Preface

The activity in the area of electronics, particularly microelectronics, has been one of the major scientific and technical frontiers for the last three decades. The complex technological development in this area emerged and appreciated only a decade later, after the invention of the transistor in 1948 and integrated circuits (ICs) soon after. Various materials and processing technologies were developed in the area of electronics during this golden age, and have made great strides in almost all facets: consumer electronics, telecommunications, medical electronics, and electronics for automotive, industrial production, and military applications. However, it is a fast-changing technology, and new materials and processes are needed for improved device applications. Many scientists and engineers are in the forefront of this area developing innovative processing and materials technologies.

The decision of the American Ceramic Society to organize several international symposia covering many aspects of advanced electronic material systems by bringing together leading researchers and practitioners of electronics and to publish the proceedings of the conferences in the Ceramic Transactions series was a major contribution to the electronics industry. As a result, symposia on (1) Materials and Processes for Microelectronic Systems, (2) Electro-Optics and Nonlinear Optic Materials, (3) Glasses for Electronics Applications, and (4) Ceramic Superconductor Materials and Processes were organized during the first and second ceramic science and technology congresses of the Society.

This volume contains a collection of papers that were presented at the International Symposium on Materials and Processes for Microelectronic Systems during the First International Ceramic Science and Technology Congress held in Anaheim, California, October 31 to November 3, 1989. The major topics of the symposium were: Dielectric Compositions; Multilayer Capacitor Systems; Glass Ceramic Composite Materials and Processing Concepts; Multilayer Interconnect Packages; Thick and Thin Film Hybrids and Wafer Scale Integration; and Device Fabrications. Over twenty-seven invited papers and many contributed papers were presented during the three-day conference. All the papers included in this volume were reviewed by experts in the field.

We, the editors, acknowledge and appreciate the contributions of the speakers, conference session chairs, manuscript reviewers, and Society officials for making this endeavor a successful one.

K.M. Nair
R. Pohanka
R.C. Buchanan

__Ceramic Transactions__ is a new proceedings series designed to meet two needs: high-quality content and rapid publication. Volumes in the series come from meetings, symposia, and forums. Each paper is reviewed by two peers, and final manuscripts are prepared by authors in a "camera-ready" format. The volumes in this series would not be possible without the hard work, dedication, and cooperation of editors, reviewers, and authors, who all deserve a great deal of thanks.

Your comments, questions, and suggestions for future __Ceramic Transactions__ volumes are welcomed and should be addressed to the Director of Publications, The American Ceramic Society, Inc.

Contents

Rational and Irrational Strategies for the Synthesis of New Materials 3
R. Roy

Recent Topics of Electronic Ceramics in Japan 11
K. Okazaki

Solution Condensed-Phase Nano-size Perovskite Particle Processing 23
K.S. Mazdiyasni

Structural Evolution of Chemically Derived Barium Tetratitanate ($BaTi_4O_9$) Powders 41
P.P. Phule, S.H. Risbud, and J.E. Pemberton

Preparation and Characterization of Ceramic Powders for Electronics 51
D.A. Warner, C.C. Hung, M.J. Munson, J.F. Wang, and R.E. Riman

Processing, Microstructure, and Properties of the Lead Scandium Tungstate Relaxor Materials 69
D. Suvorov, M. Petrič, D. Kolar, V. Kraševec

Sintering-Mechanistic Implications 77
D.L. Johnson

Effect of Nitrogen Pressure on the Sintering of AlN 91
N. Takada, M. Komatu, and T. Takahashi

Dielectric Aging of Doped PMN-PT Ceramics 105
A.E. Sutherland, S.M. Pilgrim, A.E. Bailey, and S.R. Winzer

Electric Field Effects on Crack Growth in a Lead Magnesium Niobate Ceramic 129
A.S. Raynes, G.S. White, S.W. Freiman, and B.S. Rawal

Dielectric Properties of $HfTiO_4$-$ZrTiO_4$ Solid Solutions 143

H. Ikawa, H. Narita, O. Fukunaga, and R.E. Newnham

Effects of Processing on Microstructure and Dielectric Properties in Ba ($Mg_{1/3}Ta_{2/3}$)O_3 Ceramics 153

M. Sugiyama, T. Inuzuka, and H. Kubo

Microstructure and Dielectric Properties of Mullite Ceramics 167

E.A. Giess, J.M. Roldan, P.J. Bailey, and E. Goo

Chemically Prepared PZT Films with Niobium Additions ... 179

B.A. Tuttle, D.H. Doughty, R.W. Schwartz, T.J. Garino, S.L. Martinez, D. Goodnow, C.L. Hernandez, R.G. Tissot, and W.F. Hammetter

The Effects of Raw Powder and Processing on the Properties of AlN Ceramics 193

C.-F. Chen and E. Savrun

Glass in Microelectronic Packaging: A Review 217

A.K. Varshneya and S.C. Cherukuri

The Production of Barium Hexaferrite ($BaFe_{12}O_{19}$) from High-Iron Glasses 245

B.H. Fox and A. Zybura

Low K Glass-Ceramics for Microelectronic Packaging 259

J.F. MacDowell and G.H. Beall

Ceramic-filled Glass Composite Sintering 279

K.G. Ewsuk

A Study of Laser-effected Thin Silicon Layer Recrystallization 297

C.P. Grigoropoulos and A.F. Emery

Multilayer Ceramic NPO Capacitators with Copper Electrode 313

H. Mandai, Y. Sakabe, and J.P. Canner

Parameters Important for Surface-Mount Applications of Multilayer Ceramic Capacitors 329

B.S. Rawal, K. Krishnamani, and J. Maxwell

Sintering in Silver-Glass Electrode Systems 347

T. Yamaguchi and H. Imai

A Comparison of Aqueous and Nonaqueous Slurries for Tape-casting, and Dimensional Stability in Green Tapes 355

P. Nahass, W.E. Rhine, R.L. Pober, H.K. Bowen, and W.L. Robbins

Mo-Mn Metallization on AlN Substrate 365

S. Nozaki, A. Kanda, S. Iio, Y. Nakayama, and N. Kato

Processing Multilayer Ceramics with Internal Copper Conductors 375

I. Burn and W.C. Porter

A Low-Temperature, Cofired Multilayer Ceramic L-C Filter with Copper Conductors 391

H. Hamdai, K. Wakino, H. Okamura, and J.P. Canner

High-Performance Packaging Using Laser Lithography 405

L.M. Levinson, C.W. Eichelberger, R.J. Wojnarowski, and R.O. Carlson

Ceramic Photolithography 417

W.J. Nebe

In Situ Preparation of $BaTiO_4$ Thin Film by Hydrothermal-Electrochemical Synthesis 427

M. Yoshimura, S.E. Yoo, M. Hayashi, and N. Ishizawa

Effect of Deposition Conditions on Deposition Mechanism and Surface Morphology of TiO_4 Thin Films Deposited by Chemical Vapor Deposition 437

C.S. Hwang and H.J. Kim

Nano-Indentation Hardness of Thin Films 457

M. Abouelleil, L. Conopask, W. Nighan, W. Roman, and D. Price

Sol-Gel Processing and Characterization of Ferroelectric PZT Thin Films 477

A.R. Modak and S.K. Dey

Barium Titanate Thin Films by Conventional and Rapid Thermal Annealing 489

J.R. Hanrahan, E. Sanchez, J.J. Santiago, D.H. Berry, and Q. Jiang

Microstructure-Property-Processing Relationships in Microelectronic Thin Films 503

D.P. Birnie, III, and D.R. Uhlmann

Internally Cooled Substrates for High-Power Density Packaging 527

M. Kahn and R. Nelson

New Materials for Advanced Consumer Microelectronic Applications 555
W. Borland
Evaluation of Commercial 96% Alumina Substrates and Mechanism of Paste-Substrate Interaction 575
T. Yamaguchi and S. Kato
Surface Finish Effects on Metallization of Aluminum Nitride Substrates 583
E. Savrun and W. Lougher
Evaluation of Solderability of Silver-Glass Electrode 603
Y. Wada and T. Yamaguchi
Dimensional Control in Low-Temperature Co-Fired Ceramic Multilayers 611
H.T. Sawhill, R.H. Hensen, and K.R. Mikeska
Pressure-Assisted Sintering of Multilayer Packages 629
K.R. Mikeska and R.H. Jensen
Manufacturing Piezoelectric Substrates 651
J.A. Brewer
Fast Response to Infrared Radiation in SiC Fibers 661
N. Muto, M. Miyayama, H. Yanagida, T. Kajiwara, N. Mori, H. Ichikawa, and H. Harada
Charge Transport and Reliability of Dielectric Components 675
L.C. Burton
Thermal Conductivity Considerations for AlN Interfaces 685
A.D. Westwood and M.R. Notis
Effect of Annealing on the Grain Boundary Properties of ZnO Varistors 711
T.K. Gupta
Index 729

Ceramic Transactions

MATERIALS AND PROCESSES FOR
MICROELECTRONIC SYSTEMS

RATIONAL AND IRRATIONAL STRATEGIES FOR THE SYNTHESIS OF NEW MATERIALS

Rustum Roy
Materials Research Laboratory
The Pennsylvania State University
University Park, PA 16802-4801

ABSTRACT

Using many illustrations I show that the empirical fact is that many (if not most) of the major discoveries of new materials (e.g. teflon, polyethylene, Penicillin, pyroceram lithium, rare earth magnets) have benefitted greatly from chance observations and not utilized any rational approach comparable, for example, to that used in organic synthesis.

In our own research, rational approaches based on crystal chemistry and phase equilibria have, sometimes, built upon serendipitous events to yield a whole series of new materials e.g. the rare-earth garnets, the sol-gel process, new zero-expansion families, new processes for making fine powders, etc.

THE RECORD

In a recent paper[1] I presented a review on the state of materials synthesis today describing some major events in the development of the cycle. Table 1 shows the options one may use in the discovery of new materials. The first important point to be made is that the rational approach has not and indeed cannot be effective by itself. Recent evidence to support this view comes from the enormous worldwide effort on superconductor synthesis which has led to

TABLE I. Strategies utilized in materials synthesis.

Rational	*Using reason (ratio) as principal tool.*
Irrational	*Achieved by chance. Not relying on logic or reason as principal tool.*
Non-rational or Supra-rational	*Hierarchically at a higher level. Combines chance and rational or supra-rational processes—the latter is highly dependent on experience.*

virtually no advance after ≈2000 person years of work beyond the largely serendipitous discovery of the 1:2:3 phase. The empirical data force us to conclude that:

1. Synthesis of **new** materials has been and will remain the key to both the science and technology of materials. This shows how irrational is the allocation of resources which strongly favors, analysis and detailed "autopsies" on **old** phases.
2. Step function advances in discovering new materials are always serendipitous at the molecular unit-cell level; but incremental advances at nano, micro, macro levels can be achieved systematically and are often more significant to the technology.

In contrast to the discovery of new materials, new processes provide a much more important role for reason, and less for chance. In ceramics if one lists the major new processes of the last 40 years we find the following:

TABLE II.

Process	*Developed by*	*Key events*
Glass-ceramic	*D. Stookey, Corning*[3]	*Double-chance event, and wrong explanation, yet successful systematic empirical development*
Sol-Gel (homo) and Nano-composites (hetero)	*R. Roy*[4,5]	*Rationally conceived and planned in both generations*
Rapidly solidified materials	*P. Duwez*[6]	*Rationally conceived and executed*
Float glass	*A. Pilkington*[7]	*Rationally conceived and executed*
Lanxide	*M. Newkirk*[8]	*Chance observation with rational systematic follow-up*

One can explain, simply on the basis of the complexity of the problem of predicting the properties of a complex assemblage of nuclei and electrons in a periodic or non-crystalline array, why useful new materials are often found by chance. The discoveries are a function of the large number of experiments being performed. A finite number of statistical combinations throw up hints of extraordinary properties in a material. Among this certain flux of discoverable new materials encountered by scientists, probably only a small percentage are actually recognized as being of interest. The key to this recognition is "experience." **Novices cannot make serendipitous discoveries**. (The Prince Serendip of the fable was a virtual Sherlock Holmes.) Experience in this context means an enrichment of the paradigms (or models) which are used as nets by our brains to filter out significant signals from the noise. The function of interdisciplinarity here is relevant to increasing one's repertoire of paradigms (or different shaped nets) in which the new or odd (i.e. unexpected material or fish) may be caught.

After the initial recognition (discovery) a completely different and rational process sets in. These two different stages in materials synthesis are represented in Fig. 1. The figure shows also that the second process of materials engineering optimization is just as important as the discovery.

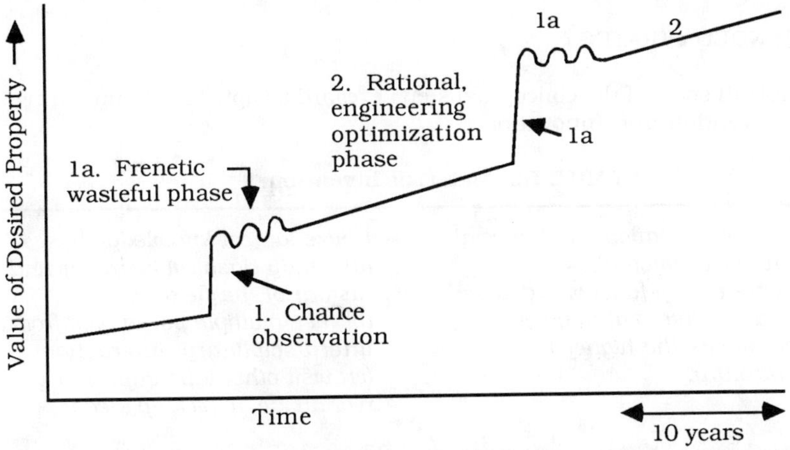

Fig. 1. New materials synthesis often consists of two steps. (1) A chance observation by an experienced observer; (1a) after a post-discovery frenetic phase, there is (2) the rational "engineering" optimization phase. The advances in desired property achieved by phase (2) are often as large as the advance in phase.

MANAGING MATERIALS SYNTHESIS: RECORD IS POOR

One can analyze the past record with respect to the amount, direction and success of research efforts expended on the most significant new materials discoveries in ceramics. I have done this[9] for the cases of ferroelectrics ($BaTiO_3$), magnetic garnets and high T_c superconductors and shown how grossly exaggerated the response was to the high T_c case. At the same time I have also shown (op cit.[9]) that among the most significant **current** discoveries if one compares the technological (and scientific) effort allocated to diamond films, "Lanxide" composites, and high T_c superconductors, the allocation is completely irrational. The source for this anomaly can also be clearly identified—the public and scientific **press**. In fact it has become increasingly clear that scientists decide on the importance of their problems <u>not</u> from the literature nor from the scientific and technological opportunity, but from the general press including the ubiquituous hyperbole contained therein. The counter example: of neglect of cement and concrete research has been developed by R. Decker[10] who has shown the enormous technical and financial opportunities in a part of materials science which gets virtually zero attention from scientists. They continue to work in the same overcrowded fields, as long as it once had media attention.

Therefore I believe that a major opportunity has now appeared in research management: **to devise the optimum strategy to respond to and capitalize on the absolutely inevitable step-function discoveries which will be made mainly outside one's own laboratories, and mainly by chance.**

RECOMMENDED STRATEGY

I list in Table III some of the conceptual steps I regard as optimum in nurturing materials invention and innovation.

TABLE III. Materials invention.

I	• *Careful observations and special attention to anomalies which indicate a step-function advance. The larger the total number of experiments the higher the probability.*	• *Utilize largest knowledge base (a) study classical literature and history of the field. (b) use multiple paradigms from interdisciplinary interaction (c) visit other labs in person worldwide, to pick up ideas*
II	• *Work systematically to improve the properties by scientific and engineering materials*	• *Use science relevant to synthesis (crystal chemistry, phase equilibrium, etc.)*
		• *Utilize panoply of synthetic tools.*

I start by noting that the truly innovative synthesis can <u>never</u> be predicted by existing theory. If it were so predictable, thousands of others would have found the materials. Hence step-function innovations in synthesis must come serendipitously. Figure 2 illustrates why. We know the relation between thermal expansion and bond strength, but a negative expansion obviously could not be predicted by such theories.

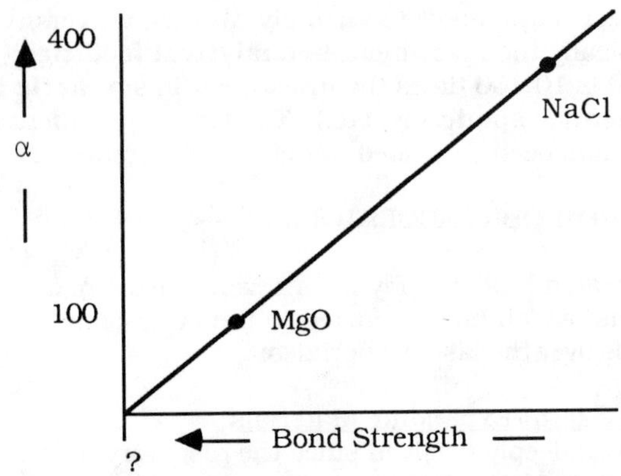

Fig. 2. The coefficient of thermal expansion (α) is well known to vary inversely with bond strength, but this cannot allow for zero or negative α. Most real innovations involve anomalous departures from known rules.

Next in Table IV I spell out the conceptual steps which go into the rational approach to new materials synthesis.

TABLE IV. Steps in actual procedure for systematic new materials search.

1. *Define structure or element of structure which is though to be correlated with desired property.*

2. *Define compositional boundary conditions if any (e.g. high m.p. eliminates halides) correlated with desired property.*

3. *Combine into possible structure-composition units using crystal chemistry structure field maps.*

4. *Use wide variety of routine synthetic tools (high p, low T; low p, high T; sputtering; sol-gel; hydrothermal; etc.)*

5. *Keep in mind phase equilibria including Me-O and P-T-X.*

6. *Scour literature for hints to guide future trials.*

7. *Increase capture cross section by large numbers (i.e. more monkeys and more word processors) and by careful analysis of results.*

Finally, in reference to step 4 in this Table, I stress the fact that the absurd imbalance in American materials research is illustrated very dramatically by the lack of synthesis equipment in virtually all U.S. university materials laboratories. Typically the investment in **analytical facilities** (TEM, XRD, SEM, Raman, NMR, etc.) is 10-100 times the investment in **synthetic facilities** (high pressure, hydrothermal, sputtering, etc.). The list of equipment I regard as a modest minimum has been described in detail in Reference 2.

CASE STUDIES FROM OUR LABORATORIES

I list here in annotated bibliography form several of the new material and process "inventions" which have occurred in the Penn State MRL and its predecessor groups over the last few decades:

• "CTP" or "NZP"—"Zero"-Expanding Materials

Penn State had been deeply involved since the discovery of the eucryptite-spodumene families in 1946-47 through the work of Hummel[11] and the phase equilibria studies by Roy.[12] This family is still the unchallenged champion among all very low expansion ceramics. Our task in 1982 was to find another structural family which could allow more crystal chemical substitution. The goal was reached successfully first in Ge-substituted cordierite. Next a serendipitous observation from the literature on fast ion conductors led us to the NZP family which has proved to be by far the most chemically versatile family imaginable. We have not only made highly refractory "zero" expansion ceramics,[13] but have even gotten rid of all anisotropy of expansion[14] and even made a ferrimagnetic zero expansion ceramic via a nanocomposite route.

• Solution-Sol-Gel II: Di-phasic Gels and Nanocomposites

In 1948 Roy developed the solution-sol-gel process for making very pure, very homogeneous ceramic powders and glasses.[4] For ten years dozens of papers extended the compositional range over which such materials could be made, before any other groups sought to exploit the process. In the seventies the sol-gel route to **homogeneous** ceramics become all the rage.

In 1982[16,17] Roy first announced the goal of maximum **heterogeneity** in ceramic materials to be attained by the SSG process. This helped advance the cause of "nanocomposites" as an entire family of new materials. It also opened the door to much lower temperatures of sintering new control of microstructure and morphology and to the confirmation of solid state epitaxy as an important phenomenon to be utilized in materials synthesis.

• Reactive Electrode Submerged Arc (RESA) Process

This was designed as an attempt at making fine crystalline oxide powders by a new process. The process modifies the Faraday method of making sols by arcing noble metals under water, by using metals and alloys which will react with the ambient fluid to make 0.1-1μ size well crystallized submicron powders of oxides, nitrides and carbides. Table V summarizes the results of many papers.[18,19]

TABLE V.

Electrodes	*Fluid*	*Products*
Al-Al	H_2O	Al_2O_3
Ti-Ti	H_2O	TiO_2
Cr-Cr	H_2O	Cr_2O_3 (+Cr_3O_4)
Cr-Cr	H_2O_2	$Cr_3O_4 + CrO_2$
C-C	*silicone*	*SiC*
C-C	*dodecane*	*diamond (minor + graphite)*
Al-Al	*Liq.* NH_3	*AlN*
Ti-Ti	*aniline*	*TiC*

• Diamond Films from the Vapor

Following the pioneer work of Derjaguin[20] which showed that in the presence of atomic H, diamond would crystallize out of the pyrolysis of hydrocarbon, Setaka et al.[21] used a microwave plasma to create coherent diamond films on a variety of substrates. In 1984 this invention, picked up in lab visits by Roy to Moscow and Tokyo, was the nucleation point for a major research effort in diamond films at Penn State and mediated therefrom to a consortium of 30 U.S. and foreign companies. The cutting edge of diamond film research has advanced now to making single crystal films of high perfection and dopable both 'p' and 'n.' The state of this art is reflected in our recent paper by Gildenblatt et al.[22] on making a Schottky diode based on a new microwave process for growing rather perfect epitaxial films.

• Low Temperature Materials for Selective Ion Fixation

By using precursor structures with strong two- and three-dimensional structural elements one can modify them near room temperature to make totally new materials with extraordinary properties. Thus starting with a fine powder of phlogopite mica one can extract the K^+ by using $NaB(Ph)_4$. This K-depleted but very highly charged sheet structure can be expanded with water or glycol, etc. The expanded phase is now a highly selective ion trap for Cs^+ with a K_d of 664,000!![23] In similar vein a nanocomposite zeolite + pillared clay has been engineered[24] to provide an almost perfect match for the thermodynamically most efficient dehydration cycle.

These examples serve to show the range of new materials which have been created—many of them during the last decade by following the procedures noted.

Acknowledgement

Materials synthesis work under the author is supported by the National Science Foundation (MRG DMR 8812824), Air Force Office of Scientific Research (AFOSR-89-0446), and Department of Energy (DE-FG02-85ER45204).

REFERENCES

1. Rustum Roy, Solid State Ionics 32/33, 3-22 (1989)
2. Rustum Roy, J. Am. Ceram. Soc. 60, 350-363 (1977).
3. S.D. Stookey, Glastechn. Berichte, 32K V, p. 1-8 (1959).
4. Rustum Roy, Science 238, 1664-1669 (1987), see references therein to early work.
5. Rustum Roy, S. Komarneni and W. Yarbrough, "Some New Advances with SSG-Derived Nanocomposites," Chapter 42, Ultrastructure Processing of Advanced Ceramics, John MacKenzie and Don Ulrich (eds.), Wiley Interscience, pp. 571-588 (1988).
6. P. Duwez and R.H. Willens, Trans. Met. Soc. AIME 227, 362 (1963).
7. A. Pilkington, Glass Technol., 12 (4), 76-83 (1971).
8. M. Newkirt et al., J. Mat. Res. 1, 81 (1986).
9. Rustum Roy, Proc. World Congress on Superconductivity, C.G. Burnham and R.D. Kane (eds.), pp. 27-41, World Scientific, Singapore (1988).
10. R.F. Decker, Proc. Conference on "Rebuilding America: Role of Materials R/D," Washington, DC, Sept. 27, 1989.
11. F.A. Hummel, Foote Prints 20, 3-11 (1948).
12. Rustum Roy and E.F. Osborn, J. Am. Chem. Soc. 71, 2086 (1949).
13. Rustum Roy, D.K. Agrawal, J. Alamo and R.A. Roy, Mat. Res. Bull. 19, 471-477 (1984).
14. Santosh Y. Limaye, Dinesh K. Agrawal, Herbert A. McKinstry and Rustum Roy, United States Patent 4,801,566 (January 31, 1989).
15. Rustum Roy and D.K. Agrawal, Proceedings, M.R.S. Symp. on Electronic Packaging, Vo. 40, pp. 83-88 (Fall 1985).
16. Rustum Roy, Bull. Am. Cer. Soc. 61, 374 (1982).
17. Rustum Roy, M.R.S. Annual Meeting Abstracts, 370 (1982).
18. Amitabh Kumar and Rustum Roy, J. Mater. Res. 3 (6), 1373-1377 (1988).
19. Amitabh Kumar and Rustum Roy, J. Am. Ceram. Soc. 72 (2), 354-356 (189).
20. B.V. Derjaguin and D.V. Fedoseev, "Growth of Diamond and Graphite from the Gas Phase," Nauka, Moscow (1977).
21. M. Kamo et al., J. Cryst. Growth 62, 642 (1983).
22. G.Sh. Gildenblatt, S.A. Grot, C.W. Hatfield, C.R. Wronski, A.R. Badzian, T. Badzian and R. Messier, Symposium F—Diamond, Boron Nitride Silicon and Related Wide Bandgap Semiconductors, Fall MRS Meeting (Nov. 28-Dec. 1, 1989).
23. S. Komarneni, A. Kijowski and R. Roy, U.S. Patent pending.
24. P.B. Malla, S. Yamanaka and S. Komarneni, Solid State Ionics 32/33, 354-362 (1989).

Recent topics of Electronic Ceramics in Japan

Kiyoshi Okazaki
Dept. of Materials Sciences and Ceramic Technology
Sagami Institute of Technology (SIT)
(Shounan Institute of Technology, Since 1990 April)
Tsujido Nishikaigan, Fujisawa 251, Japan

ABSTRACT

Essentially, ceramic material is one kind of composites consisting of multiphases such as grain boundaries, secondary phases, impurities and pores. Recently, two phase composite system is becoming more important for practical applications. In1966, Okazaki investigated the microstructure and dielectric properties of high permittivity ceramics with two Curie peaks and small temperature dependence of permittivity. This is the first attempt of two phase mixed sintering. In this paper, the recent progress of controlling microtructure and improvement of properties in electronic ceramics such as mixed sintering, relaxor/ferroelectric composite, multilayer composite, fiber reiforced ceramics, high T_c superconductor and metal composite, O_2-HIP technology, ferroelectric/ferromagnetic composite for electromagnetic wave absorber are described with the other topics of PZT repeated compressive fatique tests as a typical new result at SIT.

1 Two-Phase Mixed Sintering of Dielectric Materials

One of the characters required for ceramic capacitors with high permittivity is that the dependence of permittivity is as flat as possible. However, it is generally difficult to satisfy this requirement by using the single phase materials such as $BaTiO_3$-$SrTiO_3$ solid solution which shows a sharp Curie peak. If it is possible to make a material with two Curie peaks, it will be effective to attain the requirement mentioned above. In 1966, Okazaki et al[1] investigated the systematic experiments in order to find out the general method to obtain such dielectric ceramics. For compositions of BaO-TiO_2-CoO-SnO_2 system, markedly different characteristics were obtained by changing the calcining temperature of raw $BaCO_3$-TiO_2 material used in this composition. For example, the sample using $BaTiO_3$ powder calcined at 1400 ℃, CoO and SnO_2 showed two Curie peaks at 20 ℃ and 120 ℃, whereas the sample using the raw powders of CoO, SnO_2, $BaCO_3$ and TiO_2 showed a Curie peak at 60 ℃ as shouwn in Fig. 1(a). Thc sample with two Curie peaks seems to be consisting of two ferroelectric phases with different Curie points. The schematic model was proposed showing the coexistence of two phases as shown in Fig. 1(b). This assumption was also supported both by the mesurements of temperature dependences of remanent polarization and coercive field and by the observation using an electron microscope.

In 1977, Yonezawa et al[2] reported that $Pb(Fe_{2/3}W_{1/3})O_3$-$Pb(Fe_{1/2}Nb_{1/2})O_3$ ceramics (PFW-PFN) exhibit a high relative permittivity of 20, 000 at room temperature and they can be sintered at about 900 ℃.

Takahara et al[3] investigated on the same system of x PFW-(1-x)PFN for x = 0. 1, 0. 5, 0. 6 and 0. 7. The calcining temperatures of single phase raw materials were 1050 ℃ for = 0. 1, 750 ℃ for x = 0. 5 and 0. 6, and 700 ℃ for x = 0. 7, respectively. After calcining and grinding these

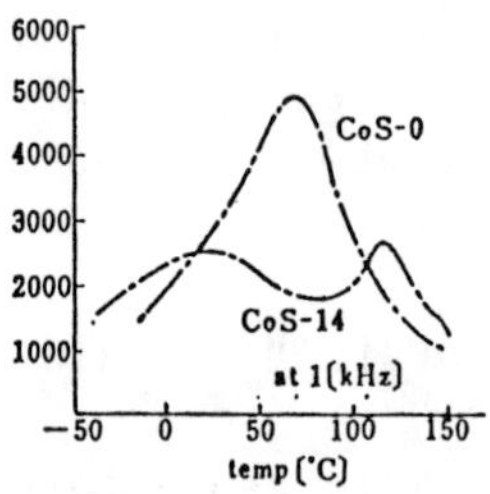

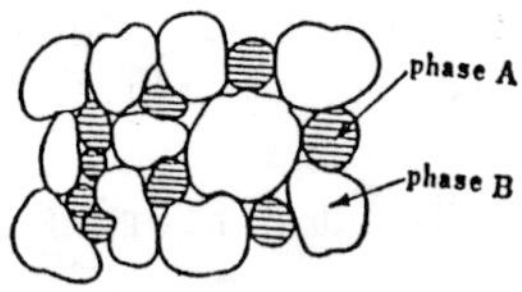

Fig. 1 (a) Relative permittivity vs temperature of 0. 95 $BaTiO_3$- 0. 05 ($Co\cdot SnO_2$)
(b) Schematic of two phase distribution

compositions, the powders with a particle size of 0.5 μm were mixed in an equimolar ratio of 0.5 $PFW_{x=0.1}$-0.5$PFN_{x=0.5}$, 0.5 $PFW_{x=0.1}$- 0.5 $PFN_{x=0.6}$, and 0.5 $PFW_{x=0.1}$-0.5 $PFN_{x=0.7}$. Figure 2 shows the relative permittivity with a broad temperature dependences.

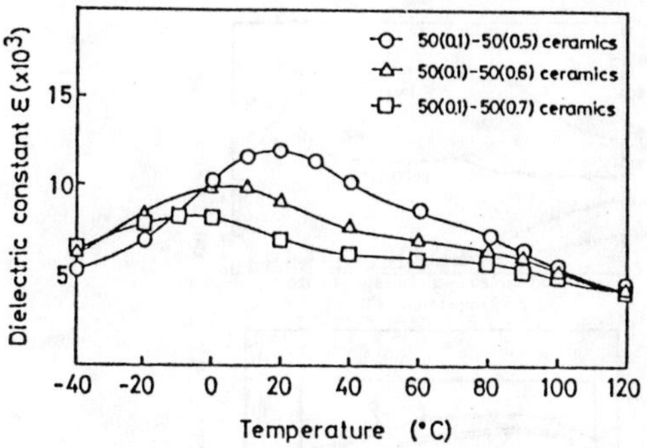

Fig. 2 Dielectric properties of the mixed-sintered PFW-PFN systems

2 A Low Firing Dielectric Material based on Relaxor/$BaTiO_3$ Composite

Yamashita et al[4] developed low firing type dielectric material based on relaxor/$BaTiO_3$ ceramic composite for multilayer ceramic capacitors(MLC). The relaxor with a composition of(Pb, Sr)($Zn_{1/3}Nb_{2/3}$)TiO_3 (Sr= 0.15~0.2, Ti= 0.15~0.3) was calcined at 800~900 ℃, and crashed into the powders with a particle size of 0.3~0.8 μm and a specific surface area of 2~5 m²/g. The ferroeletric with a composition of Ba(Ti, Zr)O_3 (Zr = 0~0.05) was calcined at a high temperature of 1200~1300 ℃ in order to reduce the reactivity with relaxor, and crashed into the powder with a particle size of 0.7~1.5 μm and a specific surface area of 1~3 m²/g. The pre-calcined relaxor and barium titanate powders were mixed and sintered at 1100 to 1140 ℃ for low firing type dielectric composite for MLCs. The high K values of 5000, 4000 and 3200 were obtained for X7T, X7S and X7R in the EIA standard, respectively, by changing the mixing ratio of relaxor/$BaTiO_3$. The advantages of this composites aretes are
(1) high K,

(2) flat temperaturedependence of K,
(3) high (capacity) X (resistance) value,
(4) low firing as shown in Fig. 4 as the concept.

They also suggested the importance of surface area[5] of the raw materi-als in order to realize the capacitor properties.

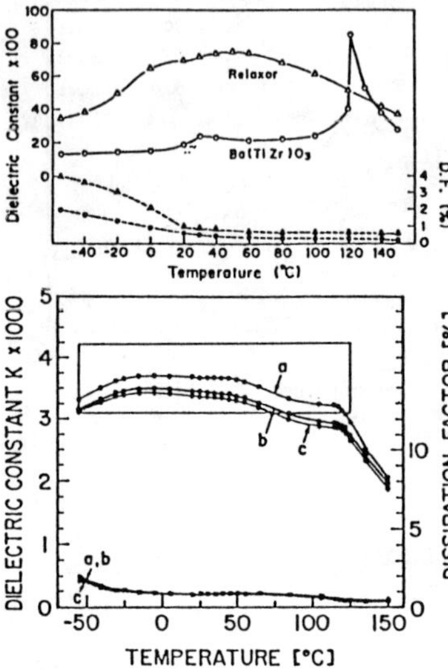

Fig. 3 (a) Dielectric properties of Relaxor and $BaTiO_3$
(b) Dielectric properties of the composites

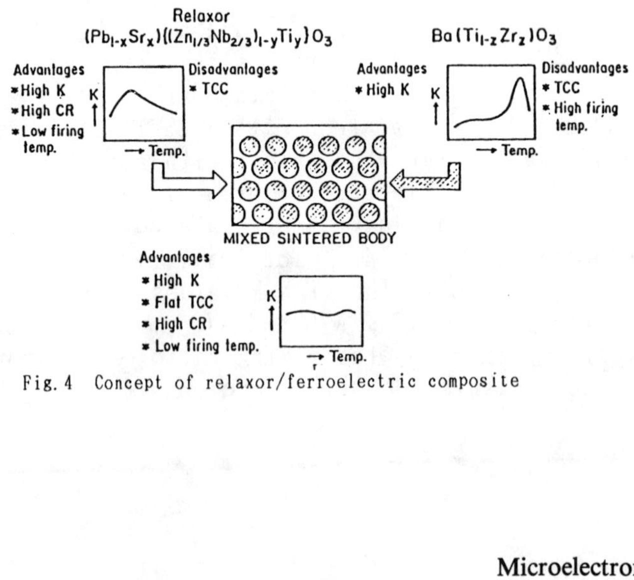

Fig. 4 Concept of relaxor/ferroelectric composite

3 Multilayer Structure of Piezoelectric and Pyroelectric Ceramics

Tashiro et al[6] investigated the piezoelectric properties of sandwich structured PZT ceramics with different Zr/Ti ratio and they ascertained that by selecting the combination of Zr/Ti ratio, the temperature coefficient can be improved.

Komiya et al fabricated[7] a multilayer-type piezoelectric and pyroelectric ceramic-ceramic composite with $PbTiO_3$(PT)-, $PbZrO_3$(PZ)- and PZT based compositions using a tape casting technique as shown in Fig. **5.** The PZ based pyroelectric composites consisting of PZ, $Pb(Zn_{1/3}Nb_{2/3})O_3$ (PZN), and $PbTiO_3$ (PT); namely $x \cdot PbZrO_3 - y \cdot (Pb_{1/3}Nb_{2/3})O_3 - z \cdot PbTiO_3$ (PZNT-100x/100z) system ($x+y+z=1, x \leq 0.1$) were investigated and the PZNT(90/1.1) and PZNT(90/2.5) showed a high pyroelectric figure of merits, F_v and F_d, which were about five times larger than those of PZT and PT-based composite. The reason is that the ferroelectric rhombohedral(LT)(F_α')-ferro-electric rhombohedral (HT)(F_α) phase transition[8] shows a very high peak of pyroelectric coefficient, p. The temperature dependence of the frequency constant, N_p, of PZT based compositewas stabilized and the best value was + 16 ppm/deg.

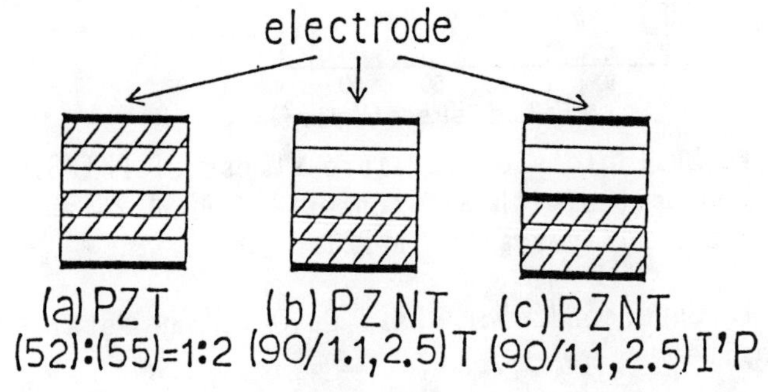

Fig. 5 Multilayer structure of piezoelectric and pyroelectric composites

4 Fiber Reinforced Piezoelectric Ceramics

As well known, the porous PZT has a large $d_h g_h$ value and a high hydrophone sensitivity. However, the mechanical strength is decreased with increasing the porosity. Yamamoto et al[9] investigated SiC reinforced porous PZT composites with the evaluation of microstructure and mechanical properties. Figure 6 shows the Weibull plots of failure stress. From the SEM, TEM with EDX observations, SiC fiber changed to SiC and reacted with PZT. The glass containing Pb, Zr and Ti surounded with the grain boundary of PZT, resulting the intergranular fracture surface. The mechanical strength of PZT with a porosity of 50 % was improved from 20 kg/cm^2 to 50 kg/cm^2 by the addition of SiC(0.25 wt %). Ohnishi et al[10] fabricated the same fiber reiforced piezo-elcctric ceramics.

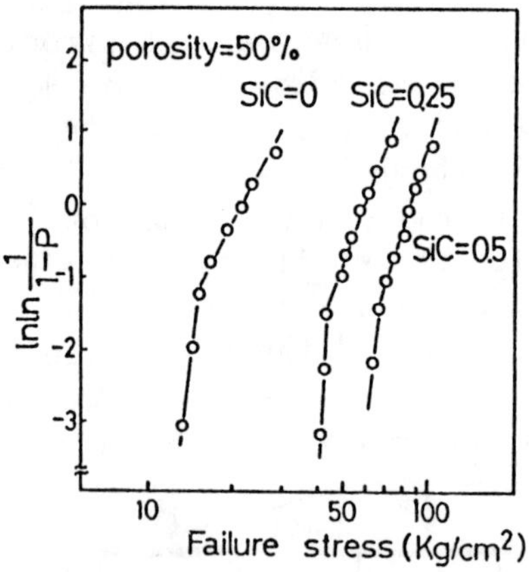

Fig. 6 Weibull plots of faillure stress of PZT/SiC wisker composites with SiC content of 0, 0.25 and 0.5 wt%, and a 50% porosity in PZT.

5 High T_c Superconductor, $Ba_2Y_1Cu_3O_{7-x}$ and Metal Ag Composite

Yamamoto el al developed a superconducting composite[11] consisting of $Ba_2Y_1Cu_3O_{7-x}$ and metal Ag by sintering a· mixture of $Ba_2Y_1Cu_3O_{7-x}$ single phase powder and Ag_2O powder. In general, Ag_2O changes into metal Ag at about 400 ℃ and the metal Ag does not react with $Ba_2Y_1Cu_3O_{7-x}$ and melts at about 800 ℃.

A high critical current density in the $Ba_2Y_1Cu_3O_{7-x}$ and Ag thick film[12] can be related as the following;

(1) The improvement of film density in the $Ba_2Y_1Cu_3O_{7-x}$ by addition of Ag_2O.

(2) The grain orientation of $Ba_2Y_1Cu_3O_{7-x}$ phase.

(3) The existence of $Ba_1Cu_1O_2$ phase

$$Ba_2Y_1Cu_3O_{7-x} \underset{\uparrow \\ \text{metal Ag}}{\rightarrow} 1/2 \cdot Ba_1Y_2Cu_1O_5 + 3/2 \cdot Ba_1Cu_1O_2 + 3/2 \cdot CuO$$

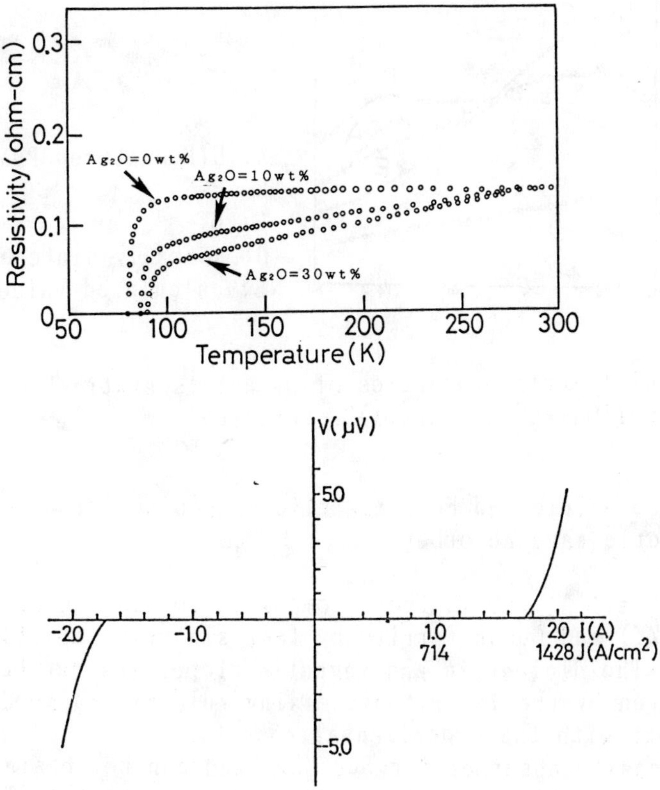

Fig. 7 Superconducting properties of $Ba_2Y_1Cu_3O_{7-x}$ and Ag composite
(a) resistivity vs temperature
(b) output voltage vs current density

6 O_2-HIP

Ohnishi et al[13] hot-isostatic-pressed(HIP) a mixture of $BaTiO_3$(BT) and $Pb(Mg_{1/3}Nb_{2/3})O_3$(PMB) in an oxygen atmosphere (O_2-HIP) at 900 and 1000 ℃ and at a pressure of 500 MPa. The sintered density was 7.01 g/cm^3 (99 % theoretical density). The dielectric properties revealed the coexistence of two phases and the dielectric permittivity coincided with the calculated values using the logarithmic mixing rule.

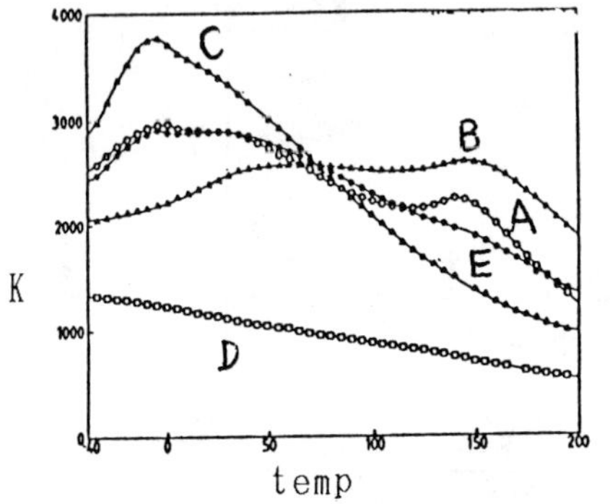

A; HIP sintered PMN-BT
B; PMN
C; BT
D; normally sintered
E; calculated value

Fig. 8 Dielectric properties of O_2-Hot-Isostatic-Pressed $BaTiO_3$-$Pb(Mg_{1/3}Nb_{2/3})O_3$ Ceramics

7 Ferroelectric and Ferromagnetic Composite for Electromagnetic Wave Absorber

Chino et al[14] fabricated a composite consisting of Pb(Zr, Ti)O_3(PZT) and Cu-Zn ferrite by fast sintering at 1100 ℃ for 0.5 hr. The dielectric and magnetic properties can be calculated by the logarithmic mixing rule with a good agreement with the experimental results.
The composite absorber for 500 MHz band can be obtained at a composition ratio of 1/10 of (PZT)/(Cu-Zn ferrite). The ratio ofthickness was 0.0417 at the central wave-length and the reflection loss at this frequency band was 20 dB.
The composite absorber for 300 MHz band can be obtained at the ratio of 6.4/10 and the thickness ratio was 0.0181. The

reflection loss was 20 dB at 100 MHz.

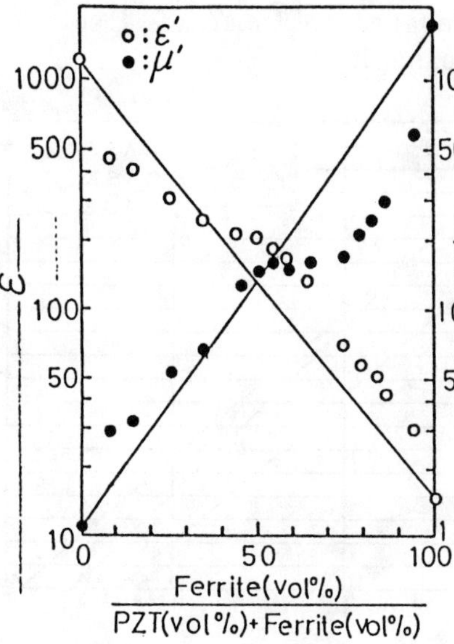

Fig. 9 Permittivity and permeability of PZT/Cu-Zn composite

8 Mechanical Properties and Fatigue of Piezoelectric Ceramics

Mechanical properties and fatigue tests[15] were carried out forpiezoelectric ceramics with a composition of PZT including the third component and pores. Both the static compression and repeated compression fatigue tests were conducted under the three different electrical conditions, namely,

case 1) electrically insulated, I = 0, (jigs Al_2O_3-Al_2O_3)

case 2) both electrodes electrically connected, E = 0, (jigs Fe-Fe)case

case 3) thermally depolarized. (jigs Al_2O_3-Al_2O_3)

Figure 10 shows the maximum cyclic stress amplitude plotted as a function of the number of cycles to the fatigue failure. As can be seen, poled samples of case 1 have a higher fatigue resistance in a longer life region than the thermally depoled samples of case 3, and the short-circuit samples of case 2 have a lower fatigue resistance about one order of magnitude compared to those of electrically insulated samples of case 1.

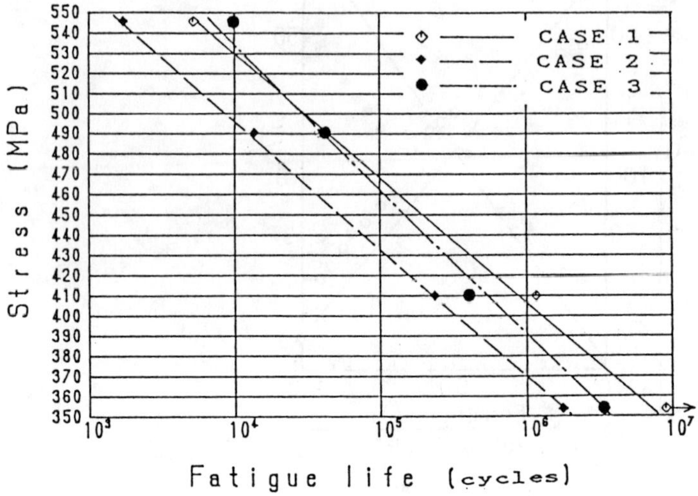

Fig. 10 Fatigue S-N Diagram for piezoelectric PZT ceramics with various poled conditions

References

1 S. Kashiwabara and K. Okazaki; High Permittivity Ceramics with Two Curie Peaks, Science Report of Defense Academy, 4, 2 (1966)pp. 136-146. (in Japanese)

2 M. Yonezawa, K. Utsumi and T. Ohno, Properties of $Pb(Fe_{2/3}W_{1/3})O_3-Pb(Fe_{1/2}Nb_{1/2})O_3$ Ceramics, Proc. 1st Meeting on Ferroelectric Materials and Their Applications(1FMA), 1977, pp. 293-301.

3 H. Takahara and K. Kiuchi; Dielectric Properties of Mixed-Sintering in the System $Pb(Fe_{2/3}W_{1/3})O_3-Pb(Fe_{1/2}Nb_{1/2})O_3$, Advanced Ceramic Materials, Vol. 1, No. 4 (1986)

4 Y. Yamashita, O. Furukawa, H. Kanai, M. Imai and M. Harada; A Low Firing Dielectric for MLC Based on Relaxor/$BaTiO_3$ Ceramic Composite, presented at Annual Meet. of Amer. Ceram. Soc., April, 1989.

5 H. Kanai, Y. Yamashita, O. Hurukawa and M. Harada; Effects of Surface Area of Starting Powders influencing on the Sinterability and Dielectric Properties of Relaxor/$BaTiO_3$ Coposites. presented at 7th FMA (1989 June) kyoto.

6 S. Tashiro, N. Arai, H. Igarashi and K. Okazaki; Piezoelectric Properties of Sandwith Structured PZT Ceramics, Ferroelectrics, Vol. 37 (1981) pp. 595-598.

7 H. Komiya, Y. Naito, T. Takenaka and K. Sakata; Multilayer Piezoelectric and Pyroelectric Composite using Tape-Casting Method, presente at 7FMA (1989, June) Kyoto.

8 T. Takenaka, A. S. Bhalla, L. E. Cross and K. Sakata; J. Amer. Ceram. Soc., 72, 6 (1989) 1016.

9 T. Yamamoto and K. Okazaki; Fabrication of Porous Piezoelectric Ceramics with high Mechanical Strength and Their Evaluation, Powder and Powder Metturgy, Vol. 35, No. 8 (1988) pp. 69-74.

10 K. Ohnishi and T. Morohashi; Studies on the Characteristics of Ceramic Fiber Reinforced Piezoelectric Ceramics, presented at 7FMA (1989, June) Kyoto.

11 T. Yamamoto, G. Terui, T. Ueyama and O. Okazaki; Fabrication of High T_c Superconductor, $Ba_2Y_1Cu_3O_{7-x}$·Ag Metal Composite Thick Film and Their Application, Ferroelectrics, Vol. 92 (1989) pp. 145-150.

12 M. Matsui, M. Hoshimura, T. Ueyama, M. Ono and T. Yamamoto; Fabrication of Screen-Ptinted Superconducting $Ba_2YCu_3O_{7-y}$·Ag

Metal Composite Thick Films, J. of the Ceram.Soc.Japan, 97, 10 (1987) pp.1053-1057.
13 K.Ohnishi, T.Morohashi and K.Uchino; Mixed-Sintering of Ferroelectric Materials using HIP, presented at 7FMA,(1989, June) Kyoto.
14 M.Chino, T.Yamamoto and A.Oshimoto; Electromagnetic Wave -Absorber of Ferroelectric/Ferromagnetic Composite by Micro-Composite Designing, presented at a Meeting of EIC Japan (1989.7.21) EMCJ 89-28.
15 T.Tanimoto, T.Yamamoto and K.Okazaki; Mechanical Properties and Fatigue of Poezoelectric Ceramics,to be appearred in Proc. of Japan-Korea Symposium (1989, Dec) Osaka.

SOLUTION CONDENSED–PHASE NANO SIZE PEROVSKITE PARTICLE PROCESSING

K. S. Mazdiyasni
General Atomics, San Diego, CA 92138–5608

Ferroelectric ceramic materials for the preparation of low loss, high permittivity capacitors and for transducer applications are much in demand for use in various kinds of microelectronic devices. Dielectric constant K and tan δ, typically of the order of K = 5000 - 6000 and loss values = 1×10^{-3} and 4×10^{-4}, have been obtained using the alkoxy solution condensed–phase $BaTiO_3$. The rare earth oxides and Nb_2O_5 doped $BaTiO_3$ exhibit much higher dielectric constant and tan δ values.

It is needless to say, that the properties of ceramic materials are dependent on microstructure. Considerable work has been done in evaluating the effects of grain size, secondary phases, chemical composition, and phase transformation on the thermal mechanical and electrical properties of various ceramic systems. With the recent interest in preparing nano size particles and their fabrication into nano crystalline sintered ceramics, the effect of decreasing grain size in the submicron region, and the contribution of interfacial or boundary layers to bulk characteristics takes on a new order of importance. Although studies in scattering characteristics of fine particles and dependence of specific magnetic and ferroelectric properties on microstructure, in particular grain size, have been studied, the precise effects of nano grain size on the electronic behavior of ceramics have not been thoroughly understood.

What is certain though, as grain size become smaller and smaller, materials properties become increasingly identified with the characteristics of surfaces and boundaries in the material due to the corresponding increase in the surface–to–volume and grain boundaries to grains ratios in particulates and sintered materials, respectively.

The dielectric properties of ferroelectrics and piezoelectrics, such as $BaTiO_3$, doped $BaTiO_3$, and lead–zirconate titanates, are strongly dependent on microstructure, as related to fine grain size.[1-9] In separate studies, Brandmayer,[3] Kniekamp and Hewang,[1] and others[10-11] observed the relative dielectric constant K′ to increase as the grain size decreases from 7 μ and reach maximum of 5500, at an average grain size of about 1 μ. The increase in dielectric constant K′, with decreasing grain size was explained based on increase in the internal stress of the material generated by decreased domain twinning in the smaller grains.[12] However, no distinction was found between the tetragonal (c/a) ratio of loose powders and sintered ceramics of similar grain size.[13]

The effects related to the existence of a surface layer on $BaTiO_3$ has not been well defined. Anliker, et al.,[14] while investigating the depolarizing effects in very fine particle size $BaTiO_3$ observed evidence of tetragonal strain well above the Curie temperature and explained this "smearing out" of the Curie transition in terms of very thin, strained surface layer. Subsequent studies of switching[15] and dielectric constant as a function of crystal thickness[16] in single crystals have concluded that a thin, nonferroelectric surface layer must exist. Also space charge layer at the $BaTiO_3$ surface have been proposed to explain anamalous pyroelectric[17] and birefringence[18] effects. Thus, in order to understand the effects of ultrafine-grain size on the magnetic and electrical properties of ceramics, an approach advanced by Mazdiyasni[19] which allow beginning with extremely fine (100 to 1000 atom) cluster powders of high purity, 99.98+%, homogeneous, and stoichiometry, controlling the growth of these cluster powder distributions into grains and observing the relationships between the resulting microstructure and the final properties of the sintered ceramics is indicated. Such aforementioned properties could potentially be further refined by solution condensed-phase source(s) to form mono sized clusters rather than cluster distributions.[20]

The present paper reviews only the application of metal alkoxide to the synthesis of perovskite in general and $BaTiO_3$ powders in particular with special emphasis on the solution condensed-phase nano size particle processing.

$$M(OR)_n \xrightarrow{H_2O} MO_n + nROH$$

$$MM'(OR)_n(OR)_{n\text{-}m} + xH_2O \longrightarrow MM'O_n + nR'OH + n\text{-}mROH$$

where M and M′ = metals, R and R′ = organic groups, and m and n = 2–5, etc.

The high surface activity associated with these powders which translate to overcoming the kinetic constraints by rapid atomic diffusion and high activities in grain boundaries, makes possible very low temperature sintering of the powder compact to near theoretical density and uniform fine grain size bodies with the desired electrical properties[7-9] for the applications in microelectronics.

The aforementioned new concept in powder synthesis is utilized to prepare mixed oxides of interest such as the rare earth oxides-doped and undoped ferroelectric and electrooptic materials. Simultaneous hydrolytic decomposition of the mixed metal alkoxide allows the mixing of a residual concentration of the dopant material to be done at something approaching the molecular level.

METHOD OF PREPARATION

For example, the barium and titanium alkoxides were prepared by the following reactions:

$$Ba + 2(CH_3)_2CHOH \xrightarrow{82°C} Ba[OCH(CH_3)_2]_2 + H_2\uparrow \qquad (1)$$

$$TiCl_4(CH_3)_2CHOH + 4NH_3 \xrightarrow{5°C} Ti[OCH(CH_3)_2]_4 + 4NH_4Cl\downarrow \qquad (2)$$

and the cocondensed-phase hydrosols were prepared in accordance with the following:

$$Ba(OC_3H_7)_2 + Ti(OC_5H_{11})_4 + 4H_2O \longrightarrow BaTiO_3{\cdot}H_2O + 2C_3H_7OH + 4C_5H_{11}OH \qquad (3)$$

$$BaTiO_3{\cdot}H_2O \xrightarrow[vac]{50°C} BaTiO_3 + H_2O$$

The reaction was carried out in a CO_2-free atmosphere. The hydrated oxide was dried in a vacuum or in a dry helium atmosphere at 50°C for 12 h. At this stage the oxide was a finely divided, stoichiometric titanate with 5 to 15 nm (maximum cluster size distributions $<1\ \mu m$) particles and was more than 99.98% pure (Table 1) with Ba/Ti $= 9.999 \pm 0.001$ as determined by both wet chemical analysis and atomic absorption technique. X-ray powder diffraction patterns of the as-prepared material revealed micro-crystallinity. Diffraction patterns on powder calcined at 700°C for $\leq$48 h showed a slight decrease in lattice parameter. Line-broadening effects prevented any differentiation between tetragonal and cubic symmetry of the as-prepared powder. The crystalite size and morphology of the agglomerated powders are shown in Fig. 1.

Table 1. Emission Spectrographic Analysis of $BaTiO_3$ from Barium and Titanium Alkoxides

Elements	Ba (ppm)	$TiCl_4$ (ppm)	$Ba(OC_3H_7)_2$ (ppm)	$Ti(OC_5H_{11})_4$ (ppm)	$BaTiO_3$ (ppm)
Si	50	100	<20	nd* <10	100
Fe	400	300	nd<10	20	<100
Sr	900	nd<5	<100	<100	nd<30
Ca	500	<100	<100	10	<100
Ni	<50	nd<3	nd<3	nd<3	nd<5
Mg	80	nd<3	<50	nd<3	nd<30
Na	<1000	500	nd<10	nd<10	nd<10
Cu	<40	nd<3	nd<5	nd<10	<30
Mo	100	nd<3	<20	nd<10	<30
Mn	100	nd<3	<50	nd<10	<30
Sn	150	nd<3	<30	nd<10	<30
Cr	<50	nd<3	<30	<30	<30
Zr	80	150	nd<10	nd<5	nd<5
Al	<80	300	nd<5	nd<5	nd<10

*nd< = Not detected less than.

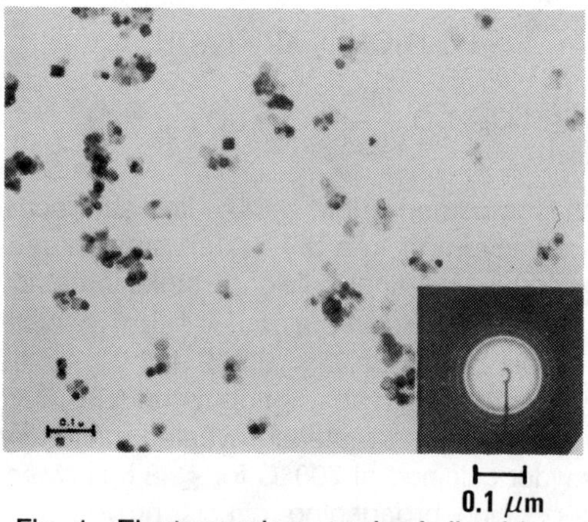

Fig. 1. Electron micrograph of alkoxide solution condensed-phase $BaTiO_3$ particulates calcined at 700°C for ≤48 h

Barium titanate powders doped with 0.1 to 0.34 mol % Ln_2O_3 were also prepared by hydrolytic decomposition of mixtures of their respective metal alkoxides. Typically, high-purity crystalline $Ba(OC_3H_7)_2$, colorless mobile $Ti(OC_5H_{11})_4$, and a crystalline lanthanide isopropoxide, $Ln(OC_3H_7)_3$, were dissolved in a mutual solvent such as isopropanol. The alkoxide mixture was hydrolyzed by adding it to a water-alcohol solution and further processed as described below:

$$Ln + n(CH_3)_2\,CHOH \xrightarrow[82°C]{H_gCl_2} Ln\,[OCH(CH_3)_2]_n + \frac{n}{2}\,H_2 \qquad (4)$$

where Ln = lanthanide metal and n = 2, 3, or 4.

$$\begin{aligned} &Ba(OC_3H_7)_2 + Ti(OC_5H_{11})_4 + La(OC_3H_7)_3 \\ &\quad + 4H_2O \longrightarrow BaTiO_3{\cdot}La_2O_3{\cdot}H_2O \\ &\quad + 2C_3H_7OH + 4C_5H_{11}OH \end{aligned} \qquad (5)$$

$$BaTiO_3{\cdot}La_2O_3 \xrightarrow[vac]{50°C} BaTiO_3{\cdot}La_2O_3 + H_2O$$

Another example of perovskite that may be cited is the synthesis of a more complex quaternary mixed oxide with nominal lead zirconate-lead titanate molar ratio of 65/35 containing 10 at percent lanthana, electrooptics, PLZT by Brown and Mazdiyasni,[21] Heartling and Land,[22] and Heartling.[23]

$$\begin{aligned} &0.9\,Pb(OR'')_2 + 0.1\,La(OR)_3 + 0.63375\,Zr(OR')_4 \\ &+ 0.34125\,Ti(OR') + 3H_2O \xrightarrow{xsH_2O} Pb_{0.9}La_{0.1}Zr_{0.063375}Ti_{0.34125}(OH)_6 \\ &\quad + 0.3\,ROH + 3.9\,R'OH + 1.8\,R'OH + xsH_2O \end{aligned} \qquad (6)$$

$$\begin{aligned} &Pb_{0.9}La_{0.1}(Zr_{0.063375}Ti_{0.34125})(OH)_6 \xrightarrow[vacuum]{60^oC} \\ &\quad Pb_{0.9}La_{0.1}(Zr_{0.63375}Ti_{0.34125})O_3 + 3H_2O \end{aligned} \qquad (7)$$

The crystallite morphology observed in these ferroelectric materials is rectangular or cubical in symmetry with particle diameters in the range of 35 to 50 nm shown in Fig. 2.

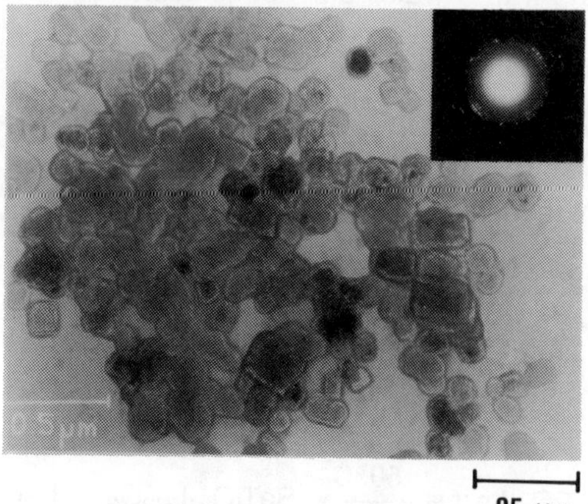

Fig. 2 Electron micrograph of PLZT crystallite morphology calcined at 500°C for 1 h

FABRICATION AND SINTERING

The as-prepared $BaTiO_3$ powders were calcined at 500°C for 30 min, ground in a B_4C mortar to effect comminution of the larger aggregates, and screened through a 200-mesh nylon screen before they were pressed. Specimens were uniaxially cold-pressed in a steel die at 1500 to 5000 psi and then isostatically pressed at 30,000 to 50,000 psi. The fine dry powder is extremely active because of the high surface area of the particulates and their clean surfaces. Contamination with H_2O and CO_2 was minimized by avoiding excessive contact with the atmosphere.

Beige specimens of 98% to 99.5% of theoretical density (6.02 g/cm^3) were readily prepared by sintering the compacted bodies at 1300° to 1350°C in air for 30 min or 1 h. Commercial practice and the references[6] cited indicate that such a density is difficult to achieve when conventional high-purity $BaTiO_3$ powders are used under identical experimental conditions.

Powders doped with 0.1 to 0.34 mol % La_2O_3, Nd_2O_3, and Nb_2O_5 were consolidated in a manner similar to that for the undoped $BaTiO_3$ and fired at 1350°C in an air or O_2 atmosphere. The sintered samples were greenish-blue with nearly theoretical density (>99%) and uniform fine grain size (1 to 1.5 μm).

The X-ray powder patterns for the doped and undoped $BaTiO_3$ powders were almost identical. Line broadening, which indicates lattice strain, was observed in the case of the as-prepared powders. A slight shrinkage of the lattice occurred with increased calcination temperature, an effect consistent with the greater degree of crystallinity resulting from increased temperature of powder calcination. All diffraction lines observed for the $BaTiO_3$ powders agreed well with the ASTM values for tetragonal $BaTiO_3$.

The lattice parameters for pure $BaTiO_3$ and $BaTiO_3$ doped with 0.1 to 0.2 mol % La_2O_3 or Nd_2O_3 were determined from Debye-Scherrer films, and the theoretical density of alkoxide solution condensed-phase $BaTiO_3$ was calculated. Table 2 compares the values obtained in the present study with existing data.

Table 2. Lattice Parameters of Tetragonal $BaTiO_3$

$BaTiO_3$	a_o (Å)	c_o (Å)	V (Å)	Density (g/cm^3)
Reference*	3.992	4.036	64.318	6.02
Alkoxy-condensed	3.9868 ± 0.0036	4.0214 ± 0.0037	63.9185	6.06
La_2O_3-doped	3.9871 ± 0.0026	0.0225 ± 0.0026	63.9457	
Nd_2O_3-doped	3.9829 ± 0.0020	4.0209 ± 0.0020	63.7855	

*Reference 24.

AGGLOMERATE AND PARTICLE SIZE EFFECTS ON SINTERING

The full potential of fine powders in terms of increased reactivity and reduced sintering cycles for final stage sintering is not realized unless agglomerates are eliminated and high green densities achieved. It is probably not possible to achieve high green density without elimination of agglomerates. Deagglomeration and increase in reactivity of an alkoxy solution condensed-phase powder can be accomplished by dry alcohol or acetone, etc., wash of the dehydrated oxide. This method of dehydration will be ineffective if the powder is heat treated beyond room temperature. The reason being OH ion fixation on the surface of finely divided and active powders.

Hoch and Nair[25] and Haberko[26] have demonstrated that calcination of gels and hydrosol washed with alcohol resulted consistently in a powder composed of agglomerates which were much porous with low powder density and mechanically weaker than agglomerates obtained from gels and hydrosol washed with water. Apparently these weak agglomerates could be easily crushed under compaction pressure, in contrast to behavior of the strong agglomerates obtained from gels and hydrosol washed with water. Haberko[26] suggests that, in compacts of the strong agglomerates two populations of pores very different in size are present; this characteristic of the strong agglomerates is perhaps related to their resistance of crushing. The presence of the two different inter- and intra-agglomerates pore-size lead to severe fluctuations of the sintering driving force across the body and the resultant densification of the body is poor. However, the collapse of weak agglomerates observed on compaction results in a very fine uniform microstructure of the compact.

MICROSTRUCTURE STUDIES

Because of the long heating and cooling cycles (300°C/h) used in sintering the $BaTiO_3$ specimens, the resultant grain size of the undoped material prepared under these experimental conditions was very large but uniform. Typical microstructures of both thermally etched and chemically etched bodies are shown in Fig. 3; the average grain size is 40 to 50 μm. In addition to the typical herringbone domain pattern, consisting of 90 and 180 deg domains,

Fig. 3. Optical micrograph of sintered at 1300°C and chemically etched coarsed-grained $BaTiO_3$

which is characteristic of $BaTiO_3$, random grains giving the appearance of a "checkerboard" or perhaps "basket-weave" pattern were also observed in the alkoxy-derived $BaTiO_3$ microstructure. An example of such a unique microstructural feature on a thermally etched surface is shown at much greater magnification in Fig. 4. At first glance it may appear that this pattern is similar to the "square-net" domain pattern observed by Forsbergh[27] and by Cook.[28] However, examination of the microstructure by hot-stage microscopy indicated that the checkerboard pattern results from the <111> ingrown twin plane which persists over the Curie temperature up to 200°C and is by no means associated with domain patterns commonly known in ferroelectric materials.

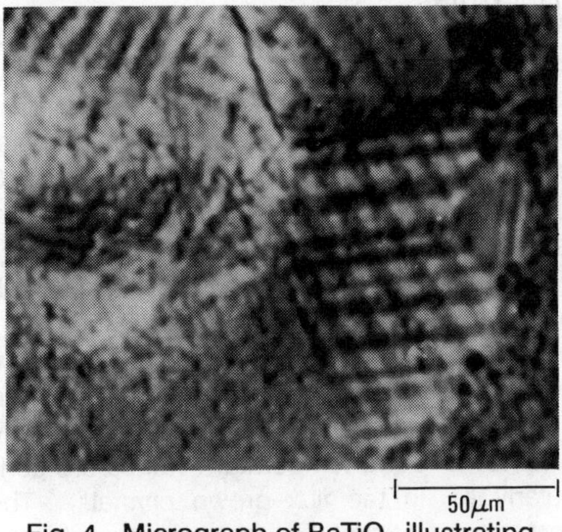

Fig. 4. Micrograph of $BaTiO_3$ illustrating checkerboard pattern of <111> ingrown twin plane

This type of microstructure is perhaps more complex than it seems and must be studied in much detail. Our explanation for the random appearance of checkerboard patterns throughout the microstructure is that mechanical strain was imposed by the surrounding grains during sintering. The strain could have resulted from the cubic-to-tetragonal phase transformation or from the already strained fine particulate $BaTiO_3$, as evidenced by X-ray diffraction line broadening. Another possibility is that twinning occurred during grain growth in the sintering process. The readily formed <111> twin plane is perhaps more pronounced because of the availability of tremendously active sites on the particles carrying surface charges which may further contribute to stresses within the fired bodies.

The polished and chemically etched surface was further examined by SEM. Well-defined 90 and 180 deg domain patterns are shown in Fig. 5(A), and a detailed view of the so-called banded structure observed by DeVries and Burke[29] is shown at high magnification in Fig. 5(B).

Typical electron micrographs of La_2O_3-doped $BaTiO_3$, polished and thermally etched and polished and chemically etched (Fig. 6), show that internal porosity is absent in the grains and on the grain boundaries. Furthermore, no segregation of a second phase and impurities is visible along the grain boundaries. This result is contrary to the findings of other investigators,[6] who have proposed that La_2O_3 and rare-earth oxides in general form a glassy phase in the grain boundaries which inhibits grain growth.

ELECTRICAL PROPERTIES

The electrical properties of $BaTiO_3$ are extremely sensitive to the conditions of sample fabrication. Dielectric measurements on most of the specimens sintered at high temperatures showed that their properties at room temperature was 5000 $\pm$ 500 with a tan δ of 4 x 10^{-3} [9] The Curie point is very sharp, with a maximum *k* of 36,000 $\pm$ 2000 occurring at 128° to 130°C. On cooling, both the dielectric constant and dissipation factor curves follow the heating curves except that the values of k are slightly higher and the Curie temperature is slightly lower ($\approx$1°C). Although a value near 120°C is generally found in commercial ceramic materials and in flux-grown crystals, Shepard[30] reported a value near 130°C for high-purity ceramic specimens and other sources[31-33] have recently reported values between 130° and 133°C for melt-grown crystals. The lower Curie temperature values are generally attributed to impurities, particularly KF in the flux-grown crystals. The chemical purity indicated in Table 1 suggests that the lower values observed in the present work probably result from strains or microstructural factors.

Some specimens sintered at high temperatures showed anomalously high apparent dielectric constant values. Room temperature and peak apparent dielectric constants were as high as 15,000 and 50,000, respectively. These values, representing an increase of a factor of $\geq$3 times the normal values, were accompanied by larger frequency dispersion of the capacitance, extremely large frequency dispersion of the conductivity, and a more pronounced discontinuity in the conductivity at the Curie temperature. Results for one specimen are shown in Fig. 7. Extrapolation of the frequency dispersion of the conductivity of these specimens to higher frequencies and cursory

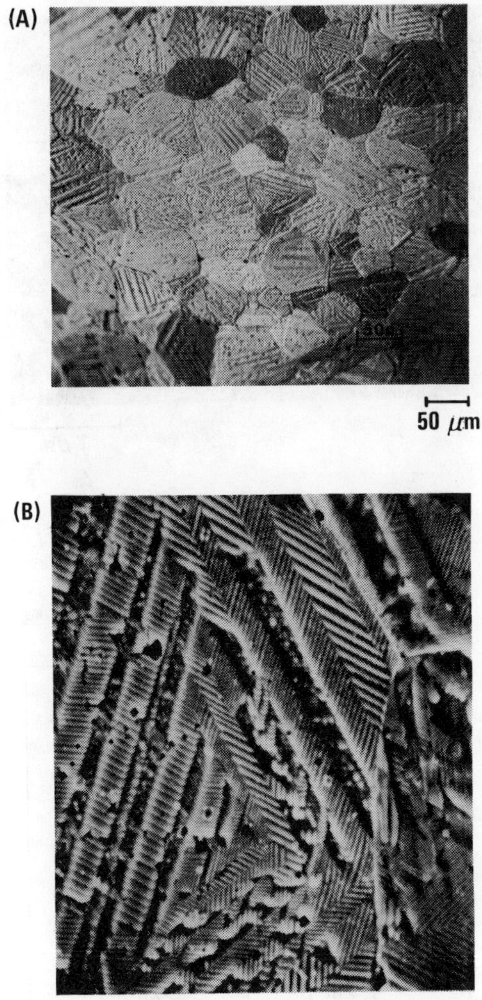

Fig. 5. Scanning electron micrographs of $BaTiO_3$ showing (A) detailed 90 and 180 deg domain twins and (B) banded structure

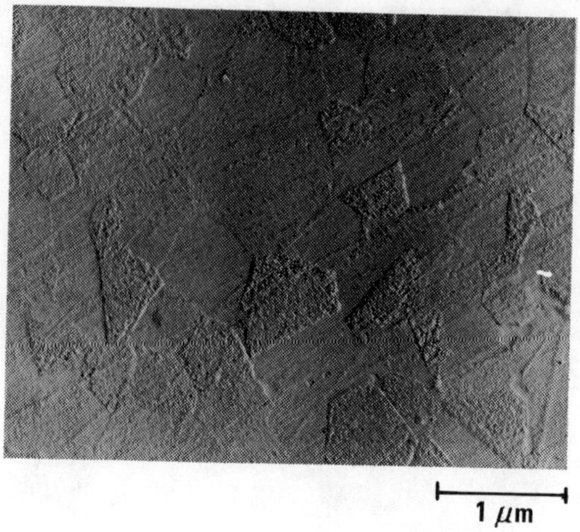

Fig. 6. Electron micrograph of La_2O_3-doped $BaTiO_3$

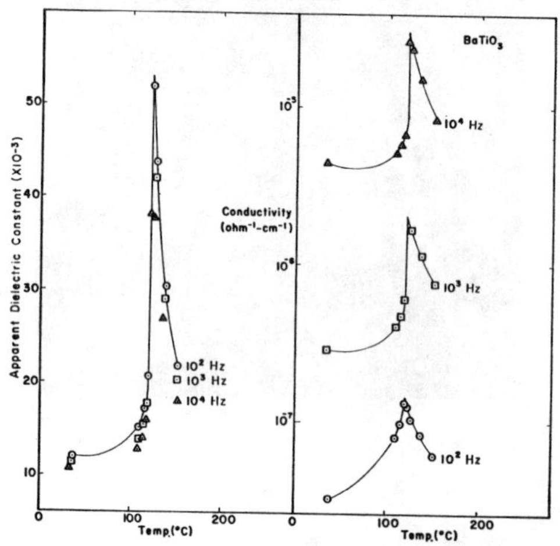

Fig. 7. Apparent dielectric constant and conductivity as a function of temperature for "enhanced" sintered specimens

measurements at 10 MHz indicate that the high-frequency bulk conductivity of the $BaTiO_3$ specimens displaying this "enhanced" behavior is of the order 10^{-1} to 10^{-2} ohm^{-1}/cm^{-1}. A possible source of such "enhanced" behavior might be Maxwell-Wagner interfacial polarization associated with the presence of thin insulating regions separating more highly conductive grains. This effect is commonly used in producing barrier layer capacitors; specimens are heated in reducing atmospheres to produce conductive substoichiometric $BaTiO_{3-x}$ and then heated for a short time in an oxidizing atmosphere to produce a thin, insulating, more nearly stoichiometric surface layer. Measurements made after grinding away the specimen surface and after prolonged heating in oxygen showed that the dielectric dispersions observed here were not associated with such oxidation-reduction effects. Because compositional variations, particularly impurity segregation, in the grain boundaries could, in principle, produce insulating surface layers, attempts were made to find these variations by electron microprobe studies on the specimen with the highest apparent dielectric constant. Within the sensitivity and resolution limits of the instrument, no significant compositional variations were detected.

Although the abnormally high apparent dielectric constants observed in this study on some specimens sintered at high temperatures were thus attributed to space charge effects, other reports have described a moderate increase in bulk dielectric constant of sintered specimens as a function of decreasing particle size.[1,33-35] As discussed earlier, it has been suggested that this grain size dependence is stress-induced by the absence of 90 deg twinning in small particles.

The samples doped with 0.2 mol % La_2O_3 or Nd_2O_3 exhibit very similar dielectric constant and tan δ values in the temperature range studied. Typical k and loss value (Fig. 8) data were obtained from samples with microstructures similar to the one shown in Fig. 7. The Curie temperature is apparently shifted to below the measured temperature. The very high dielectric constant and dissipation factor observed in these samples are in part associated with the fine grain size. Nevertheless, it should be emphasized that no second phase was observed with 0.1 to 0.2 mol % La_2O_3 and Nd_2O_3 additions to the $BaTiO_3$. The dielectric constant and loss value for 0.2 mol % Nb_2O_5 doped $BaTiO_3$ is shown in Fig. 9. These values are in agreement with the results of other workers.

SUMMARY AND CONCLUSIONS

Highly dense $BaTiO_3$ bodies with coarse-grained uniform microstructure can be prepared from alkoxy solution condensed-phase submicron powders using conventional cold-pressing and sintering techniques. The addition of residual concentrations of rare-earth oxides or Nb_2O_5 results in nearly

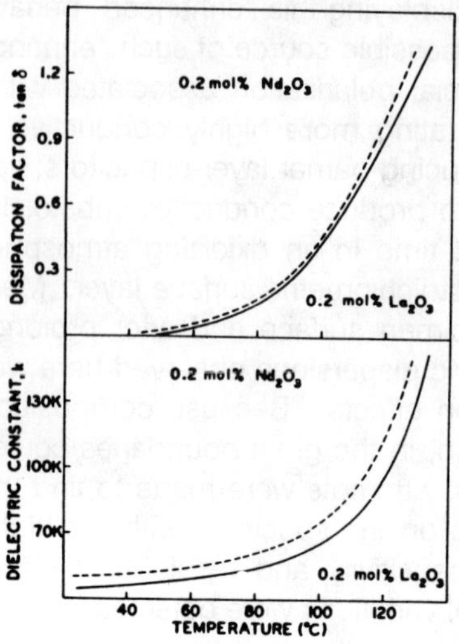

Fig. 8. Dielectric and tan δ value for La_2O_3 and Nd_2O_3 doped $BaTiO_3$

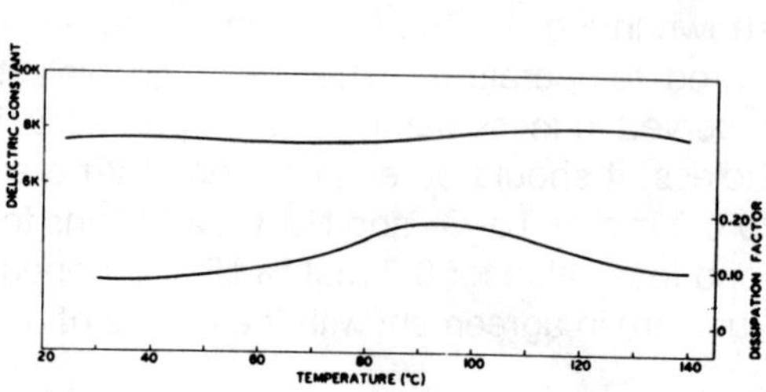

Fig. 9. Dielectric and loss value for Nb_2O_5 doped $BaTiO_3$

theoretically dense bluish-green bodies with uniform fine grain size (1 to 1.5 μm). Electrical property measurements on these bodies indicated the Curie point is depressed below room temperature. Very high room-temperature values for the dielectric constant and dissipation factor were observed and result in part from the fine grain size of the doped materials.

Based on the experimental results presented in this review paper, it is clear that much more work is needed before a quantitative understanding of the influence of submicron or nano size particles on development of the nano size structure of electronic ceramics is made possible. However, it is plausible to anticipate that if the building blocks are possible to make, their assembly to structures with unique properties are not far in the distant future. However, a crucial aspect of the stability of nano phase materials is grain growth. The driving force for normal grain growth is the chemical potential difference across a curved boundary. Since the radius of curvature of the grain boundaries is on the order of the grain size, the driving force for grain growth in a submicron grain size aggregate could be very high. This phenomenon could therefore interfere with many experiments or with processing at suitable temperatures. A systematic study of grain growth or inhibition of grain growth and grain boundary pinning in the electronic ceramics using submicron size particles are highly desirable and should be investigated.

ACKNOWLEDGMENTS

The author acknowledges all of his co-authors for their work to be included in this review paper.

REFERENCES

1. Kniekamp, H. and Heywang, W., "Depolarization Effects in Polycrystalline $BaTiO_3$," *Z. Angew, Phys.*, 6 [9], 385-90 (1954).
2. Jonker, G. H. and Noorlander, W., *Science of Ceramics,* Vol. I, Academic Press, New York, 225 (1962).
3. Brandmayr, R. J., "Properties of Ultrafine Grained $BaTiO_3$ Super-Pressed at Low Temperatures," Tech. Report ECOM-2719 (August 1966).
4. Haertling, G. H., "Hot-Pressed Lead-Zirconate Lead-Titanate Ceramics Containing Bismuth," *Bull. Am. Ceram. Soc.,* 43, 875-79 (1964).
5. Mountvala, A. J., "Hot-Pressing Piezoelectric and Ferroelectric Materials," *Bull. Am. Ceram. Soc.,* 42, 120-21 (1963).
6. Smoke, E. J., "Inorganic Dielectrics Research, 23-Year Summary Report," Tech. Rept. ECOM-0232-5, December 1969.

7. Mazdiyasni, K. S., Dolloff, R. T., and Smith, J. S., II, "Preparation of High-Purity Submicron Barium Titanate Powders," *J. Am. Ceram. Soc.,* 52 [10], 523-6 (1969).

8. Graham, H. C., Tallan, N. M., and Mazdiyasni, K. S., "Electrical Properties of High-Purity Polycrystalline $BaTiO_3$," *J. Am. Ceram. Soc.,* 52 [10], 548-53 (1969).

9. Mazdiyasni, K. S. and Brown, L. M., "Microstructure and Electrical Properties of Se_2O_3-Doped, Rare-Earth-Oxide-Doped and Undoped $BaTiO_3$," *J. Am. Ceram. Soc.,* [11], 539-543 (1971).

10. Triebwasser, S., "Space Charge Fields in $BaTiO_3$," *Phys. Rev.,* 118, 100-105 (1960).

11. Goswami, A. K., "Dielectric Properties of Unsintered Titanate," *J. Appl. Phys.,* 40 [2], 619-24 (1969).

12. Buessem, W. R., Cross, L. E., and Goswami, A. K., "Phenomenological Theory of High Permittivity in Fine-Grained Barium Titanate," *J. Am. Ceram. Soc.,* 49 [1], 33-36 (1966).

13. Bradley, F. N., "Diffraction Line Shifts Observed in Fine-Grained Barium Titanate," *ibid.,* 51 [5], 293-94 (1968).

14. Anliker, M., Burger, H. R., and Kanzig, W., "Das Verhalten von Kolloidalen Seignettealektrika III, Barium Titanat $BaTiO_3$," *Helvetica Physica Acta,* 27, 99 (1954).

15. Kanzig, W., "Space Charge Layer Near the Surface of a Ferroelectric," *Phys. Rev.,* 98, 549 (1959).

16. Fatuzzo, E. and Merz, W. J., "Surface Layer in $BaTiO_3$ Single Crystals," *J. Appl. Phys.,* 32 [9], 1685-87 (1961).

17. Schlosser, H. and Drougard, M. E., "Surface Layers on Barium Titanate Single Crystal Above the Curie Point," *ibid.,* [7], 1227-31.

18. Chynoweth, A. G., "Surface Space-Charge Layers in Barium Titanate," *Phys. Rev.,* 102 [3], 705-14 (1956).

19. Mazdiyasni, K. S., "Powder Synthesis from Metal-Organic Precursors," *Ceramic International,* 8 [2], 42-56 (1982).

20. Gherard, P. and MatiJevic, E., "Homogeneous Precipitation of Spherical Colloidal Barium Titanate Particles," *Colloid and Surfaces,* 32, 257-274 (1988).

21. Brown. L. M. and Mazdiyasni, K. S., "Cold Pressing and Low Temperature Sintering of Alkoxy-Derived PLZT," *J. Am. Ceram. Soc.,* 55 [11], 541-4 (1972).

22. Haertling, G. H. and Land, C. E., "Hot-Pressed (Pb,La)-(Zr,Ti)O_3 Ferroelectric Ceramics for Electro-Optic Applications," *J. Am, Ceram. Soc.,* 54, 1-11 (1971).

23. Haertling, G. H., "Improved Hot-Pressed Electro-Optic Ceramics in the (Pb,La)-(Zr-Ti)O_3 System," *J. Am. Ceram. Soc.,* 54, 303-309 (1971).

24. Rhodes, R. G., "Barium Titanate Twinning at Low Temperatures," *Acta Crystallogr.,* 4 [2], 105-10 (1951).

25. Hoch, M. and Nair, K. M., *Ceramurgia International,* 2, 88 (1976).

26. Haberko, K., *Ceramurgia International,* 5, 148 (1979).

27. Forsbergh, P. W., Jr., "Domain Structures and Phase Transitions in Barium Titanate," *Phys. Rev.,* 76 [8], 1187-1201 (1949).

28. Cook, W. R., Jr., "Domain Twinning in Barium Titanate Ceramics," *J. Am. Ceram. Soc.,* 39 [1], 17-19 (1956).

29. DeVries, R. C. and Burke, J. E., "Microstructure of Barium Titanate Ceramics," *J. Am. Ceram. Soc.,* 40 [6], 200-206 (1957).

30. Shepard, R., "Dielectric Properties of Lead Zirconate and Barium-Lead Zirconate," *J. Am. Ceram. Soc.,* 33 [2], 63-66 (1950).

31. Johnson, C. J., "Some Dielectric and Electro-Optic Properties of $BaTiO_3$ Single Crystals," *Appl. Phys. Lett.,* 7 [8], 221-23 (1965).

32. Wemple, S. H., Didomenico, M., Jr., and Camlibel, I., "Dielectric and Optical Properties of Melt-Grown $BaTiO_3$," *J. Phys. Chem. Solids,* 29 [10], 1797-1803 (1968).

33. Jona, F. and Shirane, G., "Ferroelectric Crystals," Pergamon Press, Inc., New York, 1962.

34. Egerton, L. and Koonce, S. E., "Effect of Firing Cycle on Structure and Some Dielectric and Piezoelectric Properties of Barium Titanate Ceramics," *J. Am. Ceram. Soc.,* 38 [11], 412-18 (1955).

35. Jonker, G. H. and Noorlander, W., pp. 255-64 in *Science of Ceramics,* Vol. 1, Edited by G. H. Stewart, Academic Press, New York, 1962.

STRUCTURAL EVOLUTION OF CHEMICALLY DERIVED BARIUM TETRATITANATE ($BaTi_4O_9$) POWDERS.

Pradeep P. Phule, Department of Materials Science and Engineering, University of Pittsburgh, Pittsburgh, PA 15261.

Subhash H. Risbud, Department of Materials Science and Engineering, University of Arizona, Tucson, AZ 85721.

Jeanne E. Pemberton, Department of Chemistry, University of Arizona, Tucson, AZ 85721.

INTRODUCTION:

Barium titanate ($BaTiO_3$) barium polytitanates (e.g. $BaTi_4O_9$, $Ba_2Ti_9O_{20}$), are technologically important electronic ceramic materials. Barium titanate, a relatively high room temperature dielectric constant (K= 1000-2500) ceramic, is widely used in manufacture of multilayer capacitors and thermistors. Titania rich compositions, such as $BaTi_4O_9$ and $Ba_2Ti_9O_{20}$ exhibit moderate dielectric constants (K= 35-40) and high quality factors (Q> 6000), in the microwave frequency range are useful for microwave resonator applications. Conventional ceramic processing route based on using oxides and/or carbonates usually results in nominal purity, inhomogeneous and coarse powders. Size reduction operations can introduce additional and unpredictable levels of impurities. For example, Mhaisalkar[1] has reported the adverse effects of Al_2O_3 impurities, introduced during ball milling, on the dielectric properties of $BaTi_4O_9$ ceramics. Chemical processing of ceramics has the potential to produce ultrafine, stoichiometric, single phase and high purity powders at lower temperatures. Different chemical processing routes for the synthesis of ceramics in the $BaO-TiO_2$ system using alkoxides, acetates, hydroxides, chlorides, citrates, nitrates, oxalates and combinations thereof have been reported in the literature[2]. In our previous work we reported development of novel routes for synthesis of such ceramics as $BaTiO_3$ in the forms of amorphous gels and crystalline powders at relatively lower temperatures[3]. Unlike $BaTiO_3$ (1:1), which can be prepared in a crystalline form at low temperatures (<100°C)[2,3], amorphous powders are obtained during sol-precipitation synthesis of titania rich polytitanate compounds [4,5,6]. Previous research on synthesis of polytitanates in the $BaO-TiO_2$ system has shown that it is difficult to prepare such compositions as $BaTi_4O_9$ (1:4) or $Ba_2Ti_9O_{20}$ (2:9) in a single phase form using conventional or chemical techniques[4-5]. We reported a sol-precipitation process, based on simultaneous hydrolytic decomposition of Ba and Ti alkoxides, for the synthesis of high purity $BaTi_4O_9$ powders in a single phase form[6].

The primary objective of this study was to investigate the structural evolution of crystalline $BaTi_4O_9$ ceramics from amorphous precursor powders using X-ray diffraction (XRD) and Raman spectroscopy. We decided to use Raman spectroscopy as a probe due to the distinct Raman spectra from the phases of interest in this work, particularly 1:1, 1:2, 1:5, 4:13 and 1:4 (Ba/Ti). Javadpour and Eror[7] have demonstrated the unique sensitivity of the Raman spectroscopy to the presence of different polytitanate phases in the $BaO-TiO_2$ system.

EXPERIMENTAL PROCEDURE:

Preparation of $BaTi_4O_9$ Powders Using Sol-Precipitation Route:

Barium metal was reacted, in-situ, with isopropanol containing appropriate amount of titanium isopropoxide (Ba/Ti 1:4). This resulted in formation of a sol containing barium and titanium alkoxides. After refluxing for 24 h, hydrolysis by distilled and carbon dioxide free water resulted in formation of spherical and ultrafine (1μm) sized amorphous powders[6]. All the reactions were conducted under a dry argon atmosphere so as to minimize formation of barium carbonate. The amorphous powders were filtered, oven dried and subsequently heat treated in an electric furnace at different temperatures ranging from 700 to 1200°C. The data presented here represent each sample separately heat treated to a given temperature.

With a view to compare the characteristics of chemically and conventionally powders, a sample of $BaTi_4O_9$ was prepared according to the following procedure and characterized. One mole of $BaCO_3$ and four moles of TiO_2 were calcined in a platinum crucible at 1050°C for 4 h. The same sample was then reground in acetone and recalcined at 1300°C for 18 h. Final calcination was conducted at 1350°c for 18 h after which the sample was reground. This resulted in what XRD data showed as a single phase crystalline $BaTi_4O_9$ powder. Note that this heat treatment was conducted on the same sample in a sequential manner with a view to generate a single phase material. The sample was provided to us by Kevin Kirby, Hughes Research Laboratory, Malibu, CA.

Structural Characterization:

Structural and morphological analyses were conducted on chemically derived and conventionally prepared powders using such techniques as X-ray diffraction and scanning electron microscopy[6]. Raman spectra were acquired on system based on a Spex 1403 double monochromator with 1800 grooves mm^{-1} holographically ruled gratings. Detection was accomplished with a high sensitivity photomultiplier tube (RCA C31034A) thermoelectrically cooled to -20° C. Excitation was provided with a Coherent Radiation Ar^+ laser operating at 514.5 nm. Laser power at the sample was typically 500 mW as measured with a Coherent Radiation model 210 broadband power meter. Spectra were acquired on these ceramic powders in melting point capillaries as single scans at 1 cm^{-1} intervals over 0.5 second integration period. The bandpass of the spectrometer was set for 6 cm^{-1} for all spectral acquisitions.

RESULTS AND DISCUSSION:

According to the XRD data (Table I), crystallization of the 1:5 phase occurred at lower temperatures (700-800°C), followed by evolution of 4:13 phase at higher temperatures (900-1000 °C). On heat treatment at higher temperatures (1100°C for 3 h), $BaTi_4O_9$ was the only phase detected. The crystallinity of the conventionally prepared powders obtained through such repeated grinding and calcination treatment was better than that of the sol-precipitated powders crystallized at 1100°C for 3 h. The crystallinity was enhanced on longer calcination time or heat treatment at 1200°C.

Table 1. Summary of XRD data for Structural Evolution of Sol-Precipitated $BaTi_4O_9$ Powders.

Heat Treatment	Identified Phases
As-prepared and Oven Dried	amorphous
700-800°C 24 h	$BaTi_5O_{11}$ (1:5)
900°c, 24 h	1:5 + $Ba_4Ti_{13}O_{30}$
1100°C, 3 h	$BaTi_4O_9$ (1:4)

Raman spectra were collected from amorphous precursors heat treated at temperatures in the range 700-1000°C for 24 h are presented in Figure 1a-d. For samples heat treated at 1100 and 1200°C (Figure 1e and 1f) the calcination times were 27 and 3 hours, respectively. The exact peak positions, which allow us to compare our data with those presented by Javadpour and Eror[7], have also been recorded but are not included here. Figure 1a shows the Raman spectrum of a precursor powder calcined at 700°C. The definitive bands at 308, 416, 537, 627 and 831 cm^{-1} confirm the presence of the 1:5 phase. This is in good agreement with our and other researchers[4,5] XRD data.

The formation of $BaTi_5O_{11}$ phase can be rationalized by considering the oxygen anion packing densities for $BaTi_5O_{11}$ (1.46) and TiO_2 (rutile, 1.71 and anatase 1.56). The Ti-O structural connectivities for TiO_2 and $BaTi_5O_{11}$ are very similar i.e. the octahedral networks of TiO_2 and $BaTi_5O_{11}$ are similar[8]. Therefore, during the synthesis of polytitanates, the first phase to nucleate seems to be $BaTi_5O_{11}$ irrespective of the original stoichiometry (Ba/Ti) of the precursor powder[4,5,8,9].

The spectrum for powder heat treated at 800°C contains Raman bands for the 1:5 phase as the major component, consistent with the XRD data. Additionally, an increase in the intensity of the band at 263 cm^{-1} and additional bands at 344, 590 and 882 cm^{-1} indicates the presence of the 1:2 phase, not identified using XRD. Its detection by Raman spectroscopy demonstrates the sensitivity of this molecular probe for characterization of such materials.

Raman spectrum presented in Figure 1c shows that the bands from the 1:2 phase are lost by heat treatment at 900°C. However, the presence of bands at 161, 188, 297, 317, 395, 514 and 699 cm^{-1} confirms existence of the 4:13 phase. Figure 1d shows Raman spectrum of precursor powder heat treated at 1000°C shows that a trace of 1:5 phase is present while as 4:13 seems to be the dominant phase. In addition to these phases, bands at 116, 137, 190, 228, 262, 288, 431, 628, 652 and 859 cm^{-1} allow us to conclusively identify the 1:4 phase as a second major component. Using XRD it was relatively difficult to ascertain the presence of 1:4 phase for precursor powders calcined at 1000°C.

Evolution of single phase $BaTi_4O_9$ from originally amorphous precursor after heat treatment at 1100°C (27h) and 1200 °C (3h) was observed using Raman spectroscopy (Figures 1e and 1f). The spectrum from the sample heat treated at 1100°C exhibits broader bands than those observed from the sample heat treated at 1200°C. This shows the improvement in the crystallinity of the $BaTi_4O_9$ on increasing the calcination temperature. The upper limit for improving crystallinity may be set by; (a) potential formation of other (non 1:4) phases and; (b) formation of undesirable agglomerates.

Raman and XRD data on our samples clearly show that the powders prepared using the precursor compounds and heat treated at 1100 and 1200°C are single phase $BaTi_4O_9$. This is consistent with the observations made by Javadpour and Eror on the Oxalate derived powders[7]. Previously reported works [4,5] did not observe single phase $BaTi_4O_9$ even after heat treatment at very high temperatures (1350°C) probably due to the formation of significant concentrations of stable non-$BaTi_4O_9$ phases.

Raman spectroscopy was also used to analyze the sample prepared using a procedure involving repeated calcination and regrinding treatments, described earlier. Comparison of the Raman spectrum from a sol-precipitated powder calcined at 1200°C (Figure 2a), with the spectrum acquired on a 1:4 powder sample produced with conventional method (Figure 2b), clearly demonstrates the superior purity and homogeneity of the sol-precipitated powders. Additional phases or impurities in the conventionally prepared 1:4 powder are easily noted by the presence of peaks at 326, 336, 462, 486, 499, 517, 546, 699 and 951 cm^{-1}.

Our experimental results presented here allow us to analyze the reaction pathways for structural evolution of $BaTi_4O_9$. Heat treatment of amorphous precursor powders results in formation of 1:5, 4:13 and traces of 1:2 phases, which subsequently react at higher (> 1000°C) temperatures to from crystalline and single phase $BaTi_4O_9$ powders. This sequence of phase evolution appears to be independent of the original stoichiometry of the chemically derived precursor powders. For studies related to such ceramic systems as $BaO\text{-}TiO_2$, Raman spectroscopy is a useful analytical technique with unique sensitivity for detection of different polytitanate phases. Precise detection of intermediate phases is important in determination of the reaction pathways for the structural evolution of ceramics. Raman spectroscopy is a powerful and versatile diagnostic tool not only for structural characterization of chemical precursors but also for determination of evolution of inorganic materials, diamond and diamond like materials and inorganic glasses[10]. We are presently applying such spectroscopic techniques for elucidating transformations of purposefully tailored organometallic precursors into amorphous gels and ceramics.

ACKNOWLEDGMENT:

We thank Kevin Kirby, Hughes Research Laboratory, Malibu, CA for providing a $BaTi_4O_9$ powder sample prepared using the oxide-mix technique. Thanks are also due to K.S. Mazdiyasni, General Atomics, San Diego, for useful suggestions and comments.

REFERENCES:

1. S.G. Mhaisalkar, M.S. Thesis, Ohio state University (1988).
2. P.P. Phule and S.H. Risbud, J.Materials Science (1990) in press.
3. P.P. Phule and S. H. Risbud, J.Mat.Sci. and Engg., **B3** (1989) 241-247.
4. J.J. Ritter, R.S. Roth and J.E. Blendell, J.Am.Ceram.Soc,.**69** (1986)155.
5. Y. Suwa, Y. Sugimoto and S.Naka, J.Jpn.Soc.Powder Metall., **25** (1978) 164 (in Japanese).
6. P.P. Phule and S. H. Risbud, in Better Ceramics Through Chemistry, (Editors Brinker, Clark and Ulrich) Mat.Res.Soc.Symp. **121** (1988) 275-280.
7. J. Javadpour and N.G. Eror, J.Am.Ceram.Soc. **71** (1988) 206.
8. J. M. Wu and H. W. Wang, J. Am. Ceram.Soc. **71** [10] 869-75 (1988).
9. S. Hirano, K. Yokouchi, M. Arai and S. Naka, in Advances In Ceramics,**19**, (1986) 139 (Editors, Blum and Cannon), The Am.Ceram. Soc., Westerville, Ohio (Publ.)
10. J.E. Pemberton, L. Latifzadeh, J. P. Fletcher and S. H. Risbud," Raman Spectroscopy of Calcium Phosphate Glasses With Varying CaO Modifier concentrations", Submitted to Chemistry of Materials.

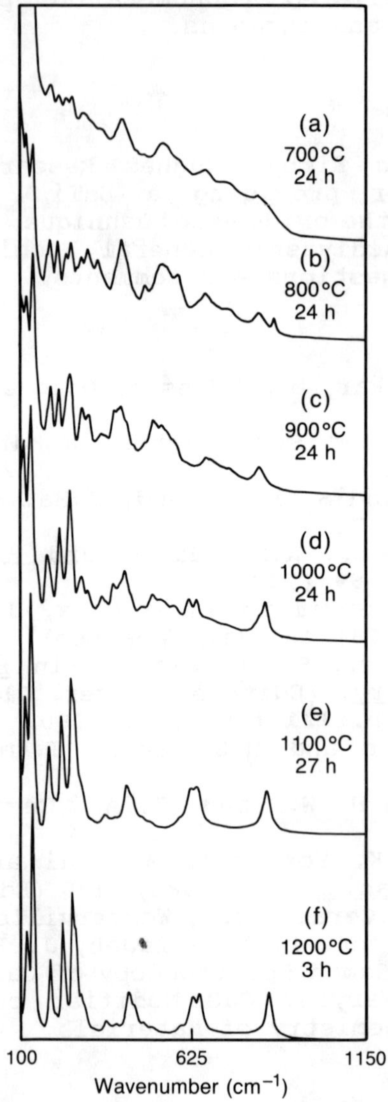

Figure 1. Raman spectra collected from $BaTi_4O_9$ precursor powders heat treated at different temperatures.

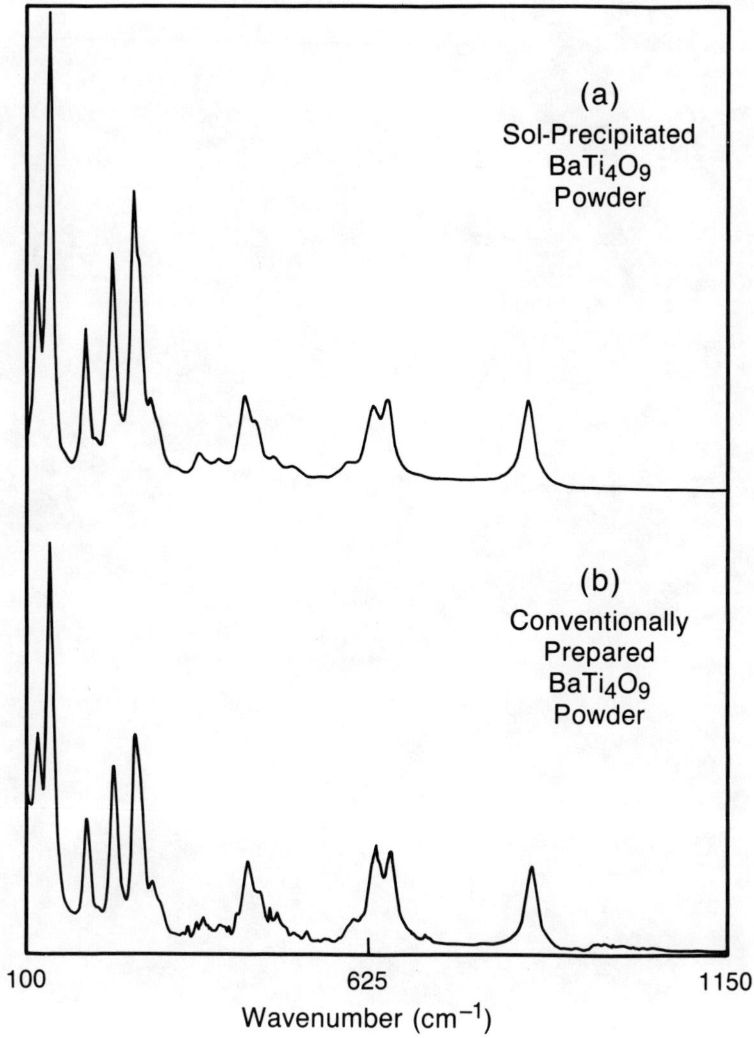

Figure 2. Raman spectra collected from crystalline $BaTi_4O_9$ powders prepared using (a): sol-precipitation and; (b) conventional oxide-mix techniques.

PREPARATION AND CHARACTERIZATION OF CERAMIC POWDERS FOR ELECTRONICS

D.A. Warner, C.C. Hung, M.J. Munson, J.F. Wang and R.E. Riman

Rutgers, The State University
Department of Ceramic Engineering
P.O. Box 909
Piscataway, NJ 08855-0909

ABSTRACT

Powder processing research at Rutgers is reviewed with respect to its importance to the preparation of electronic ceramics. Examples pertaining to powder synthesis and characterization are discussed.

INTRODUCTION

The preparation of bulk ceramic materials for electronic applications is accomplished primarily by utilizing well-established powder processing routes. Across the spectrum of applications, compositions as well as types of microstructures required vary widely. Present research challenges entail addressing how powder processing routes can be modified to prepare defect-free controlled microstructures in a reproducible fashion. From a basic science perspective, this can be accomplished through quantitative evaluation and prioritization of the effects of the powder processing variables (size, homogeneity scale, etc.) on green and fired microstructures. However, such a long-term task can only be accomplished through both the formulation of experimental designs amenable to statistical analysis and the proper selection and utilization of characterization tools.

Long-term research at Rutgers is dedicated to developing a quantitative approach to synthesis, mixing, and packing of electronic ceramic powders with controlled characteristics. This paper presents examples of ongoing research pertinent to this goal in the areas of synthesis and characterization techniques for multicomponent ceramic powder mixtures and impurity phases in multicomponent ceramic powders.

(a) POWDER SYNTHESIS

The oxides of lead and zinc play an important role in the preparation of both dielectric and piezoelectric materials. Recently, a great deal of effort has been dedicated to the controlled precipitation of powders from solution. However, most single and multicomponent powders precipitate as precursor salts that require a calcination step to form the corresponding oxide. Due to the high temperatures required for calcination, aggregates can form, rendering an otherwise useful precipitation method ineffective. For this reason, initial efforts to improve powder synthesis technology should focus more on the calcination step. Furthermore, more work is required to understand some of the fundamental processes controlling calcination-induced aggregation.

Precipitation of oxalate salts

Precipitation from homogeneous solution (PFHS) techniques[1] can be useful in the synthesis of precursor powders. This method involves the generation of an anion or pH-adjusting species in a solution containing the metal cation, and is most commonly driven by heating the solution to initiate a chemical reaction that releases species inducing precipitation. Nucleation of the particles occurs uniformly throughout the solution. By permiting adjustment of processing variables such as reagent choice, concentration, or aging conditions that affect precipitation kinetics, PFHS techniques allow for control of the physical characteristics of the powder.[2]

Using PFHS techniques being developed at Rutgers, the oxalate salts of lead and zinc were precipitated from 15 vol% H_2O in nonaqueous acetic acid solutions using diethyl oxalate as follows:

$$Et_2C_2O_4 + 2H_2O = C_2O_4^{=} + 2EtOH + 2H^+ \qquad (1)$$

$$M^{2+} + C_2O_4^{=} = MC_2O_4\ (s) \quad (2)$$

When lead ion sources (NO_3^-, $SO_4^=$, Cl^-, $OOCCH_3^-$) were employed at an aging temperature of 90°C under reaction conditions described in Figure 1a, irregularly shaped plate-like PbC_2O_4 particles formed having a scanning electron microscope (SEM) diameter of ~1-4 µm and a thickness of ~0.1 µm (photosedimentation Stoke's mean diameter = 1 µm). When the aging temperature was reduced to 70°C, and all other conditions maintained as before, spherical particles with a mean SEM diameter of ~2 µm were obtained (Figure 1b). However, the low reagent concentrations (0.025 M) produced correspondingly low yields (< 50 wt%) for further studies, while the earlier process, allowing reagent concentrations as high as 0.2 M, resulted in substantially higher yields (70-80%)

When zinc ion sources (NO_3^-, $SO_4^=$, Cl^-) were used under condition identical to those described for lead oxalate, in Figure 1a, needle-like $ZnC_2O_4 \cdot 2H_2O$ with an average SEM length of ~10 µm and an aspect ratio of 6.6 precipitated (Figure 1c). Zinc oxalate was also precipitated in an aqueous system. Using a zinc nitrate source, to avoid direct strike precipitation of the oxalate salt and maintain solution uniformity, solutions were initially acidified to pH 1. This was followed by heating to initiate a formamide decomposition to raise the pH as follows:

$$HCONH_2 + H_2O = HCOOH + NH_3 \quad (3)$$
$$2NH_3 + H_2C_2O_4 = C_2O_4^{=} + 2NH_4^+ \quad (4)$$
$$Zn^{2+} + C_2O_4^{=} + 2H_2O = ZnC_2O_4 \cdot 2H_2O\ (s) \quad (5)$$

Nonuniform cube-like particles with a diameter on the order of 10 µm were obtained (Figure 1d).

Calcination of oxalate salts

The tendency for zinc and lead oxalates to form aggregates was studied using SEM and photosedimentation. $ZnC_2O_4 \cdot 2H_2O$, in air or nitrogen, dehydrated at ~150°C and decomposed to the oxide at ~390°C via the following reaction:[3]

$$ZnC_2O_4 \cdot 2H_2O = ZnO + CO + CO_2 + 2H_2O \quad (6)$$

(a) (b)

(c) (d)

Figure 1. Precursor oxalates (a) PbC_2O_4 0.03 M M^{2+}, 0.3 M $Et_2C_2O_4$, aged at 90°C for 90 min (b) PbC_2O_4-same reaction conditions as a but aged at 70°C. (c) $ZnC_2O_4 \cdot 2H_2O$-same reaction conditions as a (d) $ZnC_2O_4 \cdot 2H_2O$ from 2.5 M NH_2CHO decomposition, 0.03 M $Zn(NO_3)_2$, 0.04 M $H_2C_2O_4$ aged at 90°C for 90 min.

Regardless of the precipitation conditions, zinc oxide exhibited no tendency to aggregate even when heated to 600°C for as long as 1 hr in air or nitrogen.

In contrast, lead oxalate decomposed to the oxide, formed aggregates under a variety of calcination conditions. In an inert atmosphere such as N_2, the following overall decomposition reaction was observed at 400°C:[4]

$$3PbC_2O_4 = Pb + 2PbO + 2CO + 4CO_2 \qquad (7)$$

In the absence of O_2, CO reduces some of the PbO present, thus producing elemental Pb and an excess of CO_2. Given that lead melts at 327°C, molten capillary bridges can form, severely aggregating the powder in addition to introducing compositional heterogeneities.

(a) (b)

Figure 2: Change in the PbO particle morphology during the litharge to massicot phase transformation. a) litharge from PbC_2O_4 calcined at 425°C for 30 min b) massicot from litharge calcined at 600°C for 1 hr.

In a flowing air atmosphere lead oxalate decomposed at 400°C as follows:[4]

$$PbC_2O_4 = PbO\ (\text{litharge}) + CO + CO_2 \qquad (8)$$

Lead oxalate showed no evidence of aggregation when calcined at temperatures below 500°C. However, at higher temperatures, aggregation was observed to such an extent that the original morphology of the powder was obscured (Figure 2).

Examining the thermochemical behavior of lead oxide in the vicinity of 500°C, it was found that a reconstructive phase transformation occurs in the 450-585°C range:[5,6]

$$PbO\ (\text{litharge}) = PbO\ (\text{massicot}) \qquad (9)$$

Study of this reaction is important since aggregation attributed to initial stage sintering could be driven by either well-established mass transport paths (i.e., evaporation-condensation, grain boundary diffusion) or enhanced diffusion paths created by the disorder during reconstructive phase transformation (the much debated Hedvall effect).[7]

To address this question, samples of PbC_2O_4 were fired at temperatures of 525, 550, and 600°C for soak times of 1,10,30, and 60 min. This provided samples with varying amounts of massicot and litharge. The samples were analyzed using x-ray diffraction, photosedimentation, and SEM. It was found that as the soak time and temperature increased, the amount of massicot, average Stoke's diameter and development of a neck structure all increased.[6,8] These data imply that enhanced mass transport caused by the reconstructive phase transformation may be responsible for aggregation.

Other evidence points to the importance of the reconstructive phase transformation to aggregation. Comparing samples from two batches, a ~50°C variation in the temperature corresponding to a specified degree of aggregation was observed. In all cases, the degree of aggregation corresponded to the expected amount of massicot present. In addition, the Θ to δ phase transformation in transitional aluminas has been known to promote neck formation in the same manner.[9]

(b) CHARACTERIZATION OF MULTICOMPONENT CERAMIC POWDER MIXTURES

High-temperature ceramic superconductors such as the 123 phase, $Ba_2YCu_3O_{7-x}$, are multicomponent materials in which cation ratios must be carefully controlled in order to achieve optimal properties. The powders are typically prepared from ternary mixtures of single-component powders, reacted in the solid state and subsequently milled. The nature of the mixing step in this conventional process, the focus of research (at Rutgers) described in this section, is of interest to any ceramist who must mix two or more components, since it controls the completeness and reproducibility of the solid state reaction. Typically, the degree of mixing is inferred qualitatively by tools such as SEM or indirectly by observations of property variations. However, quantification of the degree of mixedness in a particulate body is possible by using characterization tools such as energy dispersive spectroscopy (EDS) and statistical analysis when a proper experimental design has been employed.

In a conventional mixing operation, random mixing would be considered an optimum case. Random mixing means that the probability of withdrawing a particle of a given kind from any location in the mixture is dependent only on the overall proportion of those particles in the mixture.[10] If a number of samples containing n particles were withdrawn and their composition determined, the composition would vary according to a normal distribution. The variance in a mixture decreases as the sample size (number of particles) increases. The population of particles sampled can be thought of as occupying a finite volume and thus corresponds to a homogeneity scale. Therefore, processes resulting in random mixing should only be utilized when the corresponding variance and homogeneity scale are acceptable for subsequent processing steps which follow such as solid state reactions, sintering, etc.. For example, calculations for a 50 wt% binary mixture of monosized particles of equal density have shown that homogeneity scales are

~25 times the diameter of the starting powder given a standard deviation of 0.5 wt%.

For a three-component mixture of particles differing in density, size and size distribution such as a 123 formulation, analytical determination of the homogeneity scale is difficult. The randomness of a mixture, however, can be determined by using statistical methods to compute the probabilistic significance of variance for each component. To examine this approach, a three-component mixture of 1.0 μm $BaCO_3$ (Fisher, Springfield, NJ), 0.7 μm Y_2O_3 (PFHS-derived using the method of Sordelet and Akinc[2]) and 1.5 μm Cu (Cerac, Milwaukee, WI) was prepared by a heterocoagulation method known as polymer coacervation.[11] Employing a PVA-water-acetone system, polymer-particle flocs were formed that could be colloidally pressed into compacts at 5000 psi and subsequently analyzed in terms of chemical homogeneity.

Energy Dispersive Spectroscopy

To investigate the randomness of the heterocoagulated mixture, the chemical composition of large and small sample sizes was measured. When sampling large numbers of particles (large homogeneity scales), standardized bulk EDS (EDAX 711, Prairie, IL) was employed with a volume spanning a 1 cm diameter and depth of 50 μm. Since this was $6x10^5$ times larger than the largest particle volume, it assured that an enormous number of particles ($>3x10^5$) were sampled. For barium, copper, and yttrium spectra, statistical analysis on the 95% confidence level[12] showed no evidence for nonrandom segregation regardless of batch, sample surface (top/bottom) or method used for homogenizing the three-component mixture prior to heterocoagulation.

The above analysis yielded no information on segregation on a small homogeneity scale. This prompted development of an experimental approach capable of both sampling smaller volumes and profiling the interior of the powder compact in both the axial and the radial direction. To this end, EDS was employed on a smaller sample size (15 μm diameter, 5 μm depth) using a standardless system (IMIX System, PGT Corp., Princeton, NJ). Since this 450 μm^3 volume was 7 times larger than the largest particle volume, based on the size distributions of the various components, particle numbers in the excess of 150 were probably being sampled.

The analysis was conducted on a cross-section of the compact as shown in Figure 3. The position of the sampled volume in the compact was controlled by an adjustable sample stage and visually inspected by SEM (Amray 1200, Boston MA). Sectioning of the compacts for interior analysis was performed by first infiltrating the green compacts with a low viscosity Spurr resin,[13] curing, and then slicing with a diamond saw. SEM indicated that no microstructural distortion such as powder pullout resulted as an artifact of the sectioning procedure. Examining 2 heterocoagulated batches, 3 compacts per batch, 6 identical positions were selected for each sectioned compact for EDS analysis. A total of 36 spectra were collected for each component (barium, yttrium and copper).

Experimental Design-Analysis of Variance

Considering the large number of variables involved in this experiment, the statistical analyses for large homogeneity scales described earlier would be

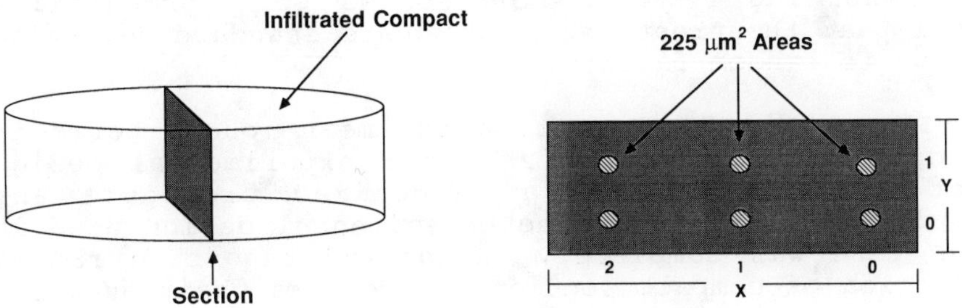

(a) Sectioning of compact (b) Sectioned area

Figure 3. Sectioning and analysis of compacts.

very time-consuming. However, the proper experimental design employed here on a small homogeneity scale allowed the analysis of variance (ANOVA) approach to be utilized.[14] In this way, many interactions could be statistically factored out to isolate variances due to particular sources.

In ANOVA, the variance of individual components can be tested for randomness using an F test, which determines the ratio of the mean square standard deviation (MS) for the treatment to the error term as follows:

$$F_p^{df_T, df_e} = [MS_T]/[MS_e] \qquad (10)$$

where

df = degrees of freedom for the treatment or error term
T = treatment or interaction term
e = error term
p = probability level corresponding to randomness

and

$$MS_i = [SS_i/df_i] \qquad (11)$$

where

SS = sum of the squares
i = treatments or the error term

The resulting F value corresponds to a probability level that the treatment mean square standard deviation is due to random factors.

The method for computing the mean square terms is contingent on utilizing a proper experimental design that allows for both a randomized procedure and statistical analysis. The experimental design used in this case was completely randomized requiring random selection of compact sections as well as their analysis order. This procedure eliminated any instrumental bias which may have occurred as the result of analysis sequence.

The use of ANOVA approach allowed the computation of position-component interaction terms such as the radial(x)-Ba, Axial(y)-Ba interactions (Table 1). A criterion was established that if the variance corresponded to a probability level greater than 5%, the interaction would be considered random. The hypothesis in each case was that the concentration of each element varied randomly with x or y. For yttrium and barium the hypothesis was supported by the F test for both x and y directions. Therefore, it can be concluded that barium and yttrium components are randomly distributed. For copper, the results indicated that, although distribution was random in the y direction, the hypothesis could not be accepted for the x direction. However, the probability of a random distribution in this direction was only slightly less than the 5% level of significance.

Based on both the direction of fluid drainage in the colloidal pressing method and the powder characteristics (size distribution, density), segregation was anticipated in the axial rather than the radial direction. Thus, further investigation into the occurrence and sources of copper metal powder radial segregation is necessary.

Table 1. Position-element interactions computed by ANOVA.

Element	Position	Mean Sq Std Dev	F	Probability Level of Significance
Ba	X	99.24	0.89	0.4572
	Y	67.41	0.65	0.3880
Y	X	492.67	4.60	0.0848
	Y	112.36	0.96	0.3726
Cu	X	148.26	7.67	0.0394*
	Y	5.44	0.04	0.8581

*Hypothesis was rejected

(c) CHARACTERIZATION OF IMPURITY PHASES IN MULTICOMPONENT CERAMIC POWDERS

For many of the advanced powders used in ceramics processing, conventional characterization methods are not adequate for the critical identification and quantification of impurity phases that impede defect-free processing. For instance, when adsorbate or amorphous phases are present in a ceramic powder, x-ray diffraction is not suitable for detection while infrared spectroscopy is unable to quantify them. X-ray photoelectron spectroscopy (XPS), on the other hand, has this capability for a variety of ceramic materials. Because this technique is novel, standard reference data is either sparse or unavailable. Therefore, correct interpretation of data requires careful experimentation.

XPS uses low energy x-ray beams to ionize inner shell electrons from the surface of the sample up to a maximum escape depth of @ 50Å.[15,16,17] Thus, it is a surface-sensitive tool. Chemical information is interpreted from the kinetic energy spectra of the photoelectrons collected, since photoelectrons of a specific kinetic energy range correspond to a specific electronic transition of a particular element. The kinetic energy of these photo-ejected electrons is converted to the binding energy of the atoms in the sample by the following equation:[16]

$$KE = hv - BE - \phi - W_C \qquad (12)$$

where
KE = kinetic energy of the ejected electron
hv = x-ray excitation energy
BE = binding energy of the atom
ϕ = work function of the instrument
W_C = work required to overcome sample charging common in insulating materials

Elements exhibit binding energy peaks whose relative positions, dependent on the electronegativity of their nearest neighbor atoms, constitute a specific bonding state. In a phase mixture in which more than one bonding state exists, the relative distribution of a particular element in each bonding state can be inferred by curve fitting the various energy shifts and computing the ratio of the various peak areas. However, care must be taken in interpreting this type of data due to errors associated with curve fitting routines and nonunique solutions that arise when energy shifts approach the peak resolution of the instrument.

When characterizing ceramic powders, rough and/or chemically nonuniform surfaces make peak assignments based on the absolute energy shift difficult. This is due to differential charging effects that shift the energy of various atoms to different degrees.[18] Furthermore, charging effects also broaden peaks, possibly leading to the false identification of multiple bonding states. These factors governing resolution, in addition to instrumental broadening inherent in any XPS spectrometer, should be considered.

Recent work at Rutgers has focused on an XPS characterization technique for barium titanate powders.[19] XPS data was collected using a Kratos XSAM-800 spectrometer (Ramsey, NJ) using referencing methods and operating conditions that facilitate a peak resolution of 1 eV. Since no Ba XPS data has been reported for barium titanate, over twenty samples of commercial and hydrothermally derived powders were examined to aid in peak assignments observed with both 3d and 4d transitions.

Ba3d spectra typical of hydrothermally derived barium titanate powder prepared at Rutgers and commercial barium titanate powders, shown in Figure 4, exhibit an asymmetry and width that could be characteristic of either charging effects or the presence of multiple bonding states. Assuming the broadening was due to multiple bonding states, Gaussian peak addition methods were used to resolve the spectra into barium titanate and barium impurity bonding states. Using 3d and 4d spectra, all peaks could be fitted as two peaks having a fixed interval of 1.4-1.6 eV. Based on electronegativity arguments used similarly for ceramic $BaY_2Cu_3O_7$ superconductors,[20] the low energy peak was assigned to the barium in the barium titanate bonding state while the higher energy peak was attributed to a barium impurity phase. Such an impurity phase could be barium carbonate, barium hydroxide and/or some type of hydroxycarbonate salt.

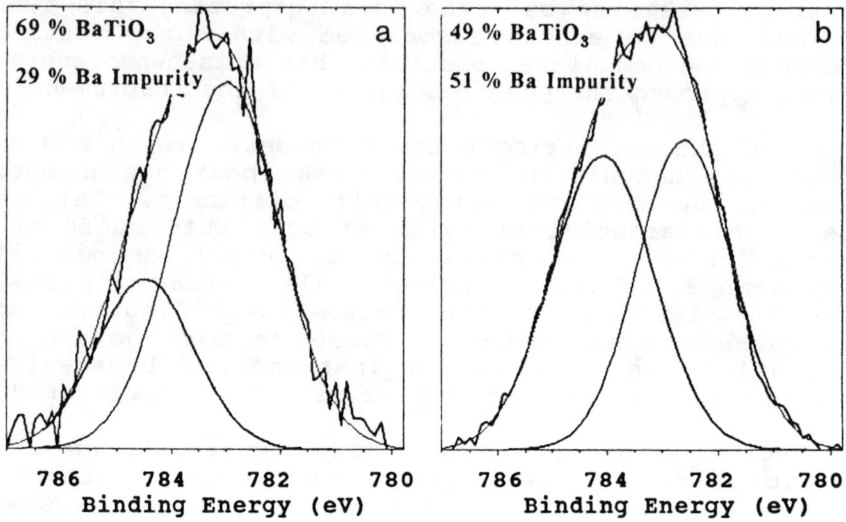

Figure 4. Fitted $Ba3d_{5/2}$ XPS peaks for: (a) experimental hydrothermally derived barium titanate powder and (b) Cabot (Boston, MA) hydrothermally derived barium titanate powder.

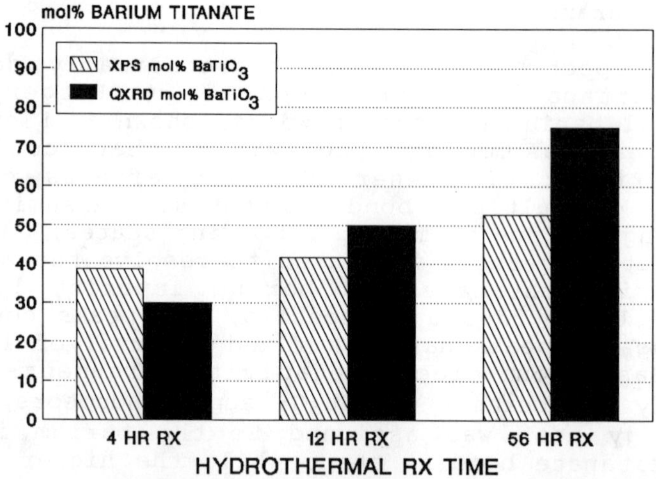

Figure 5. Formation of barium titanate as a function of time using fitted XPS $Ba3d_{5/2}$ peak data versus quantitative XRD.

To test the validity of these peak assignments, a kinetics study was performed at 90°C, 1:1 $Ba(OH)_2:TiO_2$ ratio, and 2 M concentration. Samples were taken from the reaction vessel at 4, 12, and 56 hr and washed in 1 M formic acid with a subsequent vacuum filtration step. Barium titanate and titania phases were detected by x-ray diffraction (XRD). However, no barium carbonate or amorphous phases were found. The relative proportions of barium titanate and titania were determined by using integrated peak intensities in conjunction with matrix flushing quantitative phase analysis (Model D-500, Siemens AG, Karlsruhe, FRG).[21] XRD indicated that the molar percentage of barium titanate increased with time as the titanium dioxide was consumed (Figure 5). When these samples were examined with XPS, the amount of barium titanate bonding state for the Ba 3d spectra also increased with time as the barium impurity bonding state decreased. Corroboration between the two sets of data served as evidence of a correct peak assignment.

Several points should be noted concerning the barium impurity bonding state. First, barium impurity phases were barely or not at all detected using XRD, while XPS was able to detect the impurity phase as a primary component (>50 mol%). Second, given that the escape depth of Ba3d electrons is less than 50 Å, it could be established that this impurity is concentrated at the surface. Thus, the combination of XRD and XPS data has proven useful in determining the spatial orientation of the impurity phase. As an alternative method; infrared spectroscopy could have been used to identify the impurity phase. However, the diffuse and overlapping bands of the hydroxyl and carbonate groups would have precluded quantitative determination of the relative amount of impurity phase present.

SUMMARY

Long-term research at Rutgers is dedicated to developing a quantitative approach to the powder processing of defect-free electronic ceramics. Efforts are currently focused on powder synthesis and characterization of ceramic powders, PFHS synthetic methods for preparing various types of zinc and lead oxides were presented. Solution and calcination processing variables were both found to be critical to preparing disperse oxides with controlled size and morphology. Mixedness in three-component $BaCO_3$-Y_2O_3-Cu systems was quantitatively assessed using an EDS-ANOVA experimental design approach. Heterocoagulation processing produced a random mixture with respect to barium and yttrium components. However, the mixing randomness for copper was marginal. Finally, an XPS technique was developed for examining the surface characteristics of barium titanate powders thatt cannot be adequately detected and/or quantified with conventional characterization methods. Barium titanate powder was found to have a surface which contains a considerable amount of impurity phase(s).

ACKNOWLEDGEMENTS

The authors would like to acknowledge the generous support of the New Jersey State Commission on Science and Technology, the Center for Ceramic Research, the National Science Foundation (engineering initiation award #MSM 87-07857) and Engelhard Corporation. Furthermore, the authors thank Dr. R. Carraciolo and Professor R. Lehman for their useful contributions. Finally, the efforts of Ms. K. Griffin in the preparation of this manuscript are gratefully appreciated.

REFERENCES

[1]L. Gordon, M.L. Salutsky, and H. H. Willard; Precipitation From Homogeneous Solution; pp. 61-62. John Wiley and Sons, New York, 1959.

[2]D. J. Sordelet and M. Akinc, "Preparation of Spherical Monosized Y_2O_3 Precursor Particles," *J. Col. Int. Sci.*, **122** [1] 47-59 (1988).

[3]P. E. Yankwich and P. D. Zavitsanos, "Pyrolysis of Zinc Oxalate: Kinetics and Stoichiometry," *J. Phys. Chem.*, **68** [3] 457-63 (1964).

[4]S.E. Morsi, "Kinetics of Isothermal Decomposition of Lead Oxalate in Air," *U'A.R.J. Chem.*, **13** [1] 113-20 (1970).

[5]W.B. White, F. Dachille, and R. Roy, "High-Pressure-High-Temperature Polymorphism of the Oxides of Lead," *J. Am. Ceram. Soc.*, **44** [4] 170-74 (1961).

[6]M.J. Munson, "Synthesis of Pure Reactive Zinc Oxide and Lead Oxide for Electronic Applications," M.S. Thesis. Rutgers University, New Brunwick, NJ, May 1990.

[7]P. D. Garn and T. S. Habash, "Solid State Reactions. A Rigorous Test of the Hedvall Effect," *J. Phys. Chem.*, **83** [2] 229-31 (1979).

[8]M. J. Munson and R. E. Riman, "The Chemical and Physical Properties of Oxalate Calcination Reactions," to appear in the Proceedings of the Symposium on Ceramic Dielectrics and Capacitor-Related Topics, 91st Annual Meeting and Exposition of the American Ceramic Society, April 23-27, 1989, Indianapolis, Indiana.

[9]T. Hirayama, "High Temperature Characteristics of Transition Al_2O_3 with Ultrafine Spherical Particles," *J. Am. Ceram. Soc.*, **70** [6] C-122-C-124 (1987).

[10]R. Hogg, "Grinding and Mixing of Nonmetallic Powders," *Amer. Ceram. Soc. Bull.*, **60** [2] 206-212 (1981).

[11]W. C. Moffat, P. White and H.K. Bowen, "Production of Alumina-Zirconia Composite Ceramics by Nonsolvent Flocculation of Polymer-Containing Powder Dispersion"; pp. 645-63 in Ceramic Transactions, Ceramic Powder Science II, Vol. 1, Edited by G.L. Messing, E.R. Fuller Jr., and H. Hausner. Amer. Ceram. Soc., Westerville, OH, 1988.

[12]R. Johnson, Elementary Statistics; pp. 287-90. Duxbury Press, Boston, 1984.

[13]A. R. Spurr, "A Low Viscosity Resin Epoxy Embedding Media for Electron Microscopy," *J. Ultrastructure Res.*, **26**, 31-43 (1969).

[14]C. R. Hicks, Fundamental Concepts in the Design of Experiments; pp. 39-50. Holt, Rhinehart, Winston, New York, 1973.

[15]D. M. Hercules and S. H. Hercules,"Analytical Chemistry of Surfaces, Part II. Electron Spectroscopy,"*J. of Chem. Ed.*, **61** [6] 483-89 (1984).

[16]C. Defosse and P. G. Rouxhet,"Introduction to X-ray Photoelectron Spectroscopy," pp. 169-203 in Advanced Chemical Methods for Soil and Clay Minerals Research, Edited by J. W. Stucki and W. L. Banwart. D. Reidel, Boston, 1980.

[17]F. J. Szalkowski, "The Characterization of Surfaces by Electron Spectroscopy-An Introduction," pp. 211-27 in Colloid and Interface Science. Edited by M. Kerker et al., Academic Press, New York, 1977.

[18]G. M. Bancroft, J. R. Brown, and W. S. Fyfe; "Quantitative X-ray Photoelectron Spectroscopy (ESCA): Studies of Ba^{2+} Sorption on Calcite," *Chem. Geol.*, **19**, 136 (1977).

[19]C. C. Hung, R. E. Riman and R. Carraciolo, "An XPS Investigation of Hydrothermal and Commercial Barium Titanate Powders," to appear in the Proceedings of the 3rd International Conference on Powder Processing Science, February 4-7, 1990, San Diego, CA.

[20]R. K. Brow, "An XPS Study of $YBa_2Cu_3O_x$ Surface Corrosion," pp. 598-604 in Ceramic Superconductors II. Edited by M. F. Yan. Amer. Cer. Soc., Westerville, OH, 1988.

[21]F. H. Chung, "Quantitative Interpretation of X-ray Diffraction Patterns of Mixtures, I. Matrix-Flushing Method for Quantitative Multicomponent Analysis", *J. Appl. Cryst.*, **7**, 519-25 (1974).

PROCESSING, MICROSTRUCTURE AND PROPERTIES OF THE LEAD SCANDIUM TUNGSTATE RELAXOR MATERIALS

D. Suvorov, M. Petrič, D. Kolar, V. Kraševec
"Jožef Stefan" Institute
University of Ljubljana
Yugoslavia

ABSTRACT

The system $PbO\text{-}Sc_2O_3\text{-}WO_3$ was investigated and the compound $Pb(Sc_{2/3}W_{1/3})O_3$ was synthesized. Powders were produced using different processing routes to optimize the content of the Perovskite phase. The sintered microstructure of relaxor ceramics was characterized by scanning and analytical electron microscopy and related to the electrical properties.

INTRODUCTION

Many complex lead-based Perovskite compounds with the general formula $Pb(B_xB_y)O_3$ have received considerable attention in the area of electronic ceramics. Their high dielectric constant, low sintering temperatures and temperature stability of the dielectric properties over a wide range of temperature due to characteristic diffuse phase transitions make these materials very attractive for the ceramic capacitor industry. Examples are mixed ceramics based on lead niobate, lead titanate and lead tungstate (1-4).

The synthesis of this type of material has been reported in early 1960 by Soviet authors (5,6) and since then they have been the subject of extensive research.

Phase relationships in the system $PbO\text{-}Sc_2O_3\text{-}WO_3$ have so far not been described, although the synthesis of the $Pb(Sc_{2/3}W_{1/3})O_3$ compound was performed by Tomashpolskii and Venevtsev in 1965 (6). The symmetry, dielectric constant, conductivity and transition temperature data were reported. The binary systems Sc_2O_3-WO_3 and PbO-WO_3 were described (7,8) but no information on the PbO-Sc_2O_3 system could be found in the literature. The present communication is intended to describe phase relationships in the system PbO-Sc_2O_3-WO_3 and the sequence of reaction steps leading to the $Pb(Sc_{2/3}W_{1/3})O_3$ compound with Perovskite structure. The results of dielectric measurements are also reported.

MATERIALS AND METHODS

In the present work samples were prepared by the conventional ceramic route of mixing appropriate quantities of lead oxide*, scandium oxide** and tungsten oxide***, grinding together in an agate mortar, pressing with 150 MPa into pellets and firing them covered with powder of the same composition in capped alumina crucibles. Firing temperatures up to 950°C were applied. Sintering was performed in air for various soaking times which did not exceed two hours. After cooling to 720°C ± 5°C the samples were quenched. Sections of the sintered pellets were ground for X-ray analysis (Phillips 1710 PW). Polished sections were analyzed with a scanning electron microscope (SEM, Leitz-Wetzlar AMR 1600T) and thin foils of the samples with a TEM (JEOL JEM 2000 FX). The dielectric properties of ceramic discs, 8 mm in diameter and 0.5 mm thick, were measured by an automatic capacitance bridge (HP 4192A LF) with frequencies from 1 KHz up to 100 KHz.

RESULTS AND DISCUSSION

To investigate phase relationships in the PbO-WO_3-Sc_2O_3 system, several oxide mixtures were prepared and fired at different temperatures for various soaking times. The results of the X-ray analysis are shown in Fig. 1.

Besides known peaks belonging to PbO, Sc_2O_3, WO_3, $PbWO_4$, Pb_2WO_5, Sc_6WO_{12} and $Sc_2(WO_4)_3$, a set of unidentified peaks was observed. It was determined by indexing that the reflections belonged to a compound with

---

* PbO, Ventron, 60 mesh 99.9%, Lot No. 031220402
** Sc_2O_3, Aldrich, 99.9%, Lot No. 12060-08-1
*** WO_3, Ventron, 99.7%, Lot No. 31961806

a face centered cubic unit cell with a_o = 8.141 (2) Å. Half of this value fits well with the reported parameters of a subunit cell, a_o = 4.067 Å for $Pb(Sc_{2/3}W_{1/3})O_3$ (6). Depending on the preparation temperature, Pyrochlore and Perovskite type structures were formed. The compound melts incongruently at 970°C into a lead tungstate based liquid phase and Sc_2O_3 (Fig. 2, 3). Sc_2O_3 in these samples was determined by energy dispersive spectroscopy (EDS) because Sc_2O_3 X-ray reflections were swamped by Pb_2WO_5 reflections.

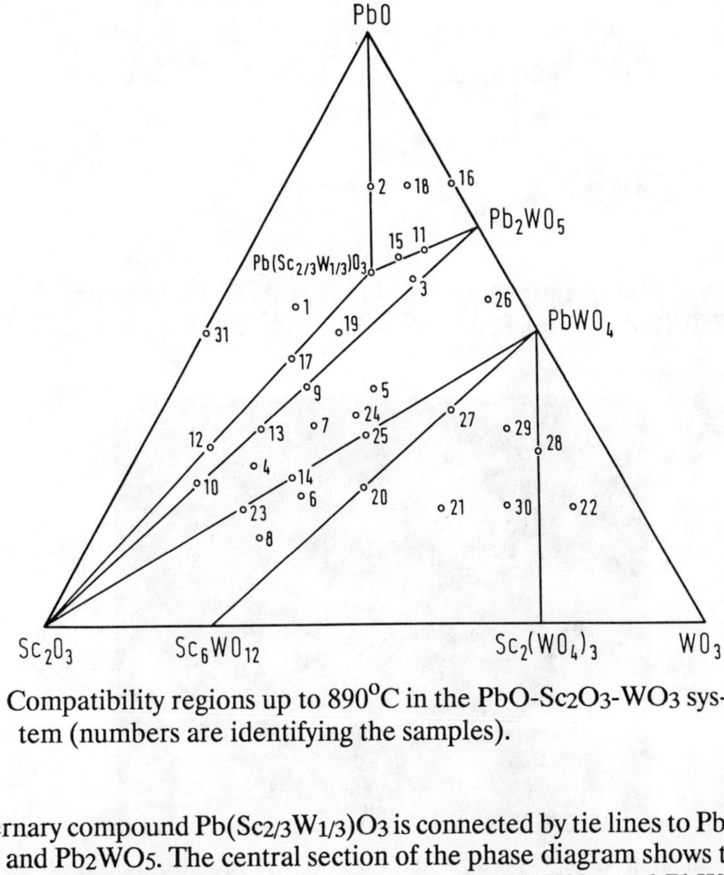

Fig. 1: Compatibility regions up to 890°C in the PbO-Sc_2O_3-WO_3 system (numbers are identifying the samples).

The ternary compound $Pb(Sc_{2/3}W_{1/3})O_3$ is connected by tie lines to PbO, Sc_2O_3 and Pb_2WO_5. The central section of the phase diagram shows the presence of the compatibility region with Sc_2O_3, Pb_2WO_5 and $PbWO_4$ corners. Also the existence of the $PbWO_4$ - Sc_6WO_{12} and $PbWO_4$-$Sc_2(WO_4)_3$ tie lines in the diagram was established. Unlike the CaO-Sc_2O_3 system, where the $CaSc_2O_4$ compound was identified (9), no binary compound was detected between Sc_2O_3 and PbO up to 880°C.

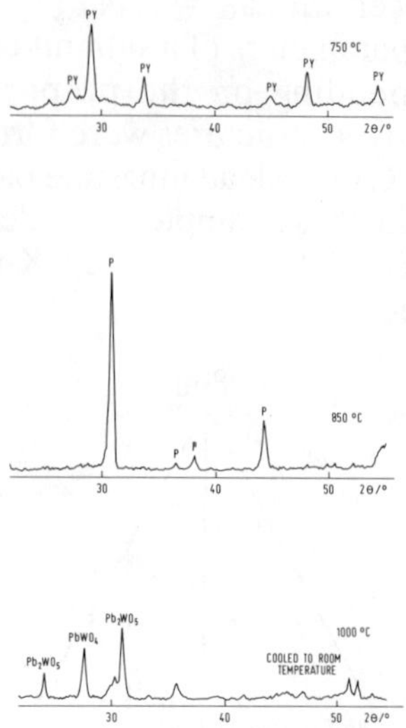

Fig. 2: X-ray powder diffractograms of $Pb(Sc_{2/3}W_{1/3})O_3$ heated at 750°C (Pyrochlore), 850°C (Perovskite) and 1000°C (decomposed).

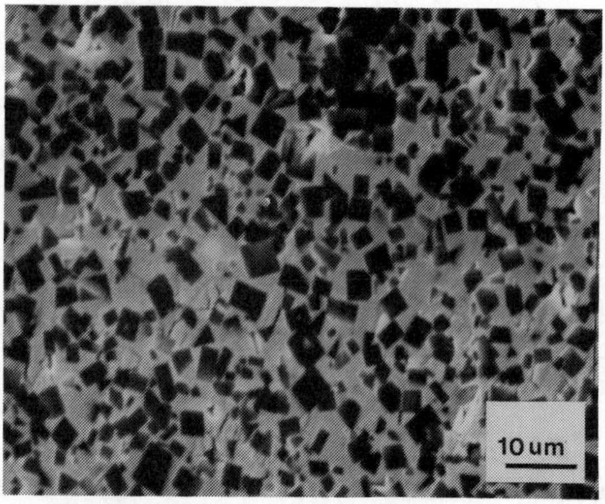

Fig. 3: Well shaped Sc_2O_3 grains in a PbO-WO_3 based liquid phase after decomposition of $Pb(Sc_{2/3}W_{1/3})O_3$ at 970°C.

Formation of $Pb(Sc_{2/3}W_{1/3})O_3$ from an initial mixture with PbO, Sc_2O_3 and WO_3 in the molar ratio 3:1:1 was found to occur by a sequence of intermediate reactions. The formation of $Pb(Sc_{2/3}W_{1/3})O_3$ was determined by high temperature X-ray analysis as follows:

$$PbO + Sc_2O_3 + WO_3 \xrightarrow{\sim 350^oC} PbWO_4 + PbO + Sc_2O_3 + WO_3$$

$$PbWO_4 + PbO + WO_3 \xrightarrow{\sim 500^oC} \alpha\, Pb_2WO_5 + PbWO_4 + PbO + WO_3$$

$$\alpha\, Pb_2WO_5 \xrightarrow{\sim 550^oC} \beta\, Pb_2WO_5$$

$$(\alpha+\beta)\, Pb_2WO_5 + PbWO_4 + PbO + WO_3 + Sc_2O_3 \xrightarrow{600^oC} \underset{\text{(Pyrochlore)}}{Pb(Sc_{2/3}W_{1/3})O_3}$$

$$\underset{\text{(Pyrochlore)}}{Pb(Sc_{2/3}W_{1/3})O_3} \xrightarrow{700^o\text{-}800^oC} \underset{\text{(Perovskite)}}{Pb(Sc_{2/3}W_{1/3})O_3}$$

$$\underset{\text{(Perovskite)}}{Pb(Sc_{2/3}W_{1/3})O_3} \xrightarrow{970^oC} \text{Liquid} + Sc_2O_3$$

High temperature X-ray analysis showed that PbO and WO_3 react at ~350°C to form $PbWO_4$. The following reaction at about 500°C is the formation of $\alpha-Pb_2WO_5$ which then partially transforms into $\beta-Pb_2WO_5$ at approximately 550°C. At around 600°C, a PSW compound with a Pyrochlore structure begins to form. At higher temperature, 700° - 800°C, Pyrochlore transforms to Perovskite and incongruently melts at 970°C into Sc_2O_3 and PbO-WO_3 based liquid phase.

These sequences partially correspond to the reaction mechanism proposed by Kassarjian and co-workers (10) for formation of $Pb(Fe_{2/3}W_{1/3})O_3$. However, they proposed the formation of two different PFW phases with Pyrochlore structure while we believe that at ~ 550°C the formation of β-Pb_2WO_5 occurs and then reacts with other constituents giving only one Pyrochlore type compound of $Pb(Sc_{2/3}W_{1/3})O_3$.

For the purpose of dielectric measurements, dense samples of nearly pure perovskite phase were prepared by hot pressing (Fig. 4).

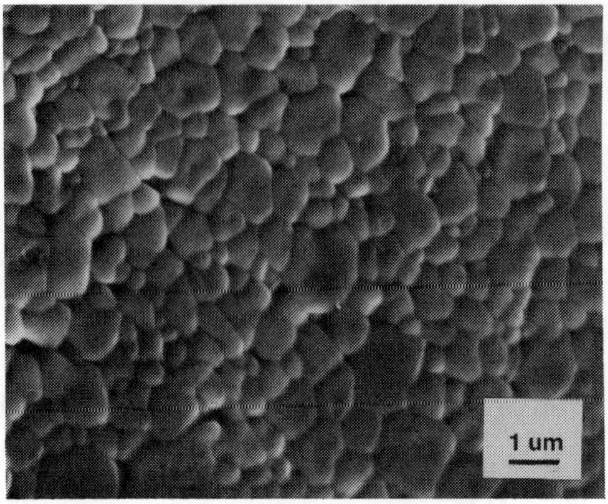

Fig. 4: Microstructure of hot pressed $Pb(Sc_{2/3}W_{1/3})O_3$ ceramic, T = 830°C/30 min, P = 20 MPa.

The dielectric constant, measured at room temperature, was 2100 and the dielectric losses 3 x 10^{-4}. Measurements of capacitance vs. temperature showed a T_c value at -50°C with ε_{max} = 2400 (at 1 kHz). T_c maximum was shifted towards higher temperatures as the measuring frequency increased (Fig. 5).

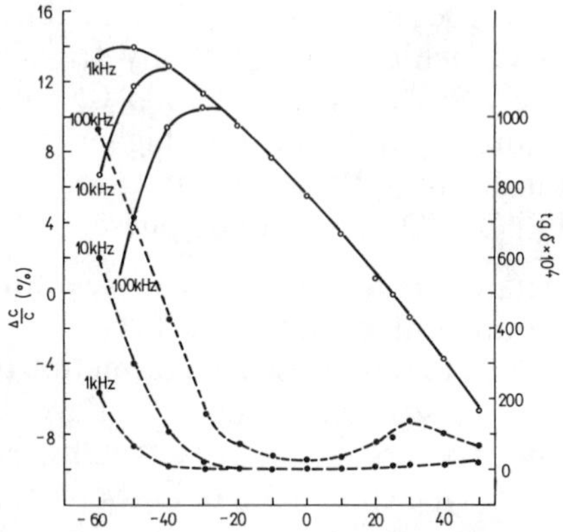

Fig. 5: Dielectric characteristics: capacitance vs. temperature and frequency

Transmission electron microscope analysis confirmed the non-homogeneity of the sample. (111) ordering spots, present on some of the [110] zone axis electron diffraction patterns indicate that the samples contain ordered as well as disordered constituents with respect to scandium and tungsten ions (Fig. 6). A prolonged time of sintering did not influence the ordering behaviour of the B-site cations as observed with $Pb(Sc_{1/2}Ta_{1/2})O_3$ by Randall et al. (11).

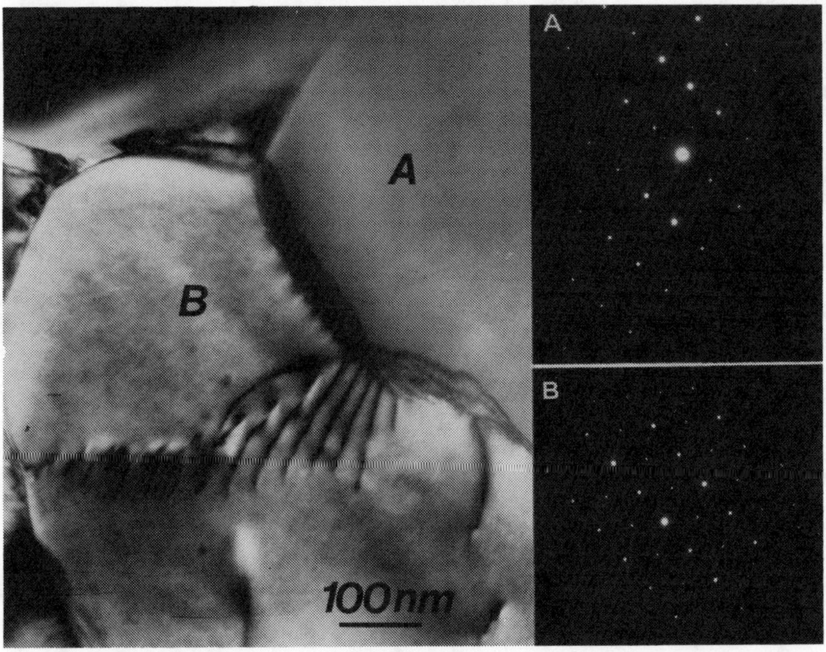

Fig. 6: A TEM image of a typical detail from the hot pressed PSW ceramic with corresponding [110] zone-axis electron diffraction patterns with ordering spots in (B).

CONCLUSIONS

A ternary compound $Pb(Sc_{2/3}W_{1/3})O_3$ was identified in the system PbO-Sc_2O_3-WO_3. Depending on the firing temperature, Pyrochlore or Perovskite type structures were formed. By heating PSW at 970°C the compound melts incongruently to Sc_2O_3 and PbO-WO_3 based liquid phase.

Dielectric measurements showed typical behaviour for compounds with diffuse phase transitions. The dielectric constant of PSW at room temperature is 2200, with 2400 as the maximum value at the Curie temperature of -50°C, measured at 1 kHz.

LITERATURE

1. M. YONEZAWA, "Low Firing Multilayer Capacitor Materials", Am. Ceram. Soc. Bull., 62, [12], 1375-1378, 1983.
2. M. YONEZAWA, "New Low-Firing Materials for Multilayer Capacitors", Ferroelectrics, 68, [1], 181-189, 1986.
3. M. HALMI, D. GESGARDIN, B. RAVEAU, "Improved Lead Perovskite compounds (PFN-PFT) for Z5U capacitor applications", Adv. Ceram. Mat., 3, [1], 32-37, 1988.
4. M. KAHN, "Electronic Ceramics: Properties, Devices and Applications", (ed. by L.M. Levinson), Marcel Dekker Inc., New York and Basel, 210, 1989.
5. G.A. SMOLENSKII, "Dielectric Polarization and Losses of Some Complex Compounds", Sov. Phys. - Tech. Phys., 3 [7], 1380-82, 1958.
6. Yu. Ya. TOMASHPOL'SKII, Yu.N. VENEVTSEV, "New Lead Containing Perovskites", Sov. Phys. Solid state, 7, [2], 412-414, 1965.
7. G.W. HALLUM, R.K. DATTA, "The System Sc_2O_3-WO_3", J. Am. Ceram. Soc., 68, [2], C-86, 1985.
8. L.L. CHANG, "Phase Relations in the System PbO-WO_3", J. Am. Ceram. Soc., 54, [7], 357, 1971.
9. A. A. SERKALO, M. YA. HODOO, A. A. FOTIEV, L. V. KRISTALLOV, "System CaO-Sc_2O_3-V_2O_5", Zh. Neorg. Khim., 31, [12], 5198-3200, 1986.
10. M.P. KASSARJIAN, R. E. NEWNHAM, J. V. BIGGERS, "Sequence of Reactions During Calcining of a Lead-Iron Niobate Dielectric Ceramic", Am. Ceram. Soc. Bull, 64, [8], 1108-1111, 1985.
11. C.A. RANDALL, D. J. BARBER, P. GROVES, R. W. WHATMORE, "A TEM Study of Ordering in the Perovskite, $Pb(Sc_{1/2}Ta_{1/2})O_3$", J. Mat. Sci. 21, [11], 4456-4462, 1986.

SINTERING - MECHANISTIC IMPLICATIONS

D. Lynn Johnson
Department of Materials Science and Engineering
Northwestern University
Evanston, IL 60208-3108

ABSTRACT

The classic sintering models have focussed upon the various stages of sintering, emphasizing their differences more than their similarities. A new approach is taken which is based upon the similarities between the stages of sintering, and a more general sintering equation is described. Simulations of sintering based upon this equation, using data in the literature for Al_2O_3, are consistent in form and magnitude with experimental data.

INTRODUCTION

The more than forty years that have passed since Frenkel[1] first proposed a model for the sintering of viscous materials have seen both progress and frustration in understanding the mechanisms of sintering. The enormous geometric complexity of even the simplest powder compact has precluded precise predictive models. Analytical models have been proposed and used with greater or lesser degrees of success to describe sintering of metals, ceramics, and glasses. The greatest successes, judged on the basis of quantitative agreement between theory and experiment, have been for the sintering of glass compacts[2-5] and the initial stage sintering of compacts of relatively large monodispersed spherical metal powders[6,7]. In the latter case, two and even three diffusion coefficients have been deduced from the sintering data, and computer simulations have closely mimicked the experimental results, even under nonisothermal cyclical heating and cooling schedules.[8] However, the sintering of more typical metal and ceramic powders has not been described adequately quantitatively, primarily because of the geometric complexities of particle shape and particle packing, simultaneous coarsening and densification, and so forth.

The sintering models have evolved in complexity and type. While all quantitative models have relied on capillarity as the driving force, a variety of responses to the capillary forces has been evoked. The earliest models described the mass flows associated with sintering in terms of the vacancy concentration gradient which arises from the concave curvature of the sharp neck between two particles. Almost at the very beginning, Herring[9] pointed out that the matter flows are driven by chemical potential gradients, rather than concentration gradients, according to the following equation:

$$j_a = -\frac{D}{kT\Omega_a}\nabla(\mu_a - \mu_v) \tag{1}$$

where j_a is the flux of atoms per unit area per unit time, D is the diffusion coefficient, Ω_a is the atomic volume, μ_a and μ_v are the atomic and vacancy chemical potentials, respectively, and kT has its usual meaning. Although the vacancy concentration models give the same answer as those based on chemical potential gradients for metals and pure ionic compounds, they fail completely for doped extrinsic ionic compounds in which the vacancy concentration is governed by the dopant concentration rather than temperature and, therefore, local hydrostatic pressure. For a pure ionic crystal, MX, exhibiting Schottky disorder, both the cation and anion vacancy concentrations are given by

$$[V_M'] = [V_X^{\cdot}] = K_s^{1/2} = K_o^{1/2}\exp\left(-\frac{\Delta H_s}{2kT}\right) \tag{2}$$

where $[V_M']$ and $[V_X^{\cdot}]$ are the cation and anion vacancy concentrations, respectively, K_s is the Schottky constant, K_o is the pre-exponential of the Schottky constant, and ΔH_s is the enthalpy of formation of a Schottky defect. In this equation the pressure appears in ΔH_s. The vacancy concentration gradients in a gradient in the hydrostatic pressure are given by

$$\nabla[V_M'] = \nabla[V_x^{\cdot}] = -\frac{\Omega_s}{2kT}K_s^{1/2}\nabla p \tag{3}$$

where Ω_s is the volume of a Schottky defect and p is the pressure. However, if an aliovalent dopant, $A_M^{\cdot}$, is added to the extent that the cation vacancy concentration is approximately equal to the dopant concentration, the cation and anion vacancy concentration gradients are given as follows:

$$\nabla[V_M'] = \nabla[V_x^{\cdot}] \sim -\frac{\Omega_s}{[A_M^{\cdot}]kT} K_s \nabla p \tag{4}$$

These equations predict that the dopant will inhibit sintering no matter whether cation or anion vacancy diffusion is rate limiting. This is not only counter intuitive but also not in accordance with observations. The problem is solved, of course, by using Eq. (1), since D will be proportional to the concentration of the vacancies of the rate limiting species, and the gradient in chemical potentials will be independent of the dopant concentration, provided the dopant is uniformly distributed.

The longest standing controversy in sintering mechanisms concerns the relative importance of dislocation motion. The level of interest in dislocation effects has been revised recently by the careful studies of Schatt and co-workers[10-12] who have demonstrated that large numbers of dislocations exist in the vicinity of large particles sintered to single crystal plates.

While the most successful models have been for the initial stage of sintering in which discreet necks can be described between pairs of sintering spheres, a number of models have been proposed for the intermediate and final stages. Here various schemes for averaging the relavent geometrical parameters have been employed. Models have been based on an assumed average grain (typically a tetrachaidecahedran with pores along the 3-grain edges or at the 4-grain corners)[13-16] or on global stereological parameters[17-19].

With the advent of fast firing, ideas have been put forth describing enhanced densification as driven by the resulting temperature gradients.[20-22]

Microstructure evolution studies and models have attempted to quantify the relationships between such parameters as density and grain growth[23-24] and pore coordination number[25] to describe the sintering behavior of compacts of real powders in the intermediate and final stages.

Recently models have been proposed which attempt to integrate kinetics and microstructure evolution.[26-28] It is too early to state with any degree of confidence the usefulness of these in ultimately predicting sintering behavior of materials. However, some interesting progress has been made.

THE SINTERING MODEL

The classic two-sphere model with adjoining neck still provides the best basis for discussing matter flows during sintering. It is well understood that densification of a compact takes place only by the flow of matter from the grain boundary between the particles to the surface of the neck. Surface diffusion of matter from the particle surface to the neck causes excess neck growth, which constitutes one manifestation of coarsening without densification. In both the initial and intermediate stages of sintering, the grain boundary intersects the pore phase around most or all of its perimeter. Here the pore phase is continuous throughout the compact. In the final stage the pores are isolated from one another and may or may not be intersected by grain boundaries. Only those intersected by grain boundaries will shrink at an appreciable rate.

In all stages of sintering the process is driven by the magnitude of the curvature of the pore surface in the vicinity of the grain boundaries that intersect the pores. Again, this is easily understood in the two sphere model, but an analogous situation obtains in the intermediate and final stages of sintering. The gradient in the chemical potential of the atoms is governed both by the magnitude of the curvature of the pore surface and the distance over which matter must be transported to reach the pore. In the two sphere model, the transport distance is on the order of the neck radius, which scales with the particle diameter. In the intermediate and early final stages, when the pores are isolated at three grain edges and four grain corners, the matter transport distance scales with the grain size. In the late final stage, as pores begin to disappear from four grain corners, the average diffusion distance scales as the mean distance between pores along the grain boundary interface. These concepts of mass transport distance will be useful in describing a universal sintering equation below.

The imposition of temperature gradients provides an additional driving force for mass transport, as has been described by Searcy[20-21] and more recently by Young and McPherson[22]. However, care must be exercised in discussing the extent to which temperature gradients can cause macroscopic densification of a compact. Consider an imposed temperature gradient across the simple two sphere model. If the gradient exists parallel to the grain boundary, mass will tend to flow from one side of the neck to the other. If the gradient exists normal to the grain boundary, mass will tend to flow from one sphere to the other. In neither case will there be a net increase in the flow of matter from the center of the grain boundary to its perimeter, which is the only flow that can cause volumetric densification in a powder compact. The situation is different at the free surface. Here matter can be driven inward from the external surfaces of the outermost particles. However, each neck represents a region of local minimum chemical potential and, therefore, a site for depositing material transported under the action of the temperature gradient. These matter flows probably take place by surface diffusion because of the high magnitude of the surface diffusion coefficient relative to grain boundary and volume diffusion. The distance that matter can be transported under a temperature gradient will be proportional to $\sqrt{(Dt)}$ and thus will be miniscule compared with the radius of a typical powder compact. Thus the most that can be expected from temperature gradients would be near surface densification. However, the hotter and cooler portions of the sample will influence the sintering of each other through mutual constraint. There probably are circumstances in which the overall densification is either enhanced or inhibited by these constraints. The possible effects of temperature gradients will not be considered further in the present discussion.

Although most sintering models have focused on the differences between the initial, intermediate, and final stages of sintering, it is useful to consider their similarities. The possible mechanisms of mass transport are identical (surface, grain boundary, and volume diffusion), densification occurs by a flow of matter from sources (grain boundaries) to sinks (pore or neck surface), additional coarsening without densification takes place by flow along this surface by surface diffusion or above the surface by vapor transport. Volume and grain boundary diffusion occur by roughly parallel paths and therefore under similar chemical potential gradients. Finally, the diffusion distance scales with the particle or grain size. Thus it is possible to write an equation relating the rate of linear shrinkage to the volume and grain boundary diffusion coefficients and the particle or grain size as follows:

$$-\frac{dL}{Ldt} = \frac{\gamma\Omega}{kT}\left[\frac{F_v(\rho)D_v}{G^3} + \frac{F_b(\rho)\delta D_b}{G^4}\right] \tag{5}$$

where L is the instantaneous length, γ is the surface tension, G is the grain or particle diameter, and $F_v(\rho)$ and $F_b(\rho)$ are dimensionless collections of scalling law parameters which depend upon the density, the relative importance of surface diffusion, and other factors not included explicitly in the other terms of the equation. The $F(\rho)$ values for the simple two sphere and tetrachaidecahedran models are known and are listed in Table I and plotted in Figs. 1 and 2. Computer simulations of the initial stage of sintering show that the curve for $F(\rho)$ is depressed by surface diffusion, particularly strongly at the earliest stages of sintering. It is likely that the values at all stages of sintering are depressed as a result of surface diffusion during the intial stage. Although the amount of depression is not known, it cannot be extreme in the intermediate and final stages. Work is underway to evaluate $F(\rho)$ for a variety of sintering conditions.

SIMULATION OF SINTERING

Equation (5) can be used to simulate sintering data for any arbitrary temperature-time profile, provided the diffusion coefficients and other parameters are known. Useful results may be obtained even in the absence of quantitative data, as long as reasonable approximations may be made. The linear shrinkage is obtained by numerical integration of Eq. (5). The density, in turn, is calculated from the initial density and the linear shrinkage, assuming isotropic shrinkage.

It can be shown from the relative magnitudes of $F_v(\rho)$ and $F_b(\rho)$ that, for particle or grain sizes of a few microns or less, grain boundary diffusion will predominate over volume diffusion for all reasonable ratios of D_b over D_v. There are several implications of this predominance of grain boundary diffusion. Since impurities and dopants segregate to grain boundaries, the diffusion coefficient will be extrinsic for most materials and, therefore, dependent upon overall purity of the material. While the effects of dopants on volumetric point defect concentrations may be indicative of those effects in the grain boundary, it would be surprising if there was a one-to-one correspondence. Volume diffusion contributions to sintering will be ignored in the present simulations.

In order to facilitate simulations of sintering based on Eq. (5), simplified $F_b(\rho)$ and grain size dependencies on density were assumed. The values of $F_b(\rho)$ must be continuous across the density range. Since the geometries of the initial and final stage models are more well defined than for the intermediate stage, the approximate form of $F_b(\rho)$ shown as a broken line in Fig. 2 was assumed. The initial stage model in Fig. 2 assumes no surface diffusion. The presence of surface diffusion will cause that portion of the curve to be lower and flatter.

Grain size is known to be a function of density, although the exact grain size-density trajectory will depend upon the presence of dopants or impurities as well as temperature and green density. For the present purposes, the grain size was arbitrarily assumed to be a second order power function on density increase, with a starting grain size of 0.25 μm at 50% dense and a final grain size at 99% dense of 3 μm, consistent with the results for plasma sintering of alumina with a surface area of 30 m^2/g* (the high surface area γ-alumina fraction transforms rapidly to coarser α-alumina, with the assumed particle size). The assumed grain diameter as a function of density is shown in Fig. 3. Alternatively, in order to explore the effects of the grain growth behavior on the simulation, a simulation was

*Baikowski CR30, Baikowski International Corp., Charlotte, NC 28210

completed in which the grain diameter was assumed to be a 6th order power function of density with the same beginning and ending point. This is shown in Fig. 3 as a broken line.

Fast firing was simulated assuming the arbitrary temperature profile shown in Fig. 4, where the maximum heating rate was 100 K/s, which occurred at 1850 K. Isothermal sintering was simulated utilizing the 0.25 μm starting particle size and the same initial temperature profile, with isothermal holds from 1500 to 1700 K.

The grain boundary diffusion coefficient reported by Cannon and Coble[29] from high temperature creep was used, as this probably is the best value of this parameter in the literature. Their value for δD_o is 8.60 x 10^{-10} m^3s, and the activation energy is 419 kJ/mol.

Figure 5 shows the linear shrinkage rates as a function of time calculated for the fast firing temperature profile for both grain growth functions. It must be kept in mind that these curves assume no surface diffusion. The presence of surface diffusion would tend to broaden the curves since it would inhibit sintering initially, which would be manifested in the present calculations as a reduction in $F_b(\rho)$ in the initial stage.

The density for isothermal sintering is plotted as a function of the log of the isothermal sintering time, according to Coble's intermediate stage sintering model,[13,14] in Fig. 6. The log of the linear shrinkage is plotted as a function of the log of the isothermal sintering time in Fig. 7.

Figures 5-7 can be compared with experimental data. First of all, it can be seen that the simulated densification rate during the the fast firing schedule (Fig. 5) closely approximates the observed results reported by Kemer and Johnson[30] for microwave plasma sintering of alumina of with the characteristics used in the simulations. Some of the features of the traditional density vs log time and log shrinkage vs log time curves used in the literature are readily apparent from the simulated curves. In particular, the log shrinkage-log time curves are, as expected, of much lower slope than the 1/3 value of the initial stage models.

It must be emphasized that the simulations presented above are only a zero order approximation. Work is under way on simultaneous measurements of grain growth and densification under various conditions to permit calculation of $F_b(\rho)$ and learn the dependence of this function on various parameters.

CONCLUSIONS

A sintering equation that extends from the initial through the late stages of sintering shows promise in providing a framework within which to describe sintering behavior. Scaling law functions, $F_v(\rho)$ and $F_b(\rho)$, relate material constants and grain size to the shrinkage rate. Simulated sintering data based upon the model is in agreement with experimental data during fast firing of Al_2O_3. Grain boundary diffusion is the predominate mass transport mechanism. Considerable work will be required to characterize the effects of surface diffusion and other sintering variables on the scaling law functions.

REFERENCES

1. J. Frenkel, "Viscous Flow of Crystalline Bodies Under the Action of Surface Tension," J. Phys. (USSR), 9 [5] 385-91 (1945).
2. G. W. Scherer, "Sintering of Low-Density Glasses: I, Theory," J. Am. Ceram. Soc., 60 [5-6] 239-43 (1977).
3. G. W. Scherer and D. L. Bachman, "Sintering of Low-Density Glasses: II, Experimental Study," J. Am. Ceram. Soc., 60 [5-6] 239-43 (1977).
4. G. W. Scherer, "Sintering of Low-Density Glasses: III, Effect of a Distribution of Pore Sizes," J. Am. Ceram. Soc., 60 [5-6] 239-43 (1977).
5. M. N. Rahaman, L. C. De Jonghe, G. W. Scherer, and R. J. Brook, "Creep and Densification During Sintering of Glass Powder Compacts," J. Am. Ceram. Soc., 70 [10] 766-74 (1987).
6. D. Lynn Johnson and T. M. Clarke, "Grain Boundary and Volume Diffusion in the Sintering of Silver", Acta Met., 12, 1173-1179 (1964).
7. D. Lynn Johnson, "New Method of Obtaining Volume, Grain Boundary and Surface Diffusion Coefficients from Sintering Data," J. Appl. Phys., 40, 192-200 (1969).
8. D. Lynn Johnson, "Methods of Determining Sintering Mechanisms," 1970 Modern Developments in Powder Metallurgy, Vol. IV, ed. H. H. Hausner, Plenum Press, N.Y. (1971), pp. 189-198.
9. C. Herring, The Physics of Powder Metallurgy, McGraw-Hill Book Co., N.Y. (1951), p. 143.
10. W. Schatt and K.-P. Wieters, "On the Nature of the Sintering Contact Zone," Powder Metall. Int., 18 [3] 135-139 (1986).
11. W. Schatt, E. Friedrich, and K. P. Wieters, "Dislocation-Activated Sintering", Reviews on Powder Metallurgy and Physical Ceramics, Freund Publ. House, Ltd., England, Vol. 3, No. 1, (1986) pp. 1-111.
12. W. Schatt and M. Hinz, "On the Generalizability of Defect-Activated Sintering," Powder Metall. Int., 20 [6] 17-20 (1988).

13. R. L. Coble, "Sintering Crystalline Solids. I. Intermediate and Final Stage Diffusion Models, "J. Appl. Phys., 32, 787 (1961).
14. R. L. Coble, "Intermediate-Stage Sintering: Modification and Correction of a Lattice-Diffusion Model," J. Appl. Phys., 36, 2327 (1965).
15. R. L. Coble and T. K. Gupta, Sintering and Related Phenomena, ed. G. C. Kuczynski, N. A. Hooton, and C. F. Gibbon, Gordon and Breach Science Publ., Inc., N.Y. (1967) pp. 423-41.
16. E. Arzt, M. F. Ashby, and K. E. Easterline, "Practical Applications of Hot-Isostatic Pressing Diagrams: Four Case Studies," Metall. Trans. A, 14a, 211 (1983).
17. D. Lynn Johnson, "A General Model for the Intermediate Stage of Sintering," J. Am. Ceram. Soc., 53 [10] 574-577 (1970).
18. R. T. DeHoff, "Stereological Theory of Sintering", Proc. VII World Round Table Conf. on Sintering, Herceg-Novi, Yugoslavia, Aug. 1989, in press.
19. H. E. Exner and E. A. Giess, "A Stereology-Based Equation for Isotropic Shrinkage During Sintering by Viscous Flow," Proc. VII World Round Table Conf. on Sintering, Herceg-Novi, Yugoslavia, Aug. 1989, in press.
20. A. W. Searcy, "Theory for Sintering in Temperature Gradients: Role of Long-Range Mass Transport," J. Am. Ceram. Soc., 70 [3] C-61-C-62 (1987).
21. D. Beruto, R. Botter, and A. W. Searcy, "Influence of Temperature Gradients on Sintering: Experimental Tests of a Theory," J. Am. Ceram. Soc., 72 [2] 232-235 (1989).
22. Robert M. Young and Reginald McPherson, "Temperature-Gradient-Driven Diffusion in Rapid-Rate Sintering," J. Am. Ceram. Soc., 72 [6] 1080-81 (1989).
23. S. J. Bennison and M. P. Harmer, "Effect of MgO Solute on the Kinetics of Grain Growth in Al_2O_3," J. Am. Ceram. Soc., 66, C90 (1983).
24. S. J. Bennison and M. P. Harmer, "Grain Growth Kinetics for Alumina in the Absence of a Liquid Phase," J. Am. Ceram. Soc., 68 [1] C-22-C-24 (1985).
25. F. F. Lange, "Sinterability of Agglomerated Powders," J. Am. Ceram. Soc., 67, 83 (1984).
26. J. Zhao and M. P. Harmer, "Effect of Pore Distribution on Microstructure Development: I, Matrix Pores," J. Am. Ceram. Soc. 71 [2] 113-120 (1988).
27. J. Zhao and M. P. Harmer, "Effect of Pore Distribution on Microstructure Development: II, First- and Second-Generation Pores," J. Am. Ceram. Soc., 71 [7] 530-39 (1988).
28. D. Lynn Johnson, "Ultra-Rapid Sintering of Ceramics," Proc. of VII World Round Table Conf. on Sintering, Herceg-Novi, Yugoslavia, August 1989, in press.

29. R. M. Cannon and R. L. Coble, "Review of Diffusional Creep of Al_2O_3," Deformation of Ceramic Materials, ed. R. C. Bradt and R. E. Tressler, Plenum Press, N.Y. (1975), pp. 61-100.
30. Eric L. Kemer and D. Lynn Johnson, "Microwave Plasma Sintering of Alumina," Am. Ceram. Soc. Bull., 64 [8] 1132-1136 (1985).

ACKNOWLEDGEMENTS

This work was supported by the National Science Foundation Grant No. DMR-8719077.

Table I. Density functions for Equation (1)

Model	Stage	$F_v(\rho)$	$F_b(\rho)$	Reference
A	Initial ($\Delta L/L < 0.05$)	$\frac{20}{\Delta L/L_o}$	$\frac{12.5}{(\Delta L/L_o)^2}$	8
B	Intermediate	126	$\frac{330}{(1-\rho)^{1/2}}$	17
C	Intermediate	$\frac{111}{\rho}$	$\frac{287}{\rho(1-\rho)^{1/2}}$	13
D	Final	$\frac{288f(\rho)}{\rho}$	$\frac{1047f(\rho)}{\rho(1-\rho)^{1/3}}$	16

$$f(\rho) = \frac{1 - (1-\rho)^{2/3}}{3\,(1-\rho)^{2/3} - [\,1 + (1-\rho)^{2/3}]\,\ell n\,(1-\rho) - 3}$$

Note: A = normalized neck surface area; X = normalized neck radius; R = normalized minimum radius of curvature of the neck surface; H = mean surface curvature; S_v = pore surface area per unit volume; L_v = pore-grain boundary perimeter length per unit volume.

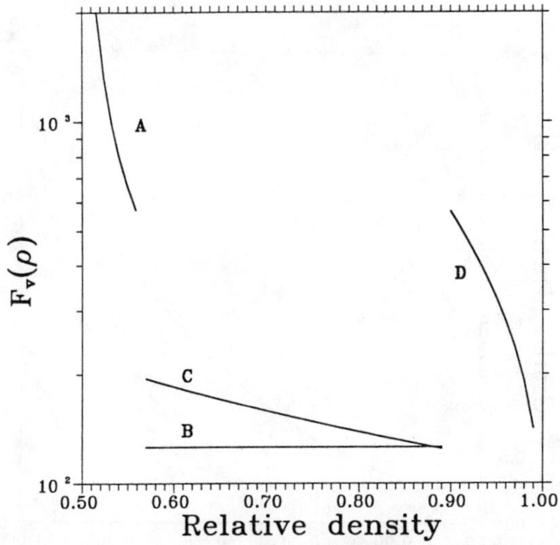

Figure 1. $F_v(\rho)$ as a function of relative density. Capital letters refer to the models in Table 1. The intermediate stage is assumed to extend from 56% to 90% dense, with the initial and final stages before and after this interval.

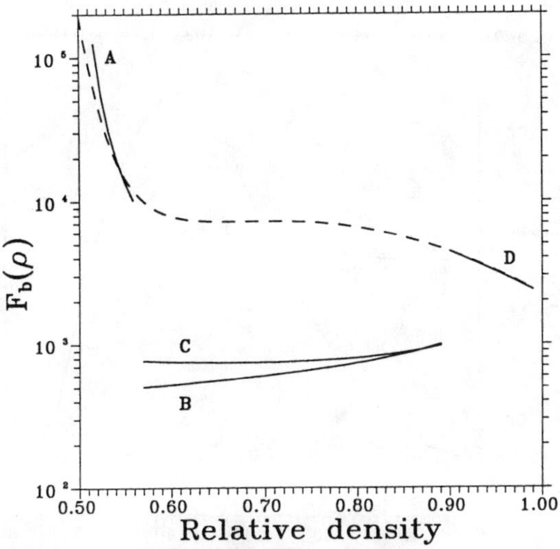

Figure 2. $F_b(\rho)$ as a function of relative density. The broken line represents values used in the simulations.

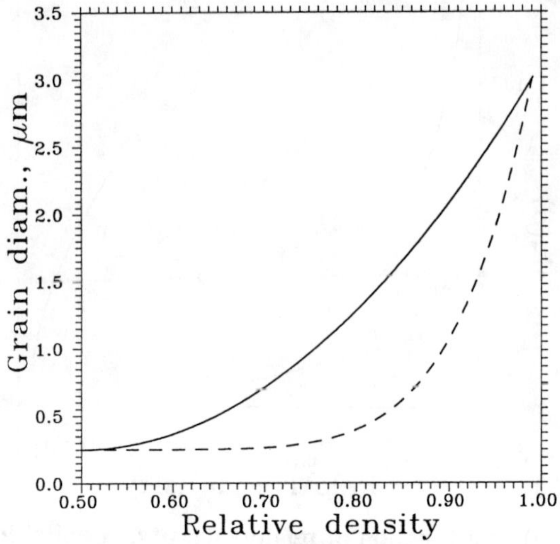

Figure 3. **Assumed grain growth trajectories. The solid and broken lines represent second and 6th order functions, respectively.**

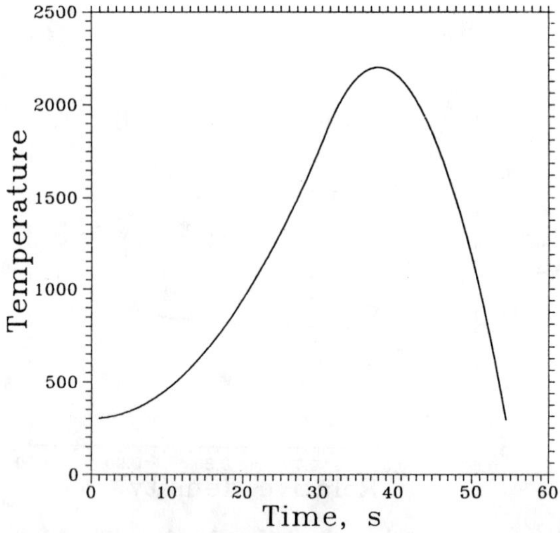

Figure 4. **Assumed temperature profile (Kelvin).**

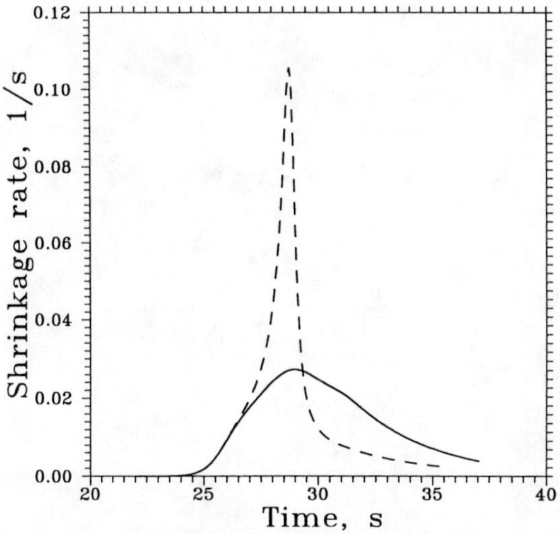

Figure 5. Simulated linear shrinkage rate for fast firing. The solid and broken lines were obtained using second and 6th order grain growth functions, respectively.

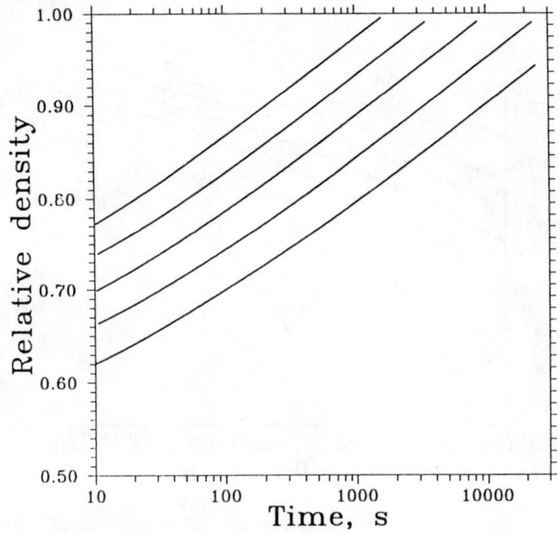

Figure 6. Simulated relative density for isothermal holds at temperatures from 1500 to 1700 K.

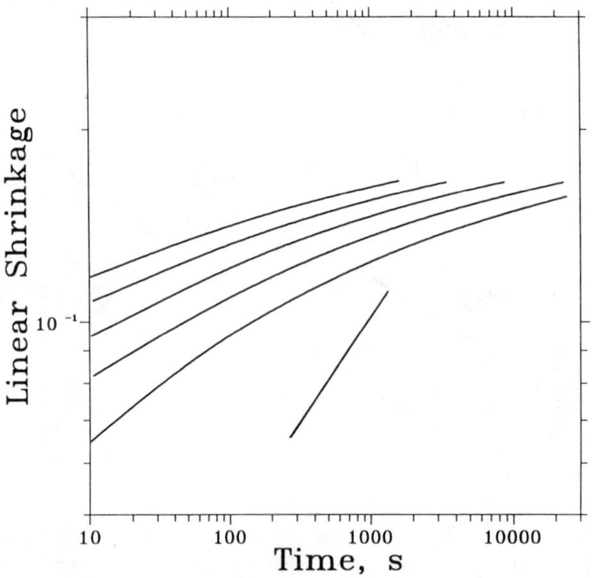

Figure 7. Simulated linear shrinkage for isothermal holds at temperatures from 1500 to 1700 K. The straight line has a slope of 1/3.

EFFECT OF NITROGEN PRESSURE ON THE SINTERING OF AlN

Naohiro Takada, Michiyasu Komatu, and Takashi Takahashi,
Toshiba Corp. New Material Engineering Laboratory
2–4.Suehiro–cho, Tsurumi–ku, Yokohama, 230 Japan

ABSTRACT

The relationship between sintering nitrogen pressure and the properties of AlN with Y_2O_3 was investigated.
Results indicate that nitrogen pressure had a strong correlation with surface topography for AlN ceramics. When nitrogen pressure was low(0.1 – 0.35 MPa), the local exudation of the boundary phases was observed at the AlN surface and made large particles. Therefore, it made the surface rougher. With an increase in nitrogen pressure(0.85 – 1.0 MPa), this phenomenon was suppressed and the surface grains had a tendency to become smaller, rather more uniform in size and rounded. Therefore, the surface roughness for AlN ceramics was improved.
On the other hand, the nitrogen pressure dependence of the thermal conductivity and mechanical strength was found to be negligibly small.

INTRODUCTION

ALUMINUM NITRIDE (AlN) is an attractive alternative candidate to BeO and Al_2O_3 in semiconductor substrates because of its high thermal conductivity, over 8 times that of Al_2O_3, its close thermal expansion coefficient to that of silicon, its high electrical resistivity and its nontoxic property.

Up to now, there have been several reports published on the sintering of AlN (1) and the development of high–thermal–conductivity AlN ceramics (2–3). However in these papers, only AlN sintered under normal nitrogen atmosphere was evaluated.

In the case of Si3N4, it was difficult to obtain dense ceramic by pressureless sintering, due to the high vapor pressure at the sintering temperature used (4).

Mitomo et al. (5) reported that the thermal decomposition of Si3N4 was suppressed by increasing nitrogen pressure during sintering.

Even though AlN is more stable at high temperature than Si3N4, AlN is a nitride ceramic, as well as Si3N4. Moreover, it seems that the interactions between the sintering atmosphere and the behavior of the boundary phases cannot be ignored for sintering AlN, because the reduction of the boundary phases during sintering was necessary for obtaining high thermal conductivity (2–3). In fact, Udagawa (6) reported that, in the AlN–CaCO3 systems, the efficiency of the formation and evaporation of the calcium aluminum oxides, which are the boundary phases, became higher when sintered under increased nitrogen gas pressure and that a high thermal conductivity (240W/mK) was obtained when sintered at 0.7MPa. The sintering pressure apparently influences not only the thermal conductivity, but also the surface properties of the ceramic.

This paper mainly reports results obtained from investigating the relations between sintering nitrogen pressure and surface topography for sintered–bodies in the AlN – Y2O3 systems, while paying attention to the behavior of the boundary phases .

EXPERIMENTATION

Sample preparation

Fig.1 shows the experimental procedure. AlN powder, with

an average particle size of approximately 0.6um and 0.85wt% oxygen content, was used for this study as a raw material. The 99.9% purity Y2O3 powder with 1.0um average grain size was used as an additive. AlN powder and 3wt% Y2O3 powder were mixed well with an organic binder and solvents. The slurry was cast into a ceramic green sheet by the doctorblade method, which was a conventional fabrication method used for alumina substrates. After punching, green sheets were heated at 773 K to remove the binder. Then, they were sintered at 2048 K for 6 – 24 hours under nitrogen atmosphere, (0.1 – 1.0 MPa). Finally, the characteristics for sintered–bodies were evaluated by several methods.

AlN surface evaluation

Surface roughness, microstructures and atomic distribution for the as–fired surface, which faced the atmosphere directly in the furnace, were evaluated by profilometry, scanning electron microscopy(SEM) and EPMA analysis, respectively. Moreover, the second phases at the AlN surface were determined quantitatively by intensity ratio of X–ray diffraction patterns for each product, calibrated in advance.

Thermal conductivity

Thermal conductivity at room temperature was measured by the laser flash method , with an InSb infrared detector. Sample size was approximately 3mm thick and 10mm in diameter. Thermal conductivity was calculated, using thermal diffusivity, density and specific heat.

Mechanical strength

Fracture strength was measured by the three point bending method. Both as–fired and ground and polished specimems were examined.

RESULTS AND DISCUSSION

AlN surface evaluation

Fig.2 demonstrates the nitrogen pressure effect on surface roughness(Ra) for the as–fired surface. The vertical axis value was normalized to the value(Ra), when sintered under 0.1MPa pressure for 6 hours. The nitrogen pressure used in the sintering process was effective in improving the surface roughness for as–fired surfaces. With increasing nitrogen pressure, the surface roughness became better. In addition, it could be recognized that the surface was roughened as a function of holding time. However, at the 1.0MPa pressure, the change was negligibly small.

AlN surface microstructures are shown in Fig.3. When the nitrogen pressure was high, surface grains had a tendency to become smaller, rather uniform in size and rounded. When sintering was done at less than 0.35MPa, large particles were observed which were larger than the normal grains. However, at more than 0.85MPa they were not observed. The microstructure observation corresponded to foregoing surface roughness tendency. It is concluded that the large particles formation and AlN grain growth were suppressed by the usage of high nitrogen pressure.

The results obtained by EPMA analysis and SEM microstructure observation on the large particle are shown in Fig.4 and 5, respectively. The large particles consisted of AlN grains and the boundary phase compounds, which were mainly yttrium and oxygen. The boundary phases were concentrated locally on and in the large particles.

According to the X–ray diffraction, boundary phases existing at the as–fired surface were Y3Al5O12(YAG) and YAlO3(YAL). The other expected phases, such as Y2O3, Al2Y4O9(YAM)and so on, couldn't be detected(Fig.6). The total content of these boundary phases (YAG + YAL) in the surface region neighborhood depended on the nitrogen pressure. A sample sintered at lower pressure had a rather larger amount of phases than one sintered at higher

pressures.
Higher nitrogen pressure not only suppressed the exudation of the large particles, but also realized more homogeneous yttrium ion distribution on the surface.
These phenomena suggested that the boundary phases behavior in the sintered–body has dependence on the nitrogen pressure during sintering. It was presumed that the large particle formation occurred due to the boundary phases exudation to the surface, AlN decomposition around the surface and grain growth of boundary phase compounds. However, at this point, in order to analyse the formation process for the large particles, it is necessary to clarify the nitrogen pressure dependence of the liquid phase formation temperature, the liquid phase viscosity, the liquid phase wettability to the AlN, the stability of boundary phase, the grain growth for boundary phase and AlN decomposition.
The amount and the compositions of boundary phases also depended on the nitrogen pressure. When the nitrogen pressure was high, YAG content was increased and YAL content was decreased (Fig.6). It was considered that this phenomena was caused by the decrease in Al2O3 evaporation or the decrease in reduction efficiency for the boundary phase, due to the suppresion of CO gas generation by high nitrogen pressure.

Thermal conductivity and mechanical strength

In this study, the nitrogen pressure dependence of thermal conductivity was found to be negligibly small. The values were from 170W/mK to 178 W/mK (Fig.7). Moreover, the same tendency appeared even if the holding time was increased up to 24hours. On the other hand, Uadagawa (6) reported in AlN–CaCO3 that Calcium Aluminate Oxides, the boundary phases, were formed and evaporated actively, when sintered under increasing nitrogen pressure. Highest thermal conductivity (240W/mK) was obtained when sintered at 0.7MPa. The authors think that the differences in these results are due to the difference in the boundary phase behavior between the AlN–Y2O3 system and the AlN–CaCO3 system. Apparently, the Calcium aluminate compound decomposes and evaporates more easily than Yttrium aluminate compound.

In the as–fired specimen which was not treated, both average and minimum mechanical strength values were increased slightly, with increasing nitrogen pressure(Fig.7). However, in the case of the ground and polished sample, the fracture strength wasnot dependent on the nitrogen pressure, A 400MPa to 500MPa value, was obtained. It is considered that the large particles were related to the origin of the fracture.

CONCLUSION

The relationship between sintering nitrogen pressure and the properties of AlN sintered with Y2O3 was investigated. Especially, the surface characteristics for the sintered–body were evaluated.

Nitrogen pressure had a direct correlation with the AlN surface topography. When the nitrogen pressure was low(0.1 – 0.35 MPa), large particles, which were larger than the AlN grain, were locally exuded onto the surface. These particles were formed from yttrium aluminate boundary phases concentrated onto the surface and AlN particles. The surface became rougher. On the other hand, when sintered at high pressure (0.85 – 1.0 MPa), the local exudation of boundary phases on the surface was suppressed and large particles were not observed. In addition, surface grains had become smaller, rather uniform in size and rounded. As a result, the surface roughness was lowered. Moreover, the composition of the boundary phase compound and the amounts in the surface layer also depended on the nitrogen pressure. These phenomena indicated that the behavior of the boundary phases in the sintered–body had dependence on nitrogen pressure during sintering.

In this AlN–Y2O3 system, the nitrogen pressure dependence

for the thermal conductivity and mechanical strength was found to be negligibly small.

REFERENCE

1)K.Komeya and H.Inoue:The influence of fibrous aluminum nitride of the strength of sintered AlN−Y2O3,Trans.& J.Br.Ceram.Soc.,70,pp.107−113(1971)

2)A.Horiguchi,F.Ueno and A.Tsuge:Development of high−thermal conductivity aluminum nitride ceramic (in Japanease),Toshiba Review 44[8] pp.616–618(1989)

3)N.Kuramoto and H,Taniguchi:Transparent AlN ceramic, J.Mater.Sci.Lett.,3[6]p.471(1984)

4)G.R.Terwilliger and F.F.Lange:Pressureless sintering of Si3N4, J.Mater.Sci.,10,pp.1169(1975)

5)M.Mitomo:Pressure sintering of Si3N4,J.Mat.Sci.,11,pp.1103 −1107(1976)

6)E.Udagawa,H.Makihara,N.Kamehara and K.Niwa:Influence of gas−pressure for thermal conductivity of AlN (in Japanease), The proceeding of the 1st symposium in autumn,pp.241−242 (1988)

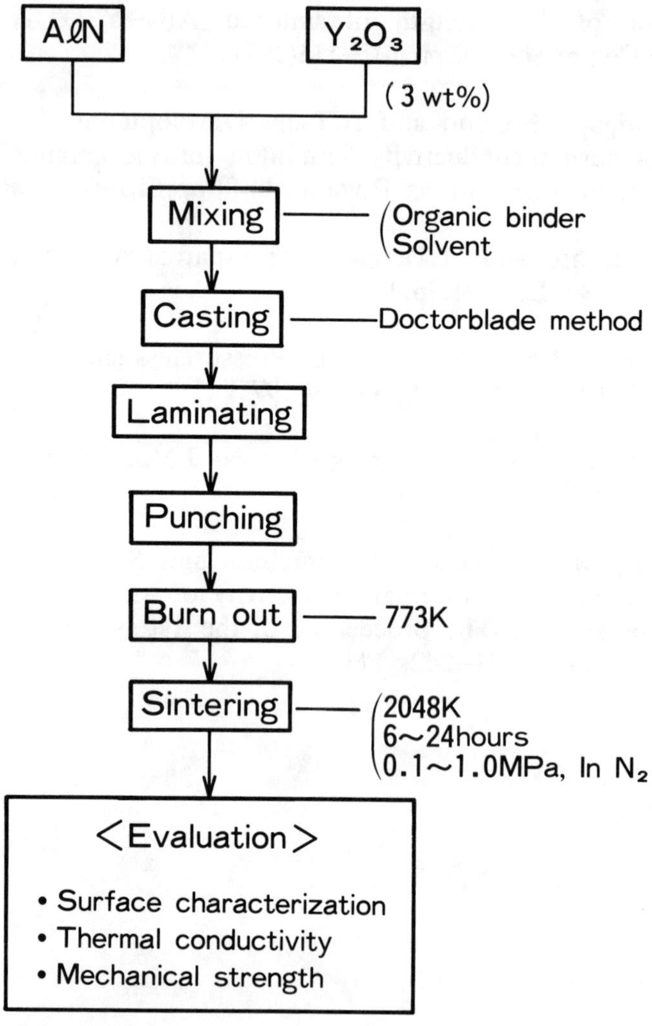

Fig. 1. Experimental procedure

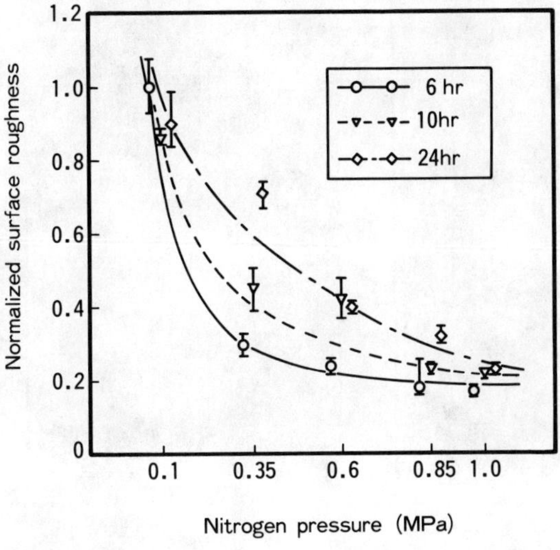

Fig. 2. Surface roughness nitrogen pressure dependence

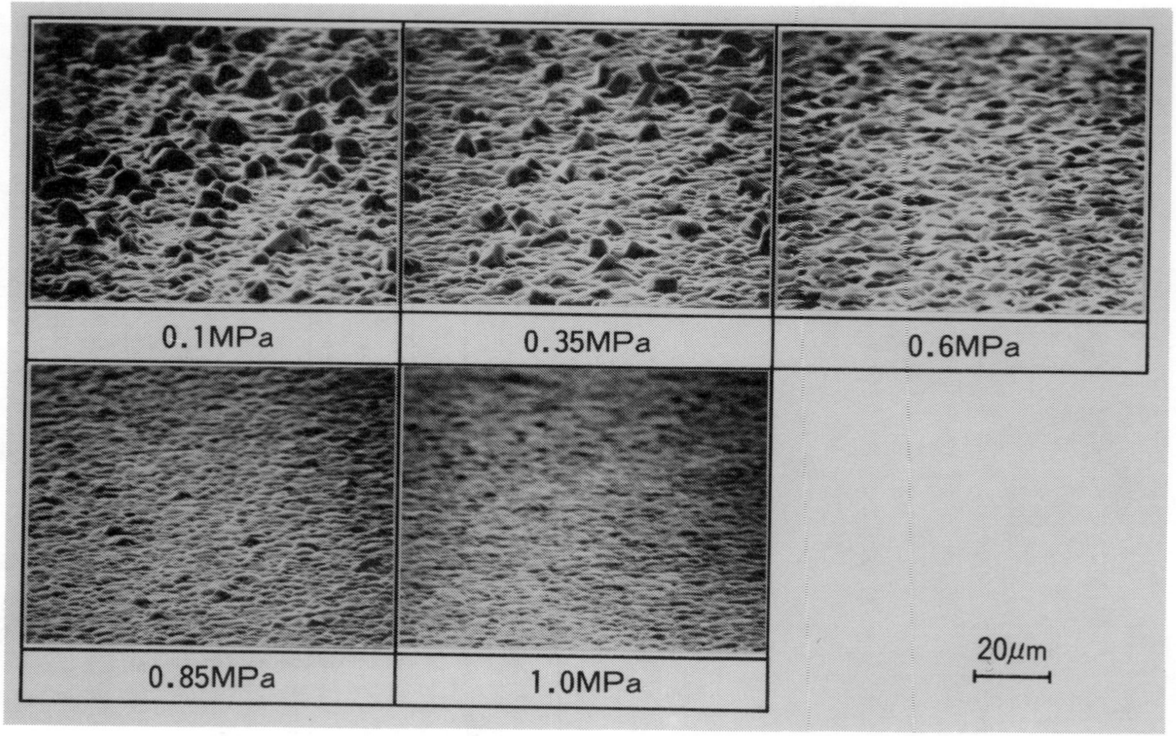

Fig. 3. Relation between surface structure and nitrogen pressure, when sintered for 24 hours

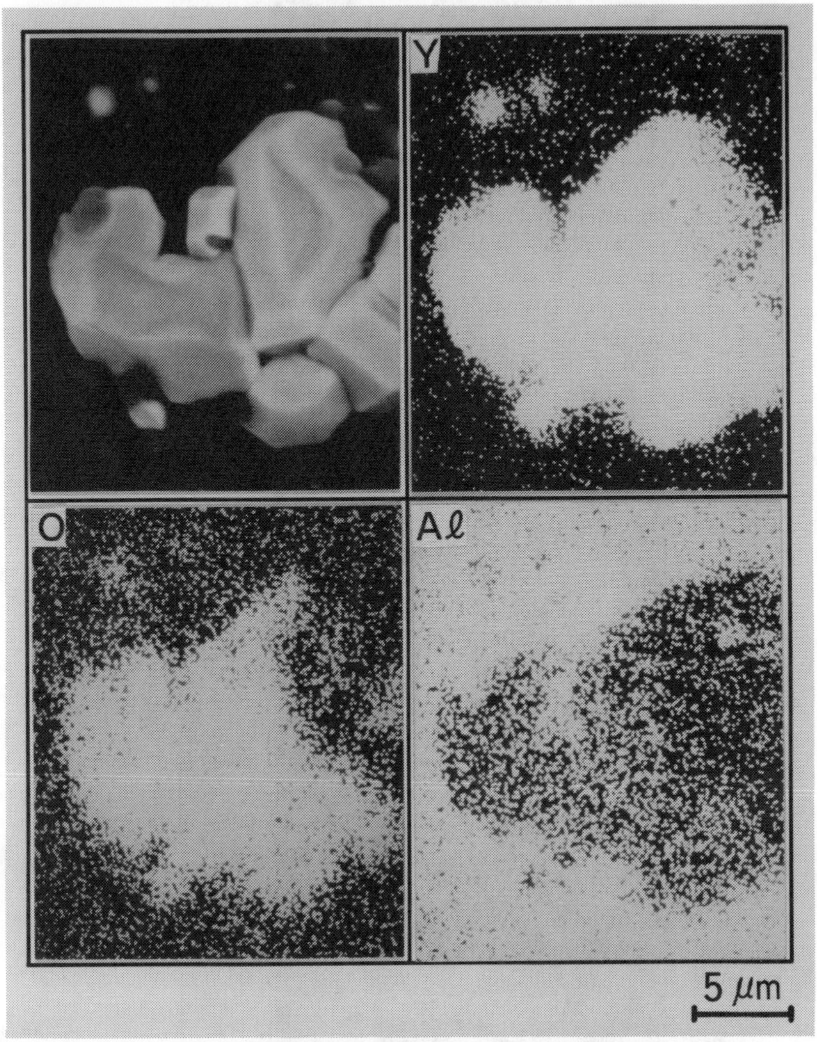

Fig. 4. EPMA analysis for large particle, when sintered at 0.1MPa for 6 hours

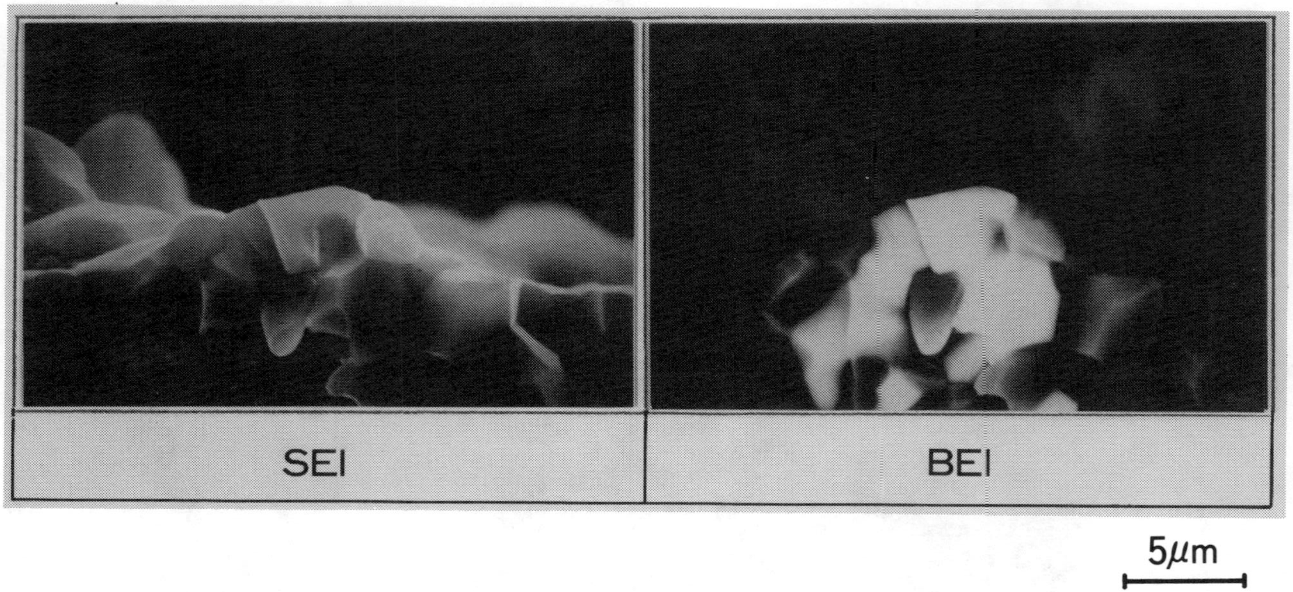

Fig. 5. Microstructure for large particle, when sintered at 0.1MPa for 6 hours

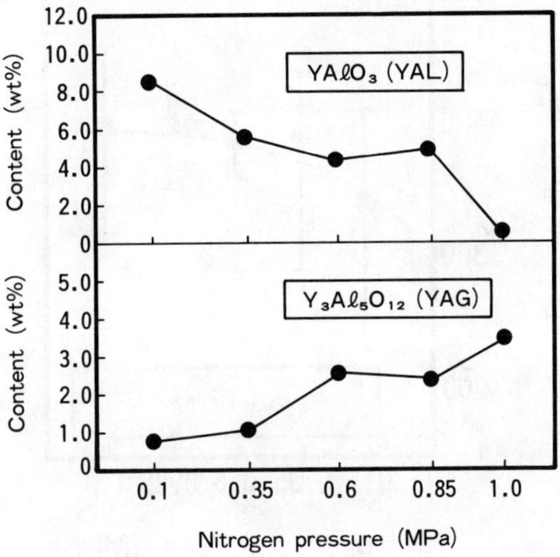

Fig. 6. Amount of boundary phases at AlN surface, determined by X-ray diffraction pattern intensity ratio

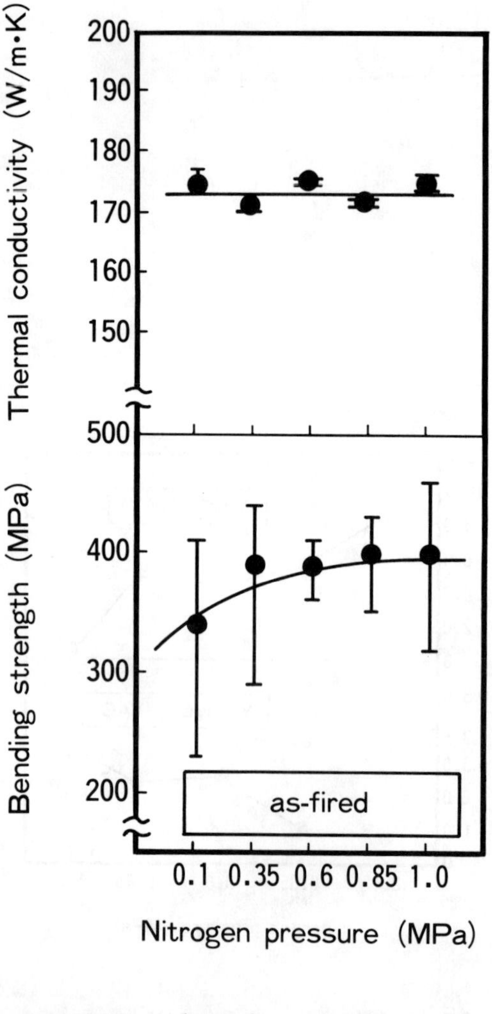

Fig. 7. Nitrogen pressure dependence of thermal conductivity and mechanical strength, when sintered for 6 hours

DIELECTRIC AGING OF DOPED PMN-PT CERAMICS

A.E. Sutherland, S.M. Pilgrim, A.E. Bailey, and S.R. Winzer
Martin Marietta Laboratories
1450 S. Rolling Road
Baltimore, Maryland 21227

ABSTRACT

Many relaxor formulations, including PLZT and SBN materials, display dielectric aging. Although much work has been done on PMN systems, no dielectric aging has been reported in stoichiometric PMN-PT. However, aging has been reported with additions of Mn or excess PbO.

We have found substantial aging effects in a number of PMN-PT and PMN-PT-ST compositions synthesized using the columbite precursor method. By varying the processing conditions of the precursor, we have been able to control aging and pyrochlore content. Aging was assessed by measurement of dielectric constant and loss as functions of temperature and by polarization-field behavior.

INTRODUCTION

Many relaxor formulations, including strontium barium niobate (SBN) and lead lanthanum zirconate titanate (PLZT) materials, display dielectric aging. Although much work has been done on lead magnesium niobate (PMN) systems, no dielectric aging has been reported in stoichiometric lead magnesium niobate-lead titanate (PMN-PT), lead magnesium niobate-lead titanate-barium titanate (PMN-PT-BT), or lead magnesium niobate-lead titanate-strontium titanate (PMN-PT-ST) systems. However, aging has been reported in PMN-PT systems with additions of Mn or excess PbO [refs. 1 and 2]. The systems that show aging all exhibit similar characteristics (Fig. 1): the formation of a saddle point in the plot of dielectric constant (K) versus temperature (T), effectively creating two dielectric constant peaks, and a dielectric loss peak at the aging temperature. Whereas in the K plot the lower temperature peak retains relaxor character, the higher temperature peak loses it. Also, the saddle point is loosely correlated with the aging temperature and is removed by heating the sample above the Curie temperature (T_C). Typically, the curve containing the saddle point is measured by increasing temperature on a sample aged for ~ 1 week at room temperature. The single-peaked curve is measured after heating the sample to 100°C for five minutes and subsequently decreasing the temperature. Such heating effectively restarts the aging clock. Three major types of models for dielectric aging have been postulated: volume, domain, and grain boundary [ref. 3]. Each of these mechanisms is based on the formation of an internal bias.

We have found aging in a number of PMN-PT, PMN-PT-BT and PMN-PT-ST systems. In this particular study, we concentrated on one composition of PMN-PT-BT (and an analogous composition in the PMN-PT-ST system) synthesized via the columbite precursor method [ref. 4]. By varying the processing conditions of the precursor, we have been able to control aging (and also pyrochlore content). In our study, aging was assessed by measurement of dielectric constant and loss as functions of temperature (only dielectric constant data will be shown for clarity) and by polarization versus field behavior.

EXPERIMENTAL PROCEDURE

In order to make the powders for our study, we synthesized two batches of the $MgNb_2O_6$ precursor (Fig. 2). Both were made from high purity MgO (99.5%) and Nb_2O_5 (99.999%), however, one was stoichiometric and the other contained 1.5 wt% excess MgO. Both batches were ball milled in Freon with zirconia media for ~ 20 hours to mix. They were then dried and divided in half. One half of each batch was then calcined at 1075°C for 4 hours while the other half was calcined at 1075°C for 15 hours. After calcining, the four batches of $MgNb_2O_6$ were ball milled in Freon with zirconia media for 20-30 hours to reduce particle size. After drying, stoichiometric amounts of high purity (99.999%) PbO, TiO_2, and $BaCO_3$ (or $SrCO_3$) were added to each. The powders were again ball milled for ~ 20 hours to ensure homogeneity, dried, and calcined at 800°C for 3 hours. They were then milled again for 20-30 hours for particle size reduction and dried. Pellets were die pressed and then isostatically pressed at 300 MPa before sintering in closed Al_2O_3 crucibles, packed in their own powder, at 1150°C for 3 hours. The batches of powder and hence pellets were labeled as in Table 1.

Table 1. Processing Differences of Compositions

Composition #	Processing of $MgNb_2O_6$
100(200)MLE*	1075°C/4 h
100(200)MLF	1075°C/15 h
100(200)MLG	1.5% excess MgO 1075°C/15 h
100(200) MLH	1.5% excess MgO 1075°C/4 h

*100 = PMN-PT-BT, 200 = PMN-PT-ST

RESULTS

X-ray diffraction (XRD) analyses were done on all sintered pellets (Fig. 3). The data indicated that samples made from the stoichiometric $MgNb_2O_6$, regardless of its calcining time, contained pyrochlore, whereas those samples made from the $MgNb_2O_6$ with excess MgO did not.

Scanning electron photomicrographs were taken of fracture surfaces of the sintered pellets to determine grain size and fracture type (Fig. 4). In the PMN-PT-BT system (the results of the PMN-PT-ST system differ, thus they will be discussed later), the 100MLE and 100MLF samples (from stoichiometric $MgNb_2O_6$) had a somewhat smaller grain size (1.5-2.0 μm) than the 100MLG sample (grain size = 3.4 μm) and a significantly smaller grain size than 100 MLH (grain size = 8.1 μm). Also note that 100MLE shows intergranular fracture, whereas 100MLF, G and H show transgranular fracture.

Quantitative analysis on polished surfaces of sintered samples by scanning electron microscopy/energy dispersive spectroscopy (SEM/EDS) showed no variation in the PbO, MgO and Nb_2O_5 contents in excess of the statistical counting error limits. The variation in TiO_2 and BaO content was outside statistical error, and 100MLE and 100MLF had a higher concentration of BaO and a lower concentration of TiO_2 than 100MLG and 100MLH. The results of the analysis are given in Table 2. Photomicrographs of the polished surfaces showing points of analysis are shown in Fig. 5.

After the physical properties of the samples were determined, the samples were measured to determine if they aged. Aging was assessed by measuring dielectric constant and loss as functions of temperature and also by measuring polarization versus field. The dielectric measurements were done on samples that had aged at least one week. The procedure involved lowering the temperature of the sample to -50^oC and then heating it at 4^oC/minute while measuring K and loss as functions of temperature at 0.1, 1, 10, and 100 kHz. This was done over the temperature range of -50^oC to 100^oC. The samples were then held at 100^oC for five minutes before cooling at a rate of 4^oC/minute to -50^oC, again measuring K and loss as functions of temperature. All measurements were done on a Hewlett Packard-controlled LCR meter system. As discussed in the Introduction, aging is evidenced by the formation of a double peak with a saddle point when the properties are measured at increasing temperatures ("up runs") and a disappearance of this saddle point on the decreasing temperature runs ("down runs"). The only PMN-PT-BT sample that showed aging of the low field dielectric constant and loss tangent was 100MLH (Figs. 6 and 7). In the other three samples, the data for both the up and the down runs were identical with respect to aging behavior. (The offset between the up and down runs seen in Fig. 6a results from a measurement artifact). The dielectric aging was confirmed by polarization-field measurements

Table 2. Average Compositions from SEM/EDS Analysis of Polished Specimens of PMN-PT-BT

Sample	PbO	MgO	Nb_2O_5	TiO_2	BaO
100MLE	71.15	2.70	21.53	2.49	2.13
+/-	0.94	0.54	1.69	0.57	0.40
100MLF	70.14	2.59	23.17	2.21	1.89
+/-	2.12	0.89	3.80	0.44	1.04
100MLG	71.11	2.98	22.32	3.42	0.16
+/-	0.77	0.40	0.80	0.36	0.34
100MLH	70.88	3.19	22.11	3.32	0.50
+/-	0.33	0.48	0.55	0.45	0.43
Nominal	68.19	3.61	23.82	3.43	0.96

(Composition (wt %))

made via an integrating capacitor technique (Figs. 8 and 9). The 100MLE and 100MLF samples aged in neither dielectric nor polarization measurements. The 100MLH sample aged according to both measurements. The 100MLG sample showed no aging of the low field K, but showed slight aging in polarization field measurements. This would indicate that polarization measurements are a more sensitive means of detecting aging than dielectric measurements. Also note that although low field dielectric constants are affected by aging when it is present, high field dielectric constant (dP/dE) seems to be independent of aging state.

It was noted in the Introduction that this study was done concurrently in a PMN-PT-BT and a PMN-PT-ST (same composition except that ST replaced BT) system. Most of the data presented thus far has been for the BT system and not the ST system. The PMN-PT-ST samples were processed in the same manner as those in the PMN-PT-BT system and the XRD patterns showed similar results (Fig. 10). However, analysis of the fracture surface indicated a definite difference in grain size (Fig. 11). The grain sizes of 200MLE, F, and G were similar to those of their BT counterparts (100MLE, F, and G); the grain size of 200MLH was 1/4 that of 100MLH. The reason for this change in grain size is not known. The only intentional difference between the two samples was the substitution of Sr for Ba. The SEM/EDS analyses of the ST samples indicated that they were all the same composition (within statistical counting error). These samples differed from the BT samples in that they did not vary in TiO_2 and SrO content as the BT samples did in TiO_2 and BaO. The results of the analysis are given in Table 3. Photomicrographs of the polished surfaces showing points of analysis are shown in Fig. 12.

Table 3. Average Compositions from SEM/EDS Analysis of Polished Specimens of PMN-PT-ST

Sample	Composition (wt %)				
	PbO	MgO	Nb_2O_5	TiO_2	BaO
200MLE	72.19	3.02	20.50	3.00	1.28
+/-	1.20	0.69	2.07	0.52	0.34
200MLF	72.36	3.09	19.97	3.03	1.55
+/-	0.56	0.50	0.85	0.31	0.33
200MLG	71.72	3.26	20.29	3.17	1.55
+/-	0.91	0.37	0.77	0.41	0.41
200MLH	71.73	3.25	20.28	3.14	1.60
+/-	0.76	0.51	0.67	0.49	0.44
Nominal	68.40	3.62	23.90	3.44	0.65

The dielectric data for these samples indicated that only 200MLG aged (Figs. 13 and 14); however, the polarization data showed that none of the samples aged (Figs. 15 and 16). It should be noted, however, that the 200MLG sample was not as dense as the others (7.4 g/cm^3 (92% theoretical density) versus 7.9 g/cm^3 (98% theoretical density)). Consequently, the dielectric data are not truly representative of the composition. Except for this sample, the material made with ST appears to be less sensitive to aging than the material made with BT. This trend was also observed in data obtained in a separate experiment. Again, a PMN-PT-BT and corresponding PMN-PT-ST material were synthesized. As can be seen from the K versus T curve, the PMN-PT-BT material showed a larger degree of aging than the PMN-PT-ST material (Fig. 17).

CONCLUSIONS

From the data obtained thus far, we can made several observations. Excess MgO in the columbite precursor acts to inhibit pyrochlore formation in the sintered pieces. Processing variables significantly affect grain size in the PMN-PT-BT system and to a lesser degree in the PMN-PT-ST system. Aging appears to be grain size dependent (only present in materials with a "larger" grain size). There is a significant variation in microchemistry with respect to Ba and Ti in the BT system, but no such variation is observed in the ST system. Pyrochlore, likely present as discrete grains since dielectric properties are not degraded, seems to inhibit aging. Calcining time and MgO concentration in the columbite precursor affect aging. (Aging was only present in the sample where the columbite precursor contained excess MgO and was calcined for 4 hours.) The polarization loops show aging consistent with the dielectric measurements and may be more sensitive to aging. Materials that contain strontium titanate seem to be less sensitive to aging that their barium titanate counterparts.

In conclusion, although pyrochlore content and small grain size correlate with the absence of aging, no causal relationship or mechanism connecting these and aging has been postulated. All may be concurrent effects of an unknown source. Systems that contain Sr seem to be less sensitive to aging. It seems that low field aging is more sensitive to processing variations and isovalent compositional changes than previously reported thus suggesting that an additional, more subtle effect than that invoked by Pan [refs. 1 and 2] can also cause aging. Regardless of aging state, high field (1 MV/m) dielectric properties remain constant.

REFERENCES

1. Pan, Wuyi, E. Furman, G.O. Dayton, and L.E. Cross, "Dielectric aging effects in doped lead magnesium niobate: lead titanate relaxor ferroelectric ceramics," Journal of Material Science Letters, Vol. 5, pp. 647-649, June 1986.

2. Pan, Wuyi, Qiming Zhuang, and L.E. Cross, "Direct current bias aging effects in lead magnesium niobate - lead titanate relaxor ferroelectric ceramics," Journal of the American Ceramic Society, Vol. 71, pp. C17-19, Jan. 1988.

3. Carl, K. and K.H. Hardtl, "Electrical after-effects in $Pb(Ti,Zr)O_3$ ceramics," Ferroelectrics, Vol. 17, pp. 473-486, March 1978.

4. Swartz, S.L. and T.R. Shrout, "Fabrication of perovskite lead magnesium niobate," Materials Research Bulletin, Vol. 17, pp. 1245-1250, Oct. 1982.

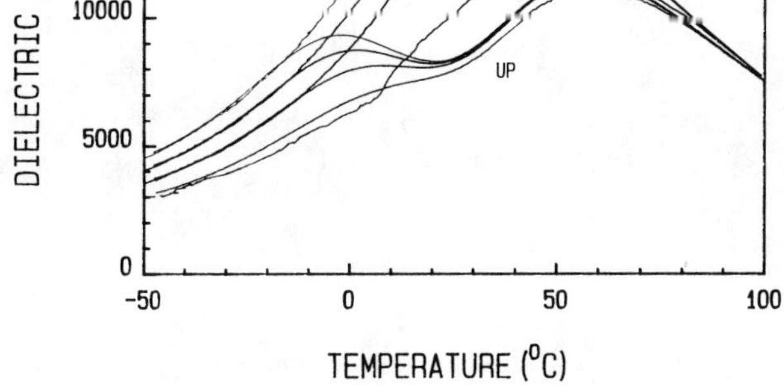

Figure 1. Typical aging curve for PMN-PT-BT

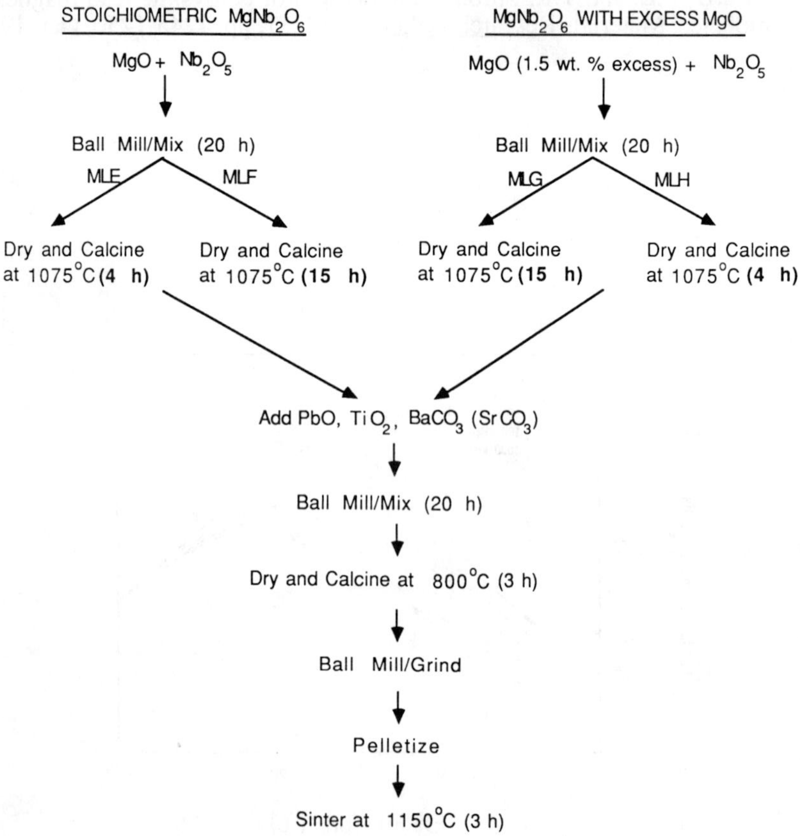

Figure 2. Processing of PMN-PT-BT(ST)

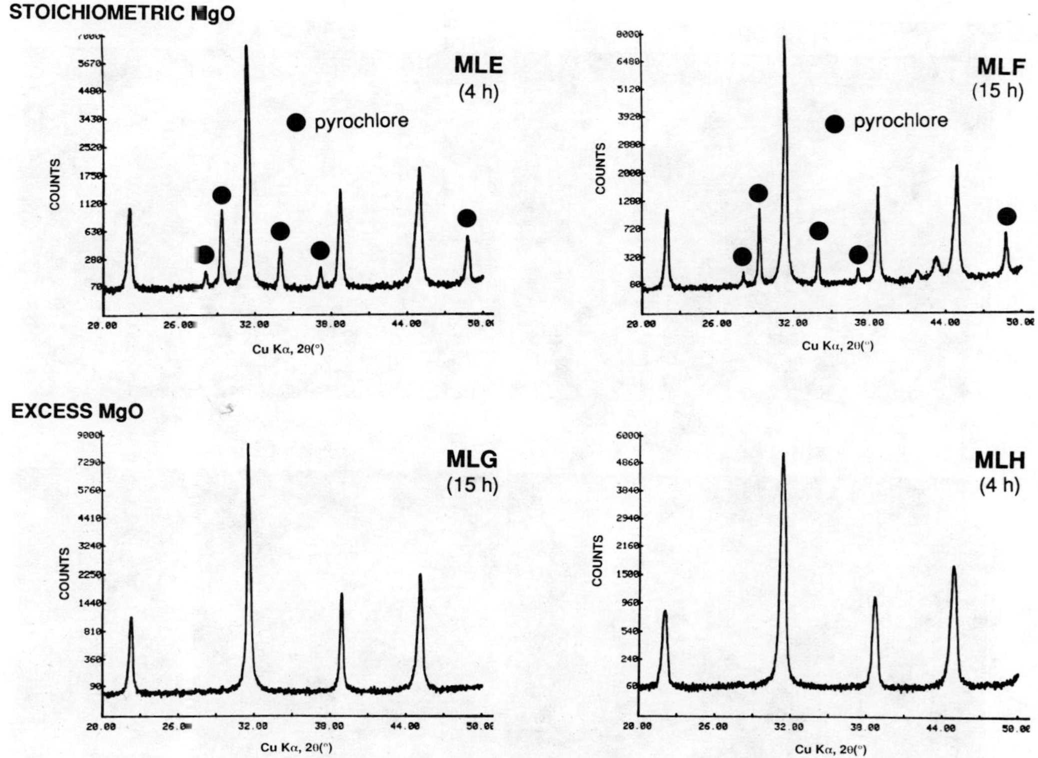

Figure 3. X-ray diffraction patterns of PMN-PT-BT: a) 100MLE, b) 100MLF, c) 100MLG and d) 100MLH

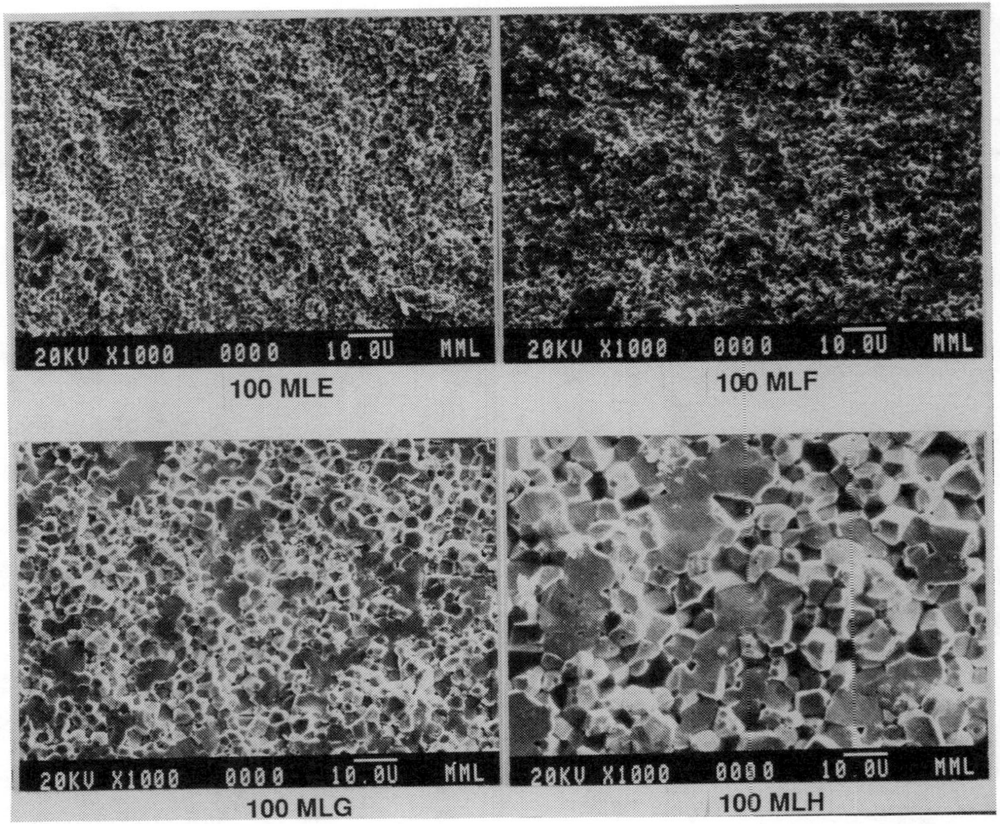

Figure 4. Effect of processing on the microstructure of PMN-PT-BT: a) 100MLE, b) 100MLF, c) 100MLG and d) 100MLH

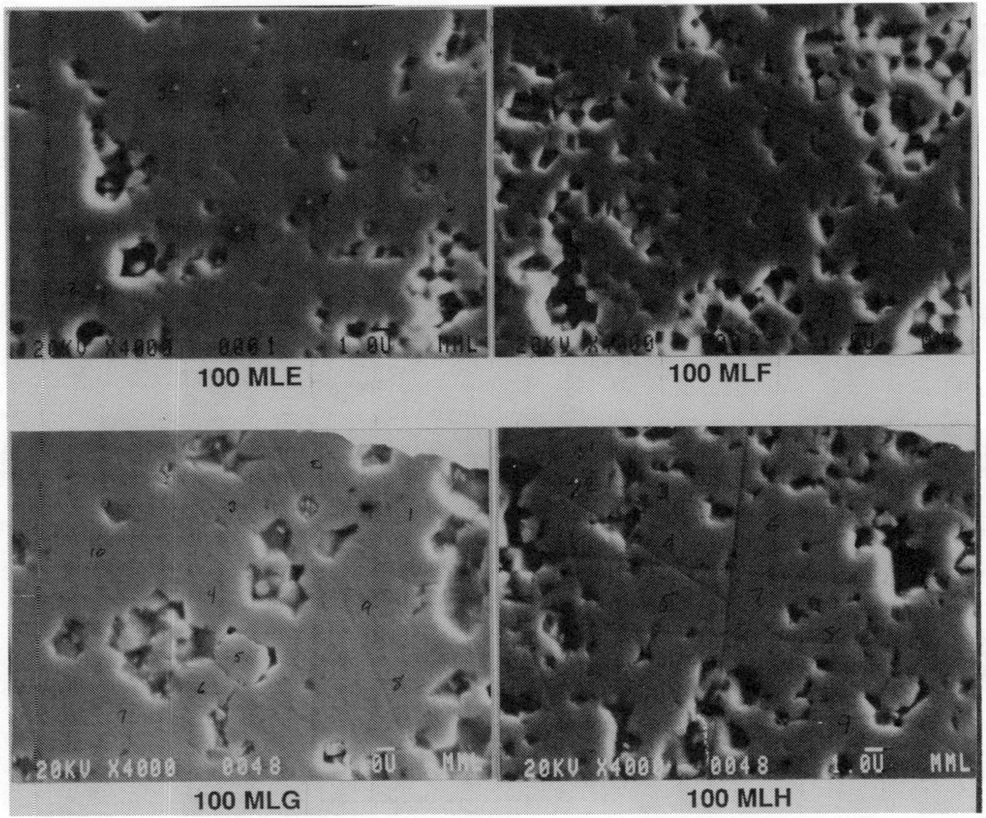

Figure 5. Scanning electron photomicrographs of polished specimens of PMN-PT-BT showing points of analysis: a) 100MLE, b) 100MLF, c) 100MLG and d) 100MLH

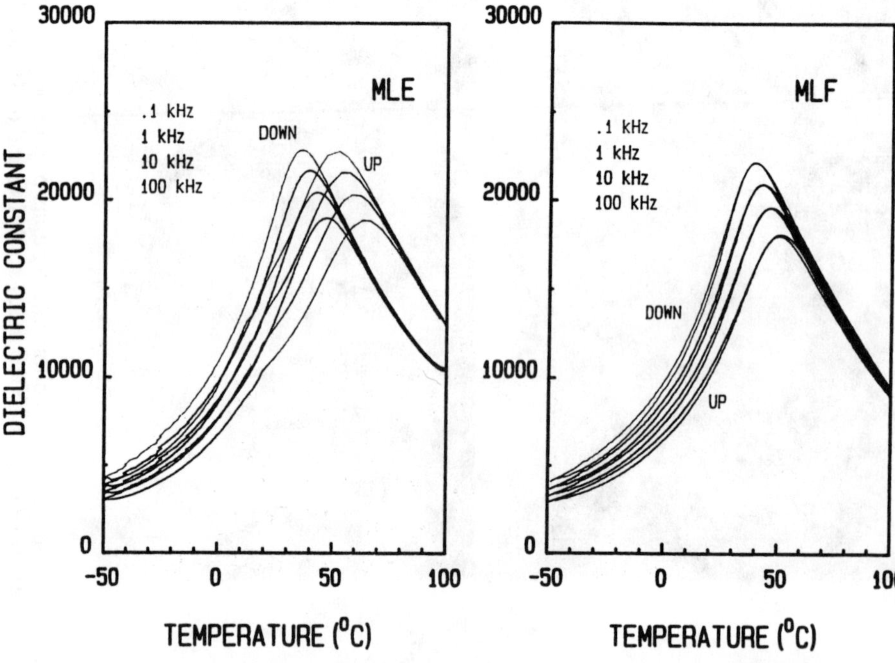

Figure 6. Effect of processing on the dielectric properties of PMN-PT-BT processed with stoichiometric MgO in the $MgNb_2O_6$ precursor: a) 100MLE and b) 100MLF

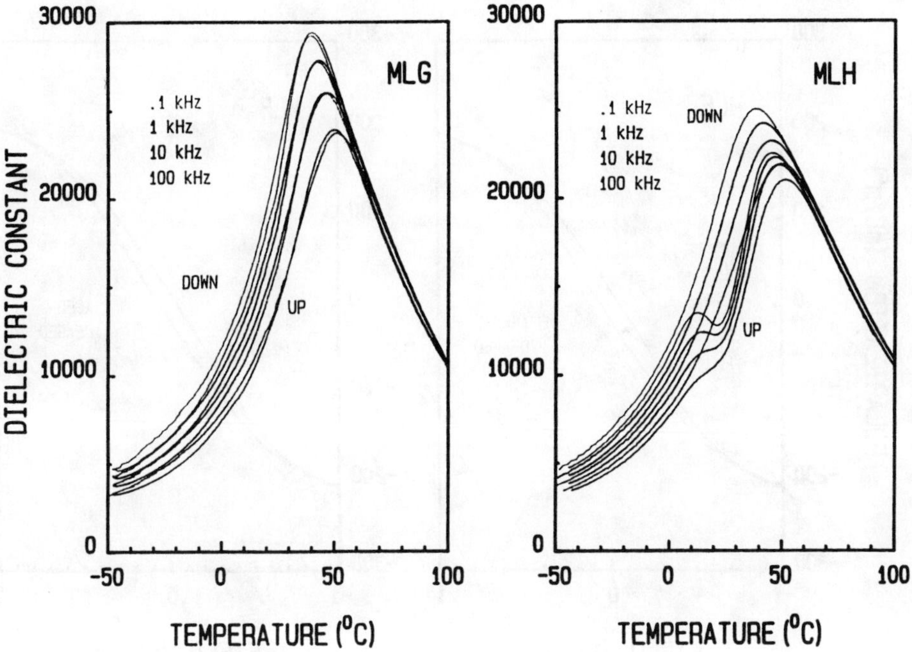

Figure 7. Effect of processing on the dielectric properties of PMN-PT-BT processed with excess MgO in the $MgNb_2O_6$ precursor: a) 100MLG and b) 100MLH

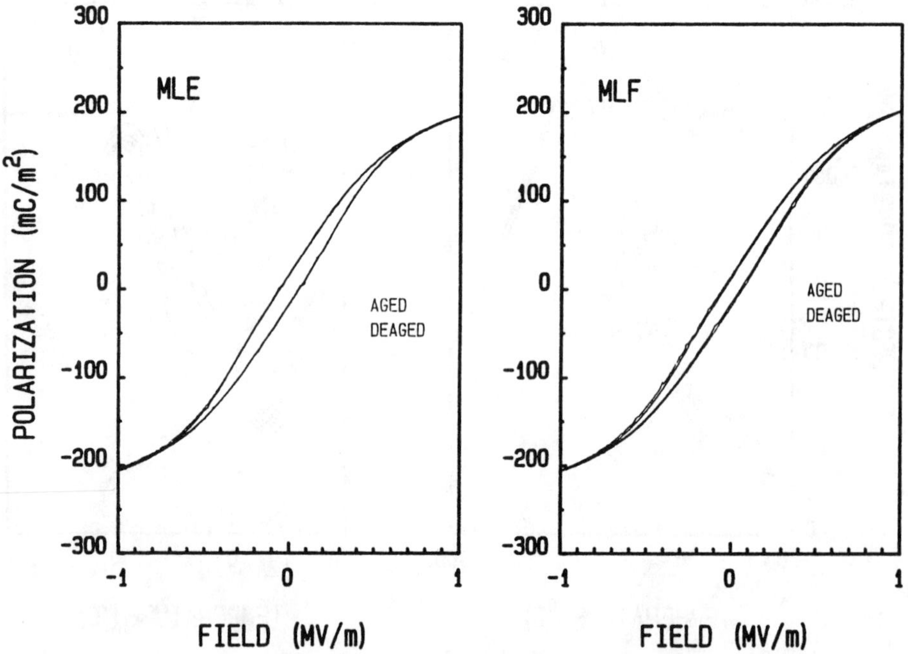

Figure 8. Effect of processing on polarization of PMN-PT-BT processed with stoichiometric MgO in the $MgNb_2O_6$ precursor: a) 100MLE and b) 100MLF

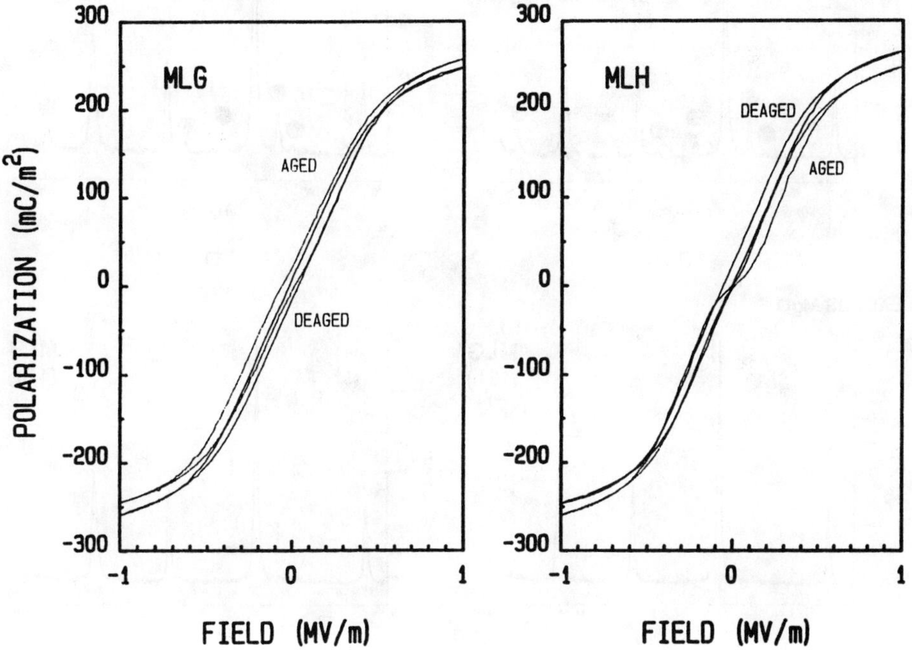

Figure 9. Effect of processing on polarization of PMN-PT-BT processed with excess MgO in the $MgNb_2O_6$ precursor: a) 100MLG and b) 100MLH

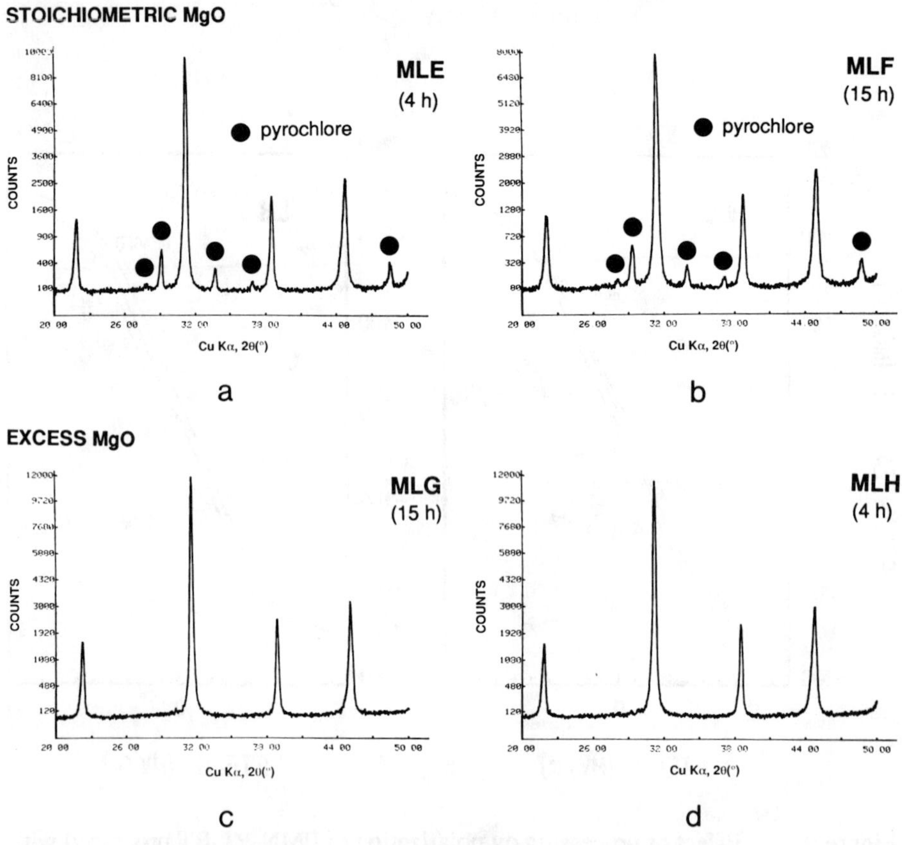

Figure 10. X-ray diffraction patterns of PMN-PT-ST: a) 200MLE, b) 200MLF, c) 200MLG and d) 200MLH

Figure 11. Effect of processing on the microstructure of PMN-PT-ST: a) 200MLE, b) 200MLF, c) 200MLG and d) 200MLH

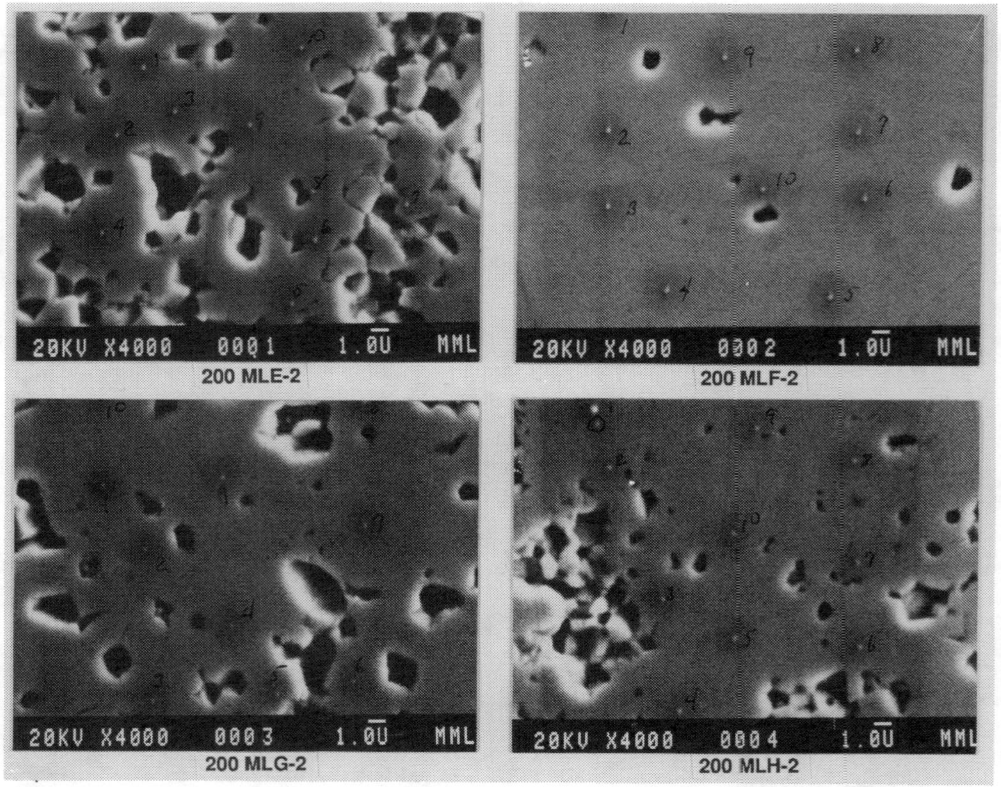

Figure 12. Scanning electron photomicrographs of polished specimens of PMN-PT-ST showing points of analysis: a) 200MLE, b) 200MLF, c) 200MLG and d) 200MLH

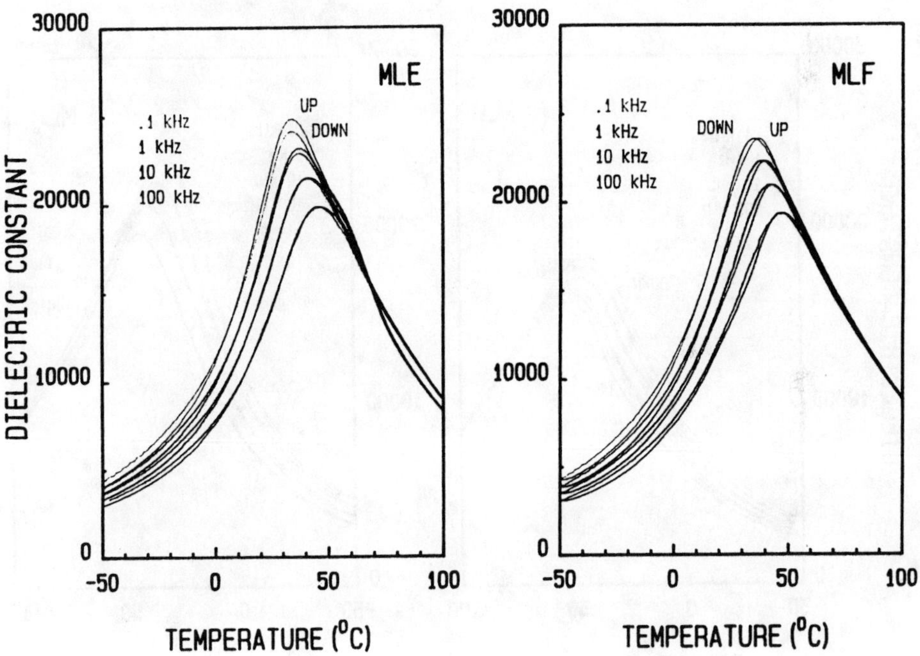

Figure 13. Effect of processing on the dielectric properties of PMN-PT-ST processed with stoichiometric MgO in the $MgNb_2O_6$ precursor: a) 100MLE and b) 100MLF

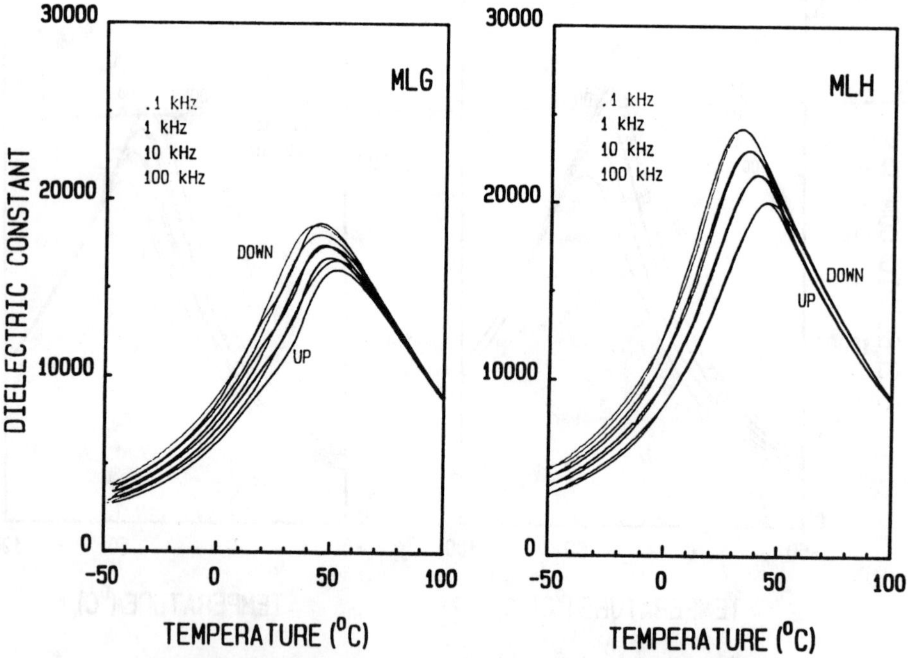

Figure 14. Effect of processing on the dielectric properties of PMN-PT-ST processed with excess MgO in the $MgNb_2O_6$ precursor: a) 100MLG and b) 100MLH

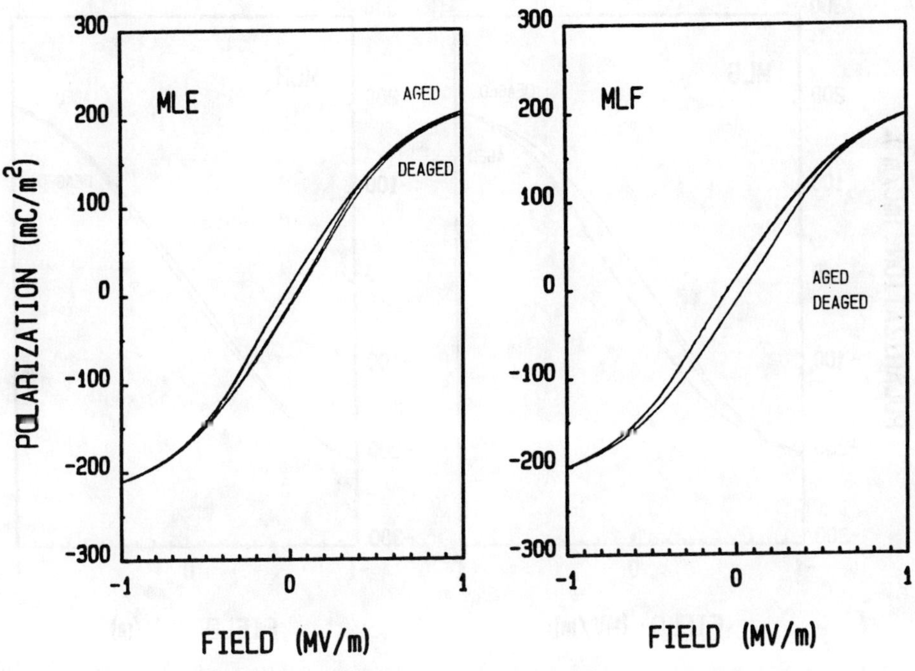

Figure 15. Effect of processing on polarization of PMN-PT-ST processed with stoichiometric MgO in the $MgNb_2O_6$ precursor: a) 100MLE and b) 100MLF

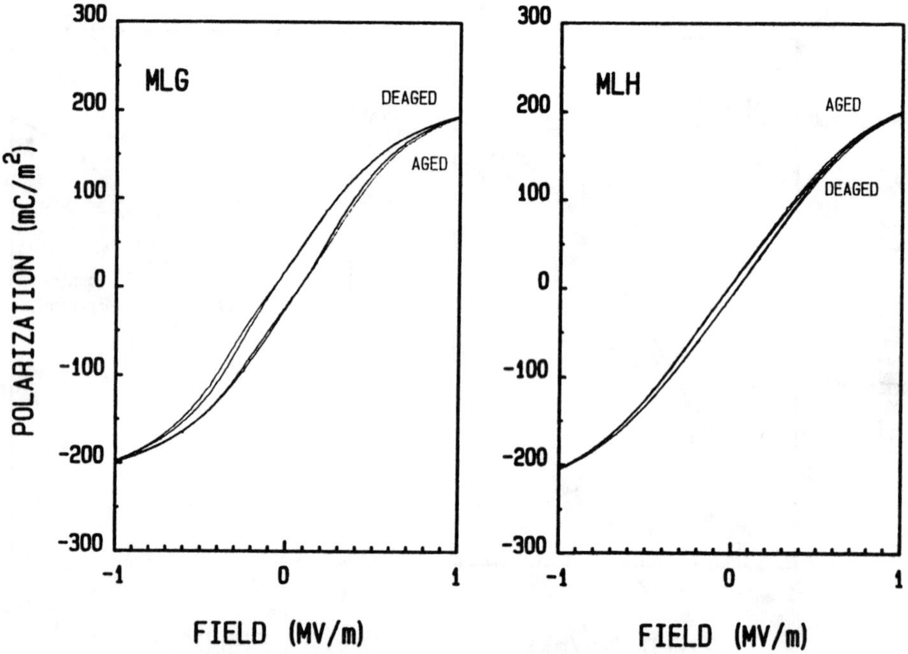

Figure 16. Effect of processing on polarization of PMN-PT-ST processed with excess MgO in the $MgNb_2O_6$ precursor: a) 100MLG and b) 100MLH

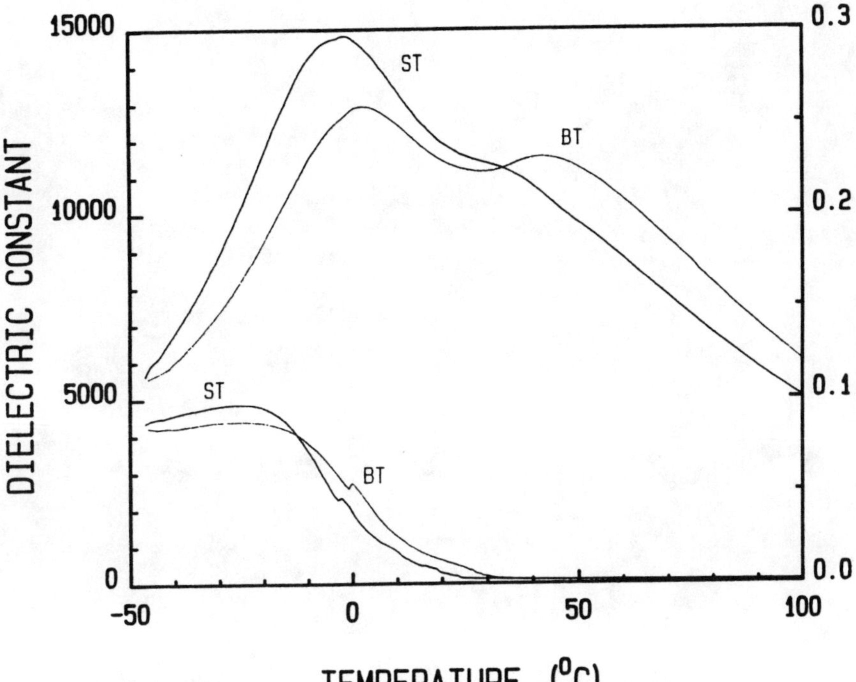

Figure 17. Aging of another composition of PMN-PT-BT and PMN-PT-ST.

ELECTRIC FIELD EFFECTS ON CRACK GROWTH IN A LEAD MAGNESIUM NIOBATE CERAMIC

A. S. Raynes, G. S. White, and S. W. Freiman
National Institute of Standards and Technology
Gaithersburg, MD

B. S. Rawal
AVX Corporation
Myrtle Beach, SC

ABSTRACT

The influence of an applied electric field on crack extension in a lead magnesium niobate ceramic was investigated in order to evaluate possible effects of electrostrictive strain on crack propagation. Cracks were generated by indenting polished specimen surfaces equidistant between imbedded parallel electrodes. The field was applied before indentation and maintained for the duration of the measurement. Results showed that cracks emanating from indentations perpendicular to the applied field were consistently shorter than those from indentations made with no field present. No effect of field was observed on cracks parallel to the field. Crack length also appeared to depend on the indentation location relative to the electrodes. A qualitative model explaining these results is proposed.

INTRODUCTION

The electrostrictive phenomenon is similar to the piezoelectric phenomenon, in that an applied electric field (e) creates a mechanical strain. However, while, in the piezoelectric effect, the strain (ϵ) is directly proportional to the electric field, in the case of electrostriction, ϵ is proportional to e^2. Materials such as lead magnesium niobate (PMN), which have relatively large electrostrictive coefficients, are of interest for both multilayer capacitors and actuators. The purpose of this work was to evaluate the effects of an electric field on crack growth.

PROCEDURE

Tape cast lead magnesium niobate specimens, containing a small percentage of Ti, underwent a firing cycle of 15 hours, including one hour at a peak temperature of 1000 °C. Multilayered specimens, ceramic and interleaved electrodes, were mounted in epoxy with five specimens to a mount. Each specimen consisted of six ceramic layers and five electrodes (see Figure 1); the ceramic layers between electrodes were 150 μm thick and the two edge layers were 190 μm thick. Samples were mounted so that the layers were perpendicular to the surface of the epoxy mount; leads attached to the electrodes extended out of the bottom of the mount. Specimens were polished with diamond paste and .05 μm alumina to expose the alternating layers of ceramic and electrodes. Porosity was evident on the polished surfaces and appeared as rows of pores with five pore rows between each set of electrodes. The rows of pores probably resulted from gaps between the tape-cast layers of ceramic used to make up the dielectric layers of the specimen, since six tapes were stacked to make each ceramic layer between electrodes.

Indentations were placed midway between electrodes in the four layers of ceramic which were bounded by electrodes on both sides. Both Vickers and Knoop indentations (see Figure 1) were introduced into the ceramic in air using a commercial microhardness tester*, with half of each type at an applied field of 17 kV/cm and half with no applied field. The Vickers indentations, made at a load of 2.45 N (250 g), were oriented with one diagonal parallel to the electrodes. Knoop indentations, made at 7.84 N (800 grams), were aligned with the long axis parallel to the electrodes. The indentation loads were chosen to generate well-developed cracks which would not intersect the electrodes. Of the Knoop indentations, 15 with and 15 without applied electric field were located within inner rows of ceramic. The remainder were placed within the two outer rows (Fig. 1). No effort was made to locate the Vickers indentations in specific sites among the four rows of ceramic. The indentations were made in several specimens; generally, groups of about 5 indentations were made in the presence of the electric field and then a similar number were made in the same specimen with no field present.

Crack length measurements (2c, i.e., measurements were made from the tip of a crack emanating from one corner of the indentation to the tip of the crack emanating from the opposite corner of the indentation) were made within 2 to 5 minutes after the load was removed from the specimen surface. Assuming an n-value of about 40, the spread in time between indentation and measurement could account for only about half the observed scatter in crack length[1].

RESULTS AND DISCUSSION

Some crushing was observed around the indentations. In the case of the Vickers indentations, individual grains or groups of grains surrounding the indentation often uplifted somewhat from the surface, with small cracks running between and around them. Knoop indentations showed similar crushing and uplift near the short diagonal regions, but displayed little or no crushing near the long diagonal regions.

Table I summarizes the results of the crack length measurements with and without an applied electric field for Vickers indentations. Corresponding data for the Knoop indentations are shown in Table II.

Table I: Vickers Indentation Crack Length Measurements

Crack/Electrode Orientation	# of Meas.	Crack Length w/o Field (μm)	Crack Length w/Field (μm)	% Change w/Field
Parallel	40	88 ± 4	82 ± 3	-7.4 ± 0.5
Perpendicular	20	82 ± 3	80 ± 4	-2.4 ± 0.6

Table II: Knoop Indentation Crack Length Measurements (long crack direction parallel to electrodes)

Location	# of Meas.	Crack Length w/o Field (μm)	Crack Length w/Field (μm)	% Change w/Field
Inner row	15	265 ± 17	236 ± 8.0	-10.9 ± 0.7
Outer row	15	282 ± 15	265 ± 11	- 6.0 ± 0.7
% Change w/ location		6.0 ± 0.9	10.9 ± 0.6	
Overall	30	273 ± 12	250 ± 8.6	- 8.4

Electric Field Effect:

The key feature of these results is that the presence of an electric field perpendicular to propagating cracks (i.e. cracks propagating parallel to the electrodes) inhibits crack growth. The effect is almost identical for both Knoop and Vickers indentations, with an overall decrease in crack length of ≈8% for an electric field of 17 kV/cm. In contrast, the effect of an electric field on cracks propagating parallel to the applied field, i.e., perpendicular to the electrodes, is much less (≈2%).

Table I shows that, for cracks parallel to the electrodes (i.e., perpendicular to the applied field), the difference in crack length between cracks with and cracks without an applied field is, on average, 6.1 μm. In contrast, the corresponding difference for the Knoop indentations is 23 μm; i.e., the Knoop indentations provide a much larger effect, relative to the measurement precision of the experiment. However, information on crack growth parallel to the applied field is restricted to Table I and relatively short cracks (≈80 μm rather than ≈250 μm).

Due to the scatter inherent in the indentation technique, the standard deviation about the average values of the measured crack lengths is fairly large. The uncertainties shown in Table I and Table II represent the 95% confidence range, rather than the standard deviation, since the confidence limits are a function of the number of measurements. Table II show that, at the 95% confidence level, the reduction in crack length for cracks perpendicular to an applied electric field is greater than the observed scatter. While the corresponding difference in Table I is not quite outside of the scatter, the same trend is evident.

The reduced crack length for cracks parallel to the electrodes and field independent crack growth for cracks perpendicular to the electrodes, can be intuitively understood in the following way. If the material is homogeneous, the presence of a uniform electric field, such as that produced by parallel plate capacitors, will generate uniform strain. However, at discontinuities in the material, the electric field will no longer be uniform and local variations in the strain will occur. In particular, discontinuities such as narrow rectangular voids have been shown[2] to result in enhanced electric fields for rectangles oriented both parallel and perpendicular to the applied field.

Crack tips, which would distort the electric field much more sharply, might be expected to enhance the electric field even more than rectangular voids. The resulting strain at the crack tip, proportional to the square of the electric field, would be much greater than the strain in the surrounding material from the unenhanced field, resulting in a stress, either compressive or tensile depending upon the sign of the electrostrictive coefficients, at the crack tips.

If the field-generated strains in the ceramic were isotropic, we would expect crack growth in all directions to be equally affected. However, we observe measurable crack growth reduction only for cracks perpendicular to the applied field. This is consistent with the directionality of the electric field[3].

$$\epsilon_i = \gamma_{ij} E_j \tag{1}$$

In Eq. 1a, ϵ is the strain tensor, E is a dyadic formed by the electric field, <u>ee</u>, and γ is the set of coefficients (fourth rank tensor) relating ϵ and E. Following standard Voigt notation, the subscripts in Eq. 1 range from 1 to 6. Because the grains of the ceramic are small, ≈5 μm, and are randomly oriented, the ceramic can be treated as an isotropic material. Therefore[3],

$$\begin{gathered} \gamma_{11} = \gamma_{22} = \gamma_{33} \\ \gamma_{12} = \gamma_{21} = \gamma_{13} = \gamma_{31} = \gamma_{23} = \gamma_{32} \\ \gamma_{44} = \gamma_{55} = \gamma_{66} = (\gamma_{11}-\gamma_{12})/2 \end{gathered} \tag{2}$$

All other components of γ are 0. However, because the electric field is generated by parallel plate electrodes, it is both uniform and unidirectional (i.e., $E_j = E_3 \delta_{3j}$, where δ_{ij} is the Kronicker delta). Therefore[a],

$$\epsilon_i = \gamma_{ij} E_3 \delta_{3j} \tag{3a}$$

or

$$\begin{gathered} \epsilon_3 = \gamma_{33} E_3 \\ \epsilon_1 = \epsilon_2 = \gamma_{32} E_3 \end{gathered} \tag{3b}$$

There are no other non-zero components to the strain. Measurements published for electrostrictive coefficients in a $Pb(Mg_{1/3}Nb_{2/3})O_3$ polycrystalline ceramic[4] show that the strain parallel to the applied field is ≈3.5 times greater in magnitude than, and opposite in direction to, the strain perpendicular to the field (i.e., $\gamma_{33}/\gamma_{32} \approx -3.5$).

Consequently, the direction of maximum strains, both with and without the presence of cracks, will be parallel to the electric field. However, due the enhancement of the applied field, the strain near the crack tips will be larger than the strain in the surrounding material, leading to compressive stresses at the crack tips as the material near the crack tips attempts to expand and is constrained by the surrounding material. These stresses are predominately in the direction of the electric field. Figure 2 is a schematic diagram of the situation; crack growth perpendicular to the applied field will be inhibited because the compressive residual stress at the crack tip acts across the crack tip opening. However, the residual stress at crack tips parallel to the applied field will act predominately in the direction of crack propagation and will, therefore, have little effect on crack extension.

The size of the region under compressive stress can be estimated if we assume that the source of the stress is similar to that in transformation toughened material[5], i.e., a region of the material near the crack tip which is attempting to expand but is constrained by the surrounding material:

$$\Delta K_I = 0.3Y\epsilon^t V_f W^{1/2} \qquad (4)$$

In Eq. 4, ΔK_I is the difference in stress intensity factor for the strained and unstrained material, Y is Young's modulus, ϵ^t is the net strain in the transformed material, V_f is the volume fraction of the material which is strained, and W is the width of the region under compressive stress. ΔK_I was determined by measuring the crack lengths[b] in the presence (0.81 MPa*m$^{1/2}$) and in the absence (0.73 MPa*m$^{1/2}$) of the applied field. Young's modulus was measured on non-electroded specimens[7] and was found to be 131.2 GPa. V_f was assumed to be 1, i.e., all of the material in the affected region was assumed to be strained[c]. Finally, to use Eq. 4, we must estimate ϵ^t, where $\epsilon^t = \epsilon(E_{crack\ tip}) - \epsilon(E_{uniform})$. From Ref. 3, we obtain the values of the electrostrictive coefficients Q_{33} and Q_{32} for a polycrystalline PMN: $Q_{33} = 2.3 \times 10^{-2}\ m^4/C^2$ and $Q_{32} = -0.64 \times 10^{-2}\ m^4/C^2$. However, these coefficients relate the polarization, $\underline{p}$, rather than electric field, to the strain, i.e.,

$$\epsilon_i = Q_{ij} P_j \qquad (5)$$

where P_j is the dyadic $\underline{pp}$. To evaluate the strain, we must evaluate the component of P_j or, due to the symmetry of the applied electric field, P_3.

As long as the measurements are made at least a few degrees (≈5 °C) above the Curie temperature[8], the expression[9]

$$p_3 = \chi \epsilon_0 e_3 \tag{6}$$

is valid where p_3 and e_3 are the nonzero components of the polarization and electric field vectors, respectively, χ is the electric susceptibility and ϵ_0 is the dielectric constant of vacuum. However, depending upon the exact composition of the PMN-based ceramic as well as upon its firing history, χ can range from[4,d,10] ≈8000 to over 20,000 with a resultant change in strain of greater than a factor of 6 for a given electric field. In addition, the amount of enhancement of the electric field at the crack tip has not been calculated and may increase substantially from the value calculated for the 25 μm wide rectangular defect quoted above. With these caveats in mind, we can attempt to obtain a rough estimate for the deformation zone. In the specimens used in this work, $\chi \approx$ 16000 rather than ≈8000 estimated from Ref. 3 for a similar material. In that work, the authors suggested that the product of the dielectric constant and the electrostrictive coefficient should remain approximately constant. In this case, the value of Q_{33} for this material should be $\approx 1.2 \times 10^{-2}$ m^4/C^2. Using these values for Q_{33} and χ and using 2.4 for the enhancement of the electric field, we calculate $\epsilon(E_{crack\ tip}) - \epsilon(E_{uniform}) \approx 3.4 \times 10^{-3}$. Solving Eq. 4, then, gives W ≈ 0.37 μm, a value much smaller than would be expected in a material with grain sizes of ≈5 μm. In fact, this value of W calls into question our initial assumption of material isotropy, since, on the scale of a micrometer, the crack tip clearly will sense grain orientation effects on Q_{ij}. However, if grain orientation strongly affected the measurements, we would expect the uncertainty values in Tables I and II to be noticeably larger for measurements made under the electric field than for those made with no field present. As the Tables show, only for Knoop indentations made on the inner levels of the specimen is there an increase in scatter for crack lengths measured in the presence of an electric field. For Knoop indentations in the outer layers and for Vickers indentations throughout the specimens, the variation in crack length measurements is identical for cracks made with and without electric fields present. Therefore, it is safe to assume that variations in grain orientation around the crack tips did not substantially alter the crack length values in the presence of an electric field and our assumption of isotropy appears to be acceptable.

Anisotropic Material Effects:

Table I indicates that the crack growth behavior in the specimens is not identical for cracks growing parallel and perpendicular to the electrodes; cracks perpendicular to the electrodes are consistently shorter than those parallel to the electrodes. In fact, the reduction in crack length for the cracks parallel to the electrodes (perpendicular to the electric field) at 17 kV/cm results in cracks which are comparable in length to the cracks perpendicular to the electrodes both with and without an electric field.

Finally, we note from Table II that cracks made within the inner rows of the specimen were shorter than those made in the outer rows. These differences are in addition to the effects of the electric field, as shown in Fig. 3. The cause of this position sensitivity in the crack lengths is not yet known, although several possibilities have been considered. Free surface effects near the edge would be expected to result in longer cracks. In fact, this is observed (Fig. 1); Knoop measurements could not be made within 75 μm of the specimen edge because the cracks would curve to intersect the surface. However, this effect is expected to be restricted to the vicinity of the surface and the phenomenon shown in Fig. 3 extended past two electrodes and ≈ 400 μm into the material.

An alternate explanation could be the presence of a composition gradient through the specimen; as the lead content decreases, the fracture of PMN should change from only intergranular to include transgranular fracture. SEM investigations of a specimen near the surface and at the inner row showed that the fracture behavior remains intergranular across the specimen. Energy dispersive x-ray analysis of the lead content at the two locations (Fig. 4) suggests that the composition of the specimen is constant. Therefore, the crack length dependence on position does not appear to result from compositional variations within the specimen.

CONCLUSION

Cracks grown perpendicular to an applied electric field in PMN have been shown to be reduced in length relative to cracks grown in the absence of the field. In contrast, cracks parallel to the field showed little or no change in length from cracks in the absence of the field.

These results are explained by a model relating enhanced electric fields at the crack tips to compressive stresses parallel to the applied field. A detailed quantitative analysis depends on precise information about the magnitude of electric fields at an atomistically sharp crack tip; this information is presently not known.

Crack lengths in each specimens were found to vary from the specimen surface to 400 μm into the interior. Cracks between inner layers of electrodes were ≈6% longer than those in the outer electrodes. The reasons for this behavior are unknown, although free surface effects and compositional variations within the specimens have largely been discounted.

ACKNOWLEDGEMENTS

The authors at NIST wish to thank the Office of Naval Research for their support of this work. The many helpful discussions with David C. Cranmer are also greatly appreciated.

REFERENCES

1. G. S. White, S. W. Freiman, E. R. Fuller, Jr. and T. L. Baker,"Effects of Crystal Bonding on Brittle Fracture," J. Mater. Res. 3 #3, 491-497 (May/June 1988).

2. D. J. Taylor, R. E. Newnham and W. B. Carlson, "The Effect of Electric Field on Mechanical Strain and Stress in Flawed Electroceramics," Ferroelectrics, 87 81-84 (1988).

3. J. F. Nye, Physical Properties of Crystals Oxford Press, Oxford, Great Britain, pp. 250-257 (1979).

4. S. Nomura and K. Uchino, "Electrostrictive Effect in $Pb(Mg_{1/3}Nb_{2/3})O_3$-Type Materials," Ferroelectrics 41 117-32 (1982).

5. A. G. Evans and A. H. Heuer, "REVIEW - Transformation Toughening in Ceramics: Martensitic Transformations in Crack-Tip Stress Fields," J. Am. Ceram. Soc., 63 (5-6) 241-48 (1980).

6. G. R. Anstis, P. Chantikul, B. R. Lawn, and D. B. Marshall, "A Critical Evaluation of Indentation Techniques for Measuring Fracture Toughness: I, Direct Measurements," J. Am. Ceram. Soc. 64 #9, 533-538 (1981).

7. G. S. White, C. Nguyen, and B. Rawal, "Young's Modulus and Thermal Diffusivity Measurements of Barium Titanate Based Dielectric Ceramics," in Conference Proceedings of Nondestructive Testing of High Performance Ceramics (A. Vary and J. Snyder, eds.) Am. Cer. Soc., Westerville, OH, pp. 371-379 (1987).

8. W. D. Kingery, H. K. Bowen and D. R. Uhlmann, Introduction to Ceramics: Second Edition, John Wiley and Sons, Inc., New York, NY, pp. 1964-72 (1976).

9. J. D. Jackson, Classical Electrodynamics, John Wiley and Sons, Inc., New York, NY, p. 109 (1962).

10. M. H. Megherhi, The Effects of Additives, Microstructure, and Macrostructure on the Fracture Behavior of Lead Magnesium Niobate-Based Ceramic Actuators, M.S. Thesis, Dept. Ceramic Science, Penn State University, University Park, PA, p. 73 (1989).

[*] Tukon Model 300 Microhardness Tester. Trade names and companies are identified in order to specify adequately the experimental procedure. In no case does such identification imply that the products are necessarily the best available for the purpose or endorsement by the National Institute of Standards and Technology.

[a] It is important to bear in mind that, in the notation of Eqs. 1-3, $E_i = e_i{*}e_i$, for $i \leq 3$, where e_i is the component of the electric field in the i direction. For $3 < i \leq 6$, E_i corresponds to products of electric field components in different directions. For this work, only E_3 is nonzero.

[b] The stress intensity factor, K_I, was calculated using the relation developed by Anstis et al.[6] relating K_I to indentation crack lengths in glass. While recognizing that this expression was derived for isotropic materials, we use it here to obtain an estimate of the apparent change in stress intensity factor due to the change in crack lengths with and without the presence of an electric field.

[c] This assumption is valid because the electrostrictive coefficients used to calculate the strain in this work are those for polycrystalline PMN. If single crystal values are used for the electrostrictive coefficients, then the orientation distribution of the grains would necessarily mean that $V_f < 1$.

[d] The values of χ listed from Ref. 3 are inferred by using eq. 6 and the reported values of P_3 and E_3.

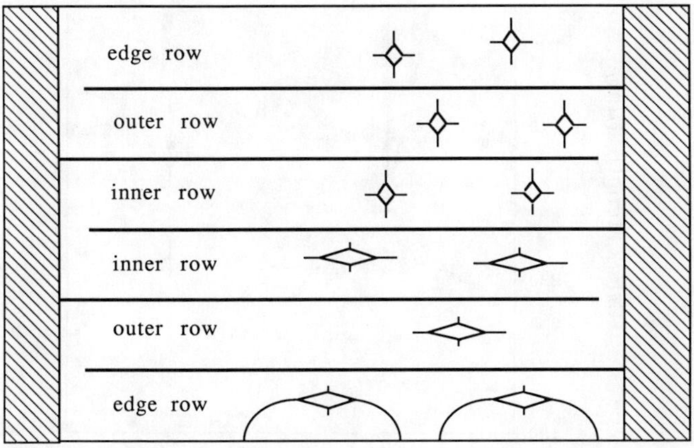

1. Schematic of specimen showing ceramic/electrode structure and typical Knoop (elongated) and Vickers (square) indentation locations and the crack patterns associated with them.

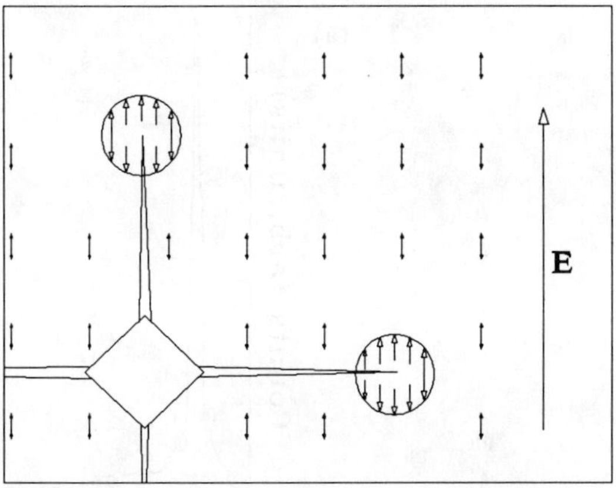

2. Schematic showing effect of applied electric field on crack-tip strains. In absence of cracks, specimen is uniformly strained. Enhanced field at crack tips results in locally greater strain. The material at the crack tip is constrained by the surrounding material, leading to a compressive stress parallel to the applied field.

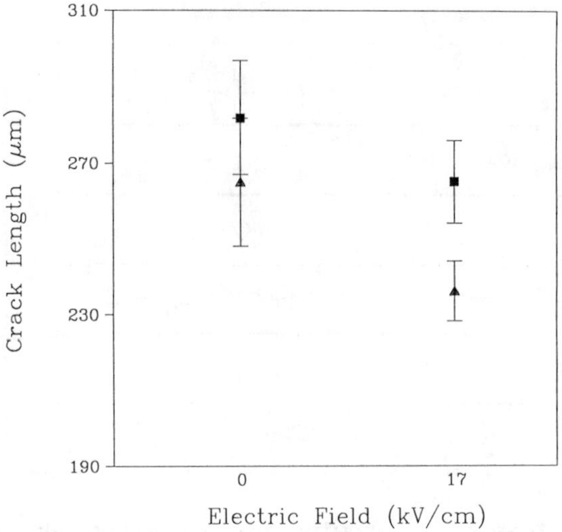

3. Crack length measurements, with and without applied electric field, as a function of indentation location (inner ▲ and outer rows of ceramic) in the specimen. Cracks are longer in the outer rows both with and without an electric field present.

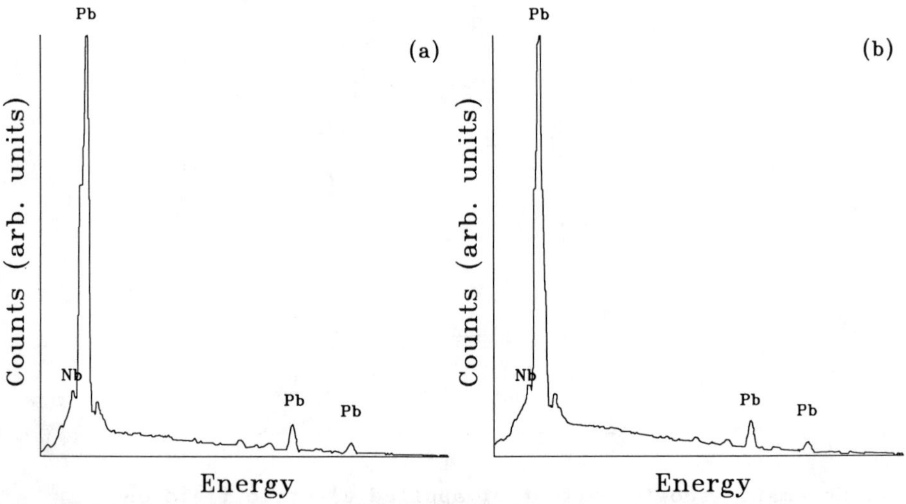

4. Energy dispersive x-ray analysis of cross-section of PMN specimen: (a) spectrum obtained from outer layer, b) spectrum obtained from inner layer of the specimen. Figure heights in arbitrary units. Identical spectra indicate that no compositional gradient exists across specimen thickness.

DIELECTRIC PROPERTIES OF $HfTiO_4$-$ZrTiO_4$ SOLID SOLUTIONS

Hiroyuki Ikawa, Hiroshi Narita and Osamu Fukunaga,
Tokyo Institute of Technology, Department of Inorganic Materials, Meguro-ku Tokyo 152 Japan
and
Robert E. Newnham
The Pennsylvania State University, Materials Research Laboratory, University Park, PA 16802

ABSTRACT

Dielectric properties of solid solutions in the system $HfTiO_4$-$ZrTiO_4$ were measured for specimens made under different heat treatments. The relative permittivity of the quenched form decreased with increasing Hf content; however, that of the furnace cooled form had a peak permittivity value for a composition with Zr content of 0.75. Similar results were achieved with the temperature coefficient of capacitance (TCc). The material with zero TCc was revealed obtainable by optimizing composition and/or heat treatment. Many changes in dielectric properties agreed with the prediction that $ZrTiO_4$ specimens containing SnO_2 exhibit properties similar to $HfTiO_4$.

INTRODUCTION

The $(A,B)_2O_4$ type compounds having the α-PbO_2 structure are of technological interests because of refractory behavior, low thermal expansivities[1-6], and excellent dielectric properties at high frequencies[7,8]. It is well known that lattice constants of $ZrTiO_4$ change significantly with heat treatments. The difference in lattice constants, however, have been attributed recently[9-12] to an order-disorder phase transformation. The high temperature form (abbreviated as high form) has the α-PbO_2 structure[13] with Zr and Ti randomly distributed in the cation sites, while the low form is an ordered arrangement showing incommensurate superstructure[9,12].
A number of anomalies, unexplainable by normal con-

cepts of crystal chemistry, are observed for these mixed oxides[11,12]. Furthermore, an interaction hypothesis has been proposed to explain the anomalies[11], and to predict that the thermal expansion of a $Zr_xTi_ySn_zO_4$ (x+y+z=2) solid solution is similar to that of $HfTiO_4$. This composition has been used as a high frequency resonator[7,8]. Accordingly, the study was initiated to further investigate the structure-property relationships of these unusual compounds.

EXPERIMENTAL

Specimen Preparation

Mixed oxides were prepared by a co-precipitation method. Reagents of extra fine grade were zirconium sulfate, titanium sulfate, ammonia water and hafnium sulfate prepared from hafnium metal. Details of the preparation are described elsewhere[11]. The calcined precipitate was shaped into pellets, and pressed isostatically at 49 MPa. The final sintering was carried out at 1600°C for 10 h to make pellets 10 mm diameter and 1 mm thickness. The pellets were furnace cooled at a rate of 10°C/min. The sintered pellets were subjected to a variety of heat treatments with the restriction that they remaind free from cracks caused by thermal shock or by anisotropic changes in lattice constants, as discussed later. To make the high form of each composition, samples were quenched in air after heating at 1300°C for 5 h; the low form was obtained by heating at 1200°C for 2 h before cooling to 900°C at a rate of 0.5°C/min followed by cooling at 10°C/min to room temperature.

Respective impurities of Hf and Zr in zirconium sulfate and hafnium sulfate were analyzed using an ICP spectrometric analyzer (Seiko Electronic, SPS7000). The apparent density and bulk density were measured by a conventional Archimedes method using 4 to 5 pellets for each preparation condition.

Lattice constants were calculated from about 25 diffraction peaks (Rigaku, RAD-II) which were corrected by using a Si internal standard.

Dielectric Measurement

Two faces of a sintered pellet were polished; care was taken to make the faces parallel within a 10 μm difference in thickness among four points. A Ag paste (Demetron, L200) was coated on both faces as electrodes. The capacitance and tan δ were measured at 1 and 10 kHz with a capacitance bridge (General Radio, 1615A) with a field of 3 V/mm. The specimen was placed in a glass tube and into which dry N_2 gas was introduced. The tube was placed in a water tank kept at temperatures ranging from 20 to 80°C. Measurements were made for four to seven pellets for each set of conditions.

Capacitance was also measured as a function of frequency using an impedance analyzer (YHP, 4192A) up to 13 MHz. However, wafting impedance and inductance caused by connector cables were not fully excluded and prevent a description on tan δ.

RESULTS AND DISCUSSION

Characterization

Table 1 lists the lattice constants and relative densities of pellets. The b lattice parameter (b_0), the unit cell volume (V) of the high form are larger than those of the low form, whereas the a_0 of the high form is slightly smaller than that of the low form. These differences decrease with Hf content, the x value. Further, the cell volume of both forms increases with x despite the smaller ionic radius of Hf^{4+} ion (0.071 nm[14]) compared to that of Zr^{4+} ion (0.072 nm). These results are consistent with reported[11]. Quasi equilibrium (quench) temperatures estimated from reported parameters[12] of the high and low forms of a composi-

Table 1. Composition, high or low temperature form, lattice constant, unit sell volume (V), relative apparent density (d_a), and relative bulk density (d_b) ofthe solid solutions of $Hf_xZr_{1-x}TiO_4$

X	Form	a_0/nm	b_0/nm	c_0/nm	V/nm^3	d_a/%	d_b/%
0.008	High	0.48024(2)	0.54746(2)	0.50311(2)	0.13227(1)	97.8	97.8
0.008	Low	0.48135(2)	0.54000(5)	0.50315(3)	0.13078(2)	98.4	98.4
0.248	High	0.47895(3)	0.55041(3)	0.50352(2)	0.13274(1)	97.7	97.7
0.248	Low	0.47971(4)	0.54533(3)	0.50315(3)	0.13162(2)	97.8	97.8
0.487	High	0.47736(3)	0.55317(4)	0.50503(3)	0.13309(2)	98.6	97.6
0.487	Low	0.47818(3)	0.54949(4)	0.50354(4)	0.13231(2)	97.9	97.8
0.727	High	0.47597(3)	0.55512(8)	0.50439(3)	0.13327(2)	97.4	96.5
0.727	Low	0.47632(2)	0.55333(3)	0.50399(2)	0.13283(1)	97.0	97.0
0.966	High	0.47489(2)	0.55727(2)	0.50482(2)	0.13360(1)	97.6	96.0
0.966	Low	0.47518(3)	0.55591(4)	0.50457(3)	0.13328(2)	98.4	97.1

tion of x=0.992 are 1200 and 1100°C, respectively. The X-ray diffraction patterns of the high forms compared to the α-PbO_2 structure. For the low forms, small halos of superstructure diffractions at 26 and 45° (2θ, CuKα) were observed for x=0.008, and weakly seen for x=0.248.
Theoretical densities of the high forms with x of 0.008 and 0.966 are 5.118 and 7.145 g/cm^3, respectively. The two values of relative densities agree well each other for pellets with small x, which means they have no open pores. However, for specimens with large x, there are incidents that bulk densities are smaller than that of the apparent. This is a reflection of higher refractory nature of $HfTiO_4$[6]. In any case those densities are larger than 95.9 %, and their dispersion is small. Accordingly, relative permittivities are thought to be comparable without corrections[15].
A well sintered microstructure with grains about 25 μm is evident from the two fractured faces in Fig. 1. Many examples of grain boundary fractures are found in Fig. 1(B). This is caused by anisotropic change in lattice constants (Table 1); however, intergranular voids are seen only limitedly.

Dielectric Property

Figure 2 shows relative permittivities (K') of the high and low forms plotted as a function of composition. The K' values measured at 10 kHz were from 99.95

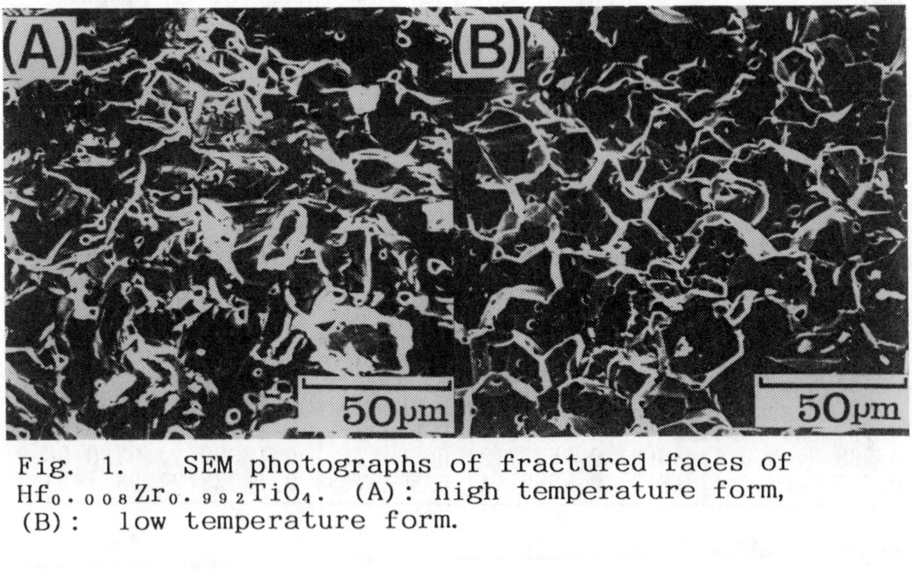

Fig. 1. SEM photographs of fractured faces of $Hf_{0.008}Zr_{0.992}TiO_4$. (A): high temperature form, (B): low temperature form.

to 99.99 % of those at 1 kHz. It is interesting that K' of the high form decreases almost linearly with x; however, K' of the low form has a peak at x=0.248. The difference in K' between the two forms is largest for x=0.008, which is consistent with the difference in b_0 (Table 1). However, no clear cut trends for the changes in K' with composition are observed between the high and low forms, as observed for the lattice constants. According to Wakino and Tamura[16], the K' of $ZrTiO_4$ is 42 and it decreases with Sn content, and that of $Zr_{0.8}TiSn_{0.2}O_4$ is 37 (7 GHz). No detail of heat treatments are given in the paper. However, the reported value agrees with our results. Moreover, the decrease in K' with Sn content in $Zr_{1-x}TiSn_xO_4$ is similar to the decrease in K' with Hf content, and this similarity between the two systems is that predicted from the hypothesis[11].

Figure 3 shows K' plotted as a function of the cube root of cell volume ($V^{1/3}$) for the two compositions. Specimens were prepared under various heat treatments, and K' and cell volume were measured for the same specimen. There is almost a linear relation between K' and $V^{1/3}$ in Fig. 3(A). The relation appears more complicated for the other composition, where K' decreases with $V^{1/3}$ for larger cell volumes. The fact that the K' vs. $V^{1/3}$ relations are opposite for the two compositions is supported by Fig. 2 also. The two lines in Fig. 2 cross at x≒0.3. Accordingly, the K' of a specimen with a composition x≒0.3 must be independent from its cell volume and the structural state of the crystal !

Loss factors (tan δ) are almost constant at 0.0001 except those of the low form of $Hf_{0.008}Zr_{0.992}TiO_4$ and both forms of the Hf rich end composition. Comparatively low relative densities are probably the reason for larger loss factors of the latter, and intergranular voids (Fig. 1(B)) may be responsible for the dielectric loss of the former.

The capacitances of each pellet were measured at four temperatures in general between 20 to 80°C. Linear relations were observed among them and temperature coefficient of capacitance (TCc) was calculated from data between 20 and 80°C. The results are shown in Fig. 4. It is clear from the results that TCc is also affected significantly by composition and heat treatment, and that we can design a material with zero TCc by controlling the two factors. Further, it is notable that the figure has many characteristics in common with Fig. 2. Additional informations concerning the TCc vs.

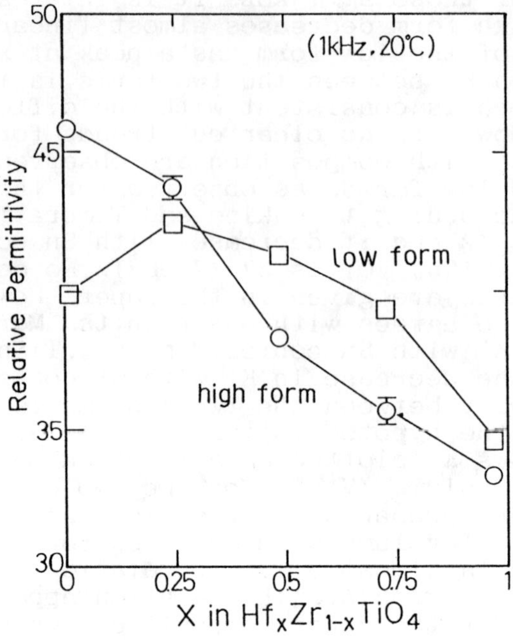

Fig. 2. Relative permittivities vs. composition of the high and low temperature forms measured at 1kHz, 20℃. Vertical line shows a standard deviation.

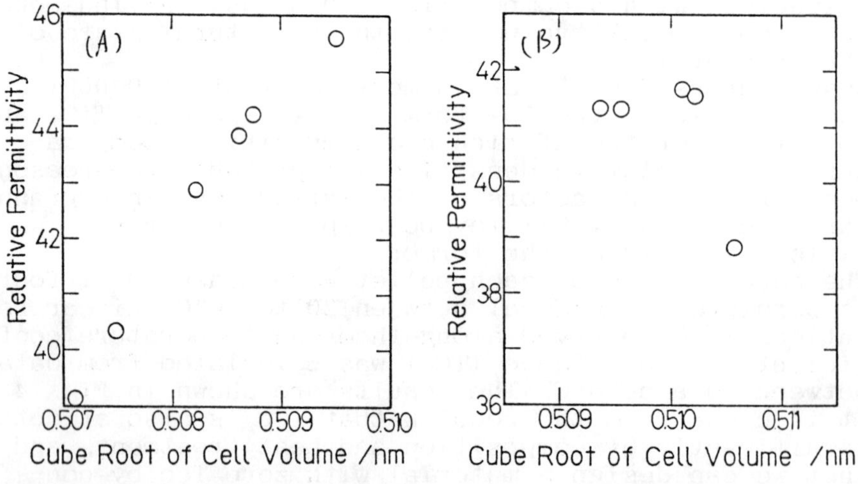

Fig. 3. Relative permittivity vs. cube root of unit cell bolume for (A) $Hf_{0.008}Zr_{0.992}TiO_4$ and (B) $Hf_{0.487}Zr_{0.513}TiO_4$

$V^{1/3}$ relation is shown in Fig. 5(A,B). In the (B), five points are dispersed crossing zero TCc, and its dispersed scheme is similar to that of Fig. 3(B). It is estimated from these similarities in figures that both K' and TCc are affected analogically by a structure related something of this compound, although the signs of dependences are opposite, and the real feature of "something" is not clear except that it is typically represented by the unit cell volume.
The TCc values of $ZrTiO_4$ in the figure agree well with reported[7,16]. Wolfram and Goubel[7], and Wakino and Tamura[16] also have reported that the value shifts to the positive values side by adding SnO_2. This change in TCc also coincides with the prediction that $ZrTiO_4$ specimens containing SnO_2 exhibit properties similar to $HfTiO_4$[11].
The K' values exhibited a slight decrease with frequency within the measurements (ca. 20 kHz to 9 MHz). Values at two frequencies are cited below in a format written next. The x value of a composition: high form (H), or low form (L), K' at 50.12 kHz- K' at 7.94 MHz.

0.008: H 46.6 - 45.3, L 40.3 - 39.0
0.248: H 44.1 - 42.9, L 42.7 - 41.5
0.996: H 34.1 - 29.7, L 35.1 - 31.1

The relative large changes in K' of the two forms of x=0.996 must be related to their larger loss factors. If we think about a stationary K' of a related compound at GHz region[8], intrinsic K' of these specimens are estimated to be constant at these frequencies.
It is evident that the ionic polarization is predominantly responsible for a large K' value of this compound. However, it is not easy to explain the reason of the changes in K' and TCc with composition and heat treatments. While, it will be worthwhile to try to explain the difference between $ZrTiO_4$ and $HfTiO_4$ by the difference in ionic polarizabilities. It is because an analytical equation is known[17] which expresses the ionic polarizability as a function of the charge on the ion, the ionic resonance frequency, the interionic distance, and masses of cation and anion. This application, however, was revealed difficult because the wavenumder differences in corresponding Raman modes of $ZrTiO_4$ and $HfTiO_4$[18] were dispersed from plus to minus. This crystal is anomalous as the high form exhibits a quenchable structural difference[11,12], and the phase change is related to the ordered arrangement of atoms accompanied with changes in coordination numbers[19]. More studies are needed to explain dielectric properties of these compounds.

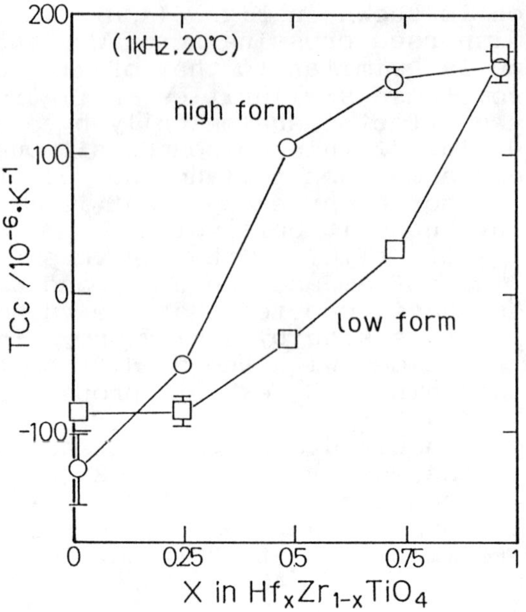

Fig. 4. Temperature coefficient of capacity (TCc) vs. composition measured at 1kHz, 20℃.

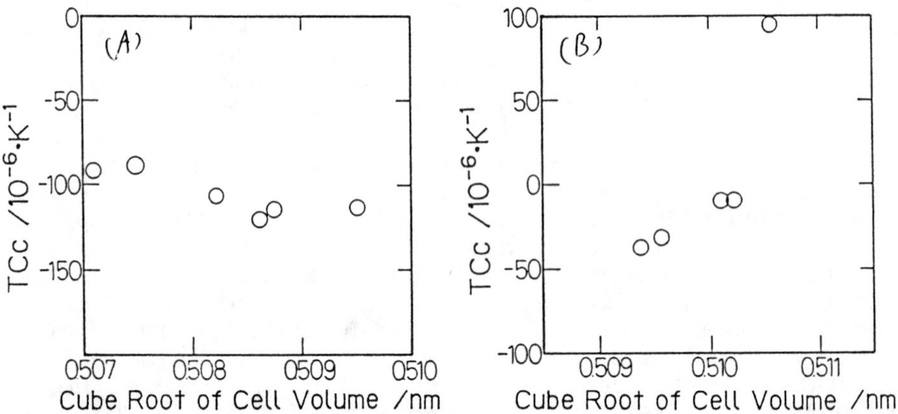

Fig. 5. Temperature coefficient of capacity (TCc) vs. cube root of unit cell volume for (A) $Hf_{0.008}Zr_{0.992}TiO_4$ and (B) $Hf_{0.487}Zr_{0.513}TiO_4$.

CONCLUSION

Dielectric properties of solid solutions in the system

$HfTiO_4$-$ZrTiO_4$ were measured for pellet specimens made under different heat treatments.
The relative permittivity of a quenched form of $ZrTiO_4$ was 45.8; the value decreased with increasing Hf content to 33.3 of $HfTiO_4$. The relative permittivities of furnace cooled forms were 39.9, 42.4 and 34.6 for $ZrTiO_4$, $Hf_{0.25}Zr_{0.75}TiO_4$ and $HfTiO_4$, respectively.
The temperature coefficient of capacity (TCc) of a quenched form of $ZrTiO_4$ was -130×10^{-6} K^{-1}; the coefficient increased with increasing Hf content to 160×10^{-6} K^{-1} of $HfTiO_4$. The material with zero TCc was revealed obtainable by optimizing composition and/or heat treatment.
The similarity between the two figures -- relative permittivity vs. composition and TCc vs. composition -- was remarked, although the reason of the similarity was unknown. Many changes in dielectric properties agreed with the prediction that $ZrTiO_4$ specimens containing SnO_2 exhibit properties similar to $HfTiO_4$.

Acknowledgment: This work is supported by grand in aid of MESC Japan; contract No. 01550590. Profs. M.Abe, K. Ema and M.Yamane are acknowledged for using ICP analyzer, capacitance bridge and SEM, respectively. Prof. L.E.Cross of MRL is acknowledged for his discussion.

REFERANCES

[1] R.W.Lynch and B.Morosin,"Thermal Expansion, Compressibility, and Polymorphism in Hafnium and Zirconium Titanates," J.Am.Ceram.Soc.,55(8) 409-13 (1972).
[2] R.Ruh, G.W.Hollenberg, E.G.Chales and V.A.Patel, "Phase Relations and Thermal Expansion in the System HfO_2-TiO_2," J.Am.Ceram.Soc.,59 (11-12) 495-99 (1976)
[3] G.A.Carlson, J.L.Anderson, R.A.Briesmeister, S.R.Skaggs and R.Ruh,"Coefficient of Thermal Expansion and Dynamic Response to Pulsed Energy Deposition in HfO_2-TiO_2 Compositions," J.Am.Ceram.Soc.,60 (11-12) 508-10 (1977)
[4] S.R.Skaggs,"Zero and Low Coefficients of Thermal Expansion Polycrystalline Oxides," Rev.Int.Hautes. Temp.Refract.,16 (3) 157-67 (1979).
[5] L'Octet Franco-Canadien,"Characterization of Some Materials in the HfO_2-TiO_2 System," High Temp.-High Pressures, 13 (1) 97-103 (1981).
[6] J.P.Coutures and J.Coutures,"The System HfO_2-TiO_2," J.Am.Ceram.Soc.,70 (6) 383-87 (1987).
[7] G.Wolfram and H.E.Goubel,"Existence Range, Structural Dielectric Properties of $Zr_xTi_ySn_zO_4$ Ceramics

(x+y+z=2),"Mar. Res. Bull., 16 (11) 1455-63 (1981).
[8]K. Wakino, K. Minai and H. Tamura," Microwave Characteristics of (Zr, Sn)TiO_4 and BaO-PbO-Nb_2O_3-TiO_2 Dielectric Resonators,"J. Am. Ceram. Soc., 67 (4) 278-81 (1984).
[9]A. E. McHale and R. S. Roth,"Investigation of the Phase Transition in $ZrTiO_4$ and $ZrTiO_4$-SnO_2 Solid Solutions," J. Am. Ceram. Soc., 66 (2) C-18-C-20 (1983).
[10]A. E. McHale and R. S. Roth,"Low-Temperature Phase Relationships in the System ZrO_2-TiO_2,"J. Am. Cera. Soc., 69 (11) 827-32 (1986).
[11]H. Ikawa, A. Iwai, K. Hiruta, H. Shimojima, K. Urabe and S. Udagawa, "Phase Transformation and Thermal Expansion of Zirconium and Hafnium Titanates and their Solid Solutions,"J. Am. Ceram. Soc., 71 (2) 120-27 (1988).
[12]H. Ikawa, H. Shimojima, K. Yamada, K. Urabe and S. Udagawa,"Polymorphism in $ZrTiO_4$,"; pp. 509-14 in Science of Ceramics 14 Edited by D. Taylor, The Instutute of Ceramics, Shelton, 1988.
[13]R. E. Newnham,"Crystal Structure of $ZrTiO_4$,"J. Am. Ceram. Soc., 50 (4) 216 (1967).
[14]R. D. Shannon,"Revised Effective Ionic Radii and Systematic Studies of Interatomic Distances in Halides and Chalcogenides,"Acta Cryst., Sect. A, A32 (9) 751-67 (1976).
[15]H. Igarashi,"Control of Microstructure of Dielectric Ceramics and their Properties,"Ceramics Japan, 12 (2) 126-131 (1977)
[16]K. Wakino and H. Tamura,"Microwave Dielectric Materials,"Ceramics Japan, 19 (11) 927-32 (1984).
[17]R. C. Buchanan,"Properties of Ceramic Insulators,"in Ceramic Materials for Electronics, Edited by R. C. Buchanan, p. 1-78, Marcel Dekker, Inc. New York and Basel (1986).
[18]M. A. Krebs and R. A. Condrate,"A Raman Spectral Characterization of Various Crystalline mixtures in the ZrO_2-TiO_2 and HfO_2-TiO_2 Systems,"J. Mate. Sci. Letters, 7 (12) 1327-30 (1988)
[19]A. Yamamoto, T. Yamada, H. Ikawa and O. Fukunaga, "Crystal Structure of Incommensurate $ZrTiO_4$," to be submitted

EFFECTS OF PROCESSING ON MICROSTRUCTURE AND DIELECTRIC PROPERTIES IN $Ba(Mg_{1/3}Ta_{2/3})O_3$ CERAMICS

Masaaki Sugiyama, Takayuki Inuzuka and Hiroshi Kubo
R&D Laboratories-I
Central R&D Bureau
Nippon Steel Corporation,
1618 Ida, Nakahara-ku,
Kawasaki 211, JAPAN

ABSTRACT

The dielectric resonator material of $Ba(Mg_{1/3}Ta_{2/3})O_3$ with high Q value, which was applied to the field of satellite communication system, was investigated by X-ray diffractometry and transmission electron microscopy. Low dielectric loss materials with Q>15000 at 10GHz could be made by conventional sintering method in air, by optimizing the process conditions. In order to improve the Q value, the factors influencing its dielectric loss were investigaeted on the view points of the microstructure of the sintered body. The domain size of the ordered phase was found to influence significantly to the Q value. On the process of ordering impingement of the ordered domains results in the formation of subboundaries in the parent grain. The subboundaries are always in a highly strained condition, and dislocations are often introduced at the interface. Since the dielectric properties are originated from the ion polarization, it is suggested that the dielectric loss at the subboundary should be large, resulting in the marked degradation of Q value.

INTRODUCTION

With the broader use of satellite systems for television and communications, special dielectric resonator materials of high permittivity with stable low loss properties at microwave frequency have been developed. At high frequencies of 4-50GHz, the electromagnetic wave of the TE_{011} mode can be utilized for the transmission mode of waves. In this case, the total Q value yielded in a dielectric resonator, which is defined to be the inverse of the dielectric loss, tan δ, is known to originate mostly in the dielectric material. Thus, dielectric resonator materials with high unloaded Q (>20000 at 10 GHz), and low temperature coefficient of the resonant frequency τ_f (0±5ppm/°C) have been developed with special attention to application as filters and/or oscillators(1). As high Q ceramics, the ceramics with complex perovskite structure on the basis of the $Ba(X_{1/3}Y_{2/3})O_3$ (X=Mg,Zn ; Y=Ta,Nb) structure have been studied(2,3). Most of them have an ordered hexagonal structure, and especially, the ordered $Ba(Mg_{1/3}Ta_{2/3})O_3$ material(BMT) sintered by the technique accompanied with the rapid heating of 1600°C/min. exhibits the highest Q value such as 35000 at 10GHz.(4) However, $Ba(Mg,Ta)O_3$ ceramics with high Q value is known as a difficult material to be sintered. Thus, the materials whose Q value at 10GHz is 20000 have been used, from the view point of the stable manufacturing technique at low cost as the dielectric resonator in the applicational field.

As the factors influencing Q value in compounds of the $A(B'_{1/3}B''_{2/3})O_3$ type, the density of the sintered body, the degree of order, stoichiometry and lattice defects have been investigated by X-ray diffractometry and measurements of dielectric properties(5,6), and Ramann spectroscopy(7). However, microstructure of dielectric resonator materials has been scarcely examined, in contrast to the studies on ferroelectric materials such as $BaTiO_3$(8) and relaxors(9). Therefore, the investigation of the microstructure of $Ba(Mg_{1/3}Ta_{2/3})O_3$ ceramics was made using transmission electron microscopy. The comparison of the microstructure of the high Q materials with the low Q materials was made to improve the Q value of the material. It is expected, through such investigations, to find the dominant factors influencing the stable production of high-Q dielectric resonator materials.

EXPERIMENTAL PROCEDURE

The starting materials were prepared by mixing $BaCO_3$, $MgCO_3$, and Ta_2O_5 weighed beforehand in the stoichiometric content of $Ba(Mg_{1/3}Ta_{2/3})O_3$ in a ball mill, followed by calcination at 1250°C for 4 hours in air. The ground mixture was granulated with PVA binders and glass compound. The powder was then formed into disks with dimensions of 8-12mm diameter, using uni-axial pressing, and sintered at temperatures between 1600 and 1700°C for 8-64 hours in air.

The samples obtained were fabricated into several sizes of different kinds of dielectric resonators, and were measured for relative permittivity, ε_r, and dielectric loss, tanδ, by the dielectric rod resonator method[10] which was based on the Hakki & Coleman's method.[11] Identification of the phases obtained were made using X-ray diffractometry, and the density of materials were measured by the Archimedes method. The specimen for transmission electron microscopy was made by cutting the sintered disks into small pieces followed by mechanical polishing and ion-milling using 4kV Ar^+ ions. The transmission electron microscope used in the present work was a Hitachi H-800 analytical microscope(200kV) equipped with energy dispersive X-ray spectroanalyzer. For high resolution observation, JEOL-4000EX microscope(400kV) was used.

EXPERIMENTAL RESULTS

Dielectric Properties and Microstructure

The change of relative densities with increasing sintering time at 1600°C is shown in Fig.1, where the theoretical density is 7.628g/cm^3 for the $Ba(Mg_{1/3}Ta_{2/3})O_3$ structure. By sintering for more than 16 hours, densified ceramics with 96% relative density were obtained. However, the relative densities have a tendency to become slightly lower for extended sintering at 1600°C. The relative permittivity, ε_r, was found to increase linearly with the relative density of the sintered body. By extrapolation to the 100% densified material, it was estimated to have the value of 25 in the sintered $Ba(Mg_{1/3}Ta_{2/3})O_3$ ceramics.

Figure 2 shows the change in Q values with the sintering time at 1600°C. To avoid the effects caused by the manufacturing conditions, the measurements were done using a series of samples made in the same batch. The length of error bars in the graph are based on values measured directly. For samples sintered at 1600°C for 64 hours, the Q values varied considerably from specimen to specimen, even though the relative densities of the specimens were similar. However, it is to be noted that the Q value decreases sharply by sintering for more than 32 hours at 1600°C, while the density of the sintered body remains constant. In order to make clear the origin of such sharp change in Q value, the structural change in the samples were examined by X-ray diffractometry. As is known(12), the $Ba(Mg_{1/3}Ta_{2/3})O_3$ ceramics has an ordered hexagonal structure. The degree of order, S, which is defined in terms of the occupation probability and the average concentration of Mg or Ta atoms, turns out to be expressed as

$$S^2 = \frac{I_{100}}{I_{110}+I_{012}} \cdot \frac{(1+\cos^2 2\theta_1)\sin^2\theta_2\cos\theta_2}{(1+\cos^2 2\theta_2)\sin^2\theta_1\cos\theta_1} \cdot \frac{(m_1|F_{110}|^2+m_2|F_{012}|^2)}{m_3|F_{100}|^2}$$

Here, I_{hkl} is the measured X-ray diffraction intensity, and θ_1 and θ_2 are difference of Bragg's angles between 110 and 012 fundamental reflections and 001 ordered reflection, respectively. (Indices are hexagonal notation) F_{hkl} is the structure factor calculated for the hexagonal ordered structure, and $m_i(i=1,2,3)$ is a multiplicity factor for each reflection. In the calculation, only the Lorentz factor was corrected and the correction by the absorption and temperature factor is not considered in the present paper. The change of the degree of order for the material during sintering was estimated, as shown in Fig.3. X-ray diffraction indicated the extra reflection peaks in the specimen sintered for more than 32hours at 1600°C. The new phase identified to be $BaTa_2O_6$(BT) was formed by precipitation after the densification of the material. It is now clear that the degradation of the Q value as well as the slight decreased density of the sample is originated in the appearance of the new phase, BT.

The amount of $BaTa_2O_6$ phase was estimated by measuring the relative intensity ratio of X-ray profiles represented by $(110)_{BMT}$ and $(420),(331)_{BT}$. Results are shown in Fig.3. In any sample sintered at 1600°C for less than 16 hours, the $BaTa_2O_6$ phase was not detected by X-ray diffraction analysis. It is to be noted , however, that the new phase does not appear in samples sintered at 1650°C for 16 hours. The $BaTa_2O_6$ phase was also observed after sintering at 1700°C for 32hours.

The morphology of the $BaTa_2O_6$ phase was examined by transmission electron microscopy, as seen in Fig.4. Figure 4(a) is a bright field image, in which long slender shaped particles indicated by arrows were identified as the $BaTa_2O_6$ phase. They were often observed at the triple point of the boundary or in the vicinity of the triple point of the sintered body. The identification of the phase was made by the electron diffraction analysis and the energy dispersive X-ray sectra analysis(EDX). The EDX spectrum taken from the new phase is shown in Fig.4(b), in which the detection of slight amount of energy peak of magnesium is caused by the overlapping effect of the $BaTa_2O_6$ and $Ba(Mg_{1/3}Ta_{2/3})O_3$ matrix phase. The dielectric property, Q was reduced by the appearance of $BaTa_2O_6$ phase.

Microstructure of Ordered Phase

Figure 5 is a series of transmission electron micrographs showing the domain structure in the ordered $Ba(Mg_{1/3}Ta_{2/3})O_3$ material. One of the dielectric properties, Q value, is strongly affected by manufacturing conditions. Even if the dielectric resonators were made by almost same experimental conditions, so that they should have almost the same densification, the Q values were extremely varied from specimen to specimen, for example, 5000 and 20000 at 10GHz. From the experimental results of observations of their microstructures,it is strongly suggested that the microstructure of the ordered phase may be responsible for the variation of Q values. The microstructure of the low Q material is constituted by lots of small ordered domains in the present grain. Fig.5(a) is a bright field image showing the typical morphology of a grain consisting of several ordered domains. As the reference, the position of a grain boundary is indicated by arrows. Selected area diffraction patterns taken from the domains, indicated by the open circles A and B in the photograph, are aligned in the right side of Fig.5. The diffraction pattern (B) exhibits a typical (001) diffraction zone in the hexagonal ordered structure. Six equivalent ordered reflections are observed in this crystal direction. On the contrary, any ordered reflection does not appear in the diffraction pattern (A), in spite of the fact that the fundamental reflections are similar to those of (B). It was made clear by tilting the sample that the domain (A) was also the ordered hexagonal phase. Consequently, those diffraction patterns of (A) and (B) were indexed systematically to be the $(221)^*$ and $(001)^*$ planes of the hexagonal structure, respectively. The photograph sited in left-down position of Fig.5 is the dark field image taken from an 010^* ordered reflection. Certain domains are satisfied with 010^* Bragg's condition, so that they should have the same crystal orientation axis. Dislocations were introduced along the interface between the domains, as indicated by arrow in the photograph. In the view point of the dielectric properties, the dielectric loss is expected to be enhanced by such distorted area. Since the degree of order of each domain was considered to be sufficiently high, judging from the intensity of order reflection taken from each domain, much attention should be paid to the interface structure between domains.

Figure 6 is a high resolution electron micrograph in the vicinity of the interface between ordered domains, which is conventionally called to be subboundary hereafter. The upper and lower area correspond to 100 and 211 zone of the crystal, respectively. As indicated by arrows, the strain contrasts of the misfit dislocations are observed along the interface at regular intervals. Figure 7 is an enlarged photograph of the area showing the strain contrast of the dislocations in Fig.6. Main two kinds of lattice fringes are observed; one corresponding to those of $(001)^*$ with lattice spacing of 7Å in the 100 zone, and the other corresponding to the fringes of $(10\bar{2})$ with the spacing of 2.9Å in the 211 zone. The misfit of lattice fringes is observed at several points as indicated by arrows in the photograph. By the structure image, the distortion of atom arrangement was observed more clearly. Figure 8 is a structure image taken from the area near the subboundary, in which the distorted interfacial region is observed. The crystal orientation relationship is represented by the conjunction of $\bar{1}01$ zone and $1\bar{4}1$ zone in the hexagonal ordered structure.

DISCUSSION

The $Ba(Mg,Ta)O_3$ ceramics with high Q values of more than 15000 at 10GHz were made by conventional sintering method in air. In order to make clear the factors influencing its dielectic loss, the microstructure of the sintered body was investigated by electron microscopy. It is elucidated that the Q value strongly depends on the microstructure of the sintered body, as shown in Fig.4-8. On the view points of the properties of the material, the factors influencing the Q value are summarized as follows;

(1) The relative density. It must be sufficiently high, that is more than 96%, to have a high Q value over 15000 at 10GHz.

(2) The degree of order inside the ordered domains. It must be high enough to be closely one to obtain the above value.

(3) The domain size of the ordered phase. It must be large enough so that the anti-phase boundaries may not affect the degree of order S.

It is found that in order to obtain a high density material, sintering at 1600°C for more than 16 hours is inevitable, as shown in Fig.1. However, it is also necessary to avoid the formation of second phase. Since the $BaTa_2O_6$ phase appeared at the triple points, as shown in Fig.4, the formation may be attributed to the diffusion of magnesium ions from a grain, resulting in the deviation of concentraion of Mg from the stoichiometry in the grain. Although the formation mechanism of $BaTa_2O_6$ phase is not made clear, its formation may be suppressed by controlling the sintering conditions.

By making the sintering at 1600°C for 32 hours, it is found to make the dielectric resonator materials with Q values more than 15000 at 10GHz. It was also found that the materials with low Q values consist of lots of ordered domains in the grain of the perovskite structure, as seen in Fig.5. Therefore, it is important to discuss about the morphology of the ordered $Ba(Mg_{1/3}Ta_{2/3})O_3$ phase. On the assumption that the ordered domains are nucleated from a disordered matrix phase, four equivalent ordered domains will be nucleated along four ‹111› directions in the cubic perovskite structure, because the ordered arrangement of magnesium and tantalum ions occurs to the ‹111› direction. As a result, it is expected that ordered domains nucleated in the same cubic perovskite disorder matrix will have different directional ordered stacking axes. In the cubic crystal, the angles between any two ‹111› directions are 70.5°. The orientation relationship between the observed ordered domains was examined for several cases. In Fig.6, the angle between (001) and $(\bar{2}21)$ is 70.5°. In other cases, the ordered domains with zone axis relationships of $(\bar{1}01)$-$(1\bar{4}1)$, and (211)-(100) were observed in Fig.8 and Fig.6, respectively. They also satisfy the above orientation relationships of the angle 70.5°. Consequently, it is concluded that the ordered domains are nucleated from the disordered matrix in the early stage of sintering at 1600°C. Because of the difference of their ordered stacking axes, the growth of each ordered domain forms the subboundary in a parent grain.

The lattice mismatch at the interface of domains appears to be small in comparison with the grain boundary of the parent phase. In the ordered phase, the lattice constant of $\langle 001 \rangle_H$ axis is smaller than the $\langle 111 \rangle_C$ axis by 0.18%. In order to accomodate such distortion, dislocations should be introduced at the subboundary as shown in Fig.5 and 6, and the strains are stored in the vicinity of the interface, as indicated by the high resolution electron micrographs of Fig.6, 7 and 8. However, so far as the lattice rearrangement in the grain occurs, the total strain could not be large. Reflecting such situation, the lattice mismatch at the interface of domains might be small. Since the electromagnetic energy is also stored as a internal energy in a grain when the crystal lattice is distorted, the dielectric loss might be enhanced by such internal stresses as observed in Fig.6-8, which are concentrated at the interface of ordered domains.

CONCLUSION

The results are aummarized as follows ; (1) Low dielectric loss materials with Q>15000 at 10GHz were made by conventional sintering in air, by optimizing the process conditions. (2) During sintering, ordered domains with hexagonal structure are nucleated and grown to the equivalent four $\langle 111 \rangle$ directions in the cubic disordered matrix. (3) Impingement of the ordered domains results in the formation of subboundaries in the parent grain. (4) The subboundaries are always in a highly strained condition, and dislocations are often introduced at the interface. It is suggested that the dielectric loss at the domain boundary is large, resulting in the marked degradation of Q value.

REFERENCES

[1]K. Wakino, "High Frequency Dielectrics and Their

Applications," Proc. of VI-IEEE Int. Symp.on Applications of Ferroelectrics, Bethlehem, PA, June, 97-106 (1986).

[2]S. Nomura, K. Toyama and K. Kaneta, "$Ba(Mg_{1/3}Ta_{2/3})O_3$ Ceramics with Temperature-Stable High Dielectric Constant and Low Microwave Loss," Jpn. J. Appl. Phys., 21[10], L624-L626 (1982).

[3]K. Wakino, M. Murata and H. Tamura, "Far Infrared Reflection Spectra of $Ba(Zn,Ta)O_3$-$BaZrO_3$ Dielectric Resonator Material," J. Am. Ceram. Soc., 69[1], 34-37 (1986).

[4]K. Matsumoto and T. Hiuga, Proc. of VI-IEEE Int. Symp.on Applications of Ferroelectrics, Bethlehem, PA, June, 118-121, 1986.

[5]S. B. Desu and H. M. O'Bryan, "Microwave Loss Quality of $Ba(Zn_{1/3}Ta_{2/3})O_3$ Ceramics," J. Am. Ceram. Soc., 68[10], 546-551 (1985).

[6]K. Endo, K. Fujimoto and K. Murakawa, "Dielectric Properties of Ceramics in $Ba(Co_{1/3}Nb_{2/3})O_3$-$Ba(Zn_{1/3}Nb_{2/3})O_3$ Solid Solutions," J. Am. Ceram. Soc., 70[9], C215-C218 (1987).

[7]H. Tamura, D. A. Sagala and K. Wakino, "Lattice Vibrations of $Ba(Zn_{1/3}Ta_{2/3})O_3$ Crystal with Ordered Perovskite Structure," Jpn. J. Appl. Phys., 25[6], 787-791 (1986).

[8]T. Mails and H. Gleiter, "Investigation of the structure of ferroelectric domain boundaries by transmission electron microscopy," J. Appl. Phys., 47[12], 5195-5220 (1976).

[9]J. Chen, H. M. Chan and M. P. Harmer, "Ordering Structure and Dielectric Properties of Undoped and La/Na-doped $Pb(Mg_{1/3}Nb_{2/3})O_3$," J. Am. Ceram. Soc. , 72[4], 593-598 (1989).

[10]Y. Kobayashi and M. Katoh, "Microwave Measurement of Dielectric Properties of Low-Loss Materials by the Dielectric Rod Resonator Method," IEEE Trans. Microwave Theory Tech, 33[7], 586-592 (1985).

[11] B. W. Hakki and P. D. Coleman, "A Dielectic Resonator Method of Measuring Inductive Capacitance in the Millimeter Range," IRE Trans, Microwave Theory Tech., 8,402-410 (1960).

[12] F. Galasso and J. Pyle, "Ordering in Compouds of the $A(B'_{0.33}Ta_{0.67})O_3$ Type," Inorg. Chem., 23, 482-484 (1963).

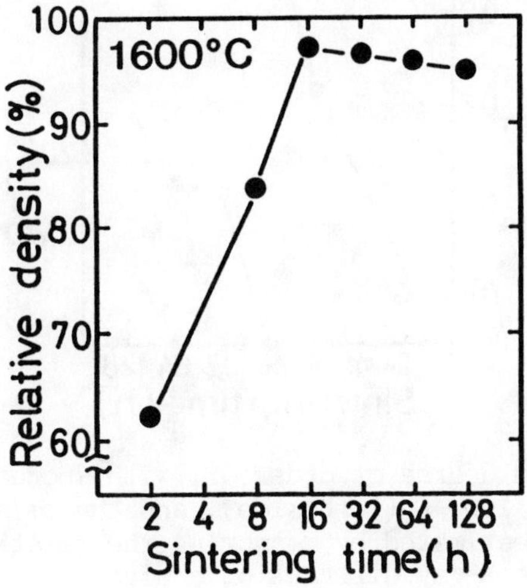

Fig. 1. Change of relative densities with increasing sintering time at 1600°C

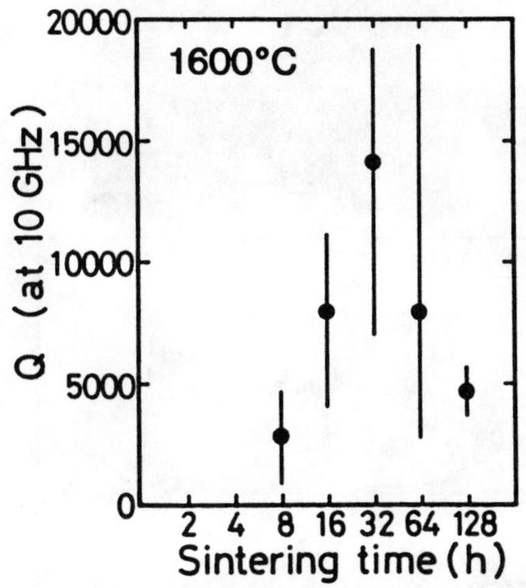

Fig. 2. Change in Q values with increasing sintering time at 1600°C

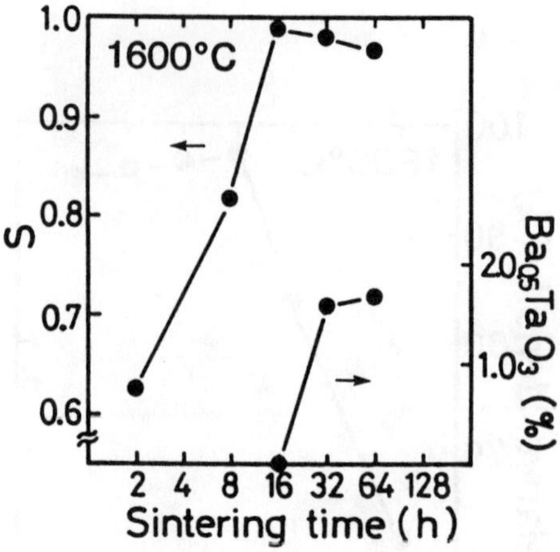

Fig. 3. Change in degree of order, S, with increasing sintering time at 1600°C (left side), and the amount of $BaTa_2O_6$ phase estimated by measuring the relative intensity ratio of X-ray diffraction analysis

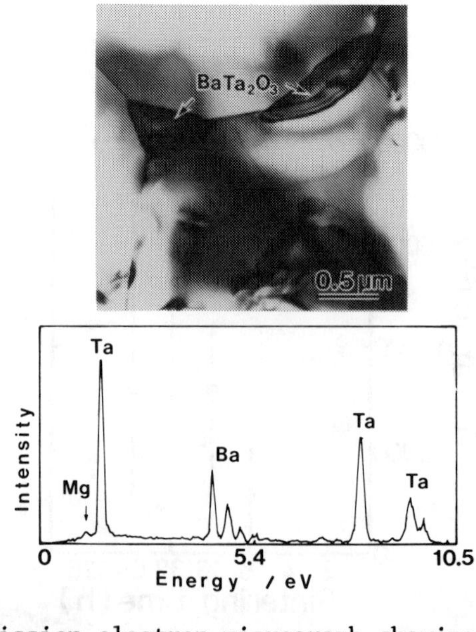

Fig. 4. Transmission electron micrograph showing the $BaTa_2O_6$ phase formed at triple point (a), and its EDX spectrum (b)

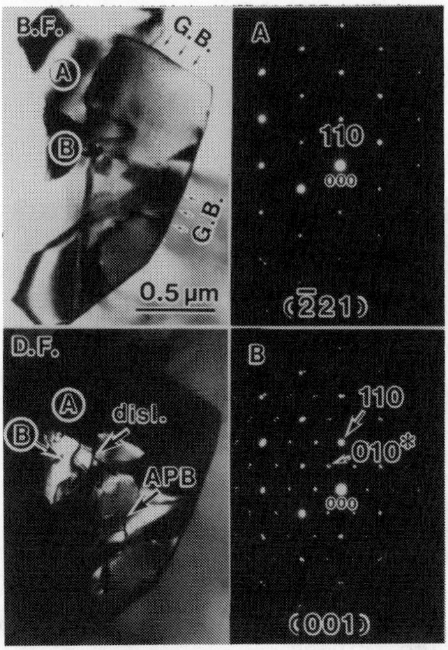

Fig. 5. A series of transmission electron micrographs showing the distribution of ordered domains in the parent grain

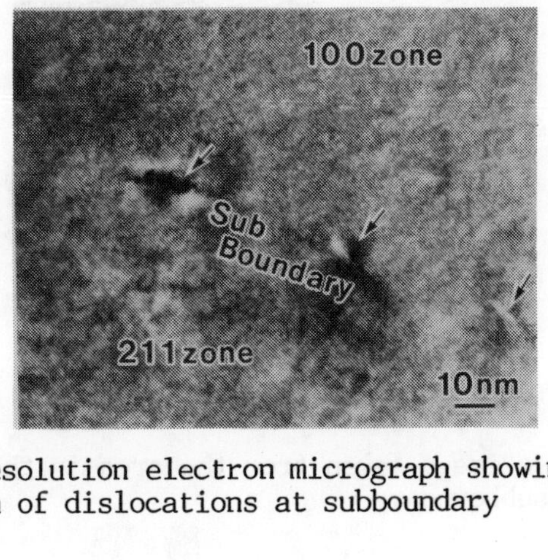

Fig. 6. High-resolution electron micrograph showing the distribution of dislocations at subboundary

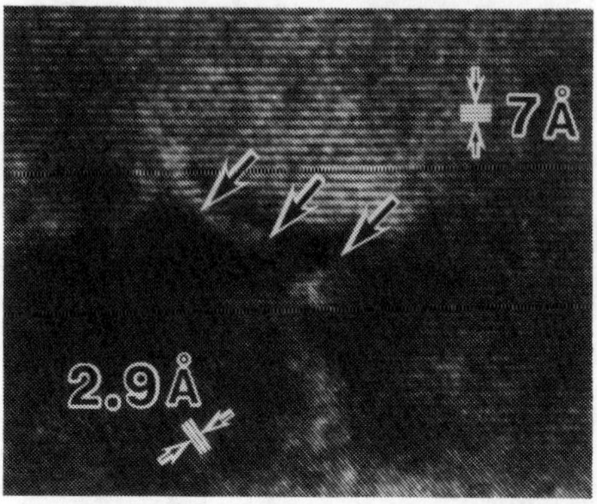

Fig. 7. Enlarged photograph of Fig. 6. Arrows indicate the misfit of lattice fringes

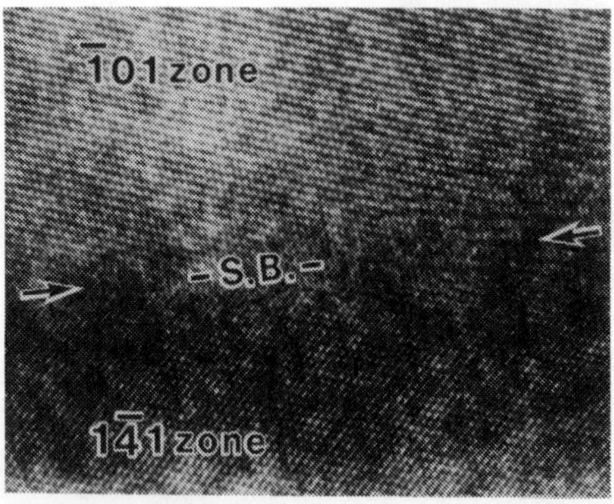

Fig. 8. Structure image taken from the area in the vicinity of the subboundary

MICROSTRUCTURE AND DIELECTRIC PROPERTIES OF MULLITE CERAMICS

Edward A. Giess, Judith M. Roldan, Philip J. Bailey,

IBM Thomas J. Watson Research Center
Yorktown Heights, New York 10598

and

Edward Goo,

University of Southern California at Los Angeles
University Park, Los Angeles, CA 90089-0241

ABSTRACT

Mullite ceramics, with 25 vol% excess silica, prepared under a variety of conditions, from both mullite-silica M-S and alumina-silica A-S with coarse- and fine-grained starting materials, produce a range of dielectric constants which are seen to vary with chemical and physical phase distributions and configurations in general accord with rules of mixtures. Besides the starting materials, porosity is one of the most important parameters influencing dielectric constant. There are major differences between the M-S and A-S composition systems beginning with the significantly higher green density of M-S bodies and the fact that no alumina is present; both factors make fired densities higher. Unreacted alumina raises the value of ε. The 1550°C firing temperature (just below the M-S eutectic) results in A-S samples with $\varepsilon \sim 4.5$, significantly lower than that of mullite alone, for firing times up to about 20 h in forming gas, but ε increases again with extended firing in forming gas. Samples from this transient low ε regime show series-type rule of mixtures behavior. Firing at 1630°C, above the eutectic, results in faster reactions and more dense bodies, but usually higher ε too. The starting material particle size has a major effect on the reaction time scale and on whether or not silica appears in the form of cristobalite. Coarse particle systems react relatively slowly and approach equilibrium slowly, especially in the A-S system. It is seen that knowing product grain morphologies alone is not sufficient to understand dielectric properties; crystallographic phases must also be identified.

I. INTRODUCTION

Electronic ceramics that support high frequency circuitry must have a low dielectric constant ε (and loss δ) to provide short signal transmission delay time

$$t_d = \frac{\sqrt{\varepsilon S}}{c} \tag{1}$$

where s is the signal path length and c is the velocity of light[1-3]. Therefore, alumina with $\varepsilon \sim 9.8$ has t_d about 21% larger than mullite with $\varepsilon \sim 6.7$, and a 40% larger t_d than other ceramics[1-7] with $\varepsilon \sim 5$.

The influence of microstructure and crystallographic second phases on dielectric properties as a result of composition and processing is known to be important in general for electronic ceramics[4-9]. Systems involving mullite and silica recently were the subject of dielectric studies[4,9] where ε values ranged from about 5 to 8. Therefore, this investigation was focussed on the microstructural and dielectric effects of starting materials, their states of particle size and aggregation, a forming gas atmosphere (found earlier[4] to favor low ε and δ), and firing temperatures of 1500, 1550, and 1630°C. The composition chosen for study was 75 vol% mullite and 25 vol% silica, in an attempt to obtain the lower ε that accompanies silica[8,9], but to retain the strength of mullite[5]. The powder raw materials included relatively coarse- and fine- grained alumina and silica, as well as coarse grained mullite. Phase equilibria studies[10-16] indicate some slight solid solubility of silica in mullite, and the existence of a eutectic at about 1590°C. Thus samples reacted at 1630°C undergo significant liquid phase reaction, while samples reacted at 1500°C represent a solid state reaction case.

II. EXPERIMENTAL PROCEDURE

(1) Starting Materials and Compositions

The specifications for starting materials are given in Table I. The high purity mullite had a molar formula $2.8Al_2O_3 \cdot 2SiO_2$, slightly rich in silica compared to the theoretical $3Al_2O_3 \cdot 2SiO_2$ mullite composition. Impurities can significantly influence the dielectric properties of electronic ceramics, so emission spectroscopic data for the starting materials also are given in Table I. The silica starting materials were amorphous, and the alumina and mullite were crystalline powders. Compositions are denoted by the powder starting materials A - alumina, M - mullite, S - silica, and by a postscript c - coarse and f - fine representing the grain size. The starting composition is equivalent to 75 vol% M and 25% S, but if it is made with fine-grained alumina (58.2 wt%) and fine-grained silica (41.8 wt%) is designated AfSf. When made with coarse-grained mullite (81.1 wt%) and coarse-grained silica (18.9 wt%), it is designated McSc.

Table I. The Starting Materials Employed

Material	Key	Particle Size	Impurities (ppm)
Mullite*1	Mc	1.2μm	Na 300 Mg 100 Fe,Ca,Ti 30 Zr 10
Alumina*2	Ac	0.96	Si 30 Na, Fe 20 Ca 13
Alumina*3	Af	0.2	Fe,Si 6 Na,Mg 2
Silica*4	Sc	3.7	Al,Fe,Ti 100 Mg,Zr,Cu 30
Silica*5	Sf	0.014	Mg,Al 3

*1 BAIKALOX MULCR Baikowski Intl. Co., Charlotte, NC28217
*2 HPA Ceralox Corp., Tucson, AZ85706
*3 AKP-50 Sumitomo Chemical America, NY, NY10154
*4 Precipitated acid washed BDH Chemicals Ltd., Poole, UK
*5 CAB-O-SIL M5 Cabot Corp., Tuscola, IL61953

(2) Processing Conditions

The process parameters investigated were: the materials and compositions discussed above; firing times of 12 h and 24 h (44 h in a few cases); a forming gas firing atmosphere; and firing temperatures of 1500°, 1550°, and 1630°C. Those starting powders with particle size >1μm were ball milled in methanol to disperse agglomerates. Five volume percent of a poly(vinylbutyral) binder and methanol were added to the powders to make a slurry that was mixed to have the consistency of a thick paste. The paste was spread to a thickness of 1 mm on a flat surface, and was dried until only ~6 wt% (~17 wt% for the submicron silica samples) solvent remained. After being ground into a powder, the material was formed into one-gram pellets that were uniaxially compacted at 74 MPa pressure in a 15.5 mm diameter die. The pellets were dried under vacuum at 110°C for 16 h.

The organic binder was removed by heating to 500°C at 100°C/h and heating for 5 h in flowing wet N_2 (bubbled through water at room temperature) followed by firing in air under the same time schedule with an additional 1 h at 1000°C. Samples were fired, in forming gas (10% H_2) flowing at 30 L/h in a 30 mm diameter tube, at a heating rate of 600°C/h to 900°C, and then at 300°C/h to one of the final firing temperatures (1500°, 1550°, or 1630°C) where they were held for 12 or 24 h.

(3) Property Measurements

Density values were determined from weights and direct micrometer measurements of pellets that had ~50 μm removed from each flat surface by diamond lapping. The theoretical specific gravities used were: alumina 3.98; mullite 3.15; and silica 2.3.

Dielectric property measurements were made on ~2 mm thick disks with their parallel flat surfaces coated by (Englehard Co.,Newark, NJ) flexible silver paint. Samples were vacuum dried at 110°C for 1 h before capacitance measurements were done on a (Hewlett Packard Co., CA) 4275A Multi-Frequency LCR Meter at 10kHz, 100kHz, 1MHz, and 10 MHz at 0.1 V.

X-ray diffraction patterns of diamond-lapped surfaces were taken with Ni-filtered Cu Kα radiation (Philips Electronics Co., NJ).

SEM micrographs were taken from pellets that were diamond polished. Samples were thermally etched in air at 1200°C for 1 h to enhance the contrast of SEM microstructures, by selectively volatilizing silica-rich phases and impurities. SEM samples were evaporation coated with about 50 nm of Au for charge dissipation. The SEM was (Hitachi Co. S-570) equipped with a LaB_6 gun and a (Gatan Co., CA) solid-state backscatter detector.

TEM specimens were prepared by mechanically grinding a thin wafer to a thickness of $\simeq$ 100 μm with 400 and 600 grit SiC papers. A 3 mm disc was cut with an ultrasonic drill and the disc was ion-milled with argon ions at an accelerating voltage of 6 kV and an incidence angle of 20°. A thin layer of carbon was evaporated on the samples to prevent charging. The TEM work was done on Philips 420 at an accelerating voltage of 100 kV.

III. RESULTS AND DISCUSSION

These experiments were based on the results of an earlier study[4] , where the lowest ε values were for samples made: with high porosity; in a forming gas atmosphere; and with firing temperature below the mullite-silica eutectic at ~1590°C. The 1630°C samples usually were the most dense and had the attendant high ε. Porosity was clearly the most important factor in determining dielectric permittivities ε which approximately followed the rules of mixtures. In Fig. 1 samples have ε values below the Maxwell rule, approaching the series rule. However, microstructures have a dispersed second phase (Maxwell) appearance instead of a layered (series) structure.

Again results can be classified according to whether the starting material phases are M-S or A-S. The starting green density of 45% (vol.% of theoretical) for the M-S system versus 31% for the A-S system is probably a major reason for the observed differences between the two precursor systems, the less complicated M-S system reaction sequences not withstanding.

The A-S system has more silica precursor powder than the equivalent M-S system and is, therefore, more sensitive to the progress of reaction toward the final equilibrium phases, mullite plus silica. This is especially true for the composition chosen for this study, viz., 25 vol% silica plus 75 vol% mullite.

While all of the McSc samples (Fig. 1) showed an increase in density and ε with both firing time and temperature, the pellets from the three different A-S systems (Figs. 2 to 4) generally increased in density with firing beyond 12 h to 44 h, but at first (12 h) ε decreased with time (to 24 h) for firing temperatures below the eutectic, before it finally increased with further firing (44 h total). In Fig. 4 the AfSf samples have a high ε at 12 h possibly because of chemisorbed anions, water, carbonate, etc. in this high surface area precursor system.

Looking at all the samples grouped according to firing temperatures, some interesting features become apparent. The 1500°C density values (Fig. 5) all increased with firing time, two systems exhibited little change in ε with time, the other two increased.

The 1550°C data assembled in Fig. 6 best show the unusual feature of ε values decreasing with firing time at first, but this trend reverses for time >20 h. It shows the transient process condition where ε is low before further reaction increases it to expected values. At 1550°C, a solid state reaction probably dominates and intergrannular liquid silica-rich glass will usually not form, so there will be more opportunity for amorphous silica starting material to crystallize. It must be recognized, however, that uncrystallized silica can undergo viscous flow at 1550°C. There were definite starting material particle size effects that were here perhaps most noticeable. For the A-S system reactions, XRD (See Table II) showed that using coarse-grained alumina with coarse-grained silica starting material resulted in significant cristobalite crystallization at lower firing temperatures.

Table II. XRD Peak Intensities for Cristobalite, Mullite, and Alumina at 2Θ values 21.7°, 26.2°, and 43.4°,

		1500°C forming gas			1550°C forming gas		
	time	cristob.	mullite	alumina	cristob.	mullite	alumina
McSc	12	373	3880	0	0	4306	0
	24	0	4020	0	0	3905	0
AcSc	12	529	4187	331	160	3510	148
	24	1040	3569	254	550	3558	0
AfSc	12	378	3133	258	0	4000	0
	24	0	3834	202	0	4268	0
	44				378	3649	0
AfSf	12				0	3102	0
	24				0	3766	0
	44				0	4316	0

1630°C 12 h and 24 h samples exhibited mullite peaks only.

The quantity of cristobalite actually increased with firing time, but disappeared with firing at 1630°C, above the eutectic. Using fine-grained starting materials seems to have avoided cristobalite formation, even below the eutectic. Not unexpectedly, under conditions where cristobalite increased with firing time, the crystalline alumina starting material decreased in concentration by reacting with silica to form mullite, because alumina is not an equilibrium product for the starting composition.

The transient low ε condition presumably was passed for even 12 h samples at 1630°C. The data in Fig. 7 show how little density and ε changed with firing from 12 h to 24 h. Firing at 1630°C may tend to enhance SiO_2/Al_2O_3 separation and segregation by liquid phase sintering, but reaction is also rapid, and final equilibrium is approached quickly relative to lower firing temperature conditions. In Fig. 7, all the 24-h arrow heads appear to be merging to an equilibrium point. At 1630°C, reaction essentially completes in 12 h for all samples, there is only a slight increase in density in a 24 h reaction.

Risbud and Zangvil[19] studied a similar A-S system by SEM and TEM, and found a microstructure with mullite reaction layers on the surface of the alumina grains with several kinds of aluminosilicate glasses present. The SEM (Fig. 8a) and TEM (Fig. 9a) results show that faceted grains are generally observed for mullite precursor systems, and these materials usually had higher density and ε than more equiaxed ones (Fig. 8b). This is not a general rule, however, because although the grains in TEM Fig. 9b appear to be equiaxed, the material also contains some alumina and, therefore, has a higher ε.

IV. CONCLUSIONS

In addition to the starting materials themselves, porosity is one of the most important parameters influencing dielectric constant.

There are major differences between the M-S and A-S composition systems beginning with the significantly higher green density of M-S bodies and the fact that no alumina is present; both factors make fired densities higher. Unreacted alumina raises the value of ε.

The starting material particle size has a major effect on the reaction time scale and on whether or not silica appears in the form of cristobalite. Coarse particle systems react relatively slowly and approach equilibrium more slowly, especially in the A-S system.

The 1550°C firing temperature (just below the M-S eutectic) results in A-S samples with ε ~ 4.5, significantly lower than that of mullite alone, for firing times up to about 20 h in forming gas, but ε increases again with extended firing in forming gas. This transient condition is a series-type rule of mixtures behavior. It is most important to recognize the existence of this transient process condition where ε is low before further firing increases it to higher, more normal values. Firing at 1630°C results in faster reactions and more dense bodies, but usually higher ε too.

Finally, knowing grain morphologies alone is not sufficient to understand dielectric properties; crystallographic phases must also be identified.

Acknowledgements: It is a pleasure to thank C. Guerci for experienced and very careful laboratory assistance. B.L. Olson conducted a variety of chemical analyses. A. Bross, J.M. Brownlow, E.I. Cooper, J.A. Casey, L.D. David, R. Gerhardt, H.E. Exner, L.W. Herron, S.H. Knickerbocker, R.O. Lussow, J.B. Niklewski, L.T. Romankiw, C.J. Sambucetti, M.W. Shafer, R.R. Shaw, and T.M. Shaw helped with discussions and advice.

REFERENCES

1. R.R. Tummala and E.J Rymaszewski, Microelectronics Packaging Handbook; Van Nostrand Reinhold, NY, 1989.

2. B. Schwartz, Electronic Ceramics; Ed. L.M. Levinson, Ch. 1, pp. 1-44, Marcel Dekker, NY, 1988.

3. W.M. Flock, Electronic Packaging Materials Science; Ed. E.A. Giess, K.-N. Tu and D.R. Uhlmann, Ch. 1, pp. 7-19, Mater. Res. Soc. Proceedings, 40 Pittsburgh, PA, 1984.

4. J.M. Roldan, E.A. Giess, S.H. Knickerbocker, L.D. David, and L.M. Litz, "Mullite and Alumina Ceramics for Low Dielectric Applications,"presented (31E89) at 91st Annual Meeting Am. Ceram. Soc., Indianapolis, IN, April 25, 1989.

5. R.M. Anderson, R. Gerhardt, J.B. Wachtman, Jr., D. Onn and S. Beecher, "Thermal, Mechanical, and Dielectric Properties of Mullite-Cordierite Composites," Proceedings Fall 1987 Am. Ceram. Soc. Denver Meeting, pp. 1-13.

6. B.H. Mussler and M.W. Shafer, "Preparation and Properties of Mullite-Cordierite Composites," J. Am. Ceram. Soc., 63 [5] 705-14 (1984).

7. W.D. Kingery, H.K. Bowen, and D.R.Uhlmann, Introduction to Ceramics ;2nd. Edn., J. Wiley and Sons, NY, 1976.

8. W.A. Yarbrough, T.R. Gururaja, and L.E. Cross, "Materials for IC Packaging with Very Low Permittivity via Colloidal Sol-Gel Processing," Am. Ceram. Soc. Bull., 66 [4] 692-98 (1987).

9. M. Horiuchi, K. Mizushima, Y. Takeuchi, and S.-i Wakabayashi, "New Mullite Ceramic Packages and Substrates," Proc. 38th Electronics Components Conf., May 1988, LA, CA, Components, Hybrid & Mfg. Soc. of IEEE, pp.574-83.

10. H.K. Bowen and J.W. Grieg, "The System $Al_2O_3 - SiO_2$," J. Am. Ceram. Soc., 7 [4] 238-54 (1924); corrections, ibid. [5]410 (1924).

11. S. Aramaki and R. Roy, "Revised Phase Diagram for the System $Al_2O_3 - SiO_2$," J. Am. Ceram. Soc., 45 [5] 229-42 (1962).

12. S.H. Risbud and J.A. Pask, "Calculated Thermodynamic Data and Metastable Immiscibility in the System $SiO_2 - Al_2O_3$," J. Am. Ceram. Soc., 60 [9-10] 418-24 (1972).

13. I.A. Aksay and J.A. Pask, "Stable and Metastable Equilibria in the System $Al_2O_3 - SiO_2$," J. Am. Ceram. Soc., 58 [11-12] 507-12 (1975).

14. W.M. Kriven and J.A. Pask, "Solid Solution Range and Microstructures of Melt-Grown Mullite," J. Am. Ceram. Soc., 66 [9] 649-54 (1983).

15. F.J. Klug and S. Prochazka, "Alumina-Silica Phase Diagram in the Mullite Region," J. Am. Ceram. Soc., 70 [10] 750-59 (1987).

16. A.J. Skoog and R.E. Moore, "Refractory of the Past for the Future; Mullite and its Use as a Bonding Phase," Bull. Am. Ceram. Soc., 67 [7] 1180-85 (1988).

17. K. Hashimoto and K. Niwa, "The Effect of MgO Addition on Mullite Substrate," J. Ceram. Soc. Jpn. Inter. Ed., 95 983-85 (1987).

18. R.J. Higgins, H.K. Bowen and E.A. Giess, "Preparation and Sintering Behavior of Fine-Grained $MgAl_2O_4 - SiO_2$ Composites," Advances in Ceramics, Vol. 21, Ceramic Powder Science and Technology, p. 691-98 (1987).

19. S.H. Risbud and A. Zangvil, "Microchemistry and Microstructure of a Multiphase Aluminosilicate Ceramic," J. Matls. Sci. 18 998-1004 (1983).

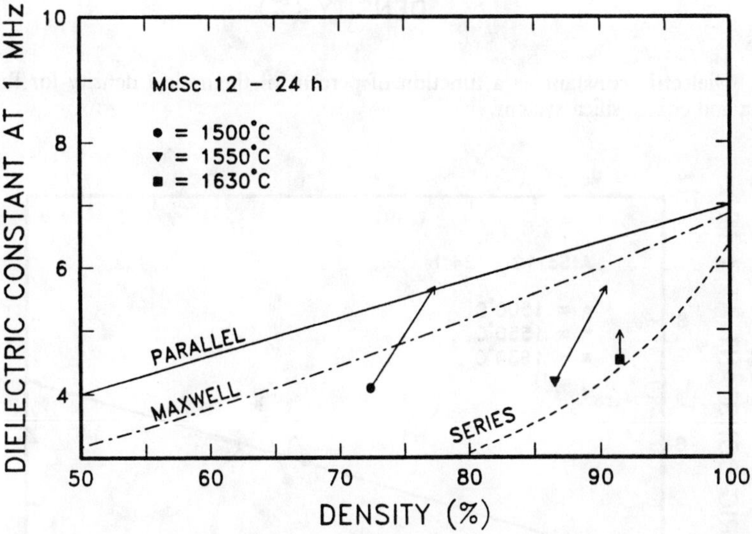

Fig. 1. Dielectric constant as a function of percent of theoretical density for the coarse mullite and coarse silica system. The solid line represents the parallel rule of mixtures with a 100% density value of $\varepsilon = 7$ (assuming mullite is ~8, silica is ~4) and open pore space value of 1. The dash-dot line, at 100% density, is a Maxwell rule value $\varepsilon = 6.8$ for mullite dispersed in silica; then for porous samples with density < 100% the plotted line is a Maxwell rule with pores dispersed in a matrix of $\varepsilon = 6.8$. The all-dashed line is a series rule. The key symbols represent 12 h data and the connected arrow heads are the respective 24 h data.

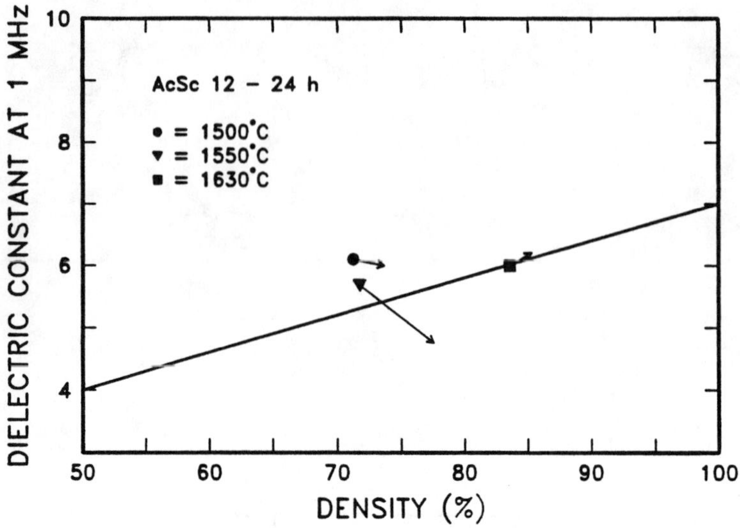

Fig. 2. Dielectric constant as a function of percent of theoretical density for the coarse alumina and coarse silica system.

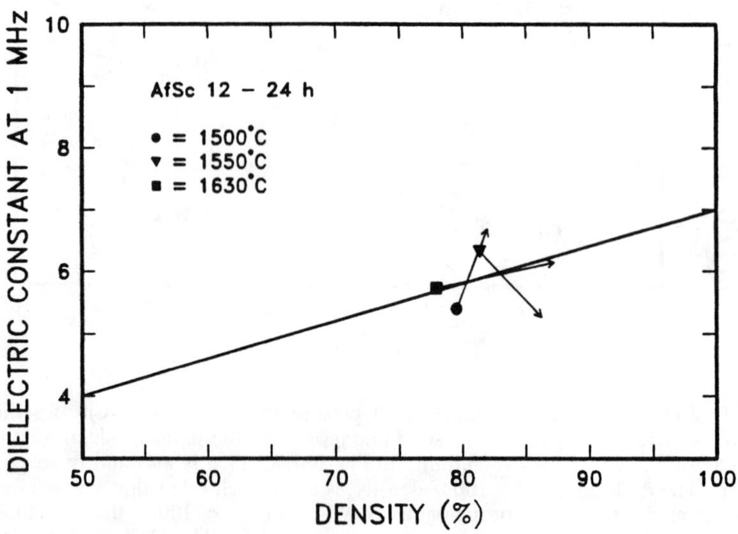

Fig. 3. Dielectric constant as a function of percent of theoretical density for the fine alumina and coarse silica system.

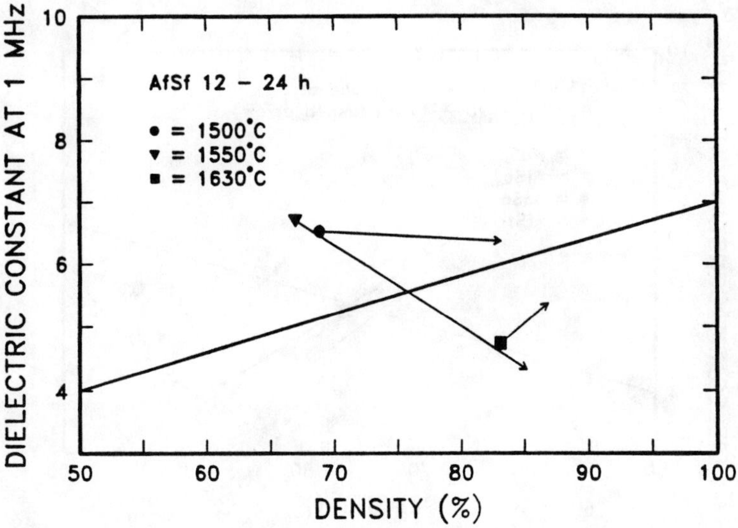

Fig. 4. Dielectric constant as a function of percent of theoretical density for the fine alumina and fine silica system.

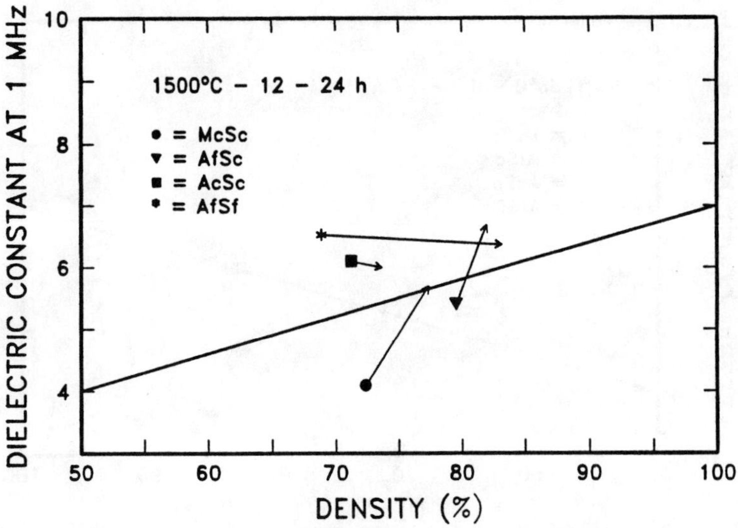

Fig. 5. The effects of a 1500°C firing temperature.

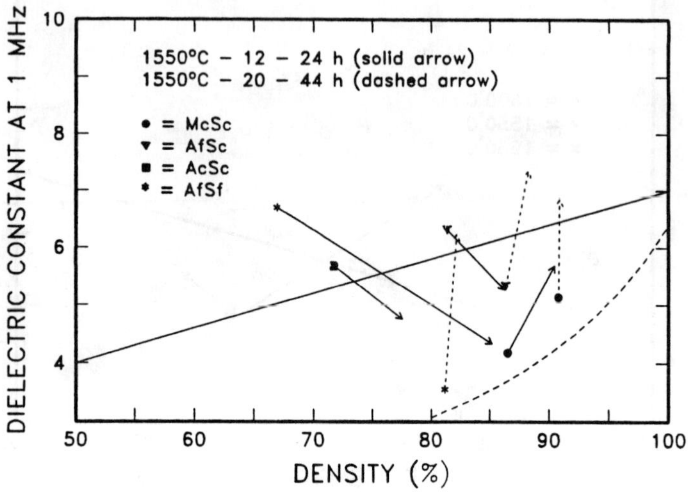

Fig. 6. The effects of a 1550°C firing temperature for samples fired 12 h and 24 h, respectively (shown by the connecting solid arrow) and 20 h and 44 h, respectively (shown by the connecting dashed arrow).

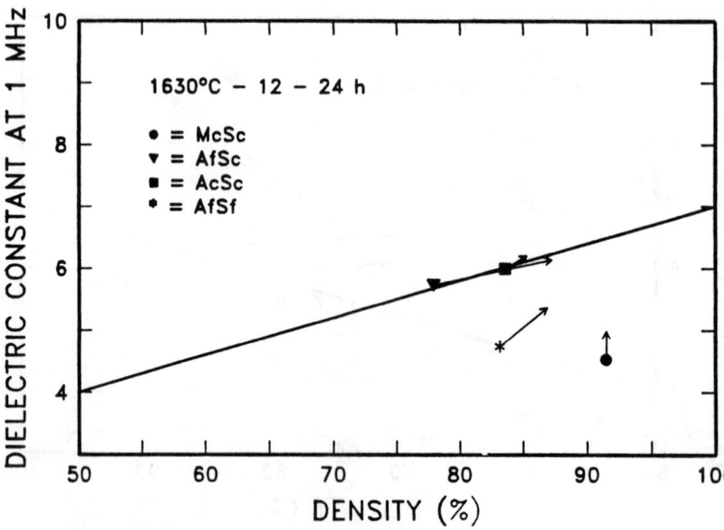

Fig. 7. The effects of a 1630°C firing temperature.

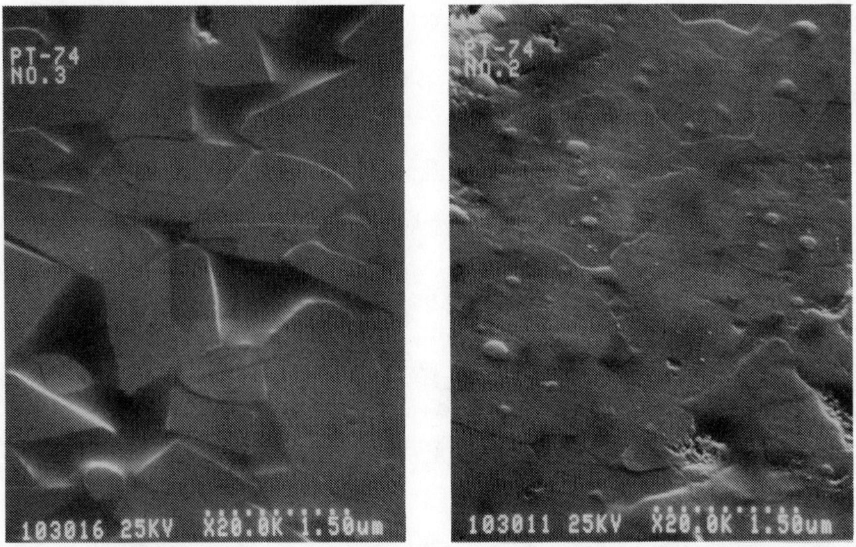

Fig. 8. SEM microstructure pictures of (a) faceted McSc grains in a 1550°C ceramic versus (b) a more equiaxed AfSc 1500° structure with lower ε

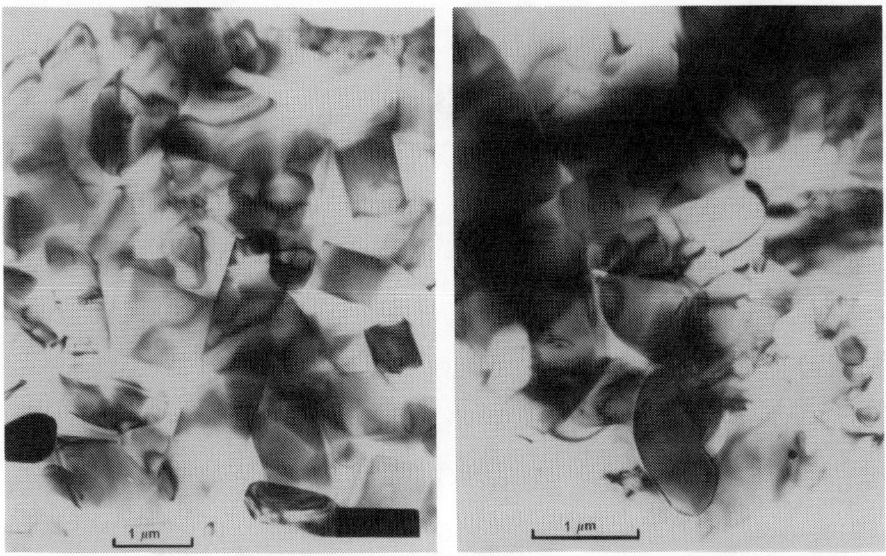

Fig. 9. TEM microstructure pictures of (a) faceted mullite grains in a McSc composition fired at 1550°C versus (b) more equiaxed AfSc grains reacted at 1550°C, but containing cristobalite and alumina, with ε~8.

Chemically Prepared PZT Films With Niobium Additions

B.A. Tuttle, D.H. Doughty, R.W. Schwartz, T.J. Garino, S.L. Martinez, D. Goodnow, C.L. Hernandez, R.G. Tissot, and W. F. Hammetter, Sandia National Laboratories, P.O. Box 5800, Albuquerque, NM 87185

ABSTRACT

We report on the first ferroelectric measurements of chemically prepared thin films of PZT containing niobium, (PNZT). Polycrystalline, perovskite PNZT thin films were fabricated by spin coating Pt coated SiO_2/Si substrates with alkoxide solutions. We systematically added niobium (0 to 10 at%) to our base composition, PZT 53/47. Further compositional modifications included the variation of Zr/Ti ratio for films containing 2 at% Nb. The dielectric constants measured for PNZT 2/56/44, 2/53/47 and 2/48/52 films were in good agreement with bulk values. Dielectric constants on the order of 700, 1000 and 500, respectively, were measured for these three films. Both linear and nonlinear current-voltage behavior was observed for our Pt/PNZT/Pt device structures, depending on the amplitude of the applied voltage and the temperature of measurement. We were able to make ferroelectric films of PNZT 2/53/47 (2 at% Nb addition) with spontaneous polarizations of 0.2 C/m^2 and coercive fields of 2.5 MV/m. PNZT films containing larger niobium additions (5 and 10 at%) had inferior ferroelectric properties because of the presence of nonperovskite phases.

INTRODUCTION

PZT ($Pb(Zr_xTi_{1-x})O_3$) thin films are of considerable interest for nonvolatile semiconductor memory applications. The ideal ferroelectric film for these applications should operate at low voltages, have a large voltage difference between read and write pulses, and exhibit maximum endurance and retention. Thus, the ideal film should have the following ferroelectric properties: low coercive field, high remanent polarization, minimal aging and minimal fatigue. Examples of other factors that may influence the compositional choice of ferroelectric films for nonvolatile semiconductor memory devices are piezoelectric effects and the dielectric constant.

Our initial efforts have emphasized the fabrication and characterization of PZT thin films with Nb additions near the morphotropic phase boundary (PZT 53/47). Bulk PZT ceramics that are compositionally near the morphotropic phase boundary have large remanent polarizations (0.4 to 0.5 C/m^2) and reasonable coercive fields (2 to 3 MV/m). Donor doped bulk PZT ceramics have lower coercive fields and higher resistivities than undoped PZT ceramics. Further, dielectric aging rates[1] in donor doped PZT ceramics are much lower than for undoped ceramics. Until now, no one has fabricated PZT thin (< 5 μm) films with niobium additions. For these reasons, we have investigated donor doping (Nb) in chemically-prepared PZT thin films near the morphotropic phase boundary.

A review of donor doping in bulk PZT ceramics is appropriate, since both resistivity and ferroelectric properties are substantially affected by dopants. Undoped PZT ceramics exhibit p-type conduction and have a resistivity on the order of 10^{10} ohm-cm. The p-type conduction in these materials is explained[1] by the creation of holes due to an intrinsic excess of lead vacancies compared to oxygen vacancies. Holes are formed in the vicinity of these Pb vacancies as shown in equation 1.

$$V_{Pb} \text{ -----> } V_{Pb}'' + 2h^{\bullet\bullet} \qquad (1)$$

Niobium doping increases the resistivity of bulk PZT. For 1 at% Nb doping, there is complete compensation[1] of intrinsic holes by the donor electrons. The resistivity of PZT ceramics containing 1 at% Nb is approximately 10^{13} ohm-cm. As the niobium doping level is increased above 1 at%, the resistivity remains constant and the coercive field decreases. Both these phenomena are explained by the creation of Pb vacancies[1] to maintain electroneutrality (equation 2).

$$Nb_2O_5 \xrightarrow{PZT} 2\,Nb_{Zr}^{\bullet} + V_{Pb}'' + PbO\uparrow + 2O_2\uparrow \qquad (2)$$

Niobium can substitute for either Zr or Ti in the lattice, only the Zr site substitution is shown in equation 2 for simplicity. The solid solubility limit for Nb in bulk PZT ceramics is approximately 3 at% Nb. As Nb doping is increased above this limit, Nb_2O_5 is formed and PNZT ceramics with inferior electrical properties result.

PbO volatility in bulk PZT ceramics fired above 1000°C facilitates the formation of lead vacancies and the accomodation of the aliovalent Nb ion into solid solution. It has not been demonstrated that chemically prepared PNZT films processed at 650°C, where PbO volatility is nonexistent or greatly reduced, will form lead vacancies. Thus, intrinsic PZT films, fired at low temperatures, may have fewer lead vacancies, and subsequently lower hole concentrations, than bulk PZT ceramics. Intrinsic PZT film resistivities would thus be larger than resistivities measured for intrinsic bulk PZT ceramics.

A lower concentration of intrinsic lead vacancies could also substantially change the effect of niobium additions to PZT. Complete compensation of holes would occur for lower niobium additions, perhaps orders of magnitude lower. For niobium additions in excess of the complete compensation limit, the following defect relationship (equation 3) may be operative for 650°C process temperatures.

$$2\ Nb_2O_5 \xrightarrow{PZT} 4\ Nb_{Zr}^{\bullet} + V_{Zr}^{''''} + Zr_{Zr} + 5\ ZrO_2(c) \quad (3)$$

For this defect mechanism, lead and oxygen are retained in the material. Niobium additions can result in formation of b site vacancies and development of a condensed phase with oxygen to cation ratio greater than 1.5. Although equation 3, for simplicity, only indicates Zr site substitution, it is understood that substitution on the Ti site can also occur and that the condensed phase would contain both Zr and Ti. For low firing temperatures, it may be thermodynamically favorable for this condensed phase to be a pyrochlore modification of PZT, such as $Pb_2(Zr,Ti)_2O_{7-x}$, where x is less than one.

Thus, it is not certain that chemically prepared PZT films with Nb additions, processed at temperatures of 650°C, will exhibit higher resistivities and lower coercive fields than undoped PZT films. Further, niobium additions to PZT films, processed at 650°C, may result in films with inferior electrical properties. These inferior electrical properties are caused by the formation of nonferroelectric phases, as shown in equation 3. Therefore, a major goal of our study is to determine if chemically prepared PZT films with niobium additions have larger resistivities and lower coercive fields than films with no added niobiuim.

Solution chemistry techniques have been used to enhance densification at low temperatures for a variety of inorganic materials. PZT and PLZT thin films, which have good ferroelectric properties, have been fabricated[2-5] using process temperatures 600°C below the sintering temperature for bulk PZT. Lipeles et al[5] developed a process, using isopropanol solutions of lead 2-ethylhexanoate, zirconium tetra-n-propoxide and titanium tetra-n-butoxide for the formation of single phase perovskite PZT 55/45 films. Extensive hydrolysis of the alkoxides in the starting solution resulted in amorphous films that contained considerable porosity. Lower water additions (mole ratios of water to metal ions equal to 0 or 0.5) were essential for the fabrication of perovskite PZT films at 525°C. We have used the Lipeles process, with slight modifications, for the fabrication of niobium doped PZT films.

EXPERIMENTAL PROCEDURE

Alkoxide derived solutions (0.25 M) were prepared from precursors of lead 2-ethylhexanoate, zirconium tetra-n-butoxide, titanium tetraisopropoxide and niobium pentaethoxide for the fabrication of PNZT films. Immediately before deposition, these solutions were filtered with a 0.2 μm syringe filter. The filtered solutions were then deposited onto our substrates at 4000 RPM using a photoresist spinner. The film was dried at 150°C for 20 minutes and then given a 400°C/20 minute heat treatment to decompose residual organics. If desired, additional film coatings could be deposited after the 400°C heat treatment. Typically, our films consisted of 3 layers and were 410 nm thick after firing. Scanning electron micrographs and NIST micron standards were used to determine film thickness. A final annealing treatment at temperatures ranging from 600°C to 725°C was used to crystallize perovskite PNZT films. For the final anneal, heating rates of 3°C/min, cooling rates of 1.5°C/min, and soak times of 30 minutes or 1 hour were used.

Our substrates consisted of 4" diameter (100) silicon wafers onto which a 600 nm thick, tetraethylorhosilicate (TEOS) derived SiO_2 layer was vapor deposited and densified at 950°C. We dc sputter deposited a 200 nm layer of Pt on top of the SiO_2. This Pt layer served as the bottom electrode for the PNZT film. For some of our PNZT films, a 50 nm Ti adhesion layer was sputter deposited under the bottom Pt film layer. The top electrode for our PNZT films consisted of a sputter deposited array of Pt dots. These dots were of three different diameters: 0.5 mm, 0.75 mm or 1.0 mm. Deposition conditions for the top electrodes were similar to those for the bottom electrode. Typically, the Pt films were 200 nm thick and deposited at a rate of 10 nm/minute.

Electrical contact to the bottom electrode was achieved by careful mechanical abrasion of the fired PNZT film to expose a selected area of the electrode. Silver epoxy was then used to facilitate connection of the bottom electrode to our microprobe. Relatively blunt microprobes ($\approx$ 100 μm diameter) were used for our electrical measurements, to minimize film damage. Our resistance measurements were performed using a two point probe technique. A sensitive ammeter (Keithley Model 619 electrometer) was required to measure the low currents (less than 1 nA) for our electrical tests.

Our device structure was chosen to be representative of ferroelectric film configurations in nonvolatile semiconductor memory devices. This device structure, consisting of a thin insulating film between two conductive layers, made meaningful 4 point probe measurements impossible. We performed resistance measurements at a variety of voltage levels to ascertain the nonlinear behavior of our films. Typically, each measurement was made 5 to 20 minutes after the initial voltage application to allow stabilization of both the electrometer and ferroelectric domain motions within the film. Resistance measurements were made as a function of temperature by placing the films on a hot chuck, with temperature control to 300°C. In addition to the thermocouple inside the chuck, an external copper-constantan thermocouple (0.125 mm diameter wire) was attached to the top of the film for temperature monitoring. Although a modified Diamant-Pepinsky bridge was used for the ferroelectric measurements, all ferroelectric measurements presented in this paper were uncompensated.

RESULTS AND DISCUSSION

Figure 1 depicts thermogravimetric and differential thermal analyses for gel particles of PNZT 2/53/47. These gel particles were formed by drying a PNZT gel at 150°C for approximately 1 hour and then crushing the dried gel into particles of less than 500 μm diameter. We interpret the sequence of DTA exotherms, with increasing temperature, to correspond to organic decomposition, pyrochlore crystallization and perovskite formation, respectively. Consistent with the

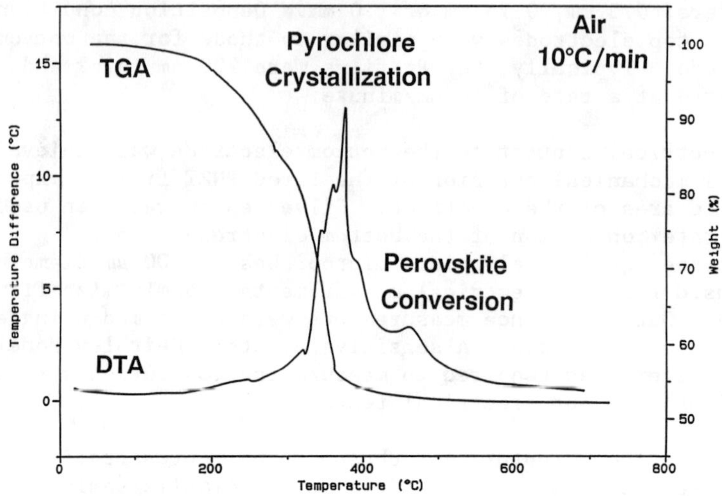

Figure 1. Thermal Analyses of PNZT 2/53/47 gel particles were performed at 10°C/min.

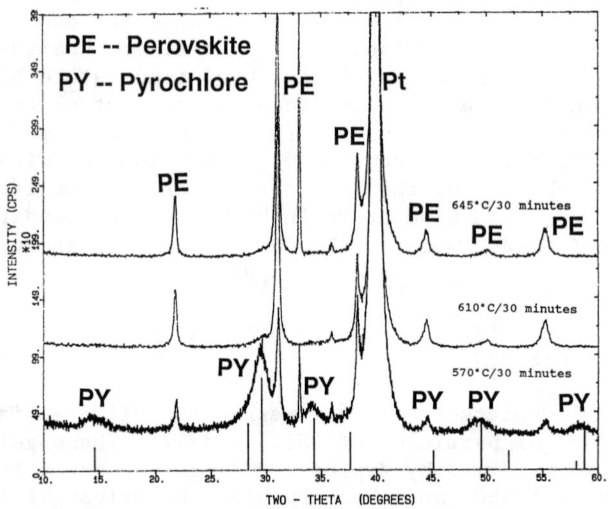

Figure 2. The X-ray diffraction patterns are of three different PZT 50/50 thin films, each fired at a different temperature: 570°C, 610°C, and 645°C.

gels, high temperature X-ray diffraction traces of our PNZT 2/53/47 films indicate that the formation of the pyrochlore phase occurs at a lower temperature than the perovskite. However, the initial crystallization of perovskite PZT may be from remanent amorphous phases rather than the pyrochlore. PZT 53/47 gel particles with other niobium additions (0 to 10 at%) had DTA and TGA traces similar to those in Figure 1. The crystallization temperature (less than 500°C) of perovskite PZT 53/47 is slightly lower for our gels than that reported (approximately 550°C) for an alternative chemical synthesis technique.[2]

Figure 2 depicts X-ray diffraction traces of three PZT 50/50 films fired for 30 minutes at different temperatures. From thermal analyses of gel particles (Figure 1) single phase perovskite PZT is expected for a thermal treatment of 570°C for 30 minutes. Although significant perovskite PZT is present, there is extensive formation of a low temperature pyrochlore phase in the the film fired at this temperature. The diffraction trace of the film fired at 570°C contains every diffraction peak ($10° < 2\theta < 60°$) of the $Pb_2Ti_2O_{7-x}$ oxygen deficient pyrochlore.[6] Interestingly, there is little evidence in the diffraction trace of this 570°C film for the existence of the PbO deficient $PbTi_3O_7$ pyrochlore.[7] Thus, PbO volatility does not appear to influence phase evolution for these thermal processing conditions. The PZT 50/50 film fired at 645°C was single phase perovskite by X-ray diffraction analysis. Our PNZT 2/53/47 films, fired at 650°C for 30 minutes, produced similar single phase perovskite X-ray diffraction patterns.

Initial measurements indicated our PNZT 2/53/47 films had a resistivity of 10^{12} ohm-cm for an applied field of 2.5 MV/m. However, current measured as a function of applied voltage at 25°C was represented by the following nonlinear relationship.

$$I(X\ 10^{10}A) = 0.84\ V^2 + 0.12\ V + .06 \qquad (3)$$

This strong second order voltage dependence was observed from 1.25 to 12.5 MV/m for our films. The implication is that a potential barrier at the Pt/PZT interface, rather than the PNZT film resistivity, dominated our measurement at 25°C. Current-voltage characteristics, measured at 100°C, for the PNZT 2/53/47 film were linear from 0 to 5.0 MV/m dc. Above 5.0 MV/m, a slight nonlinear (less than second order) dependence was observed. Current-voltage measurements at 300°C indicated that the film response was linear over a much narrower range of applied field (0 to 0.5 MV/m). Further, at 300°C, high field nonlinearity was much greater than second order.

Although we do not have definitive proof of the cause of the current-voltage behavior with temperature, we offer the following hypothesis. We propose that a potential barrier at the PZT film - Pt electrode interface influences resistivity measurements at 25°C. As temperature is increased to 100°C, a linear I-V characteristic, representative of the true PNZT film resistivity is measured. We feel the extremely nonlinear behavior at high temperatures (>150°C) and high fields is due to the nonlinear nature of PZT under these high electrical and thermal stress conditions. We have performed a calculation that indicates that the current measured at 100°C and above should not be influenced by the potential barrier. For this calculation, we assume the magnitude of the current measured at room temperature is solely due to a potential barrier greater than 2 eV. We then use the resistivity values and activation energy for bulk PZT to show that the current measured above 100°C is representative of the PZT film, rather than the potential barrier at the interface. It must be emphasized that further data is required to support our hypothesis. Electron affinities and surface state densities are among the parameters that need to be known to adequately prove or disprove the model.

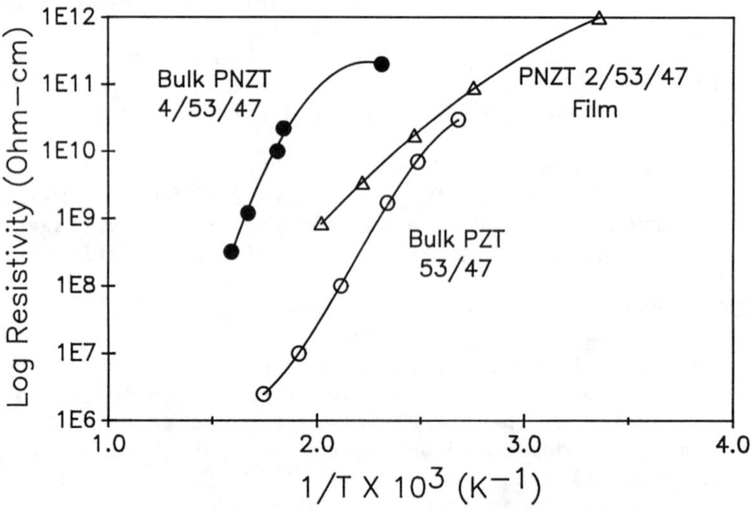

Figure 3. Log resistivity vs. inverse temperature characteristics are shown for a PZT film with 2 at% niobium addition, an undoped bulk PZT ceramic and a bulk PZT ceramic with 4% niobium addition.

Figure 3 is a plot of log resistivity vs 1000/T(K) comparing our PNZT 2/53/47 film fired at 650°C/30 min with doped and undoped bulk PZT[1] ceramics. The 2 to 3 order of magnitude difference in resistivity for doped bulk PZT compared to undoped bulk PZT is readily apparent. Our film has resistivity values that are intermediate to the bulk PZT materials. There are two possible explanations for the increase in resistivity of our film with respect to undoped bulk PZT ceramics. The first explanation, is that niobium additions create localized electrons that electrically compensate for intrinsic holes, as is observed for bulk PZT ceramics. A second explanantion is that the PZT film, fired at 650°C, has less intrinsic lead vacancies, and thus fewer holes than bulk PZT fired at higher temperatures. For this case, the PZT film fired at 650°C would have higher resistivity than the bulk PZT ceramics. Further, niobium additions would have little effect on film resistvity.

Our PNZT 2/53/47 film was ferroelectric, having a spontaneous polarization of 0.22 C/m^2, and had low field (0.25 MV/m) dielectric constants in excess of 1000. The activation energy for the bulk PZT conductivity is on the order of 0.9 eV. Disturbingly, the activation energy for our PNZT film was approximately 0.55 eV. Film conduction may be influenced by factors such as stress, porosity, space charge effects and the presence of a minor amount of pyrochlore (undetectable by X-ray diffraction analysis). Any of these factors may lower the activation energy. In addition, convection from the surface of our films may cause the true film temperature to be different than that recorded by the thermocouple.

Ferroelectric properties of PZT 53/47 films were substantially affected by niobium additions. Figure 4 compares a well saturated hysteresis loop of a PZT 53/47 film with 2 at% Nb addition with a low polarization, high coercive field hysteresis loop of an identically processed film with 10 at% Nb addition. There was little change in the polarization values of identically processed films with 2 at% and 0 at% Nb additions. Further, PZT 53/47 films with 0 at% and 2 at% Nb additions had similar coercive fields ($\approx$ 2.5 MV/m). Thus, films with 2 at% additions of Nb did not exhibit a reduction of coercive field, as is observed in bulk PZT ceramics. Films with 5 and 10 at% Nb were not single phase perovskites. This was not due to the exsolution of Nb_2O_5, but because a substantial amount of pyrochlore was present in these films.

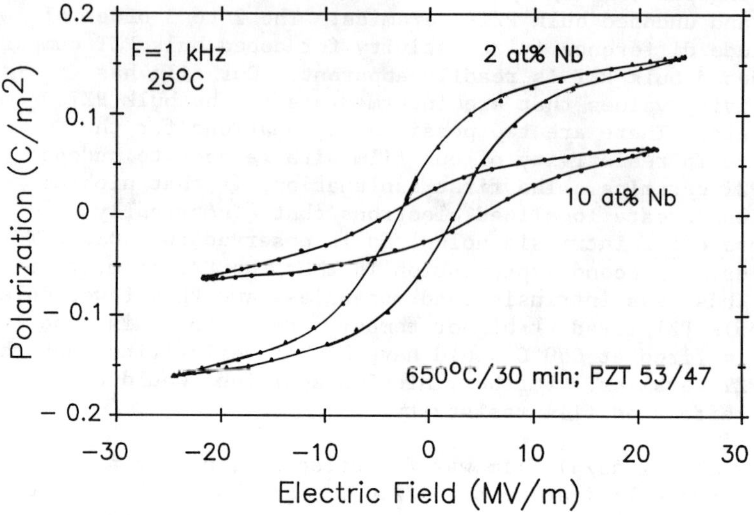

Figure 4. Ferroelectric hysteresis loops are shown for PZT thin films with 2 at% and 10 at% niobium additions.

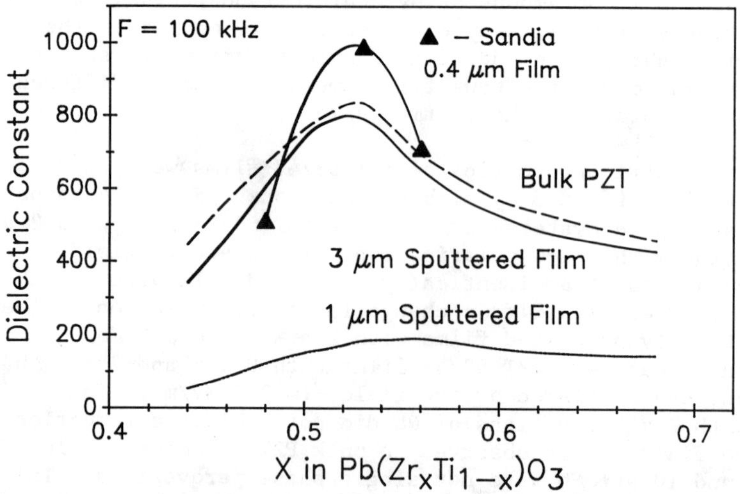

Figure 5. Dielectric constant behavior as a function of Zr/Ti ratio is depicted for chemically prepared PNZT thin films, sputtered PZT films, and bulk PZT ceramics.

An increase in the amount of pyrochlore phase with increased niobium additions is consistent with the defect mechanism described by equation 3. Attempts were made to fire these films at higher temperatures to produce single phase perovskite films. Although the percentage of perovskite phase increased with firing temperature, firing temperatures in excess of 775°C resulted in the formation of other nonperovskite phases. As expected, films with 5 and 10 at% Nb additions had lower polarizations, lower dielectric constants and higher coercive fields than films doped with 2 at% Nb or less.

A brief investigation of PNZT films with different Zr/Ti ratios was made. We fabricated films of three different Zr/Ti ratios: 1) PNZT 2/48/52, 2) PNZT 2/53/47, and 3) PNZT 2/56/44. All three films were 410 nm thick, fired at 650°C for 30 min and exhibited ferroelectric hysteresis loops. Figure 5 contains a plot of the dielectric constant of our PNZT films as a function of Zr/Ti ratio. We have also plotted data presented by Sayer et al.[8] in their paper on RF sputtered PZT films. Figure 5 indicates that our process gives good control of the Zr/Ti ratio, 1 mol% or better, since the dielectric constant peaks at the morphotropic phase boundary, as it does for bulk ceramics. The magnitude of the dielectric constant for our unpoled films agrees very well with both bulk ceramics and Sayer's 3 micron thick films. For our dielectric measurements, we used an ac field of 0.25 MV/m amplitude (0.1 volt) and measured dissipation factors in the range of 0.03 to 0.06.

Our PNZT film polarization values, measured from ferroelectric hysteresis loops, were less than the polarizations of Sayer's 3μm sputtered PZT films. However, the ferroelectric polarizations, measured for our 0.4 μm thick films, were slightly greater than those of his 1 μm thick sputtered films. Our PNZT 2/53/47 film had significantly larger polarizations than either of our two films which were compositionally removed from the morphotropic phase boundary (PNZT 2/56/44 or PNZT 2/48/52). This polarization behavior as a function of Zr/Ti ratio was also observed for Sayer's sputtered films[8] and bulk PZT ceramics.[1]

Summary:

We have fabricated and characterized chemically prepared PZT films with niobium additions. Although resistivity measurements of PNZT 2/53/47 films at room temperature indicated nonlinear behavior, linear I-V behavior was observed over reasonable ranges of applied fied at elevated temperatures. For this linear I-V region, PNZT 2/53/47 film resistivities of 10^{11} ohm-cm were measured at 100°C. The measured PNZT film resistivity is greater than that of intrinsic bulk PZT ceramics. We attribute the greater film resistivity to the formation of fewer lead vacancies, and associated holes, at the 650°C film process temperature than the ≈1200°C bulk PZT process temperature. Coercive fields for films with 0 at% Nb and 2 at% Nb additions were similar. Thus, these coercive field measurements did not provide indirect evidence of donor doping. Films with large niobium additions (5 and 10 at%) had lower dielectric constants, lower polarizations, and higher coercive fields than films doped with 2 at% Nb or less. We attribute this behavior to the significant amount of nonperovskite phase in films with large niobium additions. The above results are consistent with our proposed low temperature defect model. For this model, a condensed phase with oxygen to cation ratio greater than 1.5 is formed as niobium is added to PZT.

REFERENCES

1. B. Jaffe, W. Cook, and H. Jaffe, Piezoelectric Ceramics, Academic Press, (1971).

2. K. Budd, S. Dey, and D. Payne, "Sol-Gel Processing of $PbTiO_3$, $PbZrO_3$, PZT and PLZT Thin Films," Brit. Ceram. Soc. Proceedings, **36**, 107-20 (1985).

3. A. Shikh and G. Vest, "Kinetics of $BaTiO_3$ and $PbTiO_3$ Formation From Metallo-Organic Precursors," J. Am. Ceram. Soc. **69**, [9] 682-88 (1986).

4. J. Xu and R. Vest, "Preparation and Properties of PLZT Films From Metallorganic Precursors," Am. Ceram. Soc. Bull., **67** [3] 598 (1988).

5. R. Lipeles, D. Coleman, M. Leung, "Effects of Hydrolysis on Metallo-Organic Solution Decomposition of PZT Films," MRS Symposium Proceedings, **73**, 665-70 (1986).

6. F. Martin, "A Metastable Cubic Form of Lead Titanate Observed in Titania Nucleated Glass Ceramics," Phys. Chem. of Glasses, **6** [4] 143-7 (1965).

7. K. Iijima, Y. Tomita, R. Takayama, and I. Ueda, "Preparation of c-axis Oriented $PbTiO_3$ Thin Films and Their Crystallographic, Dielectric and Pyroelectric Properties," J. Appl. Phys., **60** [1] 361-7 (1986).

8. K. Sreenivas, M. Sayer, and P. Garrett, "Properties of DC Magnetron Sputtered Lead Zirconate Titanate Thin Films," Thin Solid Films, **172** [2] 251-67 (1989).

THE EFFECTS OF RAW POWDER AND PROCESSING ON THE PROPERTIES OF AlN CERAMICS

Ching-Fong Chen and Ender Savrun
Keramont Corp.
Tucson, AZ 85714

ABSTRACT

A complete study has been performed on two different AlN raw powders. One is synthesis through carbothermal reduction nitridation and the other is through direct nitridation. These two powders have distinctive differences in chemical properties, physical properties, and sinterability which result in different sintered properties including, color appearance, thermal conductivity, mechanical strength, microstructure, and chemical composition. This work verifies that the color of the sintered AlN is very closely related to the carbon content and the sintering temperature. The darkness of the sintered part increases as the carbon content increases. Also, for the same powder, as the sintering temperature increases, the color becomes whiter. While thermal conductivity is independent of the color appearance or the translucency of the sintered parts. These results conclude that color is not the major parameter in judging the thermal conductivity of AlN ceramics.

1.0 INTRODUCTION

It is well-known [1-3] that the properties of the raw powder can affect the densification and final properties of AlN parts. The best powder properties that have been established for AlN include: (1) spherical particle shape, (2) fine mean particle size, (3) narrow particle size distribution, and (4) low impurities content. Two types of commercially available powder can be classified based on their reaction method [1,4]; one is direct nitridation, and the other is carbothermal reduction nitridation. The carbothermal reduction AlN powder is synthesized by heating the carbon and alumina powder mixture as the raw materials. At some reaction temperature (≈1600 °C), the alumina is reduced by the carbon and reacts with the nitrogen to form AlN. The excess carbon is then removed by an oxidation process. While the direct nitride AlN powder is synthesized by direct nitridation of the aluminum metal powder and followed by crushing and milling steps.

It has been an argument between AlN manufacturers and users that the whiter the appearance of the AlN ceramics, the higher the thermal conductivity. Also, controlling the color consistency of the sintered parts has been a problem for AlN ceramics manufacturers. However, no ceramic color data has been published. In order to identify the importance of the color appearance, there is a need for a detailed study of the factors that could affect the color variation.

The color measurement of ceramics is difficult because of the complexity of total appearance and limited applicability of specific appearance scales. This issue involves potentially infinite variations in both spectral and geometric distributions of light by objects and materials. It is only feasible with instruments to measure specific attributes of appearance that are important for specific applications [5]. These method have proved to be valid and useful for identifying and controlling product appearance. In this work, the Commission Internationale de I'Eclairage (CIE) Standard Observer response, is used by the meters double-beam feedback system to measure the color variation. The advantages of this technique are (1) the assign numbers which facilitate communication about appearance and (2) providing uniform color space by plotting in rectangular coordinates [5].

2.0 EXPERIMENTAL PROCEDURES AND DATA ANALYSIS

The oxygen content was analyzed by a LECO oxygen analyzer and the carbon content was analyzed by a carbon analyzer. The metallic impurities were analyzed by the ICP emission spectroscopy method. Particle size distributions were determined using a SHIMASU particle size analyzer.

The compositions for both powders contain Y_2O_3 as the sintering aid. Weighted starting powders were ball milled in acetone with dispersant for 4 hours by using AlN grinding media. Binder and plasticizer were premixed in acetone before adding them to the slurry. The slurry was spray-dried in a Buchi mini-spray dryer. Green compacts were cold pressed in a tungsten carbides die. Compacts were then sintered in a graphite resistance furnace. A mild nitrogen gas flow rate was kept constant for all of the tests. The densities of the sintered parts were measured using Archimedes' method.

The thermal conductivity was measured by the laser flash technique [6]. The sample is coated with a thin layer of Au which is evaporated onto both faces of the sample. This is to prevent direct transmission of the laser beam through the sample. The entire sample surface is coated with graphite to aid in both the absorptivity of the front surface and the emissivity of the back surface. The laser energy is absorbed very strongly by the graphite layer and this energy, in the form of heat, is transferred through conduction to the AlN sample. The laser flash system consists of a Nd-glass laser pulse of approximately 0.25 ms duration at 1.06 um which impinges on the sample face. The IR detector is a photovoltaic Indium Antimonide system with weak spectral response in the visible. Two windows of sapphire and germanium are attached to the detector to cut off the radiation above 8 um and below 1.9 um, respectively.

Color measurement is based on numerical representations or quantifications of the three color response mechanisms of the human eye. The measurable attributes of appearance divide broadly into two categories: chromatic (having to do with color) and geometric (having to do with geometric distribution of light). The chromatic or color attributes are the most important commercially, but it is necessary, when considering overall appearance and when selecting conditions for the measurement of color, that geometric characteristics of light distribution by the objects be considered. In order to eliminate the errors from geometric factor, surface finish of the samples are limited to between 5 to 20 microns.

CIE 1976 L·a·b Uniform Color Space and Color-Difference is used for color measurement. The differences in color between a reference and a test specimen are determined from measurements made by use of a colorimeter. Lightness-difference (L), red-to-green (a), and blue-to-yellow readings from such instruments are converted by computations to color-scale values that represent the perceived color differences between the reference and the test specimen. CIE color difference were measured by MINOTA model CR-200 chroma meter. The tristimulus values Xn, Yn, and Zn, define the color of the standard white object-color stimulus. The standard stimulus is CR-A43 calibration plate. The uniform color scale is produced by plotting in rectangular coordinates.

Flexural specimens were cut from substrates to the size 25.0x5.0x1.0 mm^3 using a diamond blade . Surface finishes of the test specimens were measured using a MITUTOYO profilometer. Surface finish was reported as an average roughness, Ra. Strength distributions were determined by fracturing specimens in a 3 point bend geometry. The load span was 19 mm. All strength tests were carried out at a cross-head speed of 1 cm/ min. at room temperature on a CHATILLON Universal testing machine.

The strength data was analyzed using a two-parameter Weibull expression. The Weibull model and its applications have been described in detail by several investigators. The Weibull cumulative distribution function (CDF) is given by

$$F = 1 - \exp\left[-(\sigma/\sigma_o)^m \right] \qquad \text{--------} \quad (1)$$

where F: probability of failure,
σ: the fracture strength,
σ_o: a scale parameter,
m: Weibull modulus.

Values for F are calculated via order statistics using the relation:

$$F = n / (N+1) \qquad (2)$$

where n is order of strength, N is the total number of data points. A double logarithmic transformation of Equation (2) yields the linearized form

$$\ln \ln [1/(1-F)] = m \cdot \ln \sigma - m \cdot \ln \sigma_0 \qquad (3)$$

A least squares linear regression analysis was used to calculate the parameters σ_0 and m in equation (3). The correlation coefficient, R, was also calculated to determine the goodness of fit of the data to the two-parameter Weibull model. Statistical analyses were performed on strength data to determine the statistical significance of the observed trends. A one way analysis of variance (ANOVA) was applied to the means of distributions to determine if they were statistically identical.

3.0 RESULTS AND DISCUSSION

3.1 Characteristics of Raw Powders

The powder that was processed through carbothermal reduction will be designated as powder "CTR", while the other through direct nitridation will be designated as powder "DN". Table 1 lists the chemical characteristics of these two raw powders. The main impurities include carbon, oxygen, and metallic impurities. Comparing these two powders, one can see that both powders have almost the same content of oxygen which ranges from 1.30 to 1.40 wt%. Although both powders have the same amount of oxygen, it is believed that there is a difference in the distribution of this oxygen impurity. For powder "CTR", most of the oxygen exists on the surface of the AlN powder which is formed during the oxidation step. For powder "DN", most of the oxygen exists throughout the whole AlN powder which is formed due to the impure oxygen in the nitrogen gas. This phenomenon is further confirmed by thermal gravinometric analysis (TGA) on both powders. Figure 1 shows the weight gain from the TGA results for both powders. Comparing these two curves, it is seen that powder "CTR" has only 0.7 % of weight gain at approximately 1020 °C, while powder "DN" has a 3.0 % weight gain. These results illustrate that free oxygen on the powder "CTR" is almost saturated on the surface and is stable in the air storage.

Powder "CTR" has a less carbon, silicon, and iron impurities than powder "DN". For powder "CTR", the source of the carbon is the residual unreacted carbon while the silicon and iron impurities come from the starting raw materials. For powder "DN", the source of the carbon is a contamination from the polymer lining during the crushing step, while the silicon and iron impurities come from both the raw materials and the crushing media. It has been reported [7] that carbon, iron, and silicon impurities can diffuse into the AlN grain. The large radius impurities (e.g. iron and silicon) have the effect of creating strain in the lattice. This strain will reduce both the thermal conductivity and the translucency of the AlN parts. The small radius impurities (e.g. carbon) has the effect in reducing the translucency and increasing the darkness of the AlN parts. The strain that is created by carbon is negligible. Since powder "CTR" has smaller distortion from iron and silicon impurities, if other parameters are the same, the thermal conductivity of sample that is processed from "CTR" should be higher compared to that processed from powder "DN". Also, combined with the carbon effect, the part that is sintered from powder "CTR" should have a whiter color.

Table 2 lists the results of the physical characterizations for powders "CTR" and "DN". The main concern of the physical properties are the particle size distribution, particle morphology, and the mean particle size. The comparison of these two powders shows that powder "CTR" has a smaller mean particle size with narrower particle size distribution. The direct consequence of this small mean particle size is the high sinterability. Since powder "DN" has a wider particle size distribution, the tap density is higher. The larger particle size of powder "DN" is the major factor in (1) getting the non-uniformity in microstructure and (2) lowering the sinterability. Figures 2 and 3 show typical SEM micrographs of powders "CTR" and "DN", respectively. Comparing these two powders, one can conclude that the properties of powder "CTR" are definitely better than that of powder "DN". The mean particle size is considerably smaller. The particle size distribution is very uniform and the particle morphology is very nice and rounded. These micrographs also explain the reason why powder "DN" has a higher surface area (table 2) although its mean particle size is larger. Since powder "DN" has a very irregular morphology, this concave and convex non-uniformity increases the surface area of the powder dramatically. The sharp edge on the surface will decrease the sliding of the powder which thus decreases the rearrangement mechanism during the liquid phase sintering. The combination of these physical properties for powder "CTR" should result in high sinterability AlN parts with better thermal, optical, and mechanical properties [8].

3.2 Sinterability and Thermal Conductivity

Figure 4 shows the linear shrinkage curves for the parts that are processed from powders "CTR" and "DN", respectively. All of the parts were partially sintered for ½ hour with the same heating rate. The temperatures for the highest densification rate are approximately 1650 and 1780 °C, for powder "CTR" and powder "DN", respectively. The liquid formation temperature for powder "CTR" is approximately 130 °C lower than that of powder "DN". Due to the rounded and smooth surface morphology, powder "CTR" has a higher rearrangement ability which enhances the reaction and densification. These results support that, due to the mean particle size and particle size distribution, powder "CTR" is more reactive than powder "DN".

Table 3 shows the density results at three different sintering temperatures, 1810, 1840, and 1870 °C, for two samples that were processed from both powders. The sintering time was fixed at 6 hours for all of the sintering tests. Due to the wide distribution of the raw powder, the green density of sample "DN" is always higher than sample "CTR" at the same forming pressure. From this table, it is easy to see that powder "CTR" can reach almost full density at a much lower temperature compared to that of powder "DN". It is believed that powder "CTR" can dissolve into the liquid phase at a lower temperature which lowers the eutectic temperature of the liquid phase. Therefore, powder "CTR" has higher sinterability than that of powder "DN". Table 4 shows thermal conductivity results at three different sintering temperatures, 1810, 1840, and 1870 °C, for both powders. The thermal conductivity of powder "CTR" is always higher than that of powder "DN" throughout the temperature range. Also, as the sintering temperature increases, the thermal conductivity tends to stay at the same level. The reason for this steady thermal conductivity variation in such a wide temperature range has been suspected to be a competition among several factors, including (1) the free oxygen diffusing into AlN which tends to decrease thermal conductivity, (2) grain growth, which tends to increase thermal conductivity, (3) micro-pore elimination which tends to increase thermal conductivity.

3.3 Color Measurement

The reference sample was processed from powder "CTR" which was sintered at 1810 °C for 6 hours. Figures 5 and 6 show the red-to-green (a) and blue-to-yellow (b) hues, respectively, as a function of sintering temperature for both samples. The colors have a very small variation which is within the experimental error. Figure 7 shows the lightness (L) as a function of sintering temperature for both samples. It is evident that samples that were processed from powder "CTR" have a higher lightness scale (i.e. whiter) than powder "DN". These results further support previous arguments regarding the color variation by carbon and metallic impurities. Also, as the sintering temperature increases, the color of both sample becomes lighter (i.e. whiter) for both samples. Sample "CTR" deviates from 0 in the positive color-scale-reading while sample "DN" approaches toward 0 from the negative color-scale-reading. Although the lightness becomes whiter as the sintering temperature increases, the thermal conductivity remains at the same level for both samples (table 4). Therefore, it can be concluded that color variation of AlN parts is not directly related to the thermal conductivity. It has been suspected that the temperature dependence of the color variation is due to the microstructural variation. Several researchers [9,10] have shown that as the sintering temperature or the sintering time increases, the grain size increases. This will also decrease the number of the grain boundaries. The same microstructural changes apply to the grain boundary phase; i.e. the number of the multigrain junction decreases and the size of the multigrain junction increases. Since these microstructural variations will change the light transmission, the lightness is thus varied.

In order to determine the effect of carbon on thermal conductivity, a series of green parts were processed from powder "CTR". By controlling binder-removal, different sets of samples were obtained with different level of residual carbon content. Figure 8 shows the lightness and color variation for samples with different levels of residual carbon. All of these samples were sintered at 1810 °C for 6 hours in order to compare with the reference sample. The lightness shows a very strong dependence on the residual carbon content. The lightness scale increases from 0 to ≈14 as the residual carbon content decreases from 0.08 to ≈0.03 wt%. This increase is almost 3 times higher than the effect of increasing the sintering temperature from 1810 to 1870 °C. Regarding the color variation, both red-to-green and blue-to-yellow stay at almost the same level.

Figure 9 shows the total color scale for samples "CTR" and "DN, as a function of sintering temperature. The total color scale (E) is calculated as; $E = (L^2 + a^2 + b^2)^{1/2}$. From this figure, one can see that as the sintering temperature increases, the colors of these two samples approach each other. However, this result doesn't imply that these two samples have the same color. Since one sample with three negative values might result in the same total color scale as the other one with three positive values. Figure 10 shows the total color scale for both samples as a function of residual carbon content. Since the lightness (L) dominates the color variation, this total color scale has the same trend as the results in figure 7. Based on these two graphs, it is believed that total color scale is not a very reliable way of interpreting the color. When comparing the color, it is recommended that all three color scales (L,a,b,) are reported.

3.4 Mechanical Properties

The average strengths and Weibull parameters for both distributions are summarized in Table 5. All distributions show excellent fits to the two-parameter Weibull model as indicated by correlation coefficients (R) of 0.98 and 0.97, respectively. For powder "CTR", the average strength is 352 MPa with a standard deviation of 15 MPa. The strength ranged from 306 to 391 MPa. For powder "DN", the average strength is 233 MPa with a standard deviation of 15 MPa. The strength ranged from 205 to 265 MPa. One way analysis of variance showed that average strengths were sample "CTR" is significantly higher. Weibull moduli are about the same which are 14 and 15 for samples "CTR" and "DN", respectively. It is also worth noted that these Weibull moduli are relatively high values for a ceramic which indicate a narrow flaw population distribution for both materials. The strength distribution is presented in figure 11. It is clearly illustrated that powder "CTR" has a much stronger distribution than that of powder "DN".

4.0 CONCLUSIONS

Based upon above results and discussion, the following conclusions can be reached:

1. Both carbothermal reduction and direct nitride powders have nearly the same oxygen content which has a range of 1.30 to 1.40 wt%, while carbothermal reduction powder shows a lower level of impurity content. Carbothermal reduction powder has a narrower particle size distribution and small mean particle size.

2. The low impurities content increase the thermal conductivity, color appearance, and optical properties of the AlN ceramics that are processed from carbothermal reduction powder. The fine particle size and narrow particle size

distribution increase the sinterability and mechanical properties of the carbothermal reduction powder.

3. Color is closely related to the carbon content. Two different sources of carbon have been identified; one is from the raw powder, the other is the residual char left from the binder removal process. This implies that not only the raw powder will cause the color variation but also the different degree of binder removal.

4. For the same powder, the sintering temperature can change the color of the AlN ceramics. As the sintering temperature increases, the color becomes whiter, while thermal conductivity remains the same. Therefore, thermal conductivity is independent of color variation.

5. The parts that were sintered from carbothermal reduction powder have a higher mechanical strength than that of direct nitridation powder. However, the flaw populations appear to be similar as indicated by Weibull analysis.

ACKNOWLEDGMENTS

The authors would like to thank Alfred Ramirez for carrying out sintering experiments. Linear shrinkage curves by Michele Perisse are also acknowledged.

REFERENCES

1. N. Kuramoto, H. Taniguchi, and I. Aso, "Development of Translucent Aluminum Nitride Ceramics," Am. Cera. Bulletin, **68** [4] 883-87 (1989).

2. N. Kuramoto and H. Taniguchi, "Transparent AlN Ceramics," J. Mat. Sci. letters, 3, 471-74 (1984).

3. K. Komeya and H. Inoue, "Sintering of Aluminium Nitride: Particle Size Dependence of Sintering Kinetics," J. Mat. Sci., 4, 1045-50 (1969).

4. M. Nakamura, T. Kaga, Y. Nakajima, M. Ibukiyama, and M. Hiruta, "Some Effects of Properties of Directly Nitrided AlN powders on their Sintered Bodies," unpublished work.

5. "Calculation of Color Differences from Instrumentally Measured Color Coordinates" ASTM Standard on Color and Appearance Measurement, D2244-85, 85-93 (1987).

6. W. J. Parker, R. J. Jenkins, C. P. Butler, and G. L. Abbott, "Flash Method of Determining Thermal Diffusivity, Heat Capacity, and Thermal Conductivity," J. Appl. Phys., **32** [9] 1679-84 (1961).

7. N. Kuramoto, H. Taniguchi, and I. Aso, "Translucent AlN Ceramics Substrate," IEEE Trans. Compon., Hybrids, Manuf. Techol., **9**, 386-90 (1986).

8. C. F. Chen, "Characteristics of High Thermal Conductivity Aluminum Nitride Ceramics," pp. 1291-1304, Proceedings of Ninth Annual Int. Elec. Pack. Conf. San Diego, CA, 1989.

9. C. F. Chen and T. Y. Tien, "High Temperature Mechanical Properties of SiAlON Ceramics: Microstructural Effects," Ceramic Engineering and Science Proceedings, 8 [7-8] 778-795 (1987).

10. C. M. Hwang, "The Systems SIALON-$Y_3Al_5O_{12}$ and SiAlON-Cordierite: Sintering and Grain Growth," Ph.D. Dissertation, The University of Michigan, Ann Arbor, MI, 1988.

Table 1: Chemical Analysis of Raw Powders					
Powder	O (wt%)	C (ppm)	Fe (ppm)	Si (ppm)	Mg (ppm)
CTR	1.39	350	20	40	20
DN	1.35	1490	110	100	20

Table 2: Physical Properties of Raw Powders					
Powder	Tap Density (g/cm^3)	Surface Area (m^2/g)	D10	D50 (μm)	D90
CTR	0.65	3.20	0.50	1.22	2.47
DN	1.17	3.55	0.31	2.33	9.60

Table 3: Density for Different Powders that were Sintered at Different Temperatures for 6 Hours.

Powder	Green Density (g/cm^3)	Sintered Density (g/cm^3)		
		1810 °C	1840 °C	1870 °C
CTR	2.00	3.30	3.29	3.29
DN	2.15	3.20	3.26	3.29

Table 4: Thermal Conductivity for Different Powders that were Sintered at Different Temperatures for 6 Hours.

Powder	Thermal Conductivity (W/m·°K)		
	1810 °C	1840 °C	1870 °C
CTR	184	180	177
DN	156	153	153

Table 5: Weibull Analysis Data for AlN Ceramics

Powder	No. of Specimen	σ (MPa)	Range (MPa)	σ_o	m	R
CTR	19	352±25	306-391	368	14	0.98
DN	14	233±15	205-265	240	15	0.97

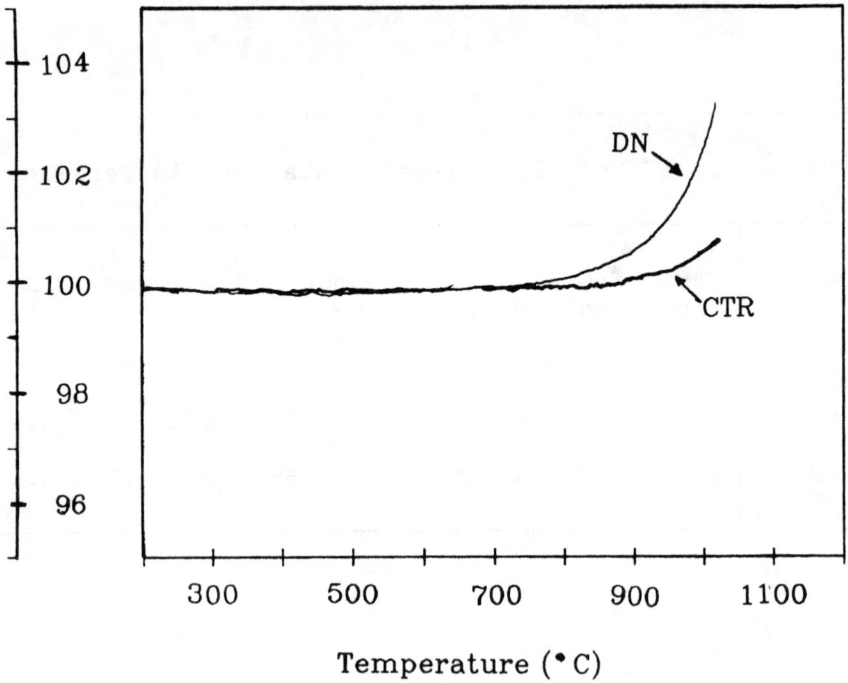

Figure 1: The weight gain from the TGA results for powders "CTR" and "DN".

Figure 2: SEM micrographs showing typical particle size, particle size distribution and particle morphology of powder "CTR" (5,000 X).

Figure 3: SEM micrographs showing typical particle size, particle size distribution and particle morphology of powder "DN" (5,000 X).

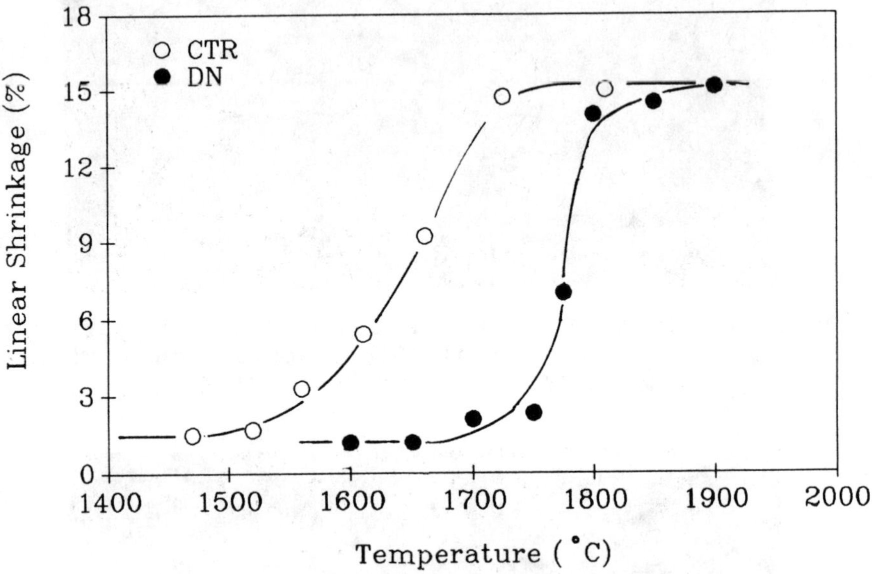

Figure 4: Linear shrinkage results for the parts that are processed from powders "CTR" and "DN".

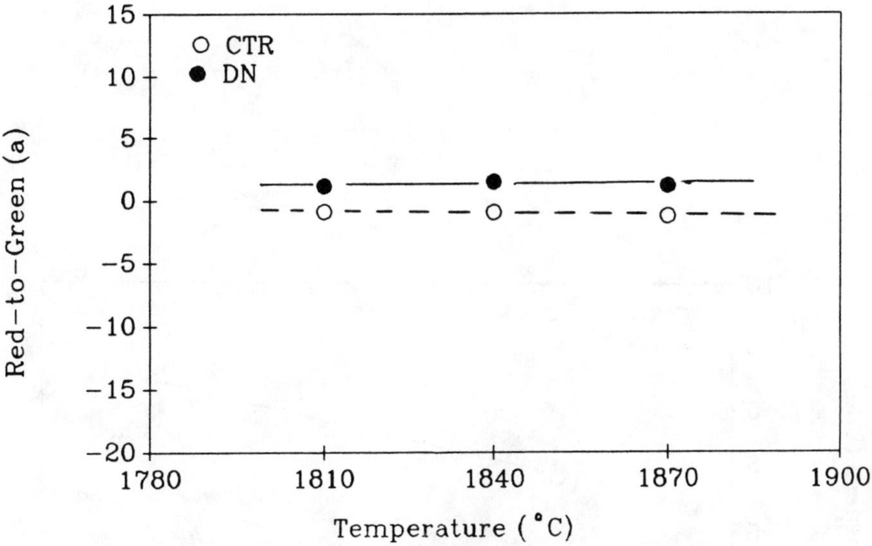

Figure 5: Red-to-green color-scale-reading as a function of sintering temperatures for samples "CTR" and "DN".

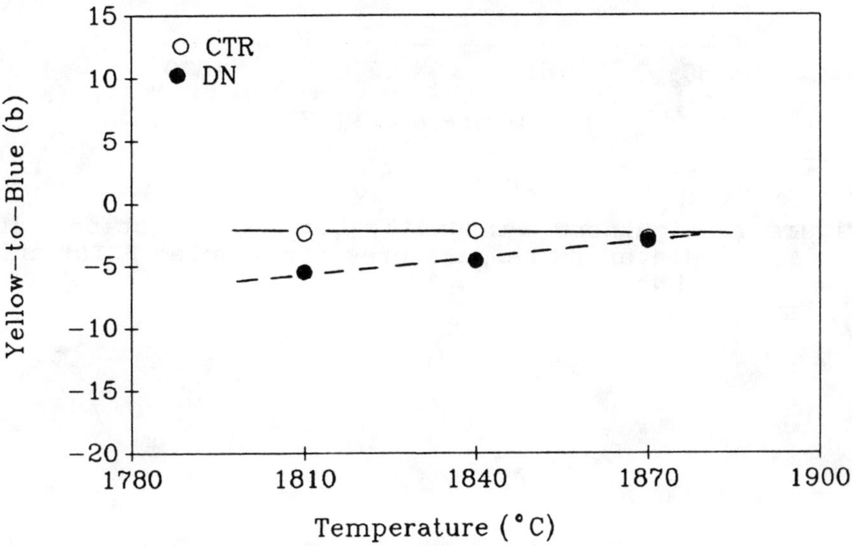

Figure 6: Blue-to-yellow color-scale-reading as a function of sintering temperatures for samples "CTR" and "DN".

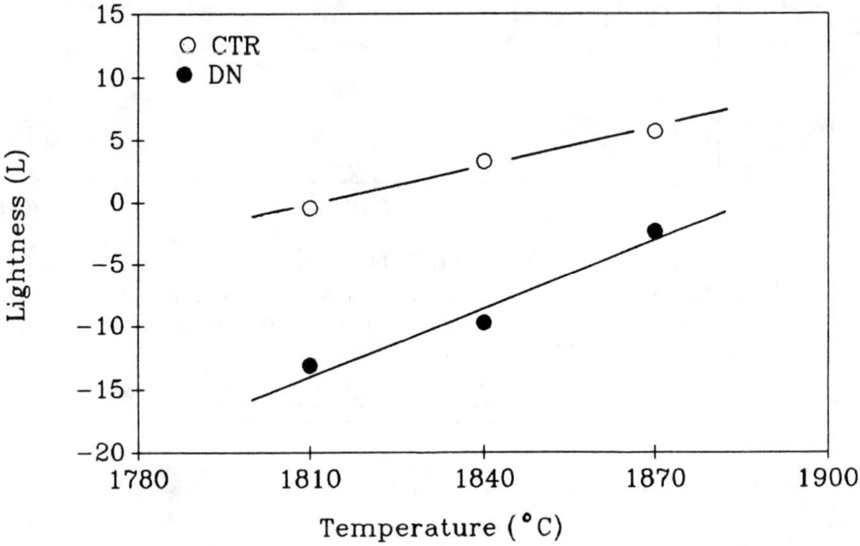

Figure 7: Lightness intensity as a function of sintering temperatures for samples "CTR" and "DN".

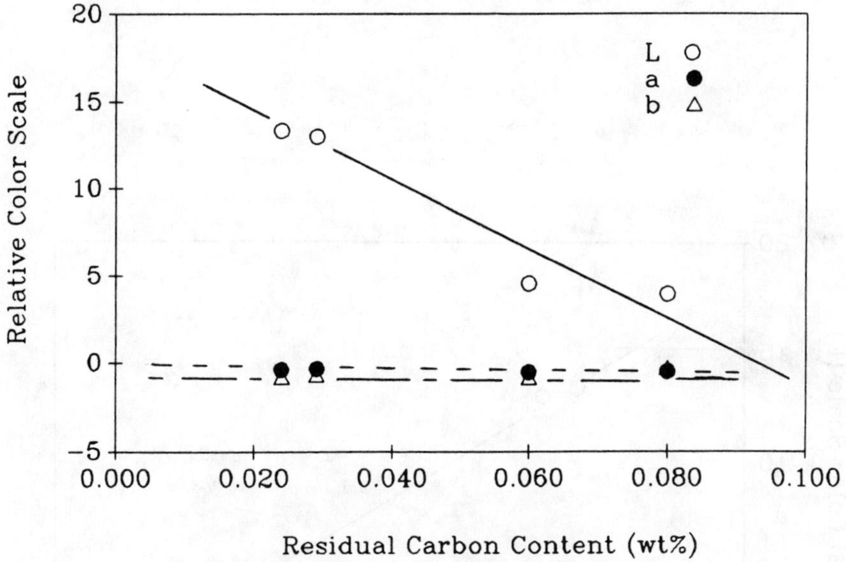

Figure 8: Color-scale-reading and lightness intensity as a function of residual carbon content for samples "CTR" and "DN".

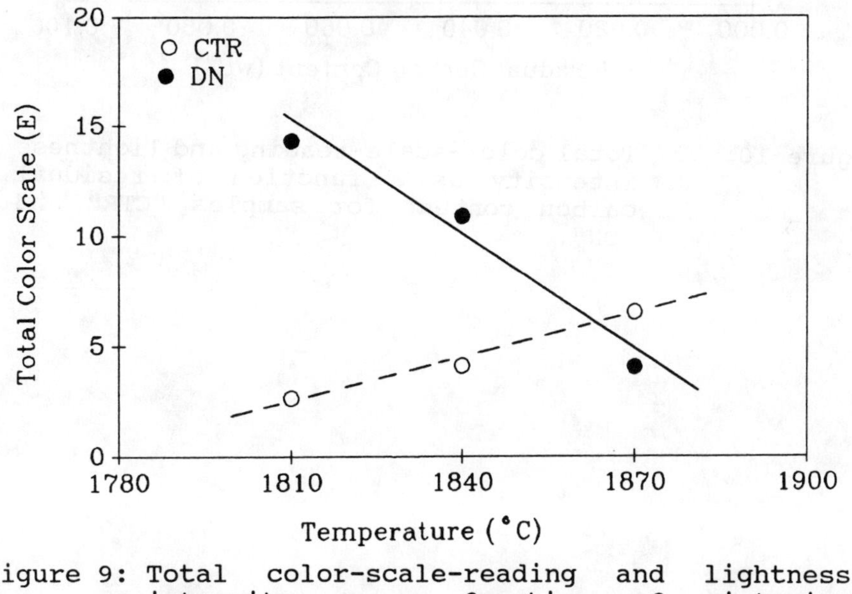

Figure 9: Total color-scale-reading and lightness intensity as a function of sintering temperature for samples "CTR" and "DN".

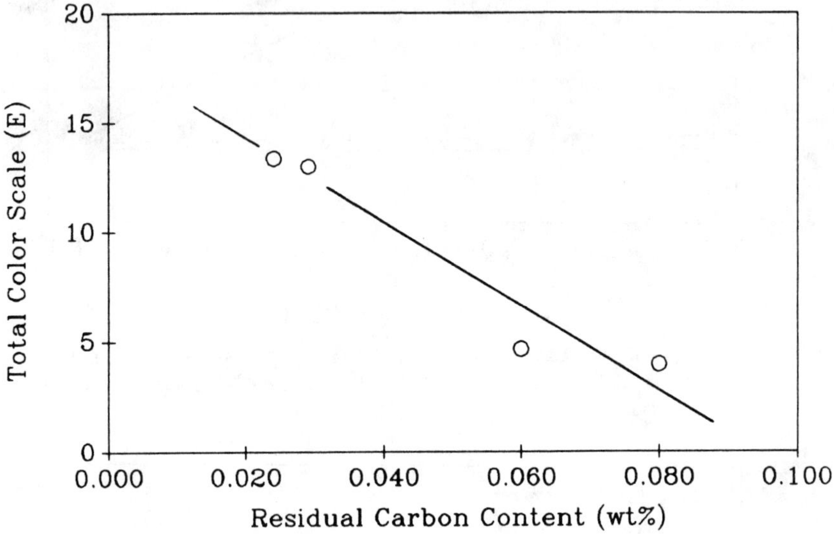

Figure 10: Total color-scale-reading and lightness intensity as a function of residual carbon content for samples "CTR" and "DN".

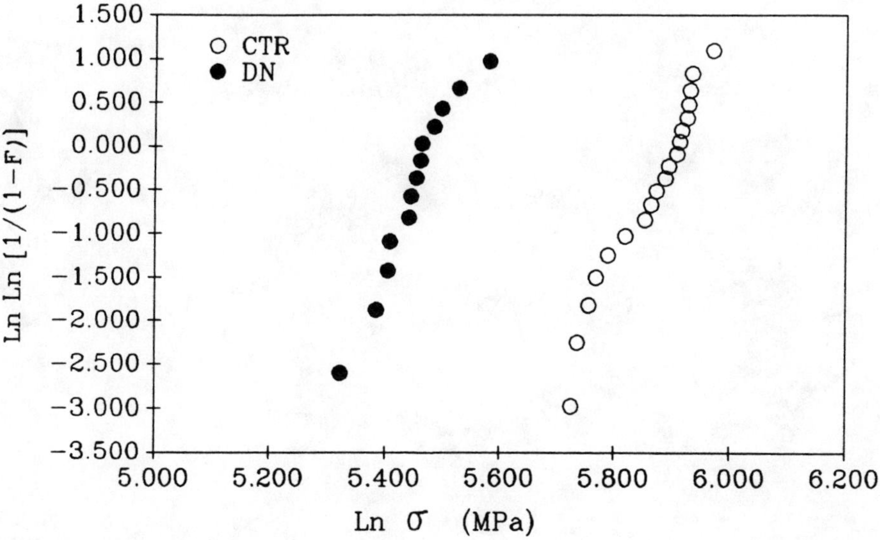

Figure 11: Linear Weibull plots of strength distribution for samples "CTR" and "DN".

GLASS IN MICROELECTRONIC PACKAGING: A REVIEW

Arun K. Varshneya, New York State College of Ceramics, Alfred University, Alfred, NY; and Satyam C. Cherukuri, Olin Corp., New Haven, CT.

ABSTRACT

The applications of glass and glass-composites in microelectronic packaging are reviewed. Emphasis is laid upon applications at the lower (chip, chip carrier and module) levels of the packaging hierarchy. The need for the understanding of material interactions during processing and application is recognized. The low permeability to atmosphere and the high durability in the working environment make glass the material of choice. Requirements of other glass properties such as viscosity, thermal expansion, dielectric constant and electrical resistivity etc. for the various applications are briefly discussed. Some suitable glass and glass-composite formulations currently used are mentioned. Development of durable glasses for high expansion metals such as Al, solder glasses with firing temperatures in the vicinity of 300°C, and glasses for bonding to advanced substrate materials such as AlN are seen as the major thrust for the near future.

INTRODUCTION

Glass plays a critical role in the function, performance, and reliability of the electronic devices and components. The evolution of microelectronics in the past four decades has been intertwined with the technological developments in ceramics, glass and glass-ceramics. The discussion of glass as a material in electronics cannot be separated from that of glass-ceramics, and glass plus fillers (usually ceramics and metals). The three types, i.e. the conventional amorphous glasses, the glass-ceramics resulting from the controlled crystallization of glasses, and glass+filler composites together form a distinct class in the microelectronics technology.

A widely accepted classification of the electronics packaging hierarchy is shown in Fig. 1. The different levels of packaging, disregarding the frequent overlap, are classified as follows:

Level 0 - chip level
Level 1 - chip capsule (carrier)
Level 2 - card or module
Level 3 - board or subassembly
Level 4 - discrete functional assemblies or gates

In addition to the above five levels, further integration can be included that involves clusters and local- or wide-area networks. The role of glass at different levels is unique; at 0-level it could be acting as a protective passivation layer whereas at the network level it might be transmitting information at high frequencies. The applications of glass at different levels of packaging are shown in Table-I.

The use of glass in microelectronics for various commercial, industrial, military, and aerospace applications is not as much a competition with the conventional ceramics and organic polymers but rather a complementary necessity. In general, glasses are brittle, require higher temperatures to form, and cannot compete with polymers in dielectric properties. A comparison of the processing issues and functional capabilities of the glasses with ceramics and polymers is shown in Table-II. The most unique property of glass that withstood the stringent competition from organic sealants is its resistance to permeation of atmospheric gases, in particular moisture. This property alone is used to classify level-1 and level-2 packages into two broad categories, 1) hermetic and 2) non-hermetic.

In this review, glass and glass-composite applications in the lower levels, in particular levels 1 and 2, are emphasized. A greater demand on the chemical and physical properties and compatibility with other materials is imposed at these levels and they directly impact the performance of the chip and other passive elements. For convenience of organization, some of the key properties of glass in relation to the requirements are briefly reviewed first.

GLASS PROPERTIES:

Key properties of interest in selecting a specific glass composition are: permeability, viscosity, thermal expansion, electrical conductivity and dielectric constant, wetting and adhesion to substrate, chemical durability, crystallization behavior and strength, and reaction with the substrate.

Permeability:

Glasses do not have high permeability paths such as grain boundaries and pores found in polycrystalline ceramics and regions of weak van der Waal force-bonded chain structures found in polymers. Some permeation of gases does occur through every glass; non-interactively for atomic and molecular species such as He, Ne, and N_2 where such gases permeate through the interstitial channels much like a sieve, and interactively for species such as H_2O where reactions with network oxygen or with modifier cations occur during the permeation process. Clearly, the rate of gaseous permeation can be reduced by filling the interstitial channels with modifier ions. Thus, a soda lime silica glass is significantly less permeable to a given inert gas compared to a fused silica glass. Permeation of moisture, on the other hand, may depend upon the exchange rate of H^+ with modifying cation at the nonbridging oxygen sites. Thus, fused silica is less permeable to moisture than soda lime glass because the former has trace levels of modifying cations only.

Viscosity:

The rapid and monotonic decrease in glass viscosity with temperature enables one to fire it on various microelectronic materials to obtain a pore-free, glazed appearance within the intended time. This is clearly a function of the actual physical shape of the glass, and the wetting conditions of the substrate.

Additions of SiO_2 and Al_2O_3 invariably raise the viscosity whereas the additions of alkalis and alkaline earths lower the same. Behavior of B_2O_3 is somewhat more complex. Considerable degree of softening is often achieved by large additions of PbO, PbF_2, CuO, Bi_2O_3, Sb_2O_3, Tl_2O, V_2O_5, and P_2O_5.

Thermal Expansion:

Thermal expansion is important primarily for the consideration of the development of permanent stresses in the finished product and, to a lesser extent, for thermal shock resistance during the processing/service. The development of stresses in glass-sealed components has been discussed by several authors. (A convenient review is by Varshneya[1].)

Most modifying cations such as the alkalis and the

alkaline earths increase the thermal expansion coefficient, as does PbO (usually). Conversely, the addition of SiO_2 and Al_2O_3 reduces the expansion coefficient. The addition of oxides such as CuO, ZnO, TiO_2, SnO_2, ZrO_2, V_2O_5 and Ta_2O_5 often has a neutral effect on the thermal expansion coefficient. CuO is interesting, in particular, because it greatly lowers the viscosity. Additivity factors for calculating thermal expansion of a variety of glasses have been given by Hormadaly.[2]

Electrical Conductivity and Dielectric Constant:

The key electrical properties of glass that influence its application in microelectronics are: volume conductivity, dielectric constant, dielectric loss factor, surface conductivity and dielectric breakdown strength. The last two properties are not only dependent on the glass composition but are also greatly influenced by surface finish, impurities and ambient conditions. The volume conductivity is primarily due to the mobility of the alkali ions. Addition of alkali and halogens in electronic glasses is usually not recommended. The presence of two or more alkali in equimolar proportions may inhibit the diffusion of alkali compared to the presence of a single alkali. This phenomenon is called "mixed alkali effect". The addition of heavy metal oxides, such as lead, improve the bulk and surface properties; however, the dielectric constant is increased due to the high polarizability of the ions. Alkali-free $PbO-B_2O_3-SiO_2$ glasses have excellent dc properties and low loss factors.

Wetting and Adhesion:

The wetting of a fluid glass on a solid substrate in a given environment is described by the contact angle, and is essentially controlled by the surface energies (Young's equation, see ref. 3). The presence of carbon and other absorbed impurities are often a deterrent to good spreading characteristics. Pre-treatment of surfaces by degreasing, or by ion-milling is often recommended, as is a decarburization treatment for metals that contain high interstitial carbon.

In bonding with glass, there are always opportunities to secure a mechanical interlock; however, chemical adhesion is generally the preferred route. Adhesion of molten glass to metals and ceramics is enhanced greatly in the presence of surface oxides.[3] Care must be taken to ensure that the oxide itself does not detach from its parent. Thus, a limited pre-oxidation treatment is recommended for adhesion to alloys such as Kovar.[3] Inclusion of small amounts of CoO and Cr_2O_3 in glass appears to improve bonding to ferrous metals.

Chemical Durability:

Although glass is more inert in most ambient environments than most metals and polymers used in the microelectronics industry, there are limits to its inertness. Durability problems in water, water vapor, or 5% brine exist in most glasses which do not have much SiO_2 or Al_2O_3. High amounts of B_2O_3, F_2, P_2O_5, V_2O_5, Tl_2O, CuO lead to gradual network dissolution.

Crystallization and Strength:

Glasses are quite weak in simple tension. In particular, the chemical reaction with the environment can cause flaws in the glass to grow and cause failure at a later date. This phenomenon, called static fatigue, guides us to keep the tension in the glass parts below about 14 MPa (~2000 psi) for trouble-free service. The strength of such glass parts can be significantly improved by introducing a second phase, generally crystalline, mixed with the glass at least on the surface. The mechanism of strengthening is essentially the arrest of an advancing crackfront by the higher strength dispersed phase. This can be accomplished by intentionally devitrifying the glass, either on the surface or throughout using nucleating agents, following the bonding operation or using a glass premixed with a desired crystalline material. The former category has led to a very broad range of sealing glasses, called devitrifiable glasses, which have much use in the electronics industry. The thermal expansion clearly changes following the devitrification. However, more importantly, large increase in viscosity allows higher service temperatures. Sealing glass for TV tubes is an outstanding example of this.

The second category, where the additives are simply admixed, is receiving increasing attention. Because many other properties besides strength are modified or enhanced, this category is discussed below separately as "Glass-Composites".

Reaction with the Substrate:

The most common concern is the dissolution or corrosion of the substrate and metallization elements by the glass constituents during firing. Such reactions can obviously be reduced by low temperature firing.

It has often been observed that some of the glass components such as CaO and MgO can migrate through the grain boundaries of the substrate and bring about increased rate of substrate sintering. This is particularly important in the cofiring of MLC from the green stage. Fluxing by glass constituents thus effects sintering at lowered temperatures. This, in turn, enables the use of Cu, Ag, Ag/Pd, etc. for conductors during cofiring instead of W, Mo, etc.

GLASS-COMPOSITES

Glass-composites are formed by admixing filler materials (granular ceramics and metals) generally to a low softening temperature glass. The admixing serves a variety of purposes. Firstly, a chemically inert, but well-defined particle-sized solid can act as a spacer during glass flow. For instance, the ferrite read/write head gap in a disk drive can be controlled using alumina particles admixed to the sealing glass. Secondly, the physical properties of the composite so produced are generally the volume average of the properties of the individual phases. Thus, for instance, the thermal expansion of a soft glass can be decreased by mixing-in β-eucryptite powder or increased by the addition of fluorspar (CaF_2). Thirdly, a glass with a ceramic filler acts much like a glass-ceramic with regards to the strength. Inclusion of ductile metal particles can toughen the material. The important point to note is that many of the property modifications and enhancements often cannot be achieved in a melted glass alone. In a reverse role, where small quantities of glass are added to larger quantities of ceramics and metals, the glass acts to adjust, for instance the resistance and its variation with temperature, etc., in addition to providing adhesion to substrates. Much of the thick-film technology is based upon the ability to tailor the firing temperatures, thermal expansion and electrical properties.

APPLICATIONS OF GLASS AND GLASS-COMPOSITES

The major applications of glass and glass-composite systems in microelectronics are broadly classified according to the function they perform, and are shown in Table I along with the packaging levels at which these functions are performed.

Passivation Glasses:

Passivation is the containment of fringing electric fields at the surfaces of P-N junctions in semiconductors. An additional purpose of the passivation coatings is to protect the surface of the chip or circuit elements from corrosive moisture, contaminants, and dust particles. One of the most common failure mechanisms in the silicon integrated circuits is the corrosion of aluminum metallization on the top surface of the chip by the atmospheric moisture. Even in the hermetically sealed packages moisture levels upto 5000 ppm may be present[4]. Typically, moisture in the presence of ionic contaminants causes the corrosion of aluminum metallization, bonding wires, pads and the interfaces.[5,6] Passivation layers are also used in thin- and thick-film circuits as protective overglazes.[7] Major requirements of glasses for passivation are shown in Table-III. Note, for instance, in the presence of aluminum metallization, the passivation glass should be fired at temperatures no higher than 577°C (Al-Si eutectic).

Common passivation glasses are: SiO_2, phosphosilicates (< 3%P_2O_5), lead borosilicates (wt%: 30-50 PbO, 0-13 B_2O_3, 3-10 Al_2O_3, 0-2 ZnO, 25-50 SiO_2, 0-5 others such as GeO_2 and PbF_2), and zinc borosilicates (wt%: 2PbO, 25B_2O_3, 60ZnO, 10SiO_2, 3CeO_2). The phosphosilicates were popular prior to the widespread use of chemical vapor deposited Si_3N_4 passivation. The phosphosilicate glass traps the moisture by chemically tying it up on the surface. The passivation coatings also act as alkali barrier layers. Phosphorous doped silica coatings are also used as a phosphorous dopant source in CMOS structures.[8] Deposition techniques often used[6,9] are thermal growth (silica only), CVD, RF sputtering, sedimentation, electron beam evaporation, electrophoretic deposition, spin-on photoresists, silk screen printing, and doctor blading.

Dielectric Pastes:

The overglaze dielectrics generally require a low dielectric constant (K) and can serve the dual function of passivation. A $PbO-B_2O_3-SiO_2$ glass with low SiO_2 and high PbO content requires <500°C for firing, and hence works adequately. Attempt should be made such that the expansion coefficient of the dielectric glass is slightly less than that of the substrate in order for the glass to be in slight compression. Some dielectric glass compositions have been given by Tummala[9] and Kobayashi[10]. Crossover dielectrics, again require low K, and can be obtained[11] by admixing $BaAl_2Si_2O_8$ ("celsian") to a crystallizable glass and firing at 850°C (industry standard). Capacitor dielectrics for 1200°C MLC firing require high temperature glasses such as the aluminosilicates. To obtain high K values, the glasses may need to be filled with $BaTiO_3$.

Metallization Binders:

Metallization of thick film conductors and resistors is often accomplished using Au, Cu, Ag, Pd, Mo, Mo-Mn, and W. These are applied as inks or pastes through silk screening and are dried and fired. For Pd/Au, Pt/Au conductor pastes, a 20-30% by volume addition of $63\%PbO\cdot25\%B_2O_3\cdot12\%SiO_2$ glass is recommended[12]. For Ag/Pd conductors, Tummala[9] indicates the use of $50PbO\cdot30SiO_2\cdot OB_2O_3\cdot5Al_2O_3\cdot2Na_2O$ plus 3 (undisclosed) oxides. In almost all the cases, glasses are used as binders which not only help the densification of the metal powders but more importantly, and often the only means of achieving adhesion between the metallization and the surrounding (or underlying) ceramic or glass composite. The glass compositions are tailored to specific metal, time, temperature and firing atmospheres. The glass powders are used in minor amounts, upto a few percent, just enough to achieve adequate adhesion. Excess use of the glass binder may impair the subsequent plating, brazing or soldering operations. Reduction of PbO to metallic Pb and entrapment of carbon from the vehicles are often a problem.

Oxide Resistor Pastes:

The advanced family of resistors in microelectronics comprise RuO_2 or $Bi_2Ru_2O_7$ (pyrochlore structure) encapsulated by $PbO-B_2O_3-SiO_2$ glasses containing small amounts of oxides such as Bi_2O_3, CdO, Cu_2O, MnO, Sb_2O_3, Nb_2O_5, V_2O_5 and ZrO_2. During firing, not only that the glass provides adhesion to the substrate, but also compounds such as $Pb_2Ru_2O_7$, $(Bi_{1-x}Pb_x)_2Ru_2O_{7-x}$, $Bi_2Ru_{2-x}Mn_xO_{7-x}$ and $Bi_2Ru_{2-x}Al_xO_{7-x}$ form which affect both the resistance as well as the thermal coefficient of resistance (TCR)[11]. A resistor glass composition given by Inokuma et al.[13] is $52PbO \cdot 35SiO_2 \cdot 10B_2O_3 \cdot 3Al_2O_3$ (softening temperature = 600°C). The pastes often contain ~30-70 vol% glass frit.

Solder Glasses:

Solder glasses are generally high lead glasses that can be worked below 500°C. The concept of using the solder glasses in a cerdip (ceramic dual-in-line) configuration was first introduced by Fairchild Semiconductor Company in the 1950's. The cerdip packages are the most commonly used low cost and high volume packages. Similar material configurations were later extended to the high lead count ceramic quad flat packs (CQFP). The substrate is usually 92-96% alumina which is used as both the base and the cap. The solder glass powder is suspended in a suitable binder and a solvent and screen printed on to the alumina piece parts. The steps involved in the burnout of organics and glass sintering are referred to as "glazing". The piece parts are assembled into a package in the following sequence: leadframe attach--die attach--wire bonding--cap frit seal. The most common leadframe material is Fe-42%Ni alloy.

Copper-Tungsten composites, with similar expansion as alumina, are also used as bases where higher thermal conductivity is required. Recently, a high expansion solder glass was developed which is matched to copper and high-copper alloys for high thermal dissipation of the base and high electrical conductivity of the lead frame.[14,15]

Solder glass compositions and processing have been reviewed by Singhdeo and Shukla[16] and Takamori[17]. Requirements for solder glasses are listed in Table-IV. Glass systems that have been studied for low temperature applications are[18-20]: $PbO-B_2O_3$, $PbO-ZnO-B_2O_3$, $PbO-B_2O_3-SiO_2$, $ZnO-B_2O_3-SiO_2$, $PbO-B_2O_3-Al_2O_3-SiO_2$, $ZnO-B_2O_3-V_2O_5-PbO$, and $PbO-B_2O_3-Tl_2O_3-SiO_2-Al_2O_3$.

The $PbO-ZnO-B_2O_3$ system is the most commonly used in electronic packaging. Both vitreous and devitrifying solder glasses can be obtained in this system. Typical thermal expansion in this system ranges from 90–120 x 10^{-7}/°C which is higher than the pressed alumina substrates used in cerdips (≈65 x 10^{-7}/°C). Fillers are used to increase or decrease the expansion of the solder glass. A list of fillers used and their thermal expansion coefficients are shown in Table-V. To obtain a glass for sealing to copper, a high expansion filler, CaF_2, is added to elevate the expansion of the base glass from 115 x 10^{-7}/°C to about 160 x 10^{-7}/°C [14,15,21]. A typical glass sealed copper alloy packaging system (CAPS^R) is shown in Fig. 2.

For higher seal strength and thermal stability of the components in subsequent high temperature excursions, $PbO-ZnO-B_2O_3$ glasses in the devitrification range are used [22,23]. These glasses, when held at the sealing temperature for sufficient length of time, precipitate $2PbO \cdot ZnO \cdot 2B_2O_3$ and $PbO \cdot 2ZnO \cdot B_2O_3$ phases [22]. A combination of both devitrifying and vitreous solder glasses is used in CQFP's. A devitrifying glass is used to embed the leadframe first, and then after die attach and wire bonding, a non-devitrifying cap is sealed at lower temperatures to avoid leadframe distortion.

Die Attach Glasses:

Silver-filled solder glass die attach has become a popular method for ceramic packages in the 80's. The major advantages with this system are:

> Die and cavity metallizations are not essential
> Less void formation than the traditional Au-Si eutectic die attach.

It has been observed that the overall thermal performance of the silver-glass die attached packages is better than the Au-Si counterpart [24]. The disadvantages are, 1) the necessity of oxidizing atmosphere during firing, and 2) special precautions for the removal of organics [25]. Nevertheless, the materials and the die attach process are more cost effective than Au-Si.

The silver-filled glass is applied in the paste form to the package cavity. A typical paste composition (JMI 4613, Johnson Matthey, Inc.) by weight is 64% Silver powder, 16.6% glass powder, 1% resin (binder), 16% solvent. After drying and organic burnout the system is approximately 80% silver and 20% glass, by weight [26]. The glasses are typically from the $PbO-ZnO-B_2O_3$ system which require firing temperatures at about 400°C. In the firing step, the glass performs three functions: 1) binds the silver particles together, 2) provides adhesion to the chip, and 3) provides adhesion to the substrate.

Sealing Glasses for Metal Housings:

A large portion of hybrid, power hybrid and power discrete devices are packaged in metal cans, tubs or similar metal housings for high reliability applications. Some of the proven techniques and materials had been developed much earlier than silicon technology. The most widely used metals are Kovar, Ni-Fe alloys, molybdenum, tungsten, copper, and nickel.

Glass-to-metal sealing of tungsten, molybdenum, Kovar and other Fe-Co-Ni alloys has been extensively studied and the techniques are well established[27-31]. Techniques for measuring and calculating stresses in various geometries have been reviewed by Varshneya[1]. Typical sealing glasses are Corning code 7052 (for Kovar), code 9010 (to make compression seals with Cu and steels), code 7720 (for W) and code 1720 (for Mo). (For various matched compositions, see reference [32]). Compositions of some of these glasses are given by Boyd and Thompson[33]. Typical seals for hybrid microelectronics are made with glass beads to form feedthroughs for the cylindrical leads. A Kovar hybrid package is shown in Fig. 3. The beads are pressed from spray-dried powders and sintered. The intergranular oxide level on the metal is critical to consistently achieving high strength hermetic seals. For Kovar, the oxide thickness is generally between 2 to 6 microns[3].

Cracking of meniscus, which is invariably formed at the glass-lead interface, during testing and use is of serious concern and is under evaluation by the package manufacturers. Using finite element analysis, Varshneya [34] has shown that the meniscus cracking arises from stress reversal, at the ends, otherwise called an end effect. Recently, to reduce the tendency of formation of meniscus, upto 5% alumina powder is dispersed in the glass powder prior to pressing; this increases viscosity at the normal working temperatures thus minimizing the meniscus [35].

Glass-Coated Metal Substrates:

The glass-coated metal substrates offer certain distinct advantages over the conventional ceramic substrates for thick-film hybrid circuit applications. The major advantages are:

- Ease of formation of relatively large boards with insertion holes by conventional metal fabrication methods.
- Mechanically rugged constructions
- Good heat spreading
- Higher temperature applications compared to laminated printed-wiring-boards
- Built-in ground plane [36].

The metals of considerable importance in this technology are low carbon steels, stainless steels, copper-Invar-copper composites, copper-molybdenum-copper composites, copper, and aluminum.

Porcelain enamels for steels were studied extensively and are widely in use for a number of applications [37]. However, almost all the compositions developed for cosmetic, sanitary, and chemical applications are not suitable for electronics due to high ionic conductivities. Alkali free glass-ceramic compositions were developed by RCA laboratories that overcome the electrical problems[38].

The copper-clad Invar and the copper-clad molybdenum composites are tailored to bring the expansion from that of copper ($175 \times 10^{-7}/°C$) to about $40\text{-}70 \times 10^{-7}/°C$ where electronic grade glasses are either commercially available or are easier to develop. Typical borosilicate and lead borosilicate systems are compatible with these composites.

Glass or glass-ceramic laminated copper is of special interest for its high thermal and electrical conductivities. Glass-ceramic compositions that are expansion-matched to copper were developed in 1960's by McMillan [39].

Typical glass coating requirements for applications in thick-film hybrid circuits are shown in Table-VI. Glass compositions used are essentially the same as those discussed above for sealing to metal housings [32].

Thin-film Substrates:

Thin film circuits are formed on smooth substrates by deposition techniques such as CVD, sputtering, and etching techniques that are commonly used in wafer fabrication. Finer pattern resolutions can be obtained with this approach as opposed to thick-film circuits where the circuits are formed by screen printing and firing of conductor, resistor or capacitor inks. The general requirements of substrates for thin-film applications are shown in Table-VII[40].

Glasses are ideal for thin-film circuits as they conform to all the requirements shown in Table-VII with the exception of thermal conductivity. For applications where thermal dissipation is not a major factor vitreous silica, borosilicate (e.g. Corning code 7740), a barium aluminosilicate (Corning code 7059), and lead borosilicate glazed alumina have been used [41,42]. The cost of the glass substrates would be much lower than of polished alumina ceramic or sapphire.

Thick-film and Cofired Multilayer Substrates:

High-alumina ceramic has been the workhorse of the post-fired and cofired single and multilayer systems; and this technology is well-established. Due to the higher processing temperatures involved (> 1500°C), the refractory metallizations W, Mo and Mo-Mn are used for cofired applications which are relatively poor electrical conductors. The ever-increasing complexity and the demand for higher performance led to the development of several low-temperature cofired (LTC*) glass composite systems that are compatible with high conductivity Ag, Ag-Pd, Cu and Au [43-49]. A variety of glass-composites with alumina, forsterite, and cordierite fillers have been studied.

A comparison of the LTC glass-composites with high-alumina ceramics is shown in Table-VIII. In addition to the low temperature firing feature, another major issue addressed in the LTC systems is the dielectric constant which is lowered for improved signal speeds. It is also possible to tailor compositions to be compatible with buried capacitors and resistors. A typical LTC multilayer process is

---

*LTC is a Trademark of Olin-Asahi Interconnect Technologies, New Haven, CT.

shown in Fig.4.

FUTURE TRENDS AND REQUIREMENTS

Electronic packaging is a complex, multidisciplinary technology. Progress in this area will depend on the understanding of material interactions during processing conditions and in application. The behavior of glass and its interfacing with metals, ceramics and organic binders and solvents at forming temperatures is not always predictable due to the lack of understanding of the mechanisms involved. The search continues to improve properties in the following manner:

-- Higher strength and fracture toughness glass-filler particulate as well as fibrous composites

-- Lower dielectric constant glass-ceramics

-- Resistance to diffusion of metals (e.g. Cu) at processing temperatures

-- Broader working range

-- Stability in reducing atmospheres during firing

In the metal housing technology, there are several missing gaps in the availability of glass and glass-ceramic systems. Even though several glass-ceramics are known for Kovar, molybdenum and other low expansion metals and alloys, processing is difficult due to long crystallization times required for obtaining desirable microstructures. Often, the weakest link is the glass-to-metal interface which requires further understanding. For instance, there are no proven chemically durable glasses having satisfactory bonding properties to aluminum, which is an ideal metal for lightweight applications.

In the available solder glass families which have satisfactory weatherability and electrical resistivity properties, the lowest sealing temperatures are around 420°C. Many of the processing steps in microelectronic packaging could be greatly simplified and the device life extended if the sealing temperature could be lowered to about 300°C. There are two interesting candidate families. One of them is the $PbO-PbF_2-B_2O_3-SiO_2$ system where the physical properties have been studied by Shelby and his students[50,51], and the other is the $Sn-Pb-NH_4$ fluorophosphate system studied by Tick[52]. In the $PbO-PbF_2-B_2O_3-SiO_2$ glasses, several compositions have Tg in the 260-300°C range and have adequately high electrical resistivity. The Tg of the $Sn-Pb-NH_4-F-P-O$ system are as low as room temperature, though several in the 100-150°C range were claimed to be quite durable[52]. The long term weatherability of these glasses remains an open question. The thermal expansion coefficients in both the systems are quite high (~100 to 200 x 10^{-7}/°C). However, it may be possible to use β-eucryptite filler, or even nylon beads to alleviate some of the stresses which may develop during sealing. High F_2-content compositions should be avoided because (i) they are less durable, (ii) F^- can conduct electricity, and (iii) F_2 can be corrosive to metallization systems.

It must be stated that the fluxing action of glasses to improve sintering of MLC and LTC, and the reactive action of glasses to improve electrical properties of resistors and capacitors had all been developments which were originally unintentional, but now are fully intentional. Glass developments shall clearly be needed for the fluxing and bonding of potential substrate materials such as AlN, SiC, and Si_3N_4, and high thermal conductivity ceramic fiber weave (incorporating anisotropic heat extraction).

And finally, with continued advances and cost-reduction in Ga-As semiconductor technology, it is not difficult to imagine that an entire series of new games involving glass will have to be played.

ACKNOWLEDGMENTS

Thanks are to W. H. Payne of Ceramic Devices, Inc., and C. J. Leedecke of Olin Corp. for helpful discussions.

REFERENCES

1. A. K. Varshneya, "Stresses in Glass-to-Metal Seals", in Treatise of Matls. Sci. and Tech. (Glass III), V.22, editors M. Tomozawa and R. H. Doremus, p. 241-306, Academic Press, 1982.

2. J. Hormadaly, "New Model for Estimating the Expansion Coefficient of Glasses Used in Microelectronics", Proc. ISHM, 77-82 (1985).

3. J. A. Pask, "Glass-Metal Interfaces and Bonding", in Modern Aspects of the Vitreous State, edotor J. D. Mackenzie, p.1-28, Butterworth & Co. Washington, DC 1964.

4. G. M. Johnson, L. Hamatier and F. Villella, "Factors Influencing Micro-circuit Package Reliability", Advanced Techniques in Failure Analysis, IEEE Proceedings, pp.108-122, Los Angeles, CA, 1978.

5. MIL-STD-883D, Test Methods and Procedures for Microelectronics, U.S. Government Printing Office, 1985.

6. G. L. Schnable, W. Kem and R. B. Comozzoli, "Passivation Coatings on Silicon Devices", J. Electrochem. Soc: Solid-State Science and Technology, 122, pp: 1092-1103 (1975).

7. A. B. Glaser and G. E. Subak-Sharpe, Integrated Circuit Engineering, pp: 376-403, Addison-Wesley, Reading, MA (1979).

8. ibid., pp: 280-284.

9. R. R. Tummala and R. R. Shaw, "Glasses in Microelectronics in the Information Processing Industry", Advances in High Technology Ceramics, 18, pp: 87-102, 1987.

10. K. Kobayashi, "DTA and TMA Compositional Dependencies in $ZnO-SiO_2-P_2O_5$ and $PbO-SiO_2-B_2O_3-GeO_2$ Glass Systems", Ceram. Bull., 66(4) 685-687 (1987).

11. L. C. Hoffman, "An Overview of Thick Film Hybrid Materials", Ceram. Bull., 63(4) 572-576 (1984).

12. R. W. Vest, "Materals Science of Thick Film Technology", Ceram. Bull., 65(4) 631-636 (1986).

13. T. Inokuma, Y. Taketa, and M. Haradome, "The Microstructure of RuO_2 Thick Film Resistors and the Influence of Glass Particle Size on Their Electrical Properties", IEEE Trans on Comp., Hybrids, and Manuf. Tech., Vol. CHMT-7(2) 166-75 (1984).

14. N. N. Singhdeo, S. C. Cherukuri and S. H. Butt, "Metal DIP with Superior Thermal Conductivity", Proceedings of the Sixth Annual IEPS conference, San Diego, CA, pp: 615-619, (1986).

15. E. F. Smith III, "Sealing Glass Composite", U. S. Patent No. 4,775,647, 1988.

16. N. N. Singhdeo and R. K. Shukla, "Solder Glass Processing", Glass Science and Technology, Vol. 2, edited by D. R. Uhlmann and N. J. Kreidl, pp.170-208, Academic Press, 1984.

17. T. Takamori, "Solder Glasses", in Treatise on Materials Science and Tech., Vol. 17, Glass II, pp.173-255, Academic Press, NY (1979).

18. R. G. Frieser, "A Review of Solder Glasses", Electrocomponent Science and Technology, Vol. 2, pp: 163-199 (1975)

19. E. M. Rabinovich, "Solder Glasses and Vitreocrystalline Cements", Izvestiya Akademii Nauk SSSR, Neorganicheskie Materialy, 7[4], pp: 545-560, (1971). Translated by Consultants Bureau, Plenum Publishing, NY, (1971).

20. Y. Shirouchi, "Devitrifying Solder Glasses", Fujitsu Scientific & Technical Journal, Vol. 5, pp: 123-165 (1969).

21. S. Smernos, "Filler for a Glass-ceramic Material Comprising CaF_2 Granules Over Coated with a Silicon Dioxide Film", U.S. Patent No. 4,185,139, (1980).

22. T. H. Ramsey, "Thermal and X-ray Analyses of Some Electronic Package Sealing Glasses", Bull. Am. Ceram. Soc., 50[8], pp: 671-75, (1971).

23. E. M. Rabinovich, "Crystallization and Thermal Expansion of Solder Glasses in the $PbO-B_2O_3-ZnO$ System with Admixtures", 58[6], pp: 595-598, 1979.

24. M. Mahalingam, M. Nagarkar and L. Lofgran, "Gold Eutectic vs. Silver Glass Die Bond in Ceramic Packages: A Relative Assessment of Thermal Performance", Proc. Sixth Annual IEPS Conf. pp.11-17, San Diego, CA 1986.

25. N. G. Kopman, T. C. Reilly and P. A. Totta, "Chip-to-package Interconnections", Microelectronics Packaging Handbook, edited by R. R. Tummala and E. J. Rymaszewski, pp: 361-454, Van Nostrand Reinhold, NY, 1989.

26. F. K. Moghadam, "Development of Adhesive Die-attach Technology in Cerdip Packages; Material Issues", Solid State Tech., Jan. 1984.

27. W. H. Kohl, Handbook of Materials and Techniques for Vacuum Devices, Reinhold Publishing, New York, 1978.

28. J. H. Partridge, Glass-to-Metal Seals, Society of Glass Technology, Sheffield, U.K. (1949)

29. J. A. Pask, "New Techniques in Glass-to-Metal Sealing", Proc. of the I.R.E.-Waves and Electrons section, Vol. 36, pp:286-289 (1948).

30. F. E. Gifford and Arthur Dolenga, "Direct Glass-Metal Seals", The Review of Scientific Instruments, 35[5], 591-593 (1964).

31. W. E. Eberly, "Glass Sealing Alloys", The Glass Industry, pp. 435-467, 1963.

32. Product Literature "Glass Sealing Combinations". Corning Glass Works, Corning, NY.

33. D. C. Boyd and D. A. Thompson, "Glass" in Encyclopedia of Chemical Technology, 3rd Edition, Vol.11, pp. 807-880, John Wiley & Sons, NY 1980.

34. A. K. Varshneya, "Application of Finite Element Analysis to Glass Processing", in Reviews of Solid State Science, Vol. 3 (3-4) pp. 561-578 (1989).

35. Product Literature, GBC Materials Corp, Latrobe,PA.

36. N. G. Aakalu and F. J. Bolda, "Coated-Metal Packaging", Microelectronics Packaging Handbook, edited by R. R. Tummala and E. J. Rymaszewski, Van Nostrand Reinhold, NY, 1989.

37. A. I. Andrews, Porcelain Enamels, Garrard Press, Champaign, IL, 1935.

38. K. W. Hang, J. Andrus and W. M. Anderson, "High-Temperature Porcelain- Coated-Steel Electronic Substrates - Composition and Properties", RCA Review, 42[2], 1981.

39. P. W. McMillan and B. P. Hodgson, British Patent No. 1,023,480, 1962.

40. Same as reference 7, pp: 346-375.

41. R. W. Berry, P. M. Hall and M. Harris, Thin Film Technology, pp: 395-418, Van Nostrand, Princeton, NJ, 1968.

42. R. R. Tummala, R. W. Keyes, W. D. Grosman and S. Kapur, "Thin Film Technology", pp: 673-726, Von Nostrand Reinhold, New York, 1989.

43. R. R. Tummala, "Ceramic Packaging", Microelectronics Packaging Handbook, edited by R. R. Tummala and E. J. Rymaszewski, pp: 455-522, Von Nostrand Reinhold, New York, 1989.

44. S. Nishigaki, J. Fakuta, S. Yano, H. Kawabe, K. Noda and M. Fukaya, "A New Low Temperature Fireable Ag Multilayer Ceramic Substrate Having Postfired Cu Conductor (LFC-2)", ISHM Proceedings, pp.429-449 (1986).

45. Y. Shimada, K. Utsumi, T. Ikeda and S. Nagasako, "Monolithic Multi-components Ceramic (MMC) Substrate", The International Journal for HYBRID Microelectronics, 7[4], 29-37 (1984).

46. W. A. Vitriol and J. I. Steinberg, "Development of a Low Temperature Cofired Multilayer Ceramic Technology", ISHM Proceedings, pp. 593-598 (1983).

47. S. Nishigaki, S. Yano, J. Fukuta, M. Fukaya and T. Fuwa, "A New Multilayered Low-Temperature-Fireable Ceramic Substrate", ISHM Proceedings, pp. 225-235 (1985).

48. Product Literature, OLIN-ASAHI Inter-connect Technologies, New Haven, CT.

49. S. Tosaka, S. Hirooka, N. Nishimura, K. Hoshi and N. Yamaoka, "Properties of a Low Temperature Fired Multilayer Ceramic Substrate", ISHM Proceedings, pp. 358-362 (1984).

50. J. Coon and J. E. Shelby, "Properties and Structure of Lead Fluorosilicate Glasses", J. Amer. Ceram. Soc., 71(5) 354-357 (1988).

51. C. A. Gressler and J. E. Shelby, "Properties and Structure of $PbO-PbF_2-B_2O_3$ Glasses", J. Appl. Phys. 66(3) 1127-1131 (1989).

52. P. A. Tick, "Water Durable Glasses with Ultra Low Melting Temperatures", Phys. Chem. Glasses, 25(6) 149-54 (1984).

TABLE-I -- Major Applications of Glass and Glass Composites in Electronic Packaging

FUNCTION	PACKAGE LEVEL
Passivation Glasses	0,2
Dielectric pastes	2
Die Attach Glasses	1,2
Metallization Binders	1,2
Oxide Resistor Pastes	2
Solder Glasses	1
Sealing Glass for Metal Housings	1,2
Thin-Film Substrates	1,2
Coated Metal Substrates	1,2
Thick-Film & Cofired Multilayer Systems	1,2,3

TABLE-II -- A Comparison of Glasses vs. Ceramics and Organic Polymers in Processing and Functions

Property	Glass, Glass Composites	Ceramics	Polymers
Forming Temperature	400-1400°C	> 1400°C	< 400°C
Thermal Conductivity	< 20 W/m·K	20-300 W/m·K	< 1W/m·K
Dielectric Constant	4-8	> 7	< 4
Flexural Strength	8-40 K Psi	20-60 K Psi	compliant
Elastic Modulus	10-20 M Psi	20-60 M Psi	< 10 MPsi
Volume Resistivity	> 10^{12} Ωcm	> 10^{14} Ωcm	> 10^{14} Ωcm
H_2O Diffusion	Impermeable	Impermeable	Permeable

TABLE-III -- Requirements of Passivation Glasses

- High Purity; must not contain mobile ions that might diffuse into the underlying circuitary (SiO_2, Si_3N_4 or metallization systems)
- Thermal expansion matched to the substrate
- Effective barrier for permeation of moisture and ionic impurities
- High insulation resistance
- Compatible with semiconductor processing (in general, low temperature deposition and densification)

TABLE-IV
Solder Glass Requirements

- Low sealing temperatures (< 450°C)
- Expansion matched to base, cap and leadframe
- Good adhesion to base, cap and leadframe
- Resistance to cyclic temperature/humidity conditions
- Resistance to salt atmospheres and sprays
- Resistance to common cleaning and tin plating solutions
- High volume and surface resistivities
- Low dielectric loss factor
- Low outgassing of moisture and other gases
- Low alpha-particle emission (target <0.1 α/cm^2/hour)

TABLE-V -- Common Solder Glass Fillers

Filler	Filler Expansion $x\ 10^7/°C$ (25-350°C)	Composite Target Expansion $x\ 10^7/°C$
β-eucryptite	- 65	60-70
Boron carbide	43	"
Silicon carbide	40	"
Zircon	45	"
Cordierite	23	"
Tin Oxide	40	"
Lead Titanate	-45	"
Fluorospar	225	150-160

TABLE-VI

Glass Requirements for Coated Metals in Hybrid Circuits

- Expansion matched to, or slightly below that of the metal substrate
- High volume and surface electrical resistivities
- Good flow properties as to form bubble and defect free coatings
- Good adhesion to the substrate
- Stability in the subsequent metallization firing steps
- Good thermal shock resistance
- High fracture toughness and strength

TABLE-VII

Substrate Requirements for Thin Film Hybrids [39]

- Smooth surface, surface roughness ~ 250A
- Expansion coefficient matched to films
- High mechanical strength and themal shock resistance
- Inertness to chemicals used in circuit processing
- Zero porosity to prevent entrapment of gases
- High electrical resistivity and low dissipation factor
- High thermal conductivity
- Uniform physical properties

TABLE-VIII
High Alumina vs. Glass-Composites for LTC Systems

Property or Process	Alumina	Glass Composites	Function/ Performance
Flexural strength (Kpsi)	40-60	20-40	Mechanical Ruggedness
Elastic modulus (Mpsi)	45-55	15-30	"
Thermal Conductivity (w/m·K°)	20-40 w/m·K	1-10	Thermal Dissipation
Cofired Conductors	W, Mo, Mo-Mn	Cu,Ag, AgPd,Au	Electrical Interconnects/ Conductance
Sintering Temps. (°C)	>1500	<1200	
Dielectric Constant (@1MHz)	≃9	4-8	Propagation Delay
Thermal Expansion ($x10^{-7}$/°C)	65-75	30-60	Compatibility with Si

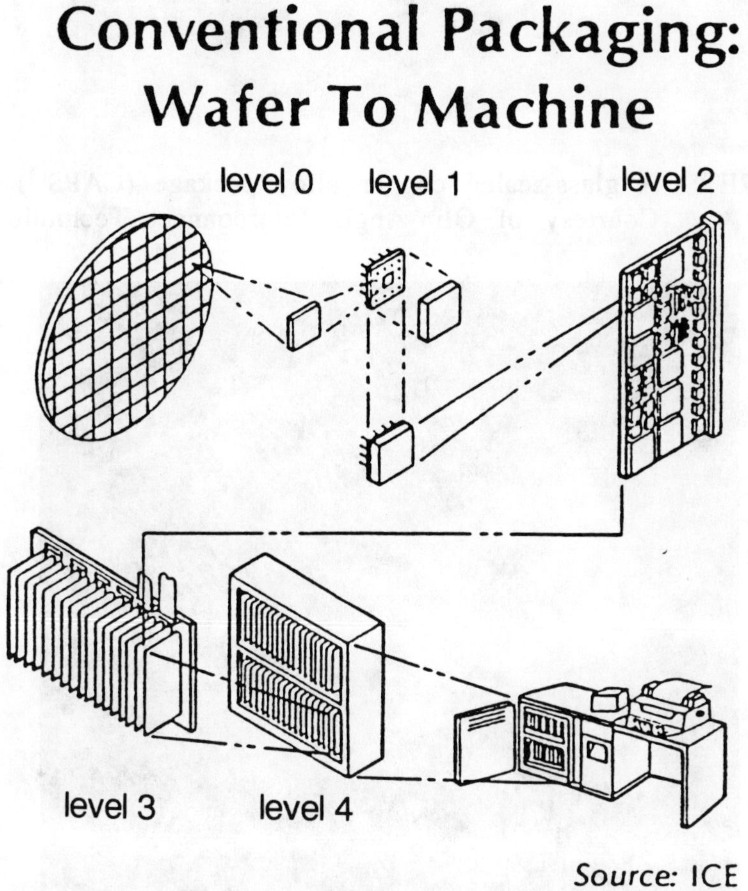

FIGURE 1: Hierarchy of microelectronic packaging levels.

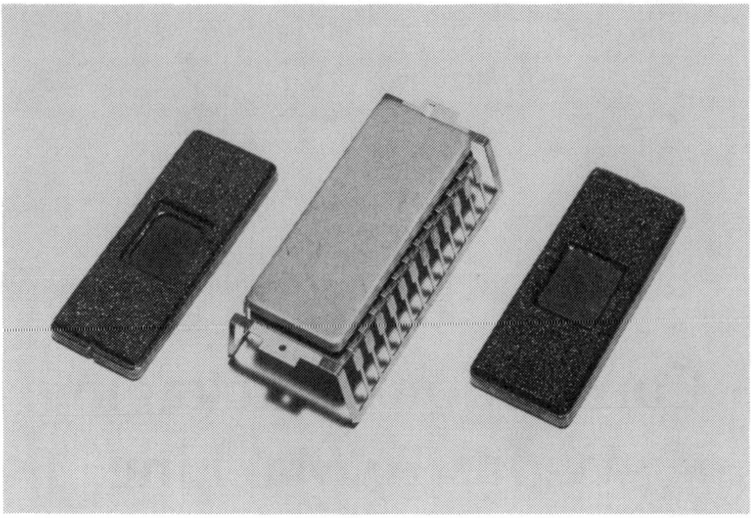

FIGURE 2: A glass-sealed copper alloy package (CAPS®) Courtesy of Olin-Aegis Interconnect Technologies.

FIGURE 3: A glass-sealed Kovar hybrid package.

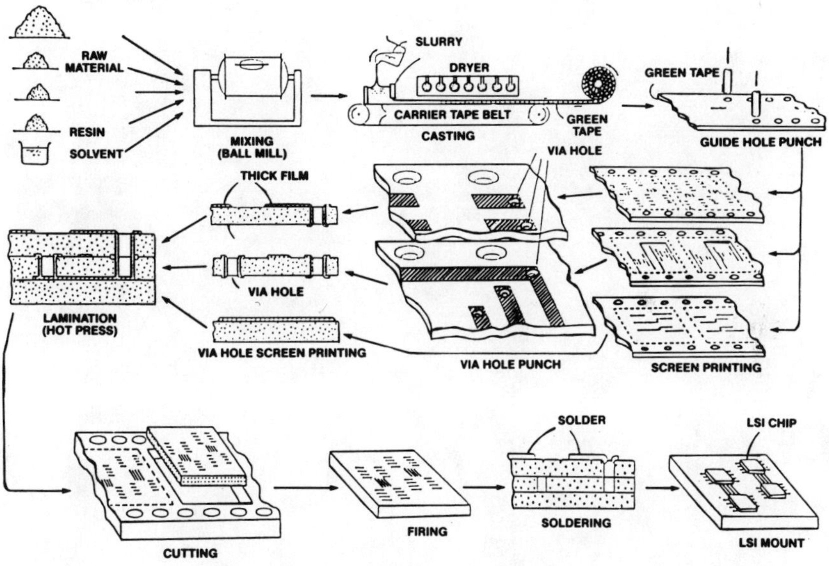

FIGURE 4: Schematics of a low-temperature cofired multilayer process. Courtesy of Olin-Aegis Interconnect Technologies.

THE PRODUCTION OF BARIUM HEXAFERRITE ($BaFe_{12}O_{19}$) FROM HIGH IRON GLASSES

B.H. Fox, ORTECH International, Mississauga, Ontario and
A. Zybura, COM DEV, Cambridge, Ontario.

Abstract

High iron content silicate and borate glass systems have been used to produce barium hexaferrite crystals by a combination of RST (Rapid Solidification Technology) and glass-ceramic processing. The materials produced were shown to have dielectric constants, dielectric losses and magnetic properties, measured under a microwave regime, to be comparable to state of the art materials produced by the traditional ferrite fabrication technique. There is every indication that these techniques can be used to produce superior microwave device materials.

Introduction

The principal purpose of this work was to produce a high quality barium hexaferrite ($BaFe_{12}O_{19}$) ferrite material for microwave device applications. The approach chosen was two-fold:

1) using a combination of RST (Rapid Solidification Technology) and glass-ceramic processing to produce fine single crystalites of $BaFe_{12}O_{19}$ within the residual glass matrix;.

2) using a flux system, which was substantially similar in composition to the original parent glasses to grow $BaFe_{12}O_{19}$ crystals .

$BaFe_{12}O_{19}$ was chosen for investigation since traditional ferrites, used in microwave devices, only meet the demands of wideband and low loss microwave components in the 1 to 30 GHz range. However, for higher frequency components, equivalent materials are not available. Above 30 GHz the magnetic splitting is restricted to around 5000 Gauss. This value renders it impossible to realize devices with a bandwidth as large as those at lower frequencies, since the ferrite volume would be small and the magnets required to provide a biased field for operation near resonance would be large. Hexagonal ferrites, such as $BaFe_{12}O_{19}$, offer a means to avoid these limitations.

Hexagonal ferrites are ferrimagnetic compounds which have a high value of magnetization (about 4000 Gauss) and a large internal anisotropy field of approximately 17000 Oersteds. The structure of barium hexaferrite was first investigated by Adelskold[1] and was shown to be that of magnetoplumbite. Barium hexaferrite and other ferrites with a magnetoplumbite structure are known as M-type ferrites.

The internal anisotropy field, which is a consequence of the $BaFe_{12}O_{19}$ structure affects the tensor permeability equivalent to an increase in magnetization. This high intrinsic magnetic field makes the compound uniquely valuable for construction of microwave devices which operate at frequencies higher than 30 GHz. The structural purity of $BaFe_{12}O_{19}$ influences the saturation magnetization as a function of the degree of substitution for the ferric ions by other elements, or iron in the wrong valence state[2]. In addition, the saturation magnetization can be adversely affected by stress and strain in the barium hexaferrite at the unit cell level.

It was the intention of this study to explore methods to control the chemical and structural purity of the material to produce a low loss (both magnetic and dielectric) material with large magnetic anisotropy for ultrahigh frequency microwave devices.

Experimental Procedures

Glass-Ceramic Process

The traditional glass-ceramic process is described by McMillan[3]. In the early work, undertaken for the production of ferrite materials from glass-ceramics (Shirk & Buessem[4] and Oda et al.[5]) it has been pointed out that the production of crystals from the amorphous glass matrix enables the crystals to grow free of any stresses and strains inherent in the traditional ferrite manufacturing process. Such inherent stresses distort the structure of the materials at a unit cell level. These distortions interact with the internal magnetic field of the crystals and cause an increase in the loss of the materials and a broadening of the uniform linewidth, both effects reducing the effectiveness of the material for device purposes.

The compositions used followed closely those given by Shirk and Buessem[4], Oda et al.[5]. The high iron content of these compositions prevented their preparation as glasses in a traditional manner, since their compositions lay outside of their glass forming region. To overcome this limitation, RST was employed. The glasses fabricated by the RST system are the parent glasses in the subsequent glass-ceramic processing. The glasses were subjected to a two stage heat treatment, first a neucleation step at 650°C for 30 min. and a crystal growth stage at 810°C for 4 hours. These treatment temperatures were established by a DTA study. Considerable stress is put into the system when it undergoes RST fabrication, the subsequent heat treatment used in the crystal formation will remove these stresses and will only leave the relatively minor stresses due to thermal expansion mismatch between the matrix and the crystal phases. Although the glass-ceramic process itself may produce minor strains in the crystals due to the thermal expansion mismatch between the crystal and glass matrix, in this case such strains are released when the crystals are separated form the matrix.

RST

Rapid Solidification Technology (also known as splat quenching), as applied to non-metallic materials, is comprehensively reviewed by Revcolevschi & Livage[6]. The principal reason for using RST on oxide material has been to extend the glass-making regions of many systems and to produce "new glasses" from systems which will not normally form glasses using traditional techniques. It is the opening up of the glass forming region that made RST the technique of choice in this case. Although Oda et al. reported that they produced glasses from their compositions by pouring the melt directly onto cooled blocks, this was not found to be the case in this work. When the melt was poured onto cooled copper blocks, only a very thin layer, approx. 1mm, was found to be non-crystalline. The RST apparatus used for this investigation is shown in Figure 1. It is a spin-melting system where the molten oxide, at a temperature of 1350°C, is squirted onto a rapidly spinning copper wheel. The cooling rates that can be achieved by this technique are in the region of 10^6°Cs^{-1}. The materials produced by this technique were a stream of non-crystalline flakes ranging in size 0.5mm to 5mm on their longest dimension, and 0.5 mm thick or less. These flakes where shown to be amorphous to XRD and therefore taken to be glassy in nature. The glasses fabricated by this RST system are the parent glasses in the subsequent glass-ceramic processing. The XRD traces showing the initial glassy nature and the crystalline development with heat treatment are shown in Figure 2a-2b.

Crystal Growth

To enhance further the understanding of the ferrite materials produced from these glass systems, and to provide comparative data to determine the effect on the dielectric and magnetic performance of the RST techniques, high quality single crystals were produced from similar compositions using a flux growth technique. Following the work of Elwell and Scheel[7], an ACRT (Accelerated Crucible Rotation Technique) system was chosen, since it had proved to be the best arrangement for the production of various forms of barium hexaferrite (see Figure 3).

The sample types produced were as follows:

Sintered Polycrystalline

- Original $BaO-Fe_2O_3-B_2O_3$ ratios wer previously optimized (designated "B1").
- Several batches ("4" - "9b") of glass were fritted and ground to 45 microns, then RST was performed on these parent materials.
- RST products were heat treated and submitted for sintering and dielectric assessment.

Single Crystal (ACRT)

- These were grown from a flux with a 3 mole% difference in Fe to B oxide ratio
- Accelerated Crucible Rotation involves the stopping and starting of the rotation of the crucible during the cooling period, from the glass melt-point to 600·C
- The solid matrix was cracked, while the single crystals were magnetically extracted, and etched,to remove any residue glass and submitted for dielectric measurement.

The results for the dielectric constant and the dielectric losses for both the polycrystalline sintered samples and the single crystal materials are shown in Figures 5 and 6.

Test Methods

The complex dielectric constant and dielectric loss tangents were carried out using the cavity perturbation technique per the ASTM standard C 525-63T. These measurements were performed at 12 GHz and 35 GHz, the Q factor measurements being performed at 35 GHz to the same ASTM standard. A diagram of a typical cavity for measurement of dielectric properties at 12 GHz and a schematic of the equipment set up are shown in Figures 4-A and 4-B.

Measurements of the gyromagnetic ratio and uniform linewidth were performed on the slow-cooled single crystals. The gyromagnetic ratio was found to be 2.0 ± 0.2 Gauss and the uniform mode linewidth 500 ± 100 Oersteds, the accuracy of these readings being low as a consequence of the relatively primitive sample preparation procedures used in this stage of the study. In each case they are compared with a leading commercially produced barium hexaferrite material.

The dielectric properties of the RST produced materials were determined on sintered blocks of crystals which where removed from the heat-treated samples, by both physical and chemical means. As a comparison, a high purity barium hexaferrite powder was obtained from a commercial supplier who had produced the material by traditional ferrite production techniques.

All the samples were sintered in as near the same manner as possible to minimize any difference due to processing. Several single crystals were produced, the largest being 15 mm across, the values given in Figs. 5 and 6 for single crystal were the best values obtained for single crystal measurement. It was observed that the better results were obtained from crystals near the centre of the melt, this was believed to be due to less crucible contamination within the crystal structure.

Results and Discussion

All the composition melts were corrosive to the crucible materials used, both in the RST and the flux growth experiments. However in the RST systems, the melt "saw" the crucible for a period of 5 to 10 minutes while the flux growth took place over 3 - 5 days. All the measured samples, RST, single crystal and commercial powder appeared to be phase pure by XRD, but the difference in dielectric properties of the single crystals grown either in a high zirconia or a platinum crucibles could also imply possible contamination from the crucibles.

When we compare the dielectric values shown in Figs 5 and 6, it can be seen that the materials produced by the RST route gave superior values of dielectric constant and loss at the measurement frequencies. In each case the RST materials were better than the traditionally produced $BaFe_{12}O_{19}$ material and could out-perform the single crystals produced in this work.

Although further work needs to be performed, it is suggested that the high performance of the materials produced by the RST route is due to the production of the barium hexaferrite crystallites in a less stressed and less contaminated manner than traditional ferrite production methods provide. The glass-ceramic process itself may produce minor strains in the crystals due to thermal expansion mismatch between the crystal and glass matrix, in this case however, such strains are released when the crystals are separated form the matrix. To further determin the relationship between the improved properties and the micro-structure, a series of SEM and TEM studies on the nature of the micro-structure and the micro-compositional nature of the RST glass and the crystals produced from it, are going to be undertaken, and will be reported in the near future.

Conclusions

The measured values which are presented in this paper show that the use of a combination of RST and glass-ceramic techniques have produced materials which are better than the best traditionally produced comparable materials, measured under the same constraints.

As a preliminary study these results strongly indicate that the further development of these techniques could produce excellent hexaferrite materials for use both in the microwave communication industry and for exploitation in the magnetic recording industry.

Acknowledgements

We would like to acknowledge NRC Space Division for their funding of this work, and T. Duncan for his assistance in preparing the text and figures.

References

1. Adelskold,V., 'X-Ray Studies on Magnetoplumbite, $Pb0.6Fe_20_3$, and other Substances Resembling Beta-Alumina, Na_20. 11 Al_20_3' Arkiv. Kemi. Minerali. Geol. **12A**, (29), 1938.

2. Von Aulock, W.H., Handbook of Microwave Ferrite Materials. New York and London: Academic Press, p 470-475,1965.

3. McMillan, P.W., Glass and Ceramics, 2nd Edition, New York: Academic Press, 1979.

4. Shirk, B.T. and Buessem, W.R. 1970, 'Magnetic Properties of Barium Ferrite Formed by Crystallization of Glass', J. Am. Ceram. Soc. **53** 192-196, 1970.

5. Oda, K., Yoshio, T., O-Oka, K., and Kanamaru,F., 'Morphology and Magnetic Properties of Ba $Fe_{12}O_{19}$ Particles Prepared by the Glass-Ceramic Method', J. Mat. Sci. Letts. **4** 876-879, 1985.

6. Revcolevschi, A., and Livage, J., 'Rapid Solidification of Nonmetals', Treatise on Materials Science and Technology **20**, 73-116, 1986.

7. Elwell, D., and Scheel, H.J., Crystal Growth from High-Temperature Solutions. New York: Academic Press, 1975.

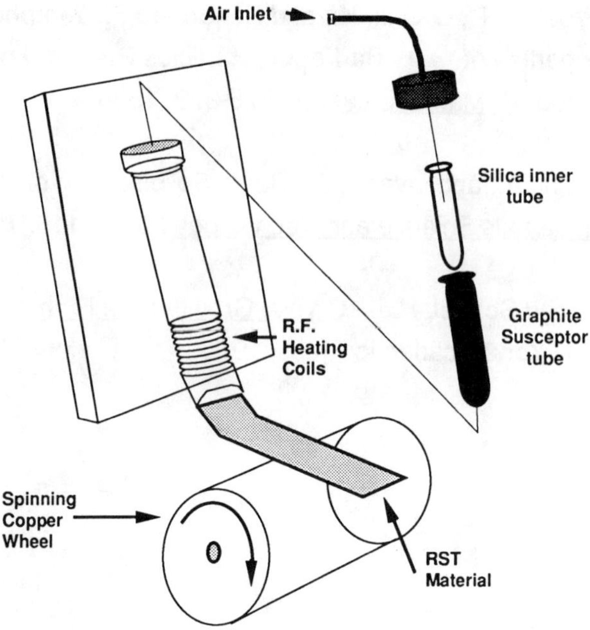

Fig. 1. Schematic diagram spin melt apparatus

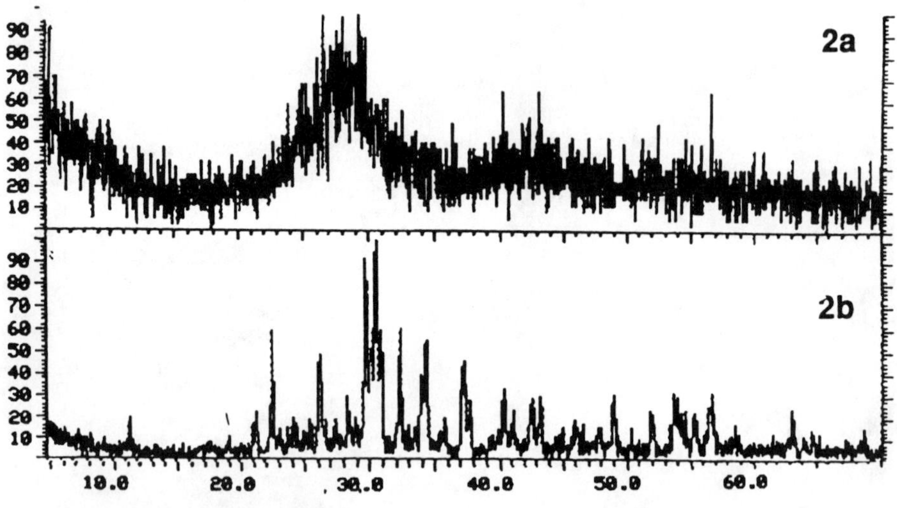

Fig. 2. XRD of RST glass (2a) and crystallized sample after heat treatment (2b)

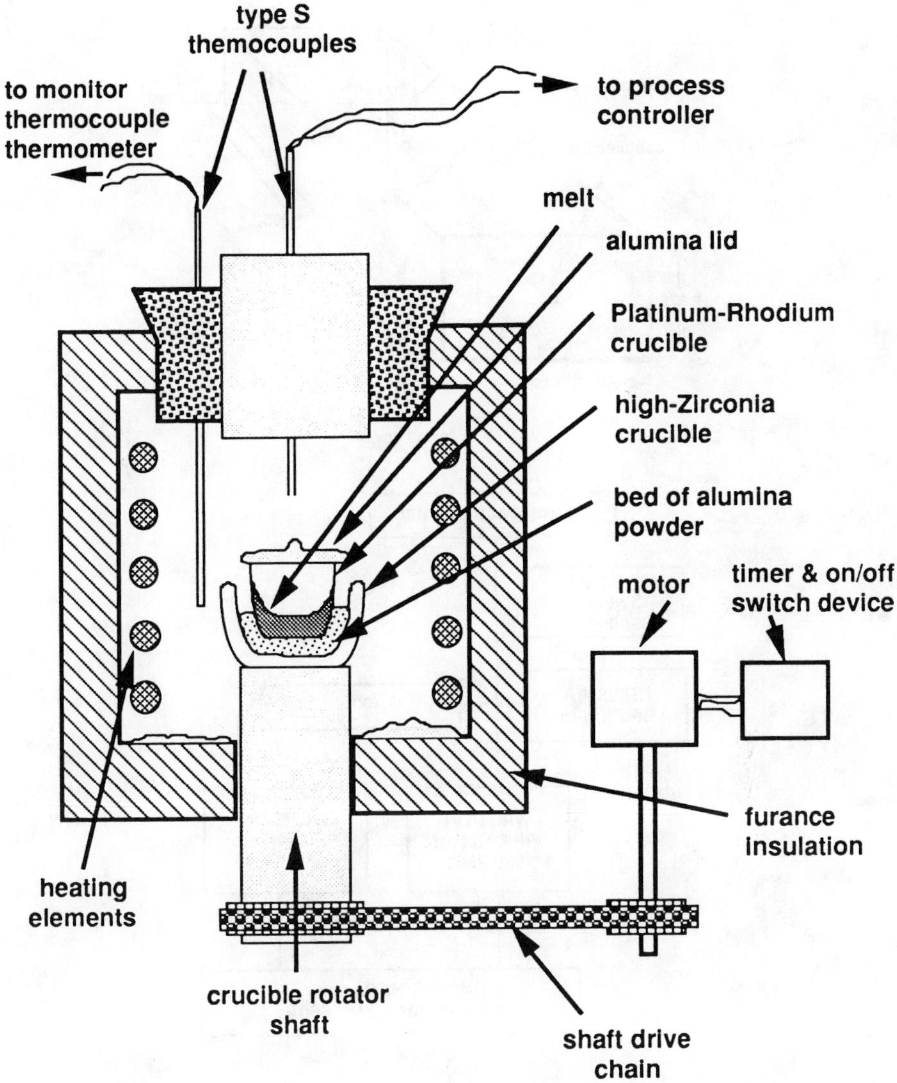

Fig. 3. Accelerated crucible rotation apparatus

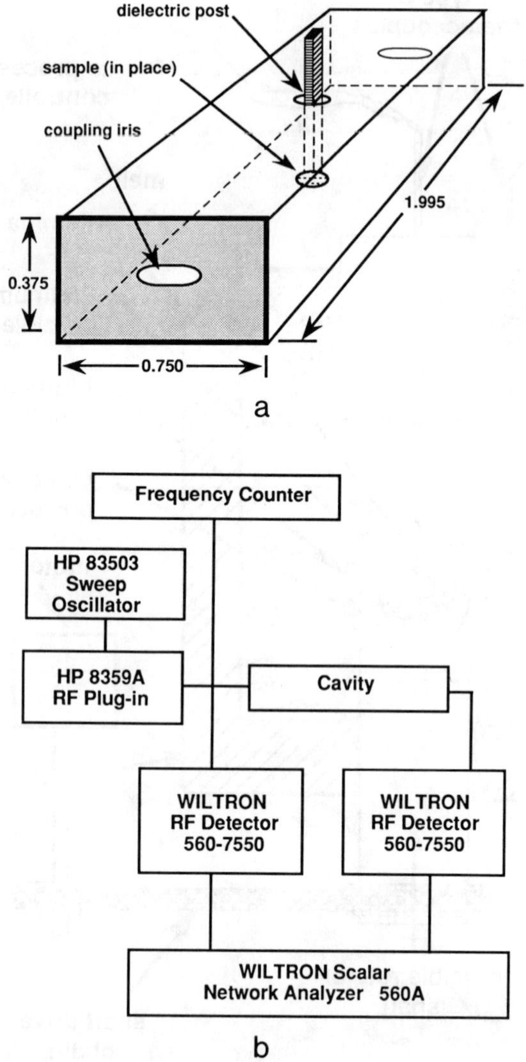

Fig. 4. (a) Resonant cavity schematic; (b) setup for dielectric measurements

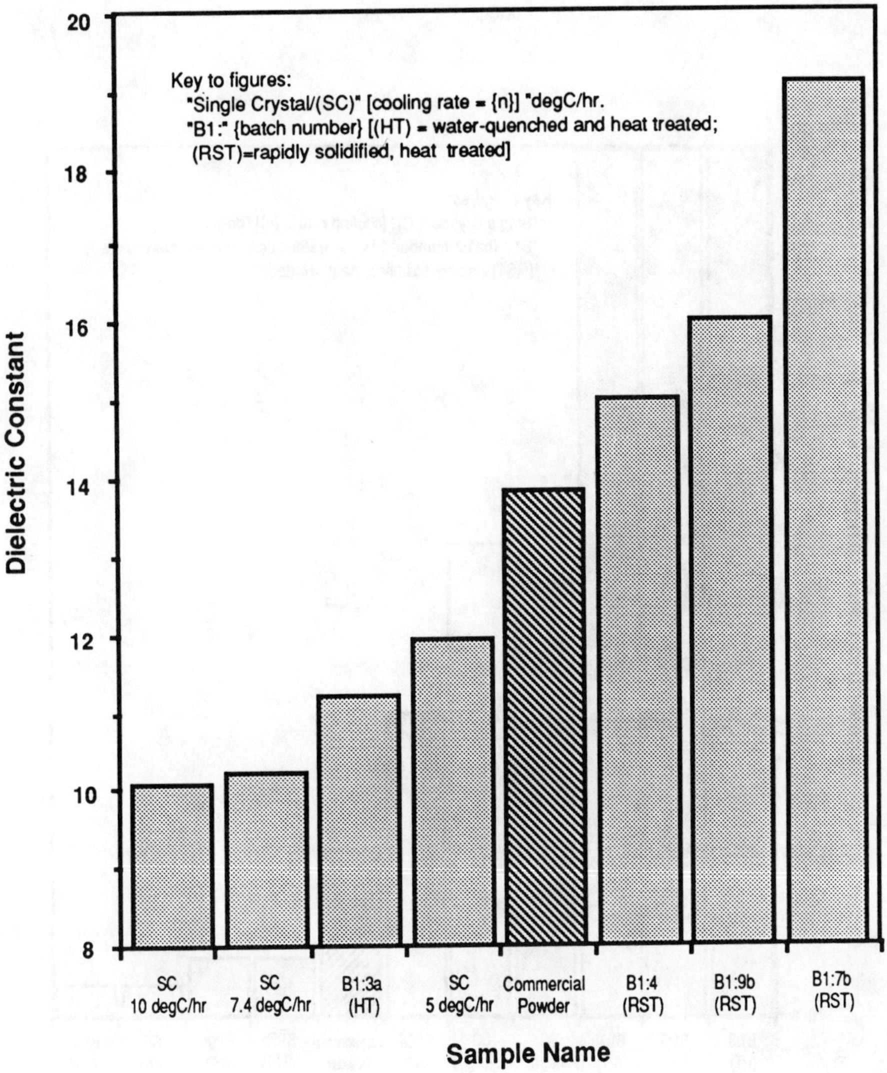

Fig. 5. Improvement of dielectric constant

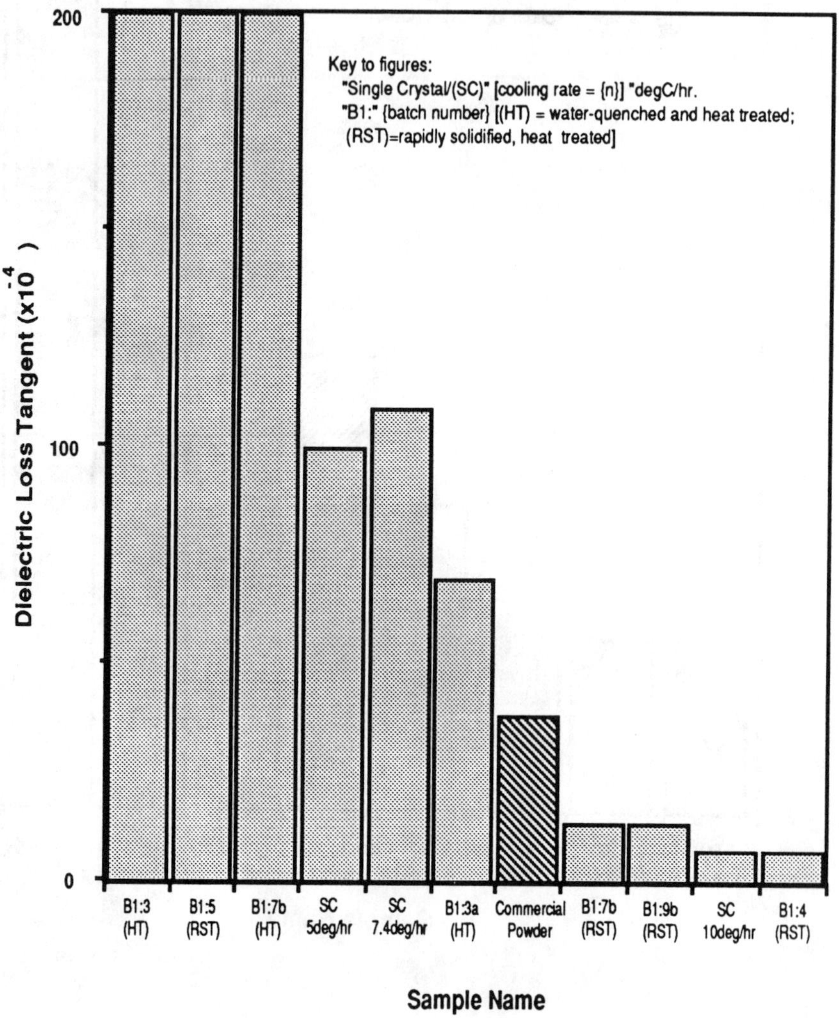

Fig. 6. Improvement of dielectric losses

LOW K GLASS-CERAMICS FOR MICROELECTRONIC PACKAGING

J. F. MacDowell, G. H. Beall

INTRODUCTION

Glass-ceramics are a class of materials that are highly crystalline, have near-zero porosity, and are uniformly microcrystalline (crystal diameters commonly around 1 micron).[1] They thus represent the most precise and reliable ceramics produced today and are found in many important applications for which neither glass nor conventional ceramics are suitable.[2,3] Perhaps more than any other property, it is their inherent predictability (constancy of properties) that qualifies glass-ceramics for use in the electronics industry.

Ceramics derived via glass crystallization can be divided into two major types: 1) those internally nucleated and crystallized from bulk glass shapes (I.N.), and 2) those made by viscous sintering and surface (externally nucleated) crystallization (E.N.) of glass powder compacts.

Before describing several new glass-ceramic materials that fit into these two categories, the authors would like to distinguish between glass-derived ceramics and glass-fluxed or glass-sintered ceramics. Recent work[4,5] in the electronic packaging field utilizes glass/crystalline powder mixtures to arrive at a dense semi-crystalline body (e.g., Al_2O_3 or α-quartz mixed with borosilicate glass). In these mixed systems, the glass phase remains intact after doing its job as a "glue" or "sintering aid" and thus becomes a permanent diluent to the crystalline phases - and to their properties. The inherent problem with this approach is viscous flow of the glass and loss of dimensional control of the substrate during firing.

Most commercial glass-ceramics fall into the I.N. (internally nucleated) category, making full use of the favorable economics of scale inherent with the viscous melting and forming of glass objects.

Internally Nucleated Glass-Ceramics

I.N. glass-ceramics require the addition of a nucleation catalyst to the raw glass batch before melting. The glass is melted at from 1400°C to

1600°C and formed during cooling, utilizing the highly predictable viscosity-temperature characteristics of a Newtonian fluid. After cooling the formed glass object, it is given a heat-treatment slightly above its transformation (annealing) range to develop microheterogeneities (amorphous or crystalline) upon which crystallization can subsequently begin (Fig. 1a).

Further heat treatment produces a major phase change from glass to crystal that begins at the low energy surfaces of the tiny nuclei and ends when the oxide raw materials of the primary crystal phase are depleted from the amorphous reservoir of the residual glass (Fig. 1c). During this process, viscous flow is drastically impeded by the growth of crystals around the widely and uniformly scattered nuclei. As a result, glass articles of intricate or precise shape can be formed by this process and crystallized without significant distortion into a uniformly mirocrystalline ceramic. The small volume decrease (or density increase) of less than 5% during formation of a glass-ceramic is in contrast to the 15 to 25% shrinkage that normally accompanies the firing of a conventional ceramic body.

Externally Nucleated Glass-Ceramics

Glass-ceramics that can be formed from particulate glass via ceramic powder processes and subsequently fired to sinter, nucleate, and crystallize have been developed recently. After melting, E.N. glasses are quenched by pouring into water, between water-cooled metal rollers, or by faster methods such as splat-cooling or atomizing. The glass is then milled (ball, impact, etc.) to an average particle diameter of between one and ten micrometers.

The glass powders can be formed into shapes using conventional ceramic processing techniques (using binders, plasticizers, etc.) and subsequently fired. Densification of the glass powder compact must occur prior to devitrification if a pore-free ceramic is desired. Although room-temperature forming of powder compacts is easier (and cheaper) than high-temperature glass forming, dimensional control of E.N. glass-ceramics is more difficult because of the shrinkage of the ceramic preforms and the low viscosity of the glass during sintering. The E.N. process is shown graphically in Figure 2. The final result is a dense structure (Figure 2c) with a uniform crystal size normally much finer than the grain size of the parent glass powder, and very similar to the product of the I.N. process shown in Figure 1c. Besides lower capital and set-up costs, the E.N. process has considerably greater compositional flexibility than internal nucleation of bulk glass. Compositions that cannot be shaped using standard glass-forming processes because of instability toward crystallization or too low a viscosity can often be quenched to glasses by simple, rapid techniques and ground to powders with relatively low energy consumption.

The thin sheets of dielectric commonly used in the electronic packaging industry are readily formed by a process called "tape-casting". The E.N. glass-ceramic process can be easily incorporated into this method of forming ceramic sheet.

Other inherent capabilities of glass-ceramics (both E.N. and I.N.) may be especially useful in the packaging field. Metastable phases (e.g., β-quartz solid solutions) unique to glass derivation can yield unusual properties (zero T.C.E. in this case). The void-free micro- (or nano-) crystalline structures of these materials can average out stresses caused by large crystal anisotropies, rendering them unable to produce strain to fracture or optical scattering. High strength and toughness and smooth surface finish are also of relevance to the microelectronic application - and readily achievable via the glass-ceramic process.

Because glass-ceramics are commonly based upon silica or aluminosilicate network structures, their dielectric properties are ideal for passive packaging. Low loss (dissipation factor), high D.C. resistivity, low dielectric constant (K), and lack of significant porosity are other key properties that can readily be obtained from glass-derived ceramics.

In this paper, the authors will describe several dielectric glass-ceramic systems that currently show promise for application in the field of microelectronic packaging.

Microelectronic Packaging Needs

The rapid progress in integrated circuit (I.C.) technology within the past decade has eclipsed minor improvements in packaging technology. As an I.C. device increases in complexity and contains more and more active elements, a higher proportion of the signal response time for the system is taken up by the dielectric packaging. Higher interconnect signal speed can best be achieved by shortening the signal path between chips and lowering the dielectric constant (K).

Although alumina currently dominates the field of ceramic packaging, materials with both lower cost and higher performance are beginning to appear on the scene in increasing numbers. Glass-ceramic packages have several inherent advantages over alumina.

Perhaps the most crucial advantage of glass-ceramic substrates is their low K (4-6) as compared to alumina's K of 9. Figure 3 shows the critical relationship between signal delay time and dielectric constant. Whereas glass-ceramics can be cofired below 1000°C and metallized with copper, silver, or gold, alumina must be cofired at >1500°C using moly or tungsten metallization. Direct die-bonding to glass-ceramics is made possible with thermal expansion coefficients matching that of the silicon chip (~30 x 10^{-7}/°C), whereas alumina's is more than double that of Si. Smoother surfaces and less costly finishing can also be achieved with glass-derived ceramics. Greater dimensional control also means larger, flatter substrates with higher registration than with alumina.

DICUSSION OF GLASS-CERAMIC SYSTEMS

1. Cordierite Glass-Ceramics

Cordierite glass-ceramics formed by the external nucleation (E.N.) or powder processing technique have been recently introduced as integrated circuit substrates. These materials have excellent dielectric properties (K 5-6, low loss), can be cofired with copper below 1000°C, and can match silicon in thermal expansion characteristics. The thermal expansion behavior of cordierite glass-ceramics can be controlled in two ways: i) by tailoring the bulk composition to produce a secondary phase, either crystalline or glass of known proportion and thermal expansion coefficient, or ii) by inducing a solid solution in the cordierite phase according to either substitutional or stuffing principles. The latter has the advantage of retaining the monophase characteristic of the glass-ceramic thereby preserving the best mechanical properties.

1.1 Cordierite with a Secondary Phase

Although cordierite can readily be prepared via solid state reaction involving mineral talc, clay, and alumina, full densification of these mineral batches requires firing temperatures of over 1400°C. Starting with particulate glass, however, full densification and crystallization to cordierite can be achieved between 900°C and 1000°C. Certain deviations from the 2:2:5 stoichiometry and/or additions of other oxides must be made to the precursor glass in order to i) increase T.C.E. from about 12 to 32 x 10^{-7}/°C for a match to silicon, and ii) allow the glass powder compact to sinter to high density before the onset of cordierite crystallization.

With some ingenuity, a sintering aid can be added that will lower T.C.E. effectively without compromising dielectric or mechanical properties. Phosphorus pentoxide[6] and boron oxide[7], alone or in combination with lead, zinc, or alkaline-earth oxides have been used extensively to this end.

The Corning Code 9641 cordierite-based composition matches silicon in thermal expansion, has been sold in small quantities as a cofirable multilayer E.N. glass-ceramic, and appears to be tape-processable (including metallization) in production.[8] The dielectric constant and loss factor at 25°C are 5.2 and 0.2% over a wide range of frequencies. The density is about 2.6 g/cc, significantly less than that of alumina, about 4.0 g/cc.

The disadvantages of cordierite are its lower strength (140 MPa M.O.R.) and thermal conductivity ($1.9\ Wm^{-1}K^{-1}$) as compared to those of alumina, (250 MPa, and $20\ Wm^{-1}K^{-1}$, respectively). Design of the packaging process to minimize mechanical and thermal stresses, however, should allow cordierite to be used at high selection rates.

1.2 Solid Solution in Cordierite

The most simple solid solutions involving low pressure cordierites consist of divalent substitutions for octahedral magnesium in the structure. Typical examples are Fe^{+2} for Mg^{+2} and Mn^{+2} for Mg^{+2}. Both these substitutions produce colored glasses which are subject to thermal oxidation in air and a reduction in the refractoriness and durability of the crystalline product. A more interesting substitution involves the substitution for two tetrahedral Al^{+3} ions by one Be^{+2} and one Si^{+4} ion.[9] This double substitution results in the stoichiometric end member $Mg_2BeAl_2Si_6O_{18}$ from normal cordierite $Mg_2Al_4Si_5O_{18}$, and produces changes in the lattice parameters as well as in its thermal expansion behavior (see Table 1). From room temperature to 800°C, the linear thermal expansion coefficient has increased from 11.7 in stoichiometric cordierite to 19.2 in the substituted Be version.

A broader form of solid solution in cordierite relies on stuffing the large hexagonal ring cavities by alkali or alkaline-earth ions.[9] This stuffing must be accompanied by a substitution of tetrahedral aluminum for tetrahedral silicon in the ring position in order to preserve charge balance. Thus, stoichiometric cordierite, $Mg_2Al_4Si_5O_{18}$, may have its hexagonal cavities stuffed with appropriate large alkali cations like potassium and cesium to give $KMg_2Al_5Si_4O_{18}$ or $CsMg_2Al_5Si_4O_{18}$. Analogous stuffing with large alkaline-earth cations such as barium will result in the stuffed end member $BaMg_4Al_{10}Si_8O_{36}$. A key effect of large stuffing ions on the cordierite composition is increased glass stability, allowing sintering prior to crystallization.

Table 1 gives the lattice parameters and the modifications in thermal expansion of various cordierite end members as crystallized from glass. It can be seen that the linear expansion coefficients cover a range from about $12 \times 10^{-7}/°C$ to $22 \times 10^{-7}/°C$ over the range of 0-800°C, with the Cs-stuffed cordierite giving the closest match to silicon.

Of some concern in glass-ceramic materials formed by the E.N. process is the stress that can develop between grains where anisotropic thermal expansion in different lattice directions is high. Clearly, this stress is a function of grain size as well as thermal expansion anisotropy. However, for a given grain size, low anisotropy such as observed with Cs-stuffed cordierite (Table 1) will produce the minimum intergranular stress. Since the anisotropy in Cs-stuffed cordierite is only one-third of that of stoichiometric cordierite, the grain size may be three times as large without exceeding the intergranular stresses present in comparable standard cordierite.

2. Fluormica Glass-Ceramics

Fluormica glass-ceramics are known for their ease of machining and their excellent dielectric properties. At least one commercial material, MACOR®, has been marketed for over fifteen years and has found wide application, particularly in dielectric and high vacuum applications.[3] The precision machinability of MACOR® with conventional metal-working tools is directly due to the fine-grained, house-of-cards microstructure where randomly oriented flexible flakes tend to either arrest fractures or cause such deflection or branching that only local damage is done.

The chemistry of fluormica glass-ceramics is rather complex,[3,10,11] involving both stoichiometric compositions and solid solutions. The early fluormica glass-ceramics were based on the fluorine phlogopite phase [$KMg_3AlSi_3O_{10}F_2$]. Recently other chemical formulations giving rise to different fluormicas have been developed. Thus, subpotassic phlogopite [$K_{1-x}Mg_3Al_{1-x}Si_{3+x}O_{10}F_2$], potassium tetrasilicic mica [$KMg_2Si_4O_{10}F_2$], alkaline-earth micas [$R_{0.5}Mg_3AlSi_3O_{10}F_2$], and biotite [$K,Na(Mg,Fe^{+2})_3(Al,Fe^{+3})Si_3O_{10}F_2$] have been reported as major phases in glass-ceramics. The dielectric properties are best in the subpotassic phlogopite and alkaline-earth mica glass-ceramics.

Table 2 lists electrical, thermal, and mechanical properties of two subpotassic mica glass-ceramics compared to standard MACOR®. Note that their log volume resistivities are around 11 ohm/cm at 500°C and room temperature dielectric constants are below 6. The loss tangents at 10 KHz can be below .001 and the dielectric strength near 2000 volts/mil. The thermal expansion coefficients from 0-400°C are around 70, close to that of alumina.

The alkaline-earth mica glass-ceramics are currently the strongest of the machinables and also boast the highest thermal stability (1100°C+) and some of the best dielectric properties (see Table 2). Abraded modulus of rupture values of 250 MPa have been measured on barium strontium phlogopite glass-ceramics with highly crystalline and interlocking microstructure.

The crystallization cycle in mica glass-ceramics includes normal nucleation and growth temperature holds, but low fluoride phases sometimes form on the surfaces of substantial fluoride species volatized prior to crystallization. Thus, a hard skin may have to be removed prior to normal etching. Machining resolution is limited only by the crystal size (10 μm dia. flakes) and unlike other machinable ceramics, there is no porosity and no postmachine firing is required.

Machinable mica glass-ceramics may have application as substrates for electronic packaging. It is relatively easy to slice thin slabs which can subsequently be welded or fritted together. Many precise, fine holes can be drilled through this glass-ceramic as has been demonstrated in the case of the EGRIT gamma ray detecting telescope where arrays of 2000 wires are anchored at precise holes driven through one-meter square frames. Copper metallization or brazing has also routinely been demonstrated in mica glass-ceramics, so plated-through holes are feasible.

3. Reconstituted Fluormica Glass-Ceramics

An entirely different type of mica glass-ceramic based on the water swelling fluorine species fluorhectorite [$LiMg_2 LiSi_4 O_{10} F_2$] has been used to make an inorganic dielectric paper.[1,2] Upon contact with water, fluorhectorite glass-ceramics swell and delaminate to such an extent that the fine platelets of mica form a stable suspension or sol-gel. If this gel is extruded through a slot into a salt solution of KCl, K^+ for Li^+ ion exchange between the solution and the mica platelets causes immediate flocculation producing a coherent and continuous film less than 25 μm in thickness. This film, when washed and dried, is flexible and creasible. It is composed of oriented crystals of a stable potassium mica approaching fluortaeniolite, $KMg_2LiSi_4O_{10}F_2$, in composition. The morphology of the mica crystals both dispersed in the original aqueous suspension and in the ion exchanged and flocculated film is interesting. Very high aspect ratio platelets and narrow strips less than 100Å thick, some approaching unit cell thicknesses, are observed in the gel (Figure 4). In the ion exchanged film, composite flakes composed of similar platelets and strips are present. The high degree of orientation and overlapping of the mica crystallites is responsible for the flexibility and creasibility of the film.

A thicker paper or board can be made by random flocculation of the gel followed by wire forming the floc on a conventional Fourdrinier machine. Glass or other fibrous material may be introduced to impart strength and tear resistance. These films, papers, and boards are about 50% porous, but the porosity is generally closed allowing for controlled permeability. The papers are as strong as quality organic paper products, can withstand temperatures of over 500°C without embrittlement, and have dielectric strengths in excess of 20 kV/mm. Additionally, they are durable over a wide range of pH.

Mica ceramic paper-based laminates can be used to make printed wiring boards with unique advantages over standard fiber-glass reinforced epoxy boards like FR-4.[13] The paper is impregnated with epoxy and BT resins, clad with copper foil which is then pattern etched and stripped, and finally stacked and pressed to form multilayer boards. The resulting laminate properties are shown in Table 3.

It is apparent from the data that the mica laminates are superior to FR-4 in such areas as bow and twist, water absorption, thermal expansion, flammability and dielectric strength. However, they are mechanically weaker and poorer in peel strength unless they contain a glass fabric facing.

Dimensional stability is probably the most important feature of ceramic mica-reinforced laminates. They are at least twice as good in this respect as the stabilized FR-4 standard material. Dimensional changes result from a complex interaction of factors such as unequal stress distribution, weave pattern, resin flow, uneven cure, etc. Lack of residual stress and lower resin flow are two of the factors that distinguish ceramic mica reinforcement from that of glass fabric.

The machinability of the ceramic mica-reinforced laminate was found to be four times better than the glass fabric laminate having the same resin. This result was based on drill bit testing where plated-thru hole quality and hole wandering were monitored. The hole walls are also straighter because protruding bundles of fiber are absent.

The capacity for higher density circuitry depends more on the nature of the fabrication process than on the type of substrate. Nevertheless, the lack of surface weave pattern and much higher dielectric strength helps in printing finer lines and in placing more layers per unit thickness in mica boards in comparison to FR-4. To determine the fine-line capability, a few six-layer boards were made with 5 mil line spacings (2.5 mil line to pad distance and 6,000 plated-through holes). The electric continuity was good and the dimensional changes were 200 ppm and 67 ppm in the **x** and **y** directions, respectively.

4. Lithium Silicate Glass-Ceramics: Fotoceram®

Photosensitive glass-ceramics[1] are produced in the lithium silicate system by homogeneous nucleation of colloidal metal particles (silver, gold) through an optical sensitizer (cerium): $[Ag,Au]^{+} + Ce^{3+} \xrightarrow{uv} Ce^{4+} + [Ag,Au]$. Dendritic lithium metasilicate (Li_2SiO_3) is then selectively nucleated by the metal colloids on heat treatment. This metastable crystal is far more easily etched in hydrofluoric acid than is the parent glass, thereby allowing an irradiated pattern to be selectively etched. The resulting photo-etched glass can then be flood exposed to u.v. and heat treated beyond the temperature region of the metastable Li_2SiO_3 phase, producing the stable crystal, lithium disilicate, Li_2SiO_5. The resulting glass-ceramic is strong, with abraded flexural strength of 150 MPa, and it negatively replicates the original photoetched pattern to a resolution of 5 μm with up to 50,000 holes/in^2.

This material is easily metallized, and has been used as a dielectric substrate in pin-grid arrays and other electronic packages. The dielectric constant and dissipation factor at 150°C are 6.0 and 0.008 respectively, while the log volume resistivity at 250°C is 8.8 (see Table 2).

5. Boron Phosphate Glass-Ceramics

It was discovered by MacDowell[14] that extremely fine-grained transparent glass-ceramics containing BPO_4 can be obtained by heat-treating B_2O_3-P_2O_5-SiO_2 glasses having molar ratios between 1:1:1 and 1:1:3. He found that, although the parent glasses are very poorly durable in water, acids, or bases, homogeneous nucleation and crystallization of boron phosphate from these glasses results in a very durable glass-ceramic with a residual glass very high in SiO_2.

Internal nucleation (I.N.) efficiency is very high in this ternary system. The fact that no distinct phase separation step with the formation of individual nuclei has been observed led the authors to conclude that homogeneous nucleation is the dominant process leading to crystallization in the BPO_4-SiO_2 system. Figures 5a - 5d are Pt pre-shadowed replica electron micrographs of a $2B_2O_3 \cdot 2P_2O_5 \cdot 3SiO_2$ glass during heating cycles ranging from no treatment to 1100°C for one hour. Boron phosphate crystallite diameters are shown to vary from less than 0.05 microns after a 900°C firing to greater than one micron at 1100°C for 2 hours.

Properties

Thermal expansivity of the boron phosphate glass-ceramics is low to moderate - between 45 and 55 x 10^{-7}/°C. As expected, compositions lowest in SiO_2 have the highest coefficients. The BPO_4 crystal has an expansion of about 75 x 10^{-7}/°C, as estimated by extrapolating a straight line relating T.C.E. to % crystallinity (via S.E.M.) to 100% BPO_4.

Perhaps the most unique feature of the boron phosphate glass-ceramics - and the key to their commercial application - is their dielectric behavior. With no effort to exclude alkali impurities, the D.C. resistivity of a B_2O_3-P_2O_5-$2SiO_2$ (900°C - 2 hr.) glass-ceramic is 10^{16} ohm-cm at 250°C - higher than 99.9% Al_2O_3! Loss tangents are generally below 10^{-3} above 1 KHz to 200°C.

Dielectric constants for these I.N. glass-ceramics are between 3.8 and 4.5, depending primarily upon the SiO_2 level, and are nearly constant as a function of temperature and frequency.

Boron phosphate glass-ceramics can also be obtained through sintering and surface crystallization (external nucleation) of glass powder compacts.[15] By increasing the SiO_2 level beyond 60 mole % and adding excess B_2O_3 to promote viscous flow, BPO_4 crystallization is inhibited sufficiently to allow for viscous consolidation of the glass powder compact. Although still nucleated internally, these compositions are lower in crystallinity and higher in SiO_2 glass, T.C.E. is about 40 x 10^{-7} /°C (a good match for silicon) and dielectric constant falls between 3.5 and 4. Using the tape-casting process, dense multilayer substrates have been obtained after firing below 1000°C for one hour. Although the modulus of rupture of the BPO_4 glass-ceramic tape is over 15,000 psi less than that of Al_2O_3, it still may be acceptable with some process modifications. Metallization using gold, silver, or copper appears feasible.

6. Hydrogen Microfoams (Gas-Ceramic)

A popular approach to achieving ultra-low dielectric constants in conventional ceramic dielectric materials is to produce a mixture of ceramics and air - commonly known as porous ceramics. However, open porosity (with high permeability) invites all manner of contamination and instabilities, especially with respect to moisture.

The authors of this paper recently discovered several glass-forming systems that nucleate spherical bubbles - in addition to crystals - to form glass-ceramic foams appropriately called "gas-ceramics".[16] These low density materials contain foam cells that are completely isolated from one another and are thus virtually impermeable to moisture and most other contaminants. We will describe herein the most promising of these glass-derived I.N. foam glass-ceramic systems.

Within the pseudobinary system BPO_4-SiO_2, already known to generate BPO_4 glass-ceramics with excellent dielectric properties, "gas-ceramics" can form when the glass batches are melted under reducing conditions, cooled, and subsequently fired to 800-1000°C. This can occur when $NH_4H_2PO_4$ or $(NH_4)_2HPO_4$ are used as a source of P_2O_5, or other reducing agents such as organics or metals are added to the glass batch in amounts up to 10 wt. % before melting.

Figure 6 is an S.E.M. of a gas-ceramic of molar composition $2B_2O_3 \cdot P_2O_5 \cdot 6SiO_2$ after firing to 1000°C, showing the closed spherical cells of the foam structure. Especially significant for use in electronic packaging is the ~100Å thick, dense glassy surface layer obvious in Figure 6a that always forms during firing of a gas-ceramic. If this skin can be kept below about 0.5 microns thick, the advantages of the open cells below the skin in preventing cross-talk between the circuit elements and in increasing the signal speed can be largely realized. The smooth glassy skin is obviously of great utility in the printing of high resolution circuit patterns that can optimize microminiaturization of the packaging system.

All foam compositions appear to generate hydrogen gas within the cells during firing. We have not been able to establish with certainty the mechanism of foam formation, but two hypotheses are currently in favor. The simplest explanation involves the breakdown of NH_4^+ ions to N_2, H_2, and a proton. The large nitrogen molecule seems to escape during the melting process, but the hydrogen molecule, about the size of a K^+ ion, is retained upon cooling - perhaps within the structural cages of the glass structure. The hydrogen could conceivably be released upon reheating of the glass as a result of changes in boron coordination which tend to densify the glass structure and shrink or eliminate the structural cages. Thus, the hydrogen may be forced to exsolve from the borophosphosilicate glass because 1) it cannot enter the more dense crystal which is forming at the expense of the glass and/or, 2) the structure of the glass itself constricts and ejects the H_2 during the boron coordination change (3 to 4) upon reheating.

Perhaps the most likely mechanism for hydrogen generation, however, involves the reduction of phosphorus from P^{+5} to P^{+3} during the melting process. During reheating of the glass, P^{+3} could oxidize back to P^{+5}, and the reduction of H_2 (from the OH^- in the glass) could occur as follows:

$$PO_3^{-3} + 2OH^- \rightarrow PO_4^{-3} + O^= + H_2$$

The importance of phosphorus in the foaming process is underscored by results from cross-polarization MASNMR spectroscopy, which indicate that hydrogen in reduced B_2O_3-P_2O_5-SiO_2 glasses is exclusively associated with phosphorus.[17] Since the phosphite redox reaction need only affect about 1% of the phosphorous present in a typical reduced glass to exsolve the observed amount of hydrogen, a very small concentration of PO_3^{-3} could produce a microfoam via this mechanism.

Retaining the excellent dielectric properties of the BPO_4 glass-ceramics, the hydrogen microfoams possess very high D.C. resistivity ($>10^{16}$ at 250°C) and low dissipation factor (<0.001). Dielectric constant of the foams obeys the law of mixtures and has been measured as low as 1.97 on a foam with a bulk density of 0.57 g/cc.

The cell diameter of the BPO_4-SiO_2 microfoams averages about 10 micrometers, but can range from less than one to greater than 100 microns, depending upon composition and firing schedule. Bulk densities below 0.5 g/cc have not been obtained - possibly limited by gas bubble nucleation efficiency in this system.

In spite of their enviable dielectric properties, two difficult problems must be overcome before the microfoams can be utilized for microelectronic packaging. First, the mechanical strength of the low density foams must either be improved or the packaging process adapted to very weak materials (perhaps through a coating technique). Though the foams crush locally, they are not subject to the catastrophic cracking failures so characteristic of glassy materials. A nail was driven through one of our lab samples and the piece stayed intact. Secondly, control of the foaming process in a bulk glass to produce a substrate of precise dimensions is difficult. The use of molds to control final dimensions is possible, but if a ceramic powder process (via external or surface nucleation of glass particulate) could be used, control would be much easier.

Table 4 summarizes the properties of the I.N. hydrogen microfoams.

Summary and Conclusions

Glass-ceramics can be formed via crystallization of bulk glass or via sintering and crystallization of a glass powder compact. Their principal attraction for microelectronic packaging is reproducibility and predictability of structure and properties. Glass-ceramics generally have superior properties, both as compared to their parent glasses, and as compared to a ceramic with the same phase assemblage produced via solid state sintering.

Five dielectric glass-ceramic systems have been described in this paper that offer several advantages over alumina in high performance microelectronic packaging.

i. Cordierite ($2MgO \cdot 2Al_2O_3 \cdot 5SiO_2$) solid solution glass-ceramics seem most likely to immediately supersede Al_2O_3 ceramics as the leading ceramic packaging material.

ii. Fluormica sheet-silicate glass-ceramics can be solubilized and processed via a unique sol-gel process to produce mica composites with excellent dielectric properties.

iii. Lithium disilicate glass-ceramics using the I.N. process can be photochemically machined to produce grooves and vias with minimum taper and near-perfect registration.

iv. Boron phosphate I.N. glass-ceramics have the best dielectric properties yet observed in a glass-derived ceramic. Composition modification can also yield glass powders that can be processed via tape-casting to form multilayer glass-ceramic sheet.

v. Hydrogen microfoams or "gas-ceramics" have densities as low as 0.5 g/cc, and dielectric constants below 2.0 have been achieved. A thin, smooth, glassy skin forms during foaming, allowing precise application of complex circuitry.

An important aspect of these glass-derived ceramics is their process versatility. Processing can be broadly tailored for compatibility with other key elements of advanced packaging systems such as copper metallization and cofiring to produce complex three-dimensional multilayer arrays. Although the inherently low dielectric constant of glass-ceramics may appear to offer only modest gains in miniaturizability, it is a powerful force when combined with processes that can produce large packages with fine surface finish and dimensional precision.

Acknowledgment

The authors would like to thank members of the corporate analytical and melting services groups for their valuable assistance in the fabrication and measurement of the recently developed glass-ceramics mentioned herein.

Thanks are also due two technologists - R. L. Andrus and J. E. Pierson - for actually accomplishing most of the experimental work in connection with these materials.

References

1. S. D. Stookey, Ind. Eng. Chem., 51 [7] 805 (1959)

2. G. H. Beall and J. F. MacDowell, "Fine Ceramics from Glass", CHEMTECH, [18], pp. 673-679 Am. Chem. Soc. (1988)

3. G. H. Beall, "Glass-Ceramics" in Advances in Ceramics, Vol. 18, Edited by J. F. MacDowell and D. C. Boyd, American Ceramic Society (1986)

4. D. M. Mattox et al, "Low Dielectric Constant, Alumina-Compatible, Co-Fired Multilayer Substrate", Ceram. Eng. Sci. Proc., 9 [11-12] pp. 1567-1578 (1988)

5. Y. Shimada et al, "Low Firing Temperature Multilayer Glass-Ceramic Substrate," I.E.E.E. Trans. CHMT [6] 382 (1983)

6. P. W. McMillan et al, "Development of the Alpha-Cordierite Phase in Glass-Ceramics for Use in Electronic Devices", Soc. Glass Tech., 6, [26], pp. 286-292 (Dec. 1985)

7. G. B. Carrier et al, "New TCE-Matched Glass-Ceramic Multichip Module", Proc. of the 39th Elec. Comp. Conf., Houston, TX (1989)

8. L. M. Holleran et al, "Glass-Ceramics for Electronic Packaging", European Patent No. 289222, Nov. 2, 1988.

9. D. L. Evans et al, "Thermal Expansions and Chemical Modifications of Cordierite", Am. Ceram. Soc. Bull, 63, 9-10, pp. 629-634 (1980).

10. S. N. Hoda and G. H. Beall, "Alkaline Earth Mica Glass-Ceramics", in Advances in Ceramics 4, pp. 287-300, Am. Ceram. Soc. (1982)

11. W. Vogel and W. Holand, "Nucleation and Crystallization Kinetics of an $MgO-Al_2O_3-SiO_2$ Base Glass with Various Dopants", in Advances in Ceramics 4, pp. 125-145, Am. Ceram. Soc. (1982)

12. G. H. Beall et al, "Inorganic Gels and Ceramic Papers, Films, Fibers, Boards, and Coatings Made Therefrom", U.S. Patent 4,239,519, issued Dec. 16, 1980.

13. S. N. Hoda, "Ceramic Mica-Reinforced Dimensionally Stable Laminates, PC FAB, pp. 109-112 (1986)

14. J. F. MacDowell, "Boron Phosphate Glass-Ceramics", Proc. XV International Congress on Glass, Edited by O. V. Mazurin, Vol 3a pp. 90-95 (1989)

15. J. F. MacDowell and R. J. Paisley, "Glass-Ceramic Substrate for Electronic Packaging", U.S. Patent No. 4,833,104, Issued May 23, 1989.

16. G. H. Beall and J. F. MacDowell, "Hydrogen-Containing Glass Microfoams and Gas-Ceramics", U.S. Patent No. 4,666,867, Issued May 19, 1987.

17. J. E. Dickinson et al, "Hydrogen Gas and Gas-Ceramic Microfoams: Raman, XPS, and MASNMR Results on the Structure of Precursor SiO_2-B_2O_3-P_2O_5 Glasses", Jour. of Non-Cryst. Sol., 102, pp. 196-204 (1988)

Table 1.

COMPOSITION, LATTICE PARAMETERS, AND THERMAL EXPANSION OF CORDIERITES

COMPOSITION	LATTICE PARAMETERS (25 °C) a(Å)	c(Å)	V(Å^3)	C.T.E. (22-800 °C) x 10^{-7}/°C	THERMAL EXPANSION ANISOTROPY (22-800 °C) $(\frac{\Delta a}{a} - \frac{\Delta c}{c})$ ppm
$Mg_2Al_4Si_5O_{18}$ (1000 °C-2 HRS.)	9.784	9.349	775.05	11.7	2600
$Mg_2BeAl_2Si_6O_{18}$ (1000 °C-2 HRS.)	9.609	9.276	741.67	19.2	2560
$KMg_2Al_5Si_4O_{18}$ (950 °C-16 HRS.)	9.811	9.470	789.42	20.6	4600
$CsMg_2Al_5Si_4O_{18}$ (950 °C-2 HRS.)	9.804	9.457	787.21	22.0	900

* ALL GLASSES MELTED AT 1650 °C, 16 HRS. GROUND, AND CRYSTALLIZED AT INDICATED SCHEDULES (Data after Evans et al, 1979)

Table 2.

PROPERTIES OF SHEET SILICATE GLASS-CERAMICS

	FLUORMICA			
	MACOR® 9658	LOW ALKALI 9652	ALKALINE EARTH	LI-DISILICATE FOTOCERAM® 8603
DENSITY	2.52	2.63	2.95	2.41
POROSITY	0	0	0	0
M.O.R. (Abraded) MPa	103	92	221	147
MACHINABILITY (MW/m^3•min)	4.5	--	15.0	Photomachinable
KNOOP HARDNESS	250	175	300	500
E-MODULUS GPa	65	64	--	86
THERMAL EXPANSION COEFFICIENT (25-400° C)	≈93	74	110	103
MAX. USE TEMPERATURE	900 ° C	1000 ° C	1100 ° C	750 ° C
LOG RESISTIVITY ohm-cm 400 ° C	7.9	12.2	10.2	6.5
DIELECTRIC CONSTANT 100 KHz, 21 ° C	5.9	5.7	7.2	5.6
DISSIPATION FACTOR 100 KHz, 21 ° C	0.0026	0.0005	0.0004	0.003
DIELECTRIC STRENGTH KV/mm, 21 ° C	4.0	8.5	--	15.0

Table 3.

LAMINATE PROPERTIES

PHYSICAL	CERAMIC MICA	FR-4
• FLEXURAL STRENGTH, psi	30,000	62,000
• FLEXURAL MODULUS, psi	2.7×10^6	3.1×10^6
• DENSITY, g/cm^3	1.70	1.82
• BOW & TWIST, %	.25	1
• BOILING WATER ABSORPTION (10 mil), %	.2	.9
• COPPER PEEL STRENGTH, pll	1.5-2	8-10
THERMAL		
• EXPANSION, ppm/ ° C		
- X-DIRECTION	14	12
- Y-DIRECTION	16	18
- Z-DIRECTION	45	60
• GLASS TRANSITION TEMP. ° C	140	120
• FLAMMABILITY, Sec (Self Extinguish)	3	15
ELECTRICAL		
• DIELECTRIC STRENGTH		
- PERPENDICULAR, Volts/Mil	2,000	750
- PARALLEL, KV	50	40
• DIELECTRIC CONSTANT (1 MHz)	4.5	4.5
• DISSIPATION FACTOR (1 MHz)	0.025	0.025
• VOLUME RESISTIVITY, ohm-cm	10^{15-16}	3.6×10^{13}
• SURFACE RESISTIVITY, ohm	10^{15}	2.5×10^{12}
• INSUL. RESISTANCE (50 ° C, 85% R.H.) ohm	10^{10}	10^8

Table 4. UNIQUE PROPERTIES OF HYDROGEN MICROFOAMS

LOW DENSITY	AS LOW AS 0.5 g/cc
HIGH STRENGTH/WEIGHT RATIO	2 TO 5 Ksi (Abraded) ON <1g/cc FOAMS
SMALL BUBBLE SIZE	1-100 μm
LOW DIELECTRIC CONSTANT	AS LOW AS 2 - POTENTIALLY LOWER
HIGH D.C. RESISTIVITY	10^{16} ohm • cm AT 250°C
LOW DIELECTRIC LOSS	<.001 IN NON-ALKALI COMPOSITIONS
LOW TO MODERATE THERMAL EXPANSION	0-70 x 10^{-7}/°C
LOW THERMAL CONDUCTIVITY	HEAT RETENTION/INSULATION VALUE FROM CLOSED CELL POROSITY
SMOOTH, DURABLE GLASSY SKIN	10 μm - 1mm THICK
MACHINABLE/FIELD WORKABLE	CAN BE SAWED, SCORED, DRILLED

Figure 1. FROM GLASS TO GLASS-CERAMIC

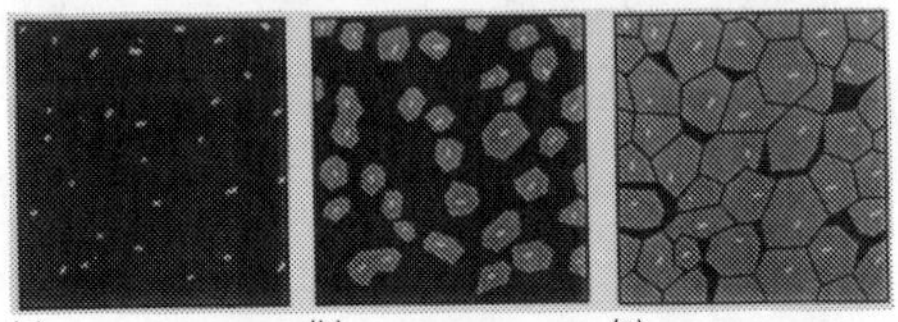

(a) NUCLEI FORMATION (b) CRYSTAL GROWTH ON NUCLEI (c) GLASS-CERAMIC MICROSTRUCTURE

Figure 2. GLASS-CERAMICS FROM POWDERED GLASS

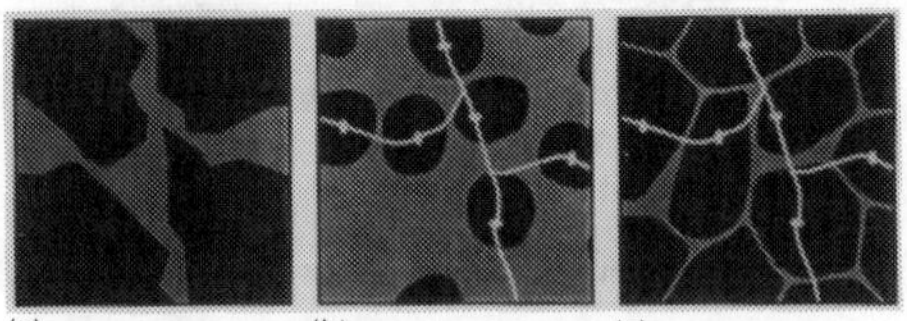

(a) POWDERED GLASS COMPACT (b) DENSIFICATION AND INCIPIENT CRYSTALLIZATION (c) FRIT-DERIVED GLASS-CERAMIC

Figure 3. PROPAGATION DELAY TIME VS DIELECTRIC CONSTANT

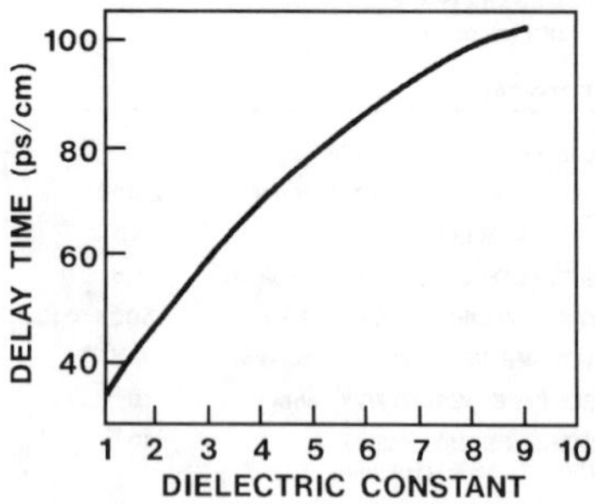

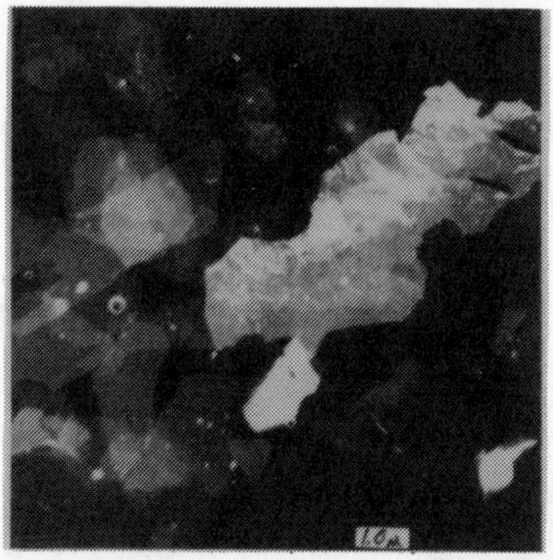

Figure 4. Transmission Electron Microphoto of Aqueous Mica Sol

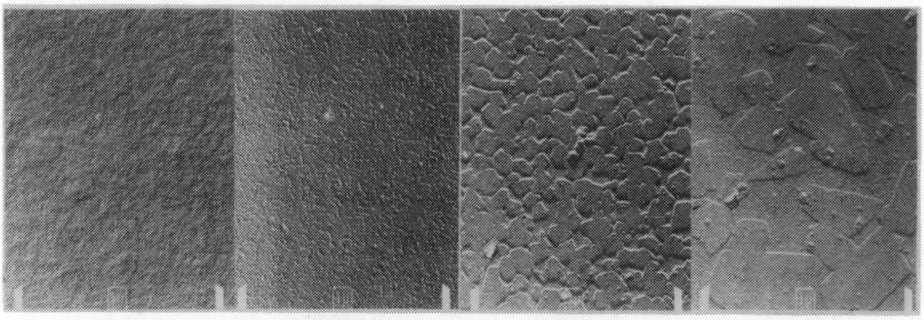

3a. untreated glass 3b. 900°C-2 hrs. 3c. 1000°C-2 hrs. 3d. 1100°C-2 hrs.

Figure 5. Crystallization of a Boron Phosphate Glass-Ceramic.

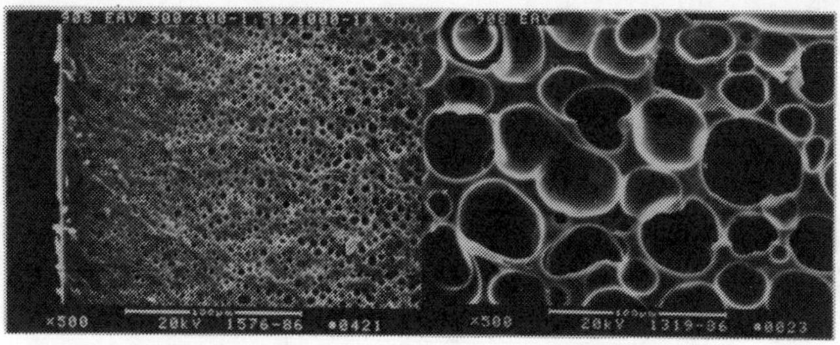

Figure 6. Fracture SEM showing range of cell sizes possible with a single gas-ceramic composition.

CERAMIC-FILLED-GLASS COMPOSITE SINTERING

Kevin G. Ewsuk, E. I. du Pont de Nemours & Co., Inc., Experimental Station, Wilmington, DE 19880-0262.

ABSTRACT

Low-temperature-sintering, ceramic-filled-glass (CFG) composites densify by a combination of glass redistribution, grain rearrangement, and viscous flow in a three-stage process described as non-reactive, liquid-phase sintering (NLPS). Pore size, filler particle size, filler concentration, and a combination of glass properties, including glass viscosity, wetting angle, and surface tension, control densification during NLPS.

The kinetics of densification during NLPS have been modeled using existing reactive liquid-phase sintering theory, and modified viscous sintering theory.

During final-stage NLPS, densification is controlled by the size of the pores in the composite microstructure, the surface tension of the glass, and the viscosity of the CFG dispersion. Sintering time increases with increasing dispersion viscosity, which increases with increasing filler concentration and/or glass viscosity. The relationship between dispersion viscosity, filler concentration, and glass viscosity in an alumina-filled, borosilicate-glass composite system has been determined using Eilers' relative viscosity model for concentrated suspensions, and experimentally derived model constants. Predicted dispersion viscosities have been used in conjunction with final-stage NLPS theory to construct sintering maps that define the critical filler concentration in a given viscosity glass below which high-density CFG composites can be produced. These sintering maps provide simple guidelines for designing and developing new, high-density, CFG composites.

INTRODUCTION

Ceramic-filled-glass (CFG) composites are becoming more widely used in microelectronic packaging as they provide materials engineers the flexibility to design and develop novel materials with the desired combination of electrical, thermal, and mechanical properties for specific applications. Additionally, CFG composites offer processing flexibility, as they can be co-fired at <1000°C with high electrical conductivity metallization (e.g., Ag and Cu).[1-4] Although low-temperature processability is a major advantage of CFG composites, processing reproducibility is a major problem. In particular, reproducibly sintering to high density (i.e., >95% theoretical density) is often a problem with composites.

Two key issues for CFG composites in microelectronic packaging are performance and reliability. To achieve the desired level of performance and reliability, high-density, hermetic microstructures must be obtained reproducibly on sintering. To achieve this reproducibility, it is necessary to understand and control CFG composite densification and microstructure development during sintering.

Low-temperature sintering, particulate, CFG composites densify by a combination of glass redistribution, grain rearrangement, and viscous flow in a three-stage process described as non-reactive* liquid-phase sintering (NLPS).[5-6] The microstructural evolution of a CFG composite during NLPS is illustrated in Figure 1.

CFG composite densification during NLPS is controlled by the size of the pores in the microstructure, the filler concentration, and a combination of glass properties, including glass viscosity, wetting angle (on the filler), and surface tension.[5-7] The driving force for material transport, and the kinetics of densification during each stage of NLPS can be defined using these parameters.

As in all sintering processes, the driving force for densification during NLPS is the reduction of the total volume free energy of the system. In a typical 55-65% dense, green powder compact composed of ceramic

---

* It is assumed that the glass (liquid) does not react with the filler to create additional liquid to aid densification during NLPS (i.e., such as solution-reprecipitation, which occurs during reactive liquid-phase sintering).

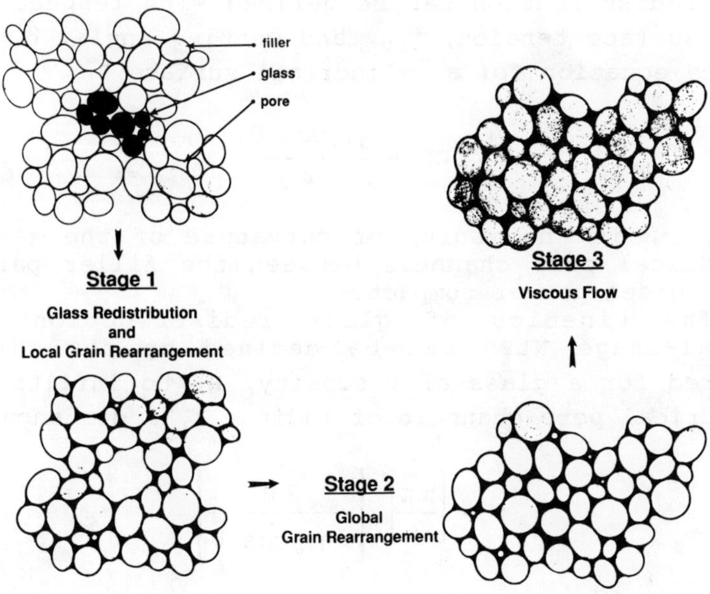

Figure 1. An illustration of CFG composite microstructure evolution during NLPS. Densification occurs by glass redistribution and global grain rearrangement during the initial and intermediate stages of NLPS, and by viscous flow during final-stage NLPS.

filler particles, glass particles, and porosity, localized ceramic-rich and glass-rich regions exist. During initial-stage NLPS, the free energy of the system decreases as lower energy solid-liquid interfaces are created at the expense of higher energy solid-solid and solid-vapor interfaces. Physically, this occurs in a CFG composite as the glass becomes fluid during heating, and infiltrates the ceramic-rich regions of the green powder compact. The redistribution of the glass to fill the capillary pore channels between the filler particles produces a dense ceramic/glass skeletal structure containing gross pores where glass-rich regions were originally present in the green powder compact. The overall volume of the composite does not change significantly during the glass redistribution process; consequently, little or no densification occurs during initial-stage NLPS.

The sintering stress or driving force, DF, for glass redistribution can be defined with respect to the glass surface tension, γ_{lv}, and wetting angle, θ, by the Laplace equation for a cylindrical surface,[8]

$$DF = \frac{\gamma_{lv}\cos\theta}{r} \tag{1}$$

where, r, is the radius of curvature of the (assumed) cylindrical pore channels between the filler particles in the green powder compact.

The kinetics of glass redistribution during initial-stage NLPS can be defined by the time, t, required for a glass of viscosity, η, to infiltrate the cylindrical pore channels of radius, r, a distance, h.[9]

$$t = \left[\frac{h\pi}{2}\right]^2 \left[\frac{2\eta}{r\gamma_{lv}\cos\theta}\right] \tag{2}$$

With a wetting glass (i.e., $\theta \leq 90°$), the infiltration rate is determined by the glass viscosity, glass surface tension, and the size of the capillary pore channels in the composite. With micron size filler particles, infiltration time decreases with decreasing viscosity and/or increasing pore size.

The coexistence of solid-vapor, liquid-vapor, and solid-liquid interfaces results in a force imbalance on the filler particles in a CFG composite body during intermediate-stage NLPS. The glass between the filler particles reduces the interparticle frictional forces, and allows the the filler particles to move relative to one another to eliminate this force imbalance. The resultant global grain rearrangement closes the gross pores created by the glass redistribution process. Intermediate-stage, global grain rearrangement occurs simultaneously with glass redistribution, and is responsible for the bulk of the densification that occurs during NLPS (i.e, from ~65 to ~90% theoretical density).

The grain rearrangement force, F, exerted on a filler particle of radius, R, can be defined in terms of the difference between the force drawing the particle into the glass, F_l, and the force drawing the particle into an adjacent pore, F_p,[10]

$$F = F_l - F_p = 2\pi\gamma_{lv} R \sin(\varphi + \theta) - 2\pi\gamma_{lv} \frac{R^2}{r} \sin\varphi \quad (3)$$

where φ is a geometric factor related to the solid-vapor-liquid intersect, and θ is the glass wetting angle.

The sintering stress, DF, during grain rearrangement can be calculated by considering the area over which the grain rearrangement force exists, which can be approximated by the solid-liquid interfacial area for a single ceramic particle. Assuming $\varphi = 90°$ this area is equal to 1/2 the surface area of a ceramic particle of radius, R, resulting in the expression,

$$DF = \frac{F}{2\pi R^2 \cos\theta} \quad (4)$$

With a good wetting glass (i.e., $\theta \leq 5°$), the maximum sintering stress during intermediate-stage grain rearrangement (i.e., for $\varphi = 90°$) is approximately the same as the sintering stress during initial-stage glass redistribution.*

The driving force for both glass infiltration and grain rearrangement disappear when all of the solid-vapor interfaces in the composite are eliminated, and only glass enclosed pores remain.

The annihilation of the residual ~10% glass enclosed porosity in a CFG composite during final-stage NLPS is controlled by viscous flow. The sintering stress during final-stage NLPS can be defined with respect to the glass surface tension, γ_{lv}, and the radius of curvature, r, of the (assumed) spherical pores by the Laplace equation for a spherical surface.[11]

$$DF = \frac{2\gamma_{lv}}{r} \quad (5)$$

---

* For micron size particles the sintering stress is ~3-4 MPa

The kinetics of densification during final-stage NLPS are defined by the kinetics of viscous flow in the composite system. For a pure glass (i.e., without filler), the time, t, required to eliminate a spherical pore of radius, r, during final-stage viscous sintering is directly proportional to the viscosity, η, of the glass.[12]

$$t = \frac{2 r \eta}{\gamma_{lv}} \tag{6}$$

However, in a CFG composite, the viscosity of the CFG dispersion (i.e., the suspension of ceramic particles in the fluid glass) must be considered in modeling the kinetics of densification during final-stage sintering. Substituting dispersion viscosity, η_s, for the glass viscosity, η, in Equation 6 yields the appropriate expression for final-stage NLPS.[6,7]

$$t = \frac{2 r \eta_s}{\gamma_{lv}} \tag{7}$$

The relationship between the relative viscosity of a CFG dispersion, η_r, glass viscosity, η, and the volume fraction of filler in the composite, ϕ, can be approximated using Eilers' relative viscosity model for concentrated suspensions,[7, 13-14]

$$\eta_r = \frac{\eta_s}{\eta} = \left(1 + \frac{\kappa \phi}{1 - \left(\phi / \phi_{max}\right)}\right)^2 \tag{8}$$

where κ is a constant related to the hydrodynamic radius of the filler particles, and ϕ_{max} is the maximum fractional packing density of the filler particles.[14]

Eilers' relative viscosity model predicts that the viscosity of a CFG dispersion increases with increasing glass viscosity and/or the volume fraction of filler; however, the exact relationship between dispersion viscosity, glass viscosity, and volume fraction filler can only be defined with the model constants κ and ϕ_{max} for a CFG composite.

The densification that occurs during final-stage NLPS ultimately determines the microstructure and properties of a CFG composite; consequently, it is particularly important to understand and control final-stage NLPS. Because final-stage sintering time is directly proportional to the viscosity of the CFG dispersion, to control and optimize densification during final-stage NLPS, it is necessary to understand the relationship between dispersion viscosity, glass viscosity, and filler concentration in a CFG composite.

The objectives of this study were to determine the relationship between dispersion viscosity, glass viscosity, and ceramic filler concentration in CFG composites, and to use this relationship to develop design and process guidelines to produce reliable, high-density, CFG composites.

EXPERIMENTAL

Composite powder samples were produced by dispersing 10-50% by volume, 1.5 µm median particle size alumina* and powdered borosilicate glass** in isopropyl alcohol in a Spex® mill for 5 min. The mixture was dried and ground, and 3-5 g powder samples were hot pressed in a 1.27 cm diameter graphite die at 675-725°C and 69 MPa for 30 min to produce 99-100% dense composite samples for viscosity measurements.

Alumina-filled, borosilicate-glass dispersion viscosity was determined as a function of filler concentration by conducting isothermal, beam bend viscosity measurements on the hot-pressed composites. Viscosity measurements were conducted using a thermomechanical analyzer*** (TMA), and a technique described by Jewell and Shelby.[15] The TMA beam bend apparatus used to determine CFG dispersion viscosity is illustrated in Figure 2.

Viscosity measurements were conducted on 1 mm by 3 mm by 11 mm bars in air at 750-770°C under a

---

* A-152, Alcoa, Pittsburgh, PA

** ES-1, Owens Illinois, Inc., Toledo, OH

*** 943 Thermomechanical Analyzer, E. I. du Pont de Nemours & Co., Inc., Wilmington, DE

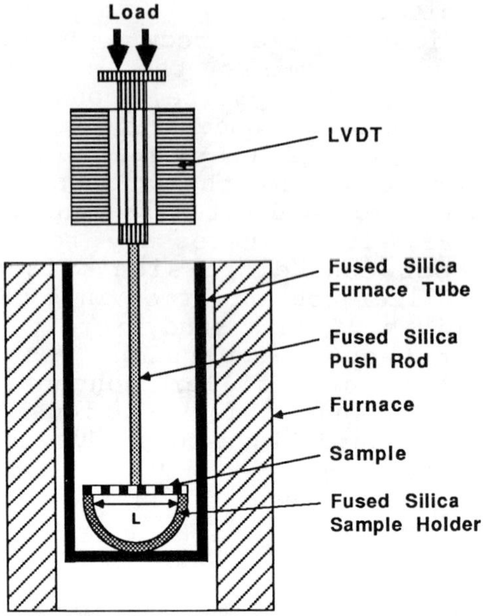

Figure 2. The TMA beam bend apparatus used to conduct the CFG dispersion viscosity measurements.

uniaxially applied load of 5-10 g. The composites were heated at 20°C/min to 20°C below the soak temperature, 10°C/min to 10°C below the soak temperature, and then 5°C/min to the soak temperature. The dispersion viscosity, η_s, of a CFG composite of density, ρ, was then determined from the measured linear deformation rate, v, of the CFG composite under an applied load, M, during the 1 hr soak, using the expression,[16]

$$\eta_s = \frac{g L^3}{2.4 I_c v} \left(M + \frac{\rho A L}{1.6} \right) \tag{9}$$

where, g, is the gravitational constant, L, is the sample span (L = 7.8 mm), and I_c and A, are the moment of inertia and cross sectional area of the sample, respectively.

The precision and accuracy of the TMA beam bend viscosity measuring technique were verified by viscosity results obtained from measurements conducted on a glass viscosity standard.* The experimental error in determining viscosity of the glass standard using the TMA technique is $\leq 10^{0.1}$ Pa·s; however, due to localized variations in filler concentration, the presence of porosity, and the possibility of devitrification, the experimental error in determining the viscosity of a CFG dispersion may be larger.

RESULTS AND DISCUSSION

The results of the TMA beam bend viscosity measurements made on the hot pressed, alumina-filled, borosilicate-glass composites are presented along with the Eilers' relative viscosity model fit to the data in Figure 3. The relative viscosity of the alumina-filled

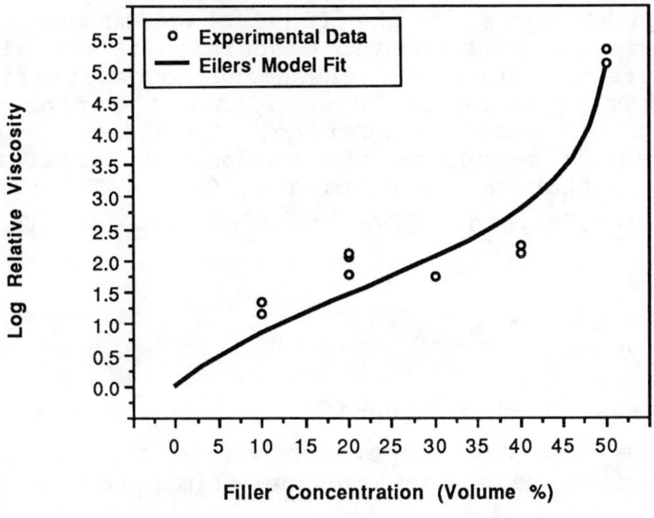

Figure 3. The measured change in relative viscosity of a borosilicate glass as function of alumina-filler concentration, and the Eilers' relative viscosity model nonlinear least squares best fit to the data generated using model constants $\kappa = 13$, and $\phi_{max} = 0.51$.

---

* NBS-711, National Institute of Standards and Technology, Gaithersburg, MD

borosilicate glass dispersion continuously increases with increasing filler concentration up to a maximum of ~$10^{5.1}$ at 50% (by volume).

The Eilers' model curve in Figure 3 was generated using model constants κ = 13, and ϕ_{max} = 0.51, which were determined using a nonlinear, least sum of squares error fit to the experimental data. A constant $\kappa \leq 1.25$ is expected for a Newtonian suspension; however, when a layer of liquid absorbs (sic) onto the surface of a filler particle, it increases its hydrodynamic radius, and the κ value increases.[13] It is hypothesized that, in the alumina-filled, borosilicate-glass dispersions examined in this study, there is some chemical interaction between the glass and alumina at 750-770°C that increases the hydrodynamic radius of the filler particles to produce the higher than expected κ value derived from the experimental data.

During final-stage NLPS, the rate of CFG composite densification is predominantly controlled by the size of the residual pores in the final-stage microstructure, and the viscosity of the CFG dispersion at the sintering temperature (Equation 7). Assuming no devitrification, and no gas pressure build-up within the closed pores during final-stage sintering, Equation 7 can be rearranged to calculate the maximum or critical size pore, r_{crit}, that can be eliminated from a CFG dispersion of viscosity, η_s, in a given final-stage sintering time.

$$r_{crit} = \frac{t\,\gamma_{lv}}{2\,\eta_s} \tag{10}$$

By rearranging Equation 10, it is possible to define the critical dispersion viscosity, $\eta_{s_{crit}}$, below which a given size pore can be eliminated in a given time.

$$\eta_{s_{crit}} = \frac{t\,\gamma_{lv}}{2\,r} \tag{11}$$

By rearranging Eilers' viscosity model for concentrated suspensions (Equation 8), critical dispersion viscosity can also be defined relative to the viscosity of the glass, and filler concentration.

$$\eta_{s_{crit}} = \eta \left(1 + \frac{\kappa\phi}{1 - \left(\phi/\phi_{max}\right)}\right)^2 \tag{12}$$

By combining final-stage NLPS theory (Equation 11) and Eilers' viscosity model (Equation 12) and rearranging, it is possible to define the maximum or critical filler concentration, ϕ_{crit}, in any glass of known viscosity and surface tension below which a given size pore can be eliminated in a given time.

$$\phi_{crit} = \frac{\left(\sqrt{\eta_s/\eta} - 1\right)\phi_{max}}{\left(\sqrt{\eta_s/\eta} - 1\right) + \kappa\phi_{max}} \tag{13}$$

By plotting the model relationship between filler concentration and glass viscosity for a given critical dispersion viscosity,* final-stage, NLPS maps can be constructed (Figure 4.). These sintering maps provide simple design and process quidelines for producing high density, CFG composites. For example, as illustrated in the final-stage sintering map in Figure 4., the filler concentration must be $\leq 22\%$ to eliminate a 10 μm radius pore from a 10^6 Pa·s viscosity glass with surface tension of 0.4 J/m^2 in 30 min. With a higher viscosity glass, the filler concentration must decrease to eliminate the same size pore in the same time. Likewise, to produce a high density CFG composite with a higher filler concentration, a lower viscosity glass must be used.

By considering the effects of changing pore size, sintering time, and/or glass surface tension, a series of critical dispersion viscosity curves can be generated, and more detailed final-stage NLPS maps can

---

* A filler concentration/glass viscosity curve is generated for a specific critical dispersion viscosity, as defined by Equation 11 for a given pore size, sintering time, and glass surface tension.

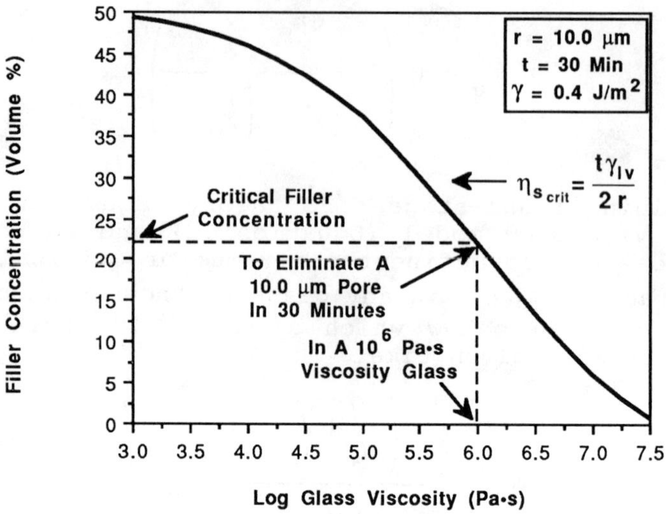

Figure 4. A final-stage NLPS map that defines the relationship between critical filler concentration and glass viscosity in a CFG composite comprised of a 0.4 J/m^2 surface tension glass. For any glass viscosity ranging from 10^3 to 10$^{7.5}$ Pa·s, the filler concentration/glass viscosity curve defines the critical filler concentration below which a 10 μm pore can be eliminated in 30 min.

be constructed. Figure 5 shows the relationship between filler concentration and glass viscosity as a function of pore size for a sintering time of 30 min, and a glass surface tension of 0.4 J/m^2. For a given glass viscosity, critical filler concentration decreases with increasing pore size, and the effect becomes more pronounced with increasing glass viscosity.

For a given glass viscosity, critical filler concentration decreases with decreasing sintering time and/or decreasing glass surface tension (Figures 6-7); however, these effects only become significant at the higher glass viscosities. In comparison to the effects of changing pore size or glass viscosity, major changes in the sintering time or glass surface tension have a relatively minor effect on the critical filler concentration.

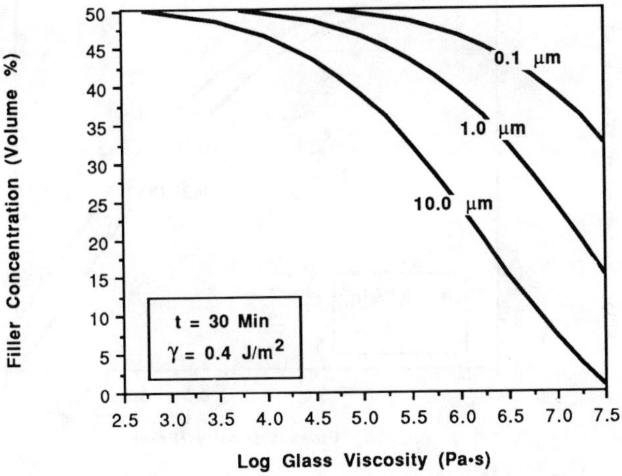

Figure 5. A final-stage NLPS map that defines the critical filler concentration in a CFG composite as a function of glass viscosity and pore size, for a sintering time of 30 min, and a glass surface tension of 0.4 J/m^2.

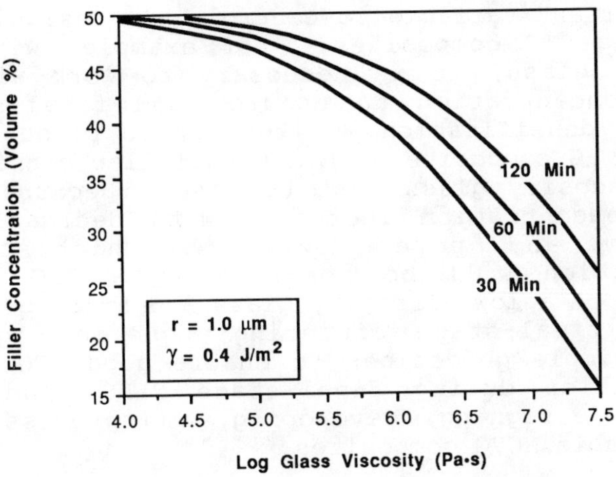

Figure 6. A final-stage NLPS map that defines the critical filler concentration as a function of glass viscosity and sintering time to eliminate 1 μm pores from a CFG composite comprised of a 0.4 J/m^2 surface tension glass.

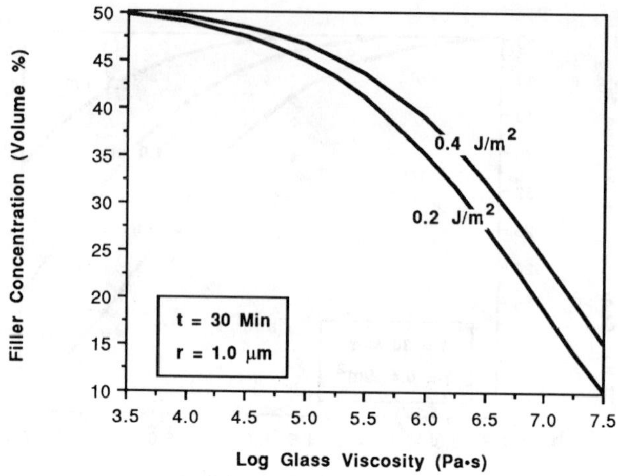

Figure 7. A final-stage NLPS map that defines the critical filler concentration as a function of glass viscosity and glass surface tension to eliminate 1 μm pores from a CFG composite in 30 min.

Glass viscosity has the most pronounced effect on critical filler concentration, and is consequently, the most important variable to control when designing and/or processing CFG composites. For example, with a high viscosity glass, it is necessary to work with a low filler concentration to ensure good final-stage CFG composite densification. Likewise, to produce a high-density, CFG composite with a high filler concentration, a low viscosity glass must be used. Because critical filler concentration increases with decreasing glass viscosity and pore size, the highest filler concentration will be possible in a CFG composite comprised of a low viscosity glass and small pores.

The final-stage sintering maps in Figures 4-7 provide simple guidelines to ensure good CFG composite densification during final-stage NLPS, and will be useful in designing, developing, and processing, new, more reliable, CFG composites.

CONCLUSIONS

Low-temperature-sintering CFG composites densify by a combination of glass redistribution, grain rearrangement, and viscous flow in a three-stage process described as non-reactive, liquid-phase sintering (NLPS). Densification during NLPS is controlled by the size of the pores in the composite microstructure, filler particle size, filler concentration, and a combination of glass properties, including glass viscosity, wetting angle, and surface tension. The kinetics of densification during NLPS have been modeled using existing reactive liquid-phase sintering theory, and modified viscous sintering theory.

A combination of pore size, glass surface tension, and CFG dispersion viscosity control densification during final-stage NLPS. Using experimentally derived model constants and Eilers' relative viscosity model for concentrated suspensions, the relationship between dispersion viscosity, filler concentration, and glass viscosity in an alumina-filled, borosilicate-glass composite system has been determined. In conjunction with final-stage NLPS theory, Eilers' model has been used to construct final-stage sintering maps that define the critical filler concentration in a given viscosity glass, below which high-density, CFG composites can be produced. These sintering maps provide simple guidelines to ensure good CFG composite densification during final-stage NLPS, and will be useful in designing, developing, and processing, new, more reliable, CFG composites.

Glass viscosity has the most pronounced effect on critical filler concentration and CFG composite densification, and is the critical variable that must be controlled when designing and processing CFG composites. With a high viscosity glass, low filler concentrations must be used to ensure good final-stage densification. To produce high-density, CFG composites with a high filler concentration, low viscosity glasses must be used. The highest filler concentration will be possible in a CFG composite comprised of a low viscosity glass and small pores.

REFERENCES

1) Y. Shimada, K. Utsumi, M. Suzuki, H. Takamizawa, M. Nitta, and S. Yano, "Low Firing Temperature Multilayer Glass-Ceramic Substrate," pp. 314-319 in The Proc. 33rd Electron. Compon. Conf. The Institute of Electrical and Electronics Engineers, Inc., New York, NY, 1983.
2) W. A. Vitriol and J. I. Steinberg, "Development of Low Temperature Cofire Multilayer Ceramic Technology," *Int. J. Hybrid Microelectron.*, **6** [1] 593-598 (1983).
3) S. Nishigaki, S. Yano, J. Fukuta, M. Fukaya, and T. Fuwa, "A New Multilayered, Low-Temperature-Fireable Ceramic Substrate," pp. 225-234 in The Proc. 1985 Int. Symp. Microelectron. The International Society for Hybrid Microelectronics, Reston, VA, 1985.
4) J. I. Steinberg, S. J. Horowitz, and R. J. Bacher, "Low Temperature Tape Dielectric Material Systems for Multilayer Interconnections," *Solid State Technol.*, 97-101 (1986).
5) K. G. Ewsuk, L. W. Harrison, and F. J. Walczak, "Sintering of Glass-Filled Ceramic Composites; Effects of Glass Properties," pp. 969-977 in Ceramic Powder Science, Part B. Edited by G. L. Messing, E. R. Fuller, and H. Hausner. The American Ceramic Society, Inc., Westerville, OH, 1988.
6) K. G. Ewsuk, L. W. Harrison, "Densification of Glass-Filled, Alumina Composites," to be published in Sintering of Advanced Ceramics. Edited by C. Handwerker, J. Blendell, and W. A. Kaysser. The American Ceramic Society, Inc., Westerville, OH.
7) K. G. Ewsuk, "Effect of Filler Loading on Final-Stage Viscous Sintering of Ceramic-Filled-Glass Composites," pp. 747-754 in Ceramic Powder Processing Science. Edited by H. Hausner, G. L. Messing, and S. Hirano. Deutsche Keramische Gesellschaft, Köln, FRG, 1989.
8) D. Dunstan and L. R. White, "A Capillary Pressure Method for Measurement of Contact Angles in Powders and Porous Media," *J. Colloid and Interface Sci.*, **11** [1] 60-64 (1986).
9) K. A. Semlak and F. N. Rhines, "The Rate of Infiltration of Metals," *Trans. Metall. Soc. AIME*, **212** 325-331 June (1958).

10) V. Smolej, S. Pejovnik, and W. A. Kaysser, "Rearrangement During Liquid Phase Sintering of Large Particles," *Powder Metall. Int.*, **14** [1] 34-36 (1982).
11) P. C. Hiemenz, Principles of Colloid and Surface Chemistry, pp. 209-251, Marcel Dekker, Inc., New York, NY, 1977.
12) A. J. Shaler, "Seminar on Kinetics of Sintering" (with discussion), *J. Met.*, **1** [11] 796-813 (1949).
13) R. F. Fedors, "Relationships Between Viscosity and Concentration for Newtonian Suspensions," *J. Colloid and Interface Sci.*, **46** [3] 545-547 (1974).
14) H. Eilers, "The Viscosity of Emulsions of Highly Viscous Materials as a Function of Concentration," *Kolloid-ZZ Polym.*, **97** 313-321 (1941).
15) J. M. Jewell and J. E. Shelby, "Transformation Range Viscosity Measurements Using a Thermomechanical Analyzer," *J. Am. Ceram. Soc.*, **72** [7] 1265-67 (1989).
16) F. T. Trouton, "Coefficient of Viscous Traction and Its Relation to that of Viscosity," *Proc. R. Soc. London, A*, **77** 426-40 (1906).
17) Hutchins, J. R., and Harrington, R. V., "Glass," pp. 533-604 in Encyclopedia of Chemical Technology, 2nd Edition, **Vol 10**. John Wiley & Sons, Inc., NY, 1966.

ACKNOWLEDGEMENTS

F. J. Walczak and L. W. Harrison prepared the composite samples for this study, and L. W. Harrison optimized the technique for and measured the CFG dispersion viscosities. W. H. Morrison and S. J. Glass provided valuable comments that contributed toward the completion of this manuscript.

A STUDY OF LASER EFFECTED THIN SILICON LAYER RECRYSTALLIZATION

by

C. P. Grigoropoulos and A. F. Emery
Department of Mechanical Engineering
University of Washington
Seattle, Washington 98195

Abstract

Laser melting and recrystallization of thin silicon semiconductor films which are deposited on amorphous substrates is capable of producing improved electrical and crystalline properties. Typically the silicon layer is traversed across a laser beam and a molten zone is produced. The process is controlled by the intensity of the laser beam, the material translation speed, the thermal and radiative properties, and the boundary conditions of the materials. This paper describes experiments using a CW argon ion laser on 0.5 micron-thick encapsulated films. Numerical calculations are presented and compared to the experimental results. These calculations assist in defining the optimal processing parameters.

Introduction

Recrystallization of thin semiconductor films which are deposited on amorphous substrates through the use of lasers has been shown to increase their efficiency in commercial VLSI applications[1]. The recrystallization is effected by shining a laser source through the substrate, Figure 1. The material is translated and recrystallizes in the wake of the molten pool. Although significant progress has been made in applying this technique, the formation of small tilt angle misalignments of the grain boundaries is detrimental to the use of the recrystallized material in electronic devices[2]. Experimental studies[3] suggest that the crystal growth is driven by the thermal gradient distribution across the moving phase change boundary and that improvements can be achieved by modifying the laser beam shape. Further improvements are likely to be dependent upon accurately modeling the thermal history of the material and the interaction between the laser beam characteristics and the temperature field produced in the thin silicon layer.

Besides leading to improved results, such thermal modeling can be used to study how the efficiency of the process depends upon different material and process parameters. Some questions which are currently unanswered are the effect of translation speed, the effect of thermal losses to the underlying substrate, the relationship between the pool size and the beam size, and the effect of the shape of the beam.

This paper describes a series of experiments from which basic information about the process has been extracted. In addition a numerical simulation of the process is presented by which details of the thermal history can be resolved. Finally, the simulation is used to provide some answers to the questions listed above.

Experimental Procedure

A sketch of the experimental apparatus is shown in Figure 1. A Spectra Physics 164-08 4W Argon laser is used as the annealing source. This laser operates in the fundamental mode, TEM_{00}, with a Gaussian intensity distribution. The multiline laser beam is formed by $n_w = 6$ distinct wavelength components, with most of the power centered around the $\lambda = 488$ and 514.5 nm wavelengths. The beam is focused to a circular spot on the silicon sample which is mounted on a precision translation state by a spherical lens of 10cm focal length. Elliptically shaped spots can be generated by a combination of a spherical lens of 7 cm focal length and a cylindrical lens of 8 cm focal length. The translation stage is driven by a piezoelectric encoder motor which is interfaced to a computer. The repeatability of the positioning is 1 μm and the maximum speed is 2mm/sec. The samples are 0.5μm-thick polysilicon deposited on bulk fused silica wafers of 7.6 cm diameter and 0.5mm thickness. The silicon layer is encapsulated by a 0.5μm-thick SiO_2 layer. This encapsulating layer prevents the generation of thermocapillary driven flows and mass transport in the molten silicon pool.

The laser beam passes through the transparent substrate and is partially absorbed in the thin layer. The transmitted beam is observed by a microscope placed directly above the sample and the observations are recorded on photographic film and video tapes. The laser beam dimensions are measured by placing a variable frequency rotating blade chopper across the filtered monochromatic line of the laser beam and focussing the chopped light on a silicon photodiode to give a waveform profile that is captured on a 11402 Tektronix digitizing storage oscilloscope. A least squares analysis is used to fit the Gaussian profile using the relationship

$$P(x_c)/P_T = 0.5\ erfc(x_c/w_{\lambda b}) \quad (1)$$

where x_c is the distance of the chopping blade from the laser beam center and P is the power measured by the detector. The beam is then characterized by the total power, P_T and the $1/e$ irradiance radius, $w_{\lambda b}$ for the 488 nm line. The corresponding irradience distribution is given by

$$Q_{\lambda b}(x, y) = Q^o_{\lambda b}\ exp[-(r/w_{\lambda b})^2] \quad (2)$$

In the above expression r is the radial distance from the center of the beam and $Q^o_{\lambda b}$ is the peak irradiance intensity. Figures 2a and 2b show a schematic of the sample and identify some of the dimensions used to characterize the process.

Experimental Results

In the experiments, three different power levels and four material translation speeds were applied. Use of the spherical lens focussing system yielded nearly circular molten pools. With the cylindrical lens, the laser beam was expanded in the transverse to the motion direction and elliptical pools of varying aspect ratio, up to about 10:1, were obtained. Measurements showed that the molten pool size is increased as the power is increased and that it is nearly insensitive to speed in the range of 0 to 2 mm/sec. Speeds of this low range have been found to be effective in melting and recrystallizing these thin layers[4].

As the laser beam size is decreased (and the beam intensity is increased) the molten pool temperatures increase. For very high intensities, the molten silicon is pulled into a "keyhole" shape, leaving a narrow strip of bare substrate behind the center of the beam. As the beam diameter is increased and the intensity is decreased, the keyhole disappears and the molten pool is nearly circular. Finally as the intensity is sufficiently reduced, the outer periphery of the pool begins to evidence partial melting. This partially molten zone is recrystallized into a chaotic structure as it moves through the polysilicon layer.

Figure 3a, 3b, and 3c show the history of partial melting in a silicon layer. Figure 3a shows a slight pertubation of the the solid/liquid phase boundary that develops into dendritic penetration of the solid crystallites into the molten pool, Figure 3b, and thereafter to what we call a "mushy" zone, Figure 3c, of solid/liquid silicon coexistence. The origin of this mushy zone has been studied by Grigoropoulos[5,6,7]. An extended account of the experimental results was given by Grigoropoulos et al.[8].

Analysis

Referring to Figure 2a, a silicon layer of thickness d_{si} is deposited on a SiO_2 substrate of thickness d_{ss} and encapsulated by a SiO_2 layer of thickness d_{enc}. The temperature distribution in each of the different layers is represented by an equation of the form

$$\begin{aligned} \rho_j(T_j)c_{pj}(T_j)(\frac{\partial T_j}{\partial t} + V\frac{\partial T_j}{\partial x}) = \frac{\partial}{\partial x}(k_{xj}(T_j)\frac{\partial T_j}{\partial x}) + \frac{\partial}{\partial y}(k_{yj}(T_j)\frac{\partial T_j}{\partial y}) \\ + \frac{\partial}{\partial z}(k_{zj}(T_j)\frac{\partial T_j}{\partial z}) \\ + \text{Source}_j - \text{Losses}_j \end{aligned} \tag{3}$$

The subscript j refers to the different materials, substrate, thin silicon layer, and the encapsulating layer. The term labelled *Losses* represent the convective and radiative losses from the surfaces of the substrate and encapsulating layers. Continuity of temperature and heat flux is applied at the interfaces of the different materials. It is assumed that thermal contact between the adjacent layers is perfect. Throughout these calculations bulk properties are assumed for the silicon, although these properties are assumed to vary with temperature.

The term *Source* refers to the laser beam energy which passes through the transparent substrate and is absorbed by the thin silicon layer. This energy is given by

$$Q_{ab}(x, y, T_{si}) = \sum_{i=1}^{n_w} [1 - R_{\lambda i}(T_{si}) - \tau_{\lambda i}(T_{si})] Q_{\lambda i}(x, y) \tag{4}$$

where $Q_{\lambda i}(x, y)$ is the laser beam energy distribution and $R_{\lambda i}$ and $\tau_{\lambda i}$ represent the thin film reflectivity and transmissivity for the ith wavelength of the laser.

In performing these calculations, the substrate was divided by a three dimensional grid. The grid spacing was kept at $w_{\lambda b}/10$ near the center of the beam. This spacing was allowed to increase as one moved away from the center of the beam. The rate at which the grid spacing increased was adjusted to ensure that the steep temperature gradients near the melted silicon layer were adequately modelled. The silicon layer and the encapsulating layer were assumed to be thin enough so that the temperature did not vary in the z direction. Thus Equation 3 was integrated through the thickness of these regions and only the mean temperature in each layer determined. The computations were begun by assuming that the layers were at room temperature and continued until the temperature profile, as seen from the point of view of an observer located at the center of the beam, became constant in time. Computations were also performed with the grid spacing reduced by a factor of 2; no significant changes were observed.

During the computations, if the beam power was sufficient, the temperature under the beam reached the melting point. When this happened, the energy equation was rewritten in terms of enthalpy given by

$$H_{si}(T) = \int_0^T \rho_{si}(T) c_{p-si}(T) dT, \qquad T < T_m \tag{5a}$$

$$H_{si}(T) = \int_0^T \rho_{si}(T) c_{p-si}(T) dT + L, \qquad T > T_m \tag{5b}$$

The details of the numerical scheme were described by Grigoropoulos et al.[8] (1989).

Numerical Results

The calculations were carried out for a range of parameters that corresponds to the experimental conditions. Figure 4 shows how the temperature in the silicon layer at the center of the laser beam varies with time as the intensity of the beam is changed by changing the size of the beam. It is seen that after a processing time of about 30 msec that a quasi-steady temperature distribution is established.

Results are presented in Table 1 for laser beam total power of 1.4 and 1.8 watts and a material translation speed of 2 mm/sec. Examining the values in the table reveals that the numerical predictions show the same characteristics as the experimental results. For a power of 1.8 watts, as the intensity is reduced, the experimentally measured pool diameter increases strongly at first, and then becomes nearly constant. The predicted values also initially rise, but instead of remaining constant, decrease. The experimental results for 1.4 watts are similar. Keeping in mind that the silicon layer absorbtivity drops by a factor of 2 upon melting, these behaviors can be explained as follows. Initially the pool is larger than the beam. Since the system is at steady state, all of the energy which is absorbed from the beam must be lost to the substrate and to the ambient surroundings. This loss is proportional to the temperature of the molten pool and its area. Since the temperature drops as the intensity is reduced, in order to maintain a constant loss the area of the pool must increase. Hence, the increase of the pool diameter. Finally, as the beam diameter increases sufficiently, the Gaussian distribution results in a portion of the beam falling outside of the pool; thus the energy absorbed in the pool decreases. The amount of energy which is absorbed in the pool, and which must be lost, drops very rapidly as the beam size is increased. Since the pool temperature remains reasonably constant, approximately equal to the melting temperature, this reduced loss can only be consistent with a reduced pool area. Otherwise the loss would exceed the energy absorbed and the pool would not melt. Finally, the losses do increase so much that the silicon does not reach the melting temperature, and we have what is called the mushy zone. Further reductions in intensity lead to an unmelted silicon layer.

Table 1

Molten Spot Diameter (μm)				
$w_{\lambda b}$ (μm)	P_T=1.8 W		P_T=1.4 W	
	Experiment	Comput.	Experiment	Comput.
40	108	89	91	86
50	118	107	95	87
60	120	106	100	90
70	126	102	Partial melting	Partial melting

In the case of the computations for 1.8 watts, the results follow this trend at first. However, since the predicted pool starts out too small, i.e., it is the case that the beam is already large compared to the pool, the results quickly show a reduction of pool diameter with increasing beam diameter. It is suspected that the substrate thermal conductivity or the surface convective coefficient was chosen too high, thus leading to an underpredictions of the pool size. Rather than pursue the computations to match exactly the experimental results, it was felt that the results obtained thus far are sufficient to lend support to the fundamental correctness of the approach. Thus the numerical method was used to investigate the effect of changing different parameters on the process.

Sensitivity Studies

The calculations described above are extremely time consuming because of the large three dimensional region which must be treated and because of the very slow approach to a steady state. Accordingly, we chose to model a process using an elliptical laser beam whose aspect ratio was such that over the central part of the beam the molten pool boundary was nearly straight and the thermal field was two dimensional[9]. As mentioned above, the first question is the effect of a changing substrate conductivity. Figure 5 illustrates this effect. It can been seen that the pool size is strongly dependent upon the conductive loss to the substrate. If the intensity is sufficient, a 50% rise in conductivity reduces the pool size by about 60%. If the intensity is low, the effect is to prevent melting completely. It is apparent that a small modification to the conductivity would bring the predicted results in line with the measured values listed in Table 1.

Figure 6 shows the effect of processing speed. For speeds in the range of 0 to 3 mm/sec, there is essentially no effect. This is in agreement with our experimental measurements. Starting with a speed of 5 mm/sec, we see that the pool size begins to reduce and is significantly smaller for 10 mm/sec. This suggests that for wafers of similar dimensions, that the processing speeds should be no more than 3 mm/sec.

Finally Figure 7 shows how the pool size varies with beam intensity. First notice the relatively good agreement between experiments carried out with a beam of aspect ratio of 7:1 and the predictions. The lines represent the least squares fit to the data and the predictions. The differences are felt to be due to inaccuracies in estimating the radiation lost from the pool and in accounting for the temperature dependent properties. Second is the agreement with the behavior shown in Table 1. In Figure 7, Q_o^* was chosen as the intensity necessary to yield a molten pool size of 10 μm and varies with $w_{\lambda x}$. Replotting for a constant power of 1.4 watts over a 250 μm wide elliptical pool for different intensities yields the results shown in Figure 8. These results support the arguments made in comparing the measured and predicted pool size given in Table 1.

Improvements

As mentioned in the introduction, it has been found that concave solidification fronts may yield improved recrystallized films. Kawamura et al.[2] used a doughnut shaped laser beam to yield such concave fronts. They operated the laser in a linear combination of low order rectangular Hermite-Gaussian modes, the TEM_{01} and TEM_{10} modes. Figure 9 shows the calculated temperature field distribution for a laser beam of total power of 8 watts, material processing speed of 2 mm/sec and a beam 1/e width of 20 μm. It can be seen that the predicted solidification trailing edge is curved inwards.

Conclusions

This paper has described the melting and recrystallization of thin silicon layers effected by a 4W, multiline CW Argon ion laser beam. The experiments measured the molten pool size as a function of the laser beam power, shape of intensity distribution and the material translation speed. The phase change process was modelled using conductive heat transfer algorithms based on the enthalpy formulation. Given the uncertainty for the material properties, good agreement was found between the computed and the experimental results. Computational experiments yielded improved thermal gradient distributions across the phase change boundary and demonstrated the usefulness of numerical modelling towards process optimization.

Acknowledgements

Support to the first of the authors by the National Science Foundation, Division of Mechanics, Structures, and Materials, under Grant MSM-8708757 is gratefully acknowledged.

References

1 Tsaur, B.Y., "Assessment of Silicon on Insulator Technologies for VLSI," Proceedings, Materials Research Society, A. Chiang et al. eds., MRS, Pittsburgh, PA, Vol. 53, pp. 365-373, 1986.

2 Celler, G.K., "Laser Crystallization of Thin Si Films on Amorphous Insulating Substrates," J. Crystal Growth, Vol. 63, pp. 429-444, 1983.

3 Kawamura, S. et al., "Recrystallization osf Si on Amorphous Substrates by Doughnut Shaped CW Ar Laser Beams," Appl. Phys. Lett., Vol. 40, No. 5, pp. 394-395, 1982.

4 Pfeiffer, L., Gelman, A.E., Jackson, K.A., and West, K.A., "Growth Mechanisms During Thin Film Crystallization From the Melt," Proceedings, Materials Research Society, M.O. Thompson et al., eds., MRS, Pittsburgh, Vol. 74, pp. 543-553, 1987.

5 Grigoropoulos, C.P., Buckholz, R.H., and Domoto, G.A., "A Thermal Instability in the Laser-Driven Melting and Recrystallization of Thin Silicon Films on Glass Substrates," J. Heat Transfer, ASME, Vol. 109, pp. 841-847, 1987.

6 Grigoropoulos, C.P., Buckholz, R.H., and Domoto, G.A., "An Experimental and Theoretical Study on Laser Driven Silicon Phase Boundary Stability," ASME paper 87-WA/HT-3, 1987.

7 Grigoropoulos, C.P., Buckholz, R.H., and Domoto, G.A., "An Experimental Study on Laser Annealing of Thin Silicon Layers," J. Heat Transfer, Vol. 110, pp. 416-423, 1988.

8 Grigoropoulos, C.P., Emery, A.F., and Wipf, E.P., "Heat Transfer in Thin Silicon Film Melting by Scanning Laser Line Sources," Int. J. Heat-Mass Transfer (in press), 1989.

9 Grigoropoulos, C.P., Long, S.E., Emery, A.F., and Dutcher, W.E., "Experimental and Computational Analysis of Laser Melting of Thin Silicon Films," Presented at the 1989 National Heat Transfer Conference, ASME, HTD - Vol.113, pp. 45-53, 1989.

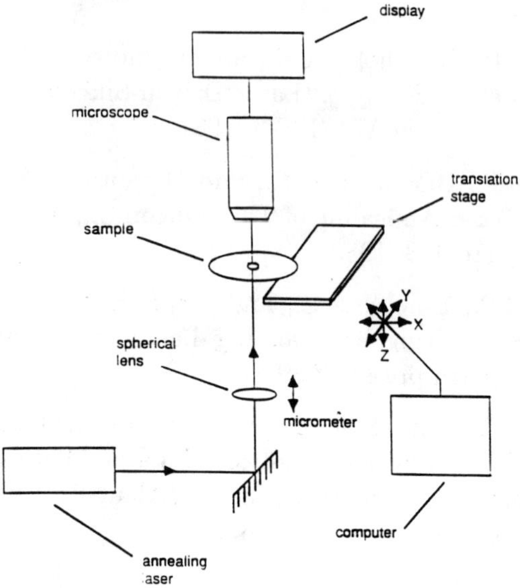

Figure 1 Schematic of the Laser Annealing Experimental Apparatus

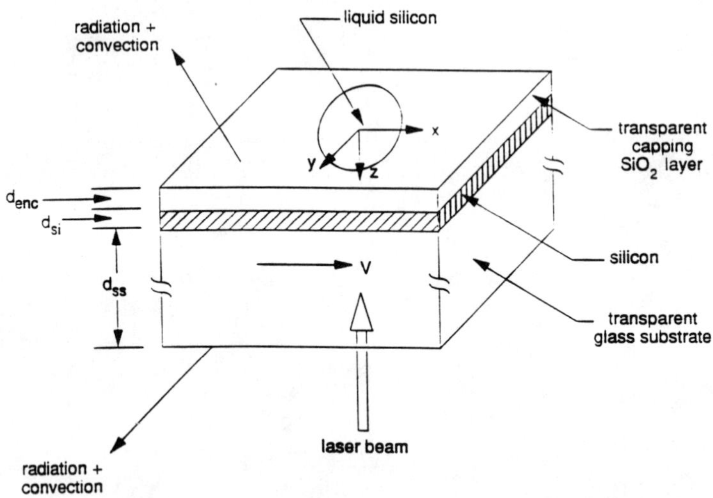

Figure 2a The Sample is Irradiated by a Laser Beam of Circular Cross-sectional Shape

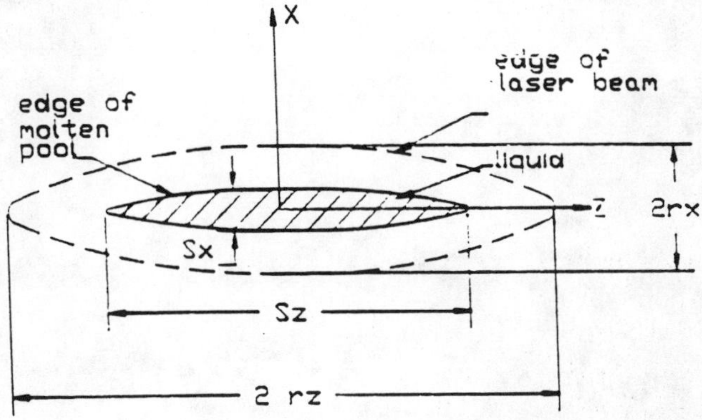

Figure 2b A Laser Beam of Elliptical Cross-Sectional Shape Produces a Molten Spot of Size S_x, along the Direction of Motion, x.

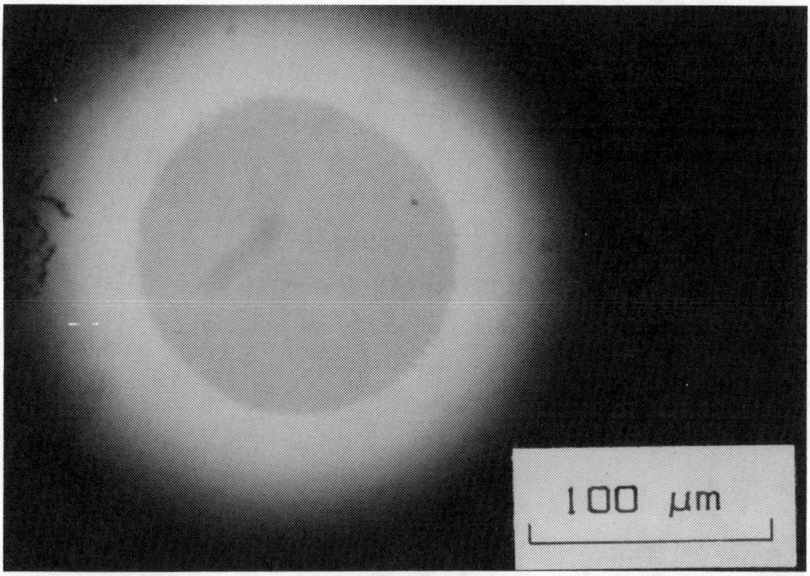

Figure 3a Micrograph of Light Transmitted through the Silicon Layer, $P_T = 1.8W$ and $w_{\lambda b} = 78\mu m$

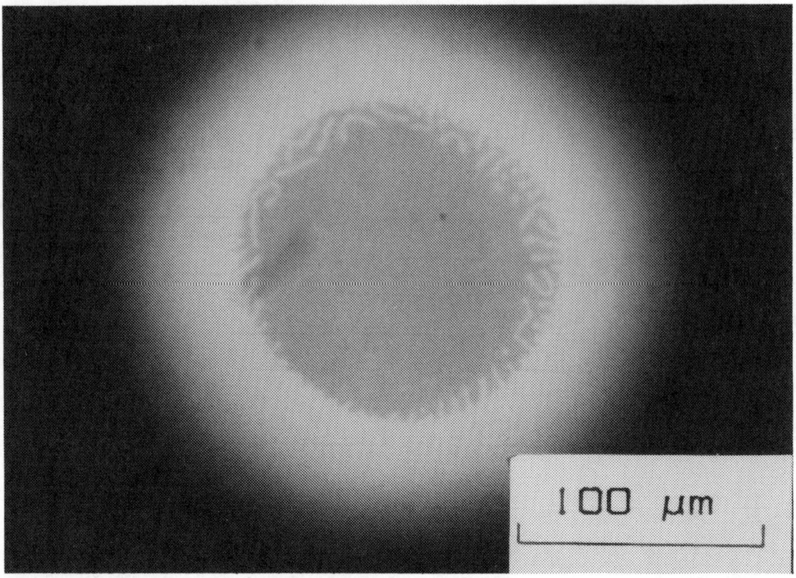

Figure 3b Micrograph of Light Transmitted through the Silicon Layer, $P_T = 1.8W$ and $w_{\lambda b} = 84\mu m$

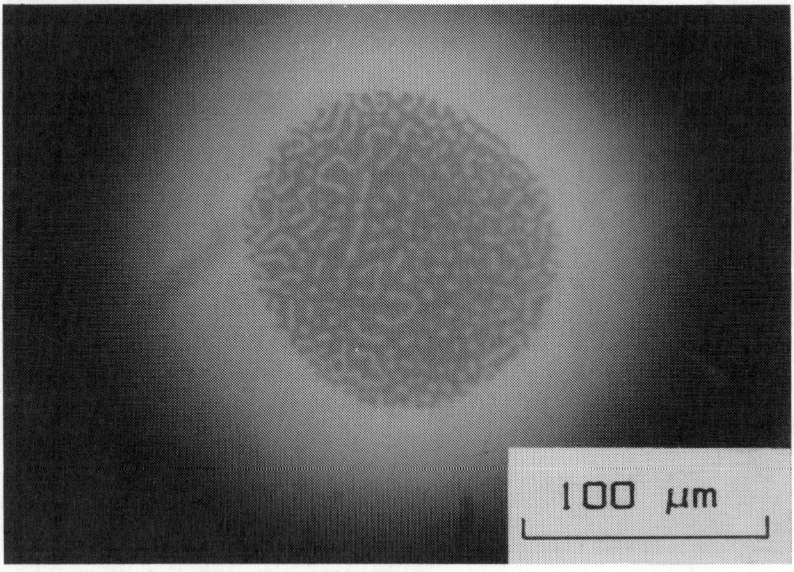

Figure 3c Micrograph of Light Transmitted through the Silicon Layer, $P_T = 1.8W$ and $w_{\lambda b} = 90\mu m$

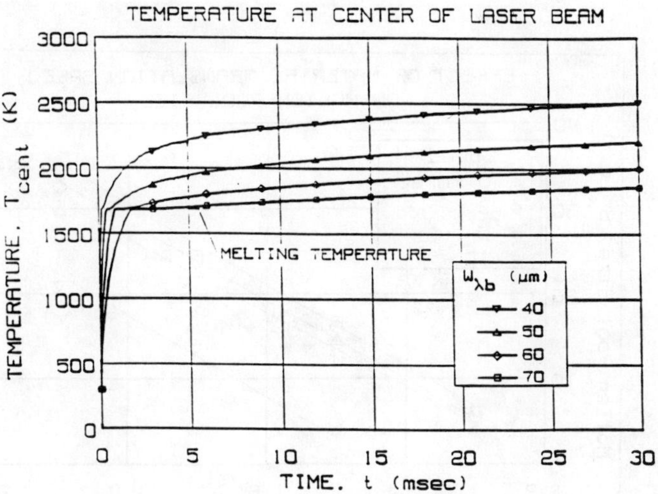

Figure 4 Temperature History of the Silicon Layer at the Center of the Laser Beam whose Cross-sectional Shape is Circular. The Laser Beam Power, $P_T = 1.8W$ and the Material Translation Speed, v=2mm/sec

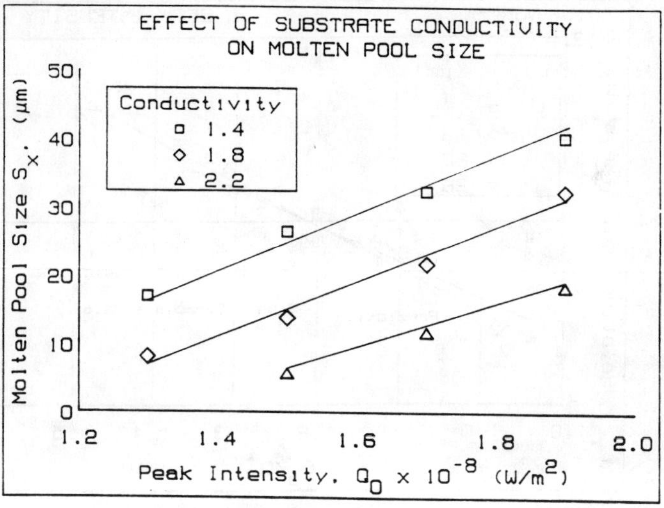

Figure 5 Effect of the Substrate Thermal Conductivity on the Molten Pool Size for a Laser Beam of $w_{\lambda x} = 19\mu m$. The Material Translation speed, v=2mm/sec

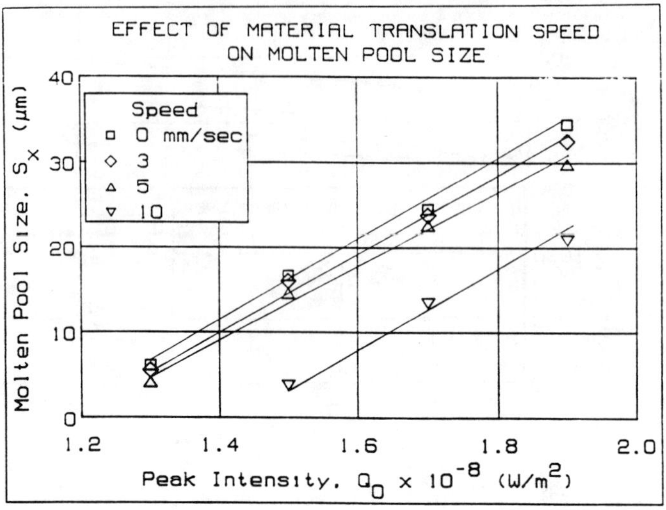

Figure 6 Effect of the Material Translation Speed on the Molten Pool Size for a Laser Beam of $w_{\lambda x} = 19\mu m$

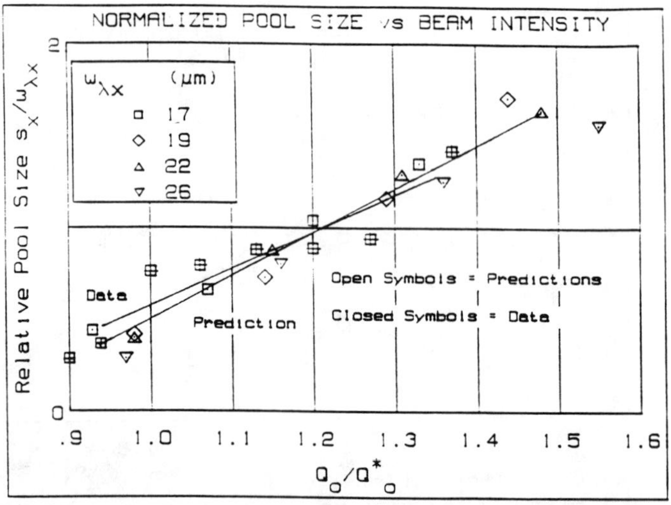

Figure 7 Normalized Molten Pool Dimensions, $S_x/w_{\lambda x}$ versus Normalized Laser Beam Peak Intensities, Q_o/Q_o^*. Q_o^* is the Calculated Peak Irradiance Value for a Molten Pool Size of $S_x = 10\mu m$

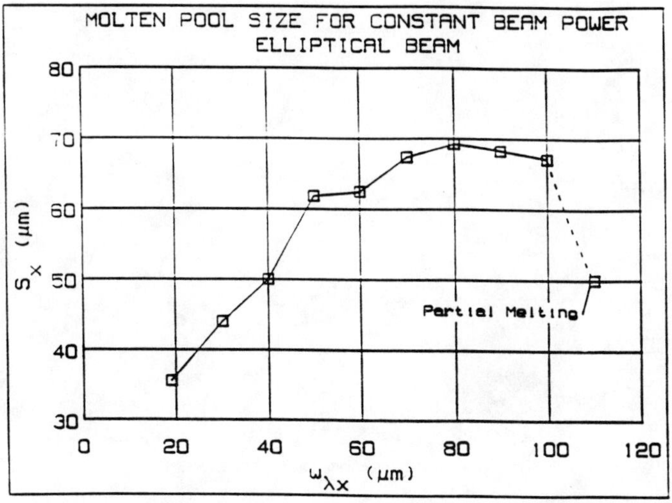

Figure 8 Computed Molten Pool Size, S_x, for an Elliptical Beam of Total Power, $P_T = 1.4W$. The Power is Uniformly Distributed over a Distance of $250\mu m$. The Material Translation Speed is v=2mm/sec

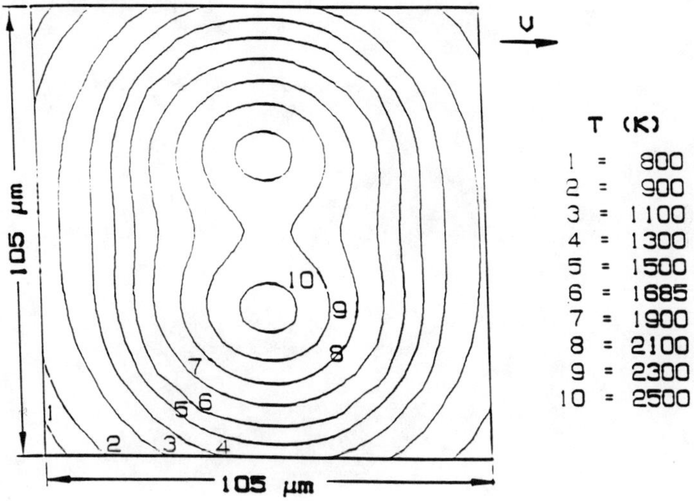

Figure 9 Predicted Temperature Field in the Silicon Layer at t=20msec. The Laser Beam operates on the TEM_{01} mode, the Total Power, $P_T = 8W$, $w_{\lambda b} = 20\mu m$ and the Material Translation Speed, v=2mm/sec

MULTILAYER CERAMIC NPO CAPACITORS WITH COPPER ELECTRODE

Harufumi Mandai, Yukio Sakabe
Murata Mfg. Co., Ltd.
2-26-10 Tenjin
Nagaokakyo-shi
Kyoto, 617 Japan

James P. Canner
Murata Erie N.A., Inc.
1900 West College Ave.
State College, PA 16801

ABSTRACT

The new Multilayer Ceramic Capacitor (MLC) is composed of a low temperature fireable ceramics (BSSZ) consisting of BaO, SrO, SiO_2, and ZrO_2, which is co-fireable with copper electrodes. The dielectric constant is 15 and the temperature coefficient is NPO. This ceramic is fired at 900 - 1000°C in a slightly reducing atmosphere under precise control. The high frequency characteristics (Q, ESR) are superior to high frequency MLC's using palladium electrodes because of the reduced electrical resistivity of copper. These new products will reduce the cost of high frequency MLC's also.

INTRODUCTION

The demand for high Q chip capacitors for high frequency and high power circuits is growing with the growth of the high frequency communication system. Conventional high Q multilayer capacitors (MLC) are composed of palladium (Pd) metal electrode and $MgTiO_3$- $CaTiO_3$ dielectric.

At the frequency region higher than 100MHz, the Q value of the capacitor depends on the resistance of the inner electrode, i.e. $Q = 1/(\omega CR_e)$. The Equivalent Series Resistance (ESR) of the capacitor is equal to the electrode resistance, R_e. The allowable electric power capability is also dependent on the ESR value.

To design MLC's with higher electrical performance, resistivity of the electrode must decrease. One of the possible metals for a lower resistance electrode is copper (Cu) (Cu = 1.55×10^{-6} Ohm-cm; Pd = 9.77×10^{-6} Ohm-cm).(1)

The copper electrode must be fired under low oxygen partial pressure since copper is oxidized completely if fired in air. The melting temperature of Cu is 1083°C. As a result, MLC's with Cu electrode must be fired at temperatures lower than 1000°C in an atmosphere of lower oxygen partial pressure (PO_2) than the equilibrium oxygen partial pressure of Cu_2O/Cu.

A new NPO dielectric having high insulation resistance and firing temperature lower than 1000°C is required. The purpose of this study is to evaluate the ternary dielectric compositions of (Ba Sr) O-SiO_2-ZrO_2, in the making of MLC's with copper electrodes.(2)

DIELECTRIC MATERIALS

Preparation of Single Layer Dielectric Sample

The compositions of the ternary system (Ba Sr) O-SiO_2-ZrO_2 with additives, Al_2O_3 and TiO_2 were prepared from reagent grade $BaCO_3$, $SrCO_3$, SiO_2, ZrO_2, Al_2O_3, and TiO_2. All batches were wet mixed for 12 hours using a plastic jar mill with agate (SiO_2) ball media. The mixed powder was calcined at 825°C for two hours.

The test samples were prepared in square plate form (1mm thick) by laminating cast green sheets. Samples were sintered over the range of 900°C through 1100°C for two hours under different reducing atmospheres. The reducing condition can be obtained by mixtures of H_2 and H_2O gases with N_2 carrier gas. Indium - gallium alloy was applied as a conducting electrode on the sintered plates to make the electrical measurements.

Composition

A. Basic Compositions

The dielectric materials used for the new MLC's consist of BaO, SrO, SiO_2, ZrO_2, and small amounts of additives - Al_2O_3 and TiO_2.(3),(4) In the region SiO_2 <30% (wt) and (Ba Sr)O <20% (wt), the ceramics materials are not sintered under 1050°C. In order to use the copper electrodes, the firing temperature must be lower than the melting point of copper (1083°C), preferably under 1000°C. Consequently, the compositional range where the ceramics materials can be sintered is limited. The firing atmosphere $PO_2 = 10^{-9}$ to 10^{-10}atm (at 1000°C) was used because it was found to be optimum for Cu_2O/Cu metal system.
The dielectric constant increases with increasing the amounts of (Ba Sr)O and

ZrO_2. The Q value becomes higher with decreasing amounts of SiO_2. The TCC becomes more negative with increasing amounts of (Ba Sr)O.

The most suitable composition is 35% (Ba Sr)O, 35% SiO_2, and 30% ZrO_2. This composition, called BSSZ, will be used through the remainder of this paper.

B. Additives - Al_2O_3, TiO_2

Fig. 1 shows the relation between TCC and the amounts of TiO_2 added to BSSZ. The addition of TiO_2 makes the TCC become negative. On the other hand, the addition of Al_2O_3 makes the TCC curve straight and more positive. Consequently, we can control the TCC by the addition of TiO_2 and/orAl_2O_3.

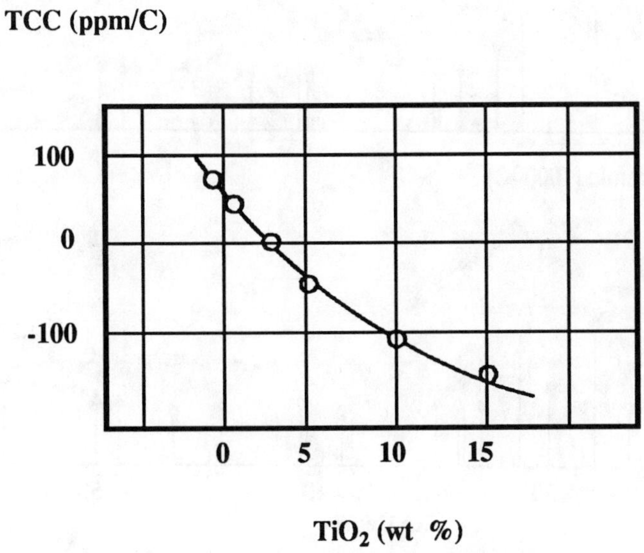

Fig. 1 Change in temperature characteristics with addition of TiO_2.

Microstructure

A. Crystal Phase

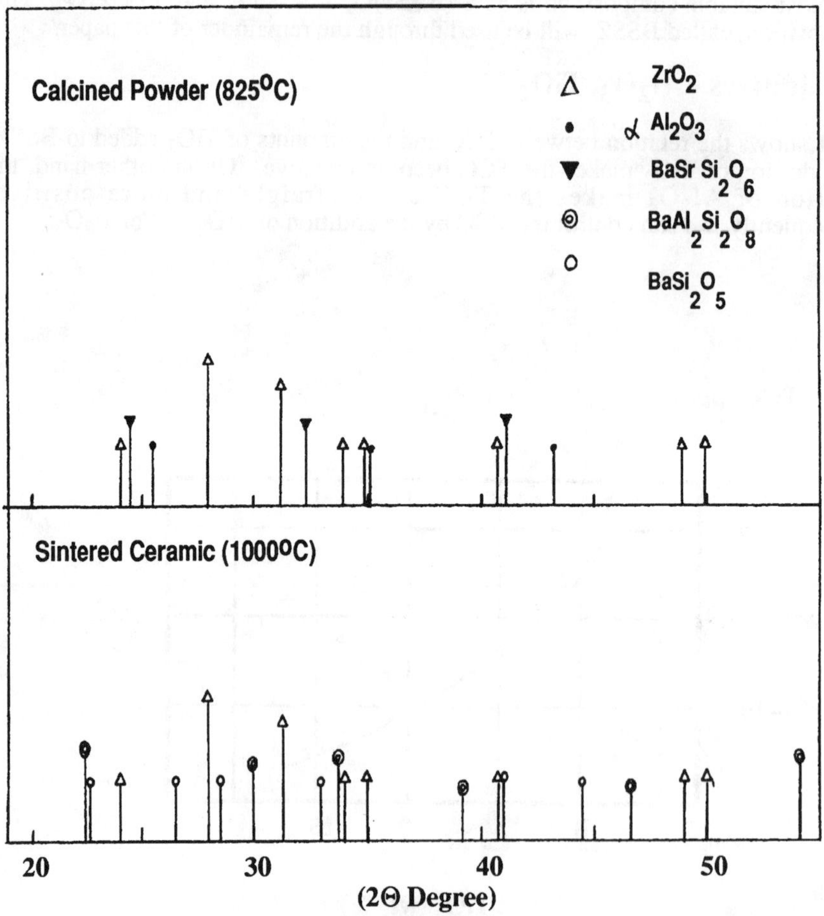

Fig. 2 X-ray diffraction pattern of BSSZ calcined at 825°C 9(a) and BSSZ fired at 1000°C (b).

The raw material powders for BSSZ are calcined at 800-850°C in air for two hours. Fig. 2 shows the X-ray diffraction (XRD) pattern of the calcined powders. In the calcined powder, three crystalline phases (ZrO_2, α-Al_2O_3, $BaSrSi_2O_6$) are observed. The X-ray diffraction pattern of the sintered ceramics at 1000°C is also shown in Fig. 2. The crystal phases of ZrO_2, $BaAl_2Si_2O_8$ (celsian)[5], and $BaSi_2O_5$ exist.

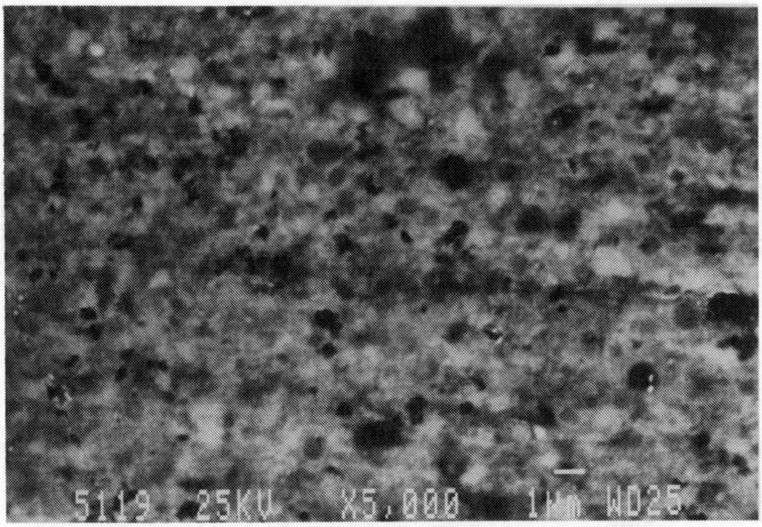

Fig. 3 SEM photo of BSSZ ceramic fired at 1000°C

B. X-ray Microanalysis (XMA)

The cross section SEM photograph of the BSSZ materials fired at 1000°C is shown in Fig. 3. At least three kinds of phases are observed. We analyzed the forming composition of these phases using XMA. Figs. 4 - 6 show the results. The following phases are assumed based on XMA along with the result of XRD.

1. Phase 1 (light) is a Zr - rich phase. See Fig. 4. It is presumed to be ZrO_2 dispersed in $BaSi_2O_5$ glass phase.

2. Phase 2 (dark) is a Al - rich phase. See Fig. 5. It is presumed to be $BaAl_2Si_2O_8$.

3. Phase 3 (medium) is a Ti - rich phase. See Fig. 6. However, the crystal phase including Ti is not detected with the XRD analysis. Consequently, it is assumed that Ti is dissolved in the amorphous phase (glassy - phase).

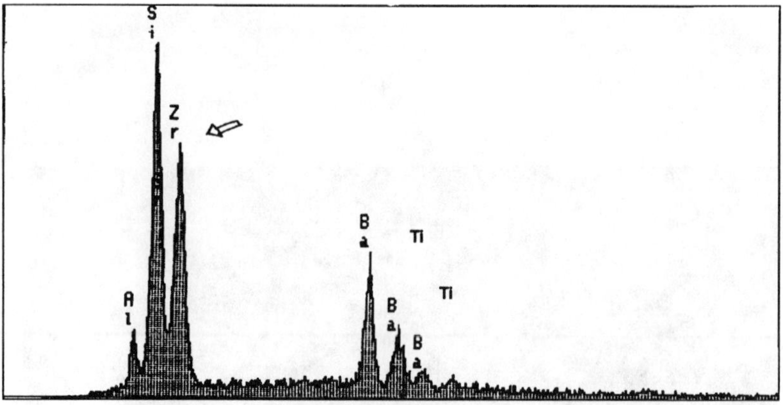

Fig. 4 XMA of light phase - ZrO2 rich

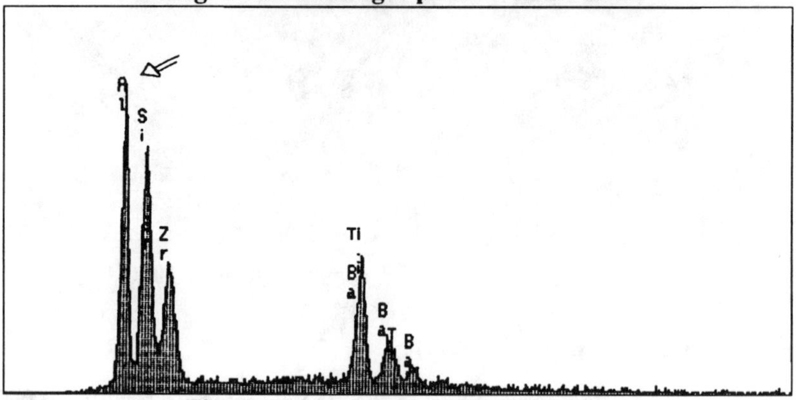

Fig. 5 XMA of dark phase - Celsian

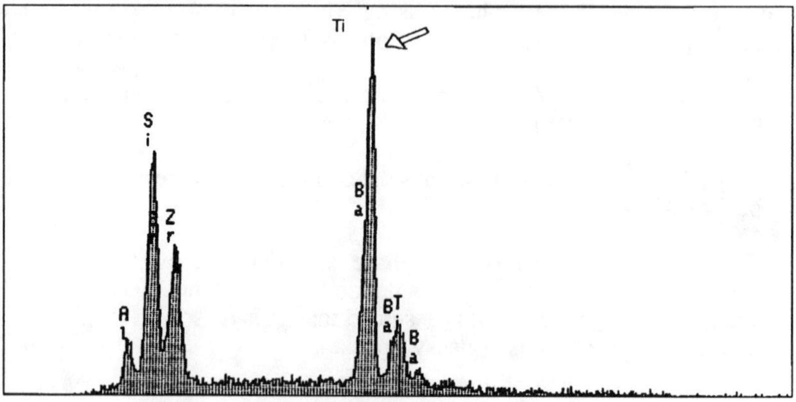

Fig. 6 XMA of medium phase - Ti rich glass

FIRING BEHAVIOR

A. Non-reductivity of the Dielectric

Table 1

Log PO_2 (atm, 1000°C)	Dielectric Constant	Q Log	Resistivity (ohm-cm)
-18	12.8	800	14.6
- 9	12.6	1040	14.0
- 0.7(in air)	12.7	1230	14.5

Table 1 shows the relationship between dielectric characteristics (dielectric constant, Q value, and resistivity) and the oxygen partial pressure (PO_2) of the firing atmosphere. The resistivity does not depend on the PO_2 (at 1000°C) from 0.2atm (in air) to 10^{-18} atm. This data suggests that the BSSZ ceramics is a non-reducing dielectric material. The cause of this non-reducing stability is assumed to be due to the following:

1. The valency of the elements (such as Ba, Sr, Si, Zr) constituting the main compositions are stable.

2. The transition metal elements are not used in the main compositions except for Zr.

3. ZrO_2 has ionic conductivity because of the oxygen vacancies, but the electronic conductivity is very low.

B. Sintering Behavior

The shrinkage curve is shown in Fig. 7. The BSSZ ceramics begins to shrink at 800°C and the shrinkage is virtually complete at 20% above 980°C.

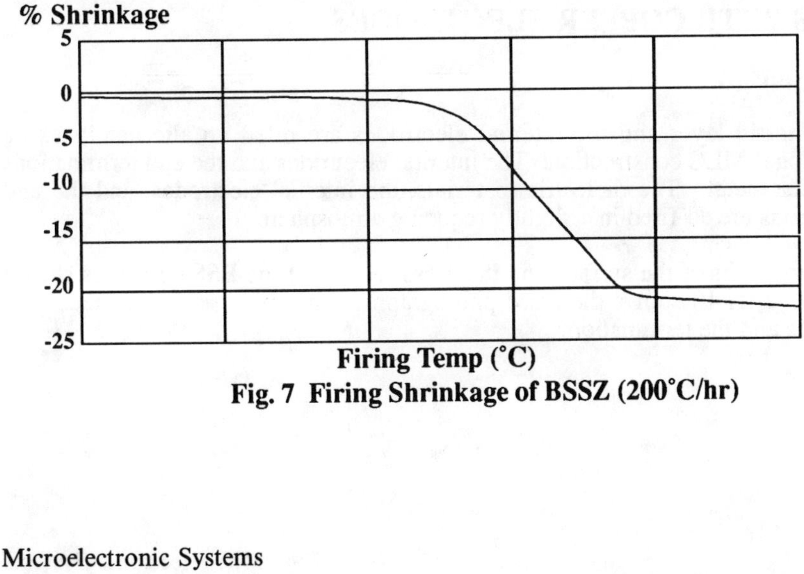

Fig. 7 Firing Shrinkage of BSSZ (200°C/hr)

C. TCC

TCC of the BSSZ material depends on the sintering temperature. As shown in Fig. 8, the TCC curve becomes negative with increasing sintering temperature. It is necessary to keep the sintering temperature in the range from 980°C to 1020°C in order to satisfy the TCC specification for NPO of ±30ppm/°C.

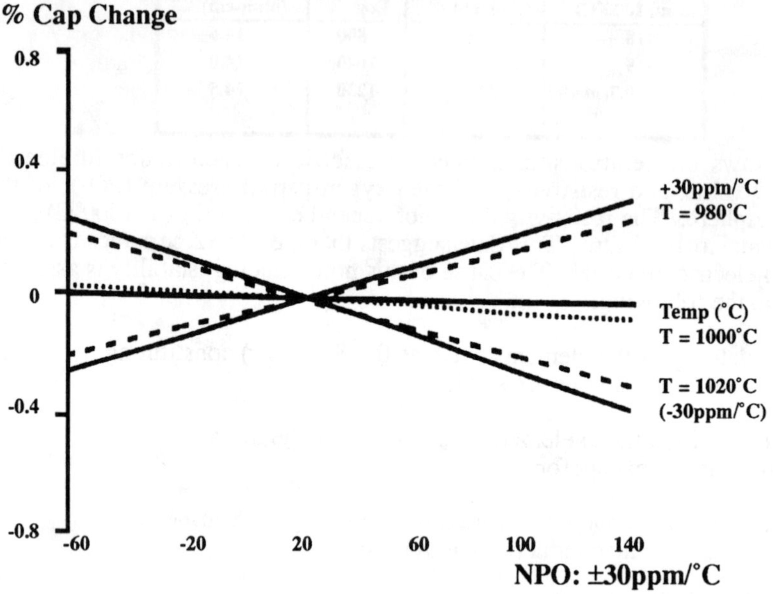

Fig. 8 TCC change with BSSZ firing temperature

MLC'S WITH COPPER ELECTRODES

Construction

The dielectric layer and the internal electrodes are piled up alternately using conventional MLC construction. The internal electrodes and the end terminations are copper metal. The dielectric materials, the internal electrodes, and the end terminations are co-fired in a slightly reducing atmosphere.

The photographs of the surface and the cross section of the BSSZ MLC is shown in Fig. 9. Fig. 10 shows the SEM photograph of the cross section of the inner electrodes and the termination.

Fig. 9a Exterior view of BSSZ MLC

Fig. 9b Interior view of BSSZ MLC

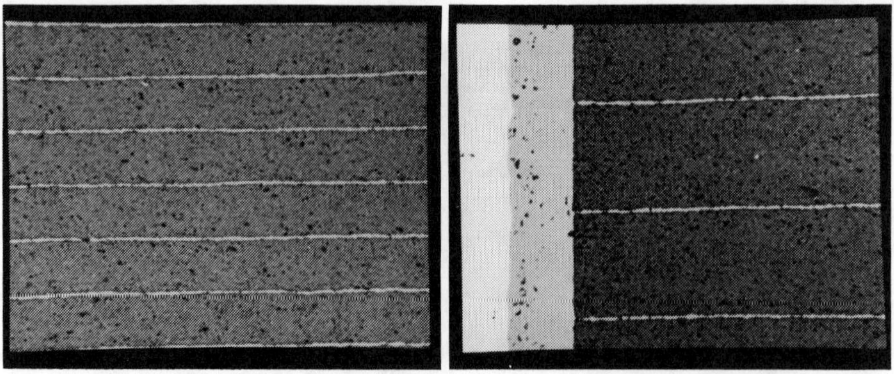

a b

Fig. 10 SEM photo of BSSZ inner electrodes (a) and termination (b)

Process

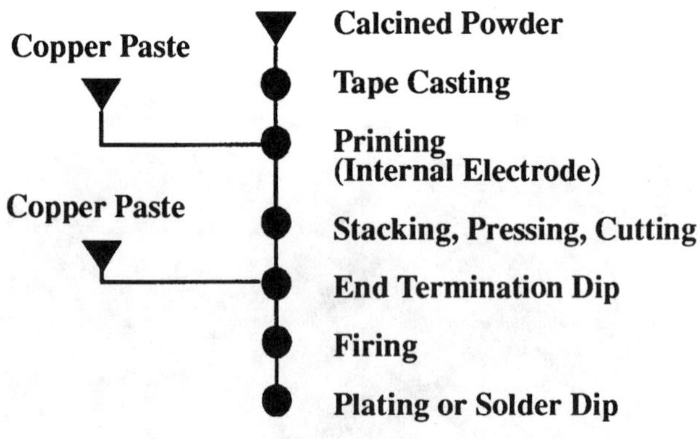

Fig. 11 Process Flow Chart, BSSZ multilayer capacitor

The fabrication process flow chart for the MLC's is shown in Fig. 11. Slurry consisting of dielectric material powders and organic binder (polyvinyl butyral) is cast by a doctor blade method. The green sheets are 40μm in thickness. The internal electrode copper ink is printed on these sheets. Then the sheets are stacked and pressed at 40°C and 2000Kg/cm^2. The pressed blocks are cut into chips and end terminations are applied by dipping the chip into a copper ink. The MLC chips are then fired at 1000°C in a reducing atmosphere using the forming gas mixture of N_2, H_2, and H_2O.

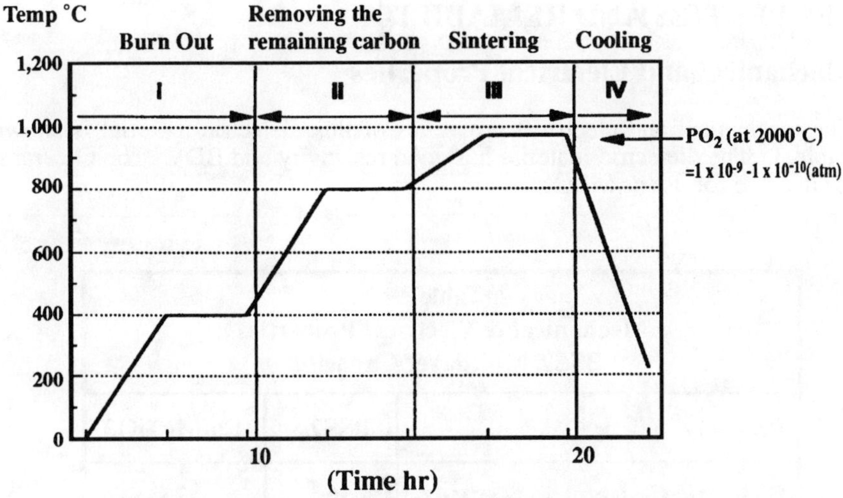

Fig. 12 Time-temperature firing profile for BSSZ

As shown in Fig. 12, the time - temperature profile consists of 4 zones. Each zone uses forming gas mixture to control partial oxidation pressure.

Zone 1 Binder Burnout Zone
Zone 2 Removal of the Remaining Carbon Zone
Zone 3 Sintering Zone
Zone 4 Cooling Zone

The firing atmosphere is established under the condition that the oxygen partial pressure is $10^{-9} \sim 10^{-10}$ atm at 1000°C. In such an atmosphere, copper is maintained in the metallic state and carbon is removed as CO and CO_2. Table 2 shows the residual carbon of the MLC's fired in this firing condition. The MLC chip before firing has 5wt% carbon. The magnitude of the remaining carbon after firing at 1000°C is 10ppm.

Table 2
Residual Carbon
BSSZ Multilayer Capacitor

Temp (°C)	Residual Carbon(wt%)
Before firing	5%
450°C	0.4%
900°C	120 ppm
1000°C	10 ppm

PROPERTIES AND RELIABILITY

Mechanical and Electrical Properties

The mechanical and electrical properties of this dielectric material is shown in Table 3. The dielectric material has good resistivity and BDV. The Q parameter is adequate for use as MLC's.

Table 3
Mechanical & Electrical Properties
BSSZ Multilayer Capacitor

Property		BSSZ	Ca/MgTiO3
Dielectric Constant	@1MHZ	15	25
Q	@1MHz	2000	10,000
Temperature Coefficient of Capacitance	(ppm/°C)	NPO±30	NPO±30
Log Resistivity	(ohm-cm)	>14	>14
Dielectric Strength	(kV/mm)	>50	>50
Density	(g/cc)	3.6	3.7
Flexural Strength	(kg/cm²)	2300	2500
Thermal Expansion Coefficient (25-600°C)	(ppm/°C)	6.8	10

High Frequency Characteristics

High frequency Q and ESR was measured on a Boonton Model 34A resonant coaxial line system. Fig. 13 shows the comparison of high frequency characteristics (Q, ESR) between the MLC's with palladium electrodes using the dielectric materials of the $MgTiO_3$ - $CaTiO_3$ system and the BSSZ MLC's with copper electrodes. The ESR of the latter is about half that of the former because of the low resistivity of copper.

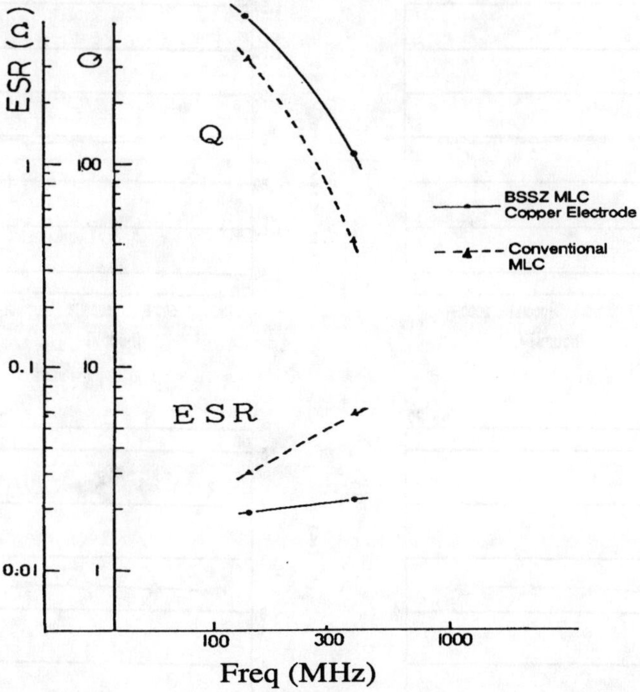

Fig. 13 ESR and Q of BSSZ and conventional MLC

Reliability

Fig. 14 shows the results of a load life test and a load humidity test. The new MLC's shows no degradation under these tests through 2000 hours.

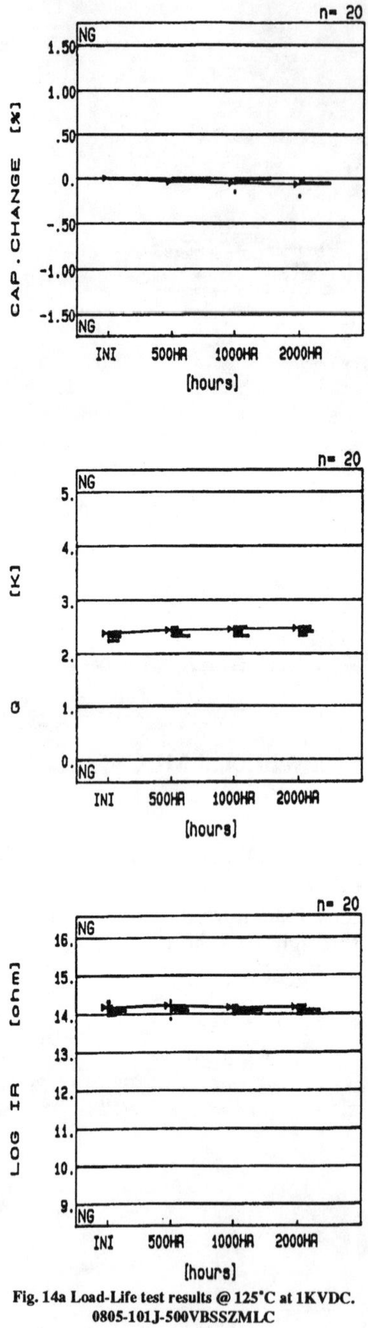

Fig. 14a Load-Life test results @ 125°C at 1KVDC. 0805-101J-500VBSSZMLC

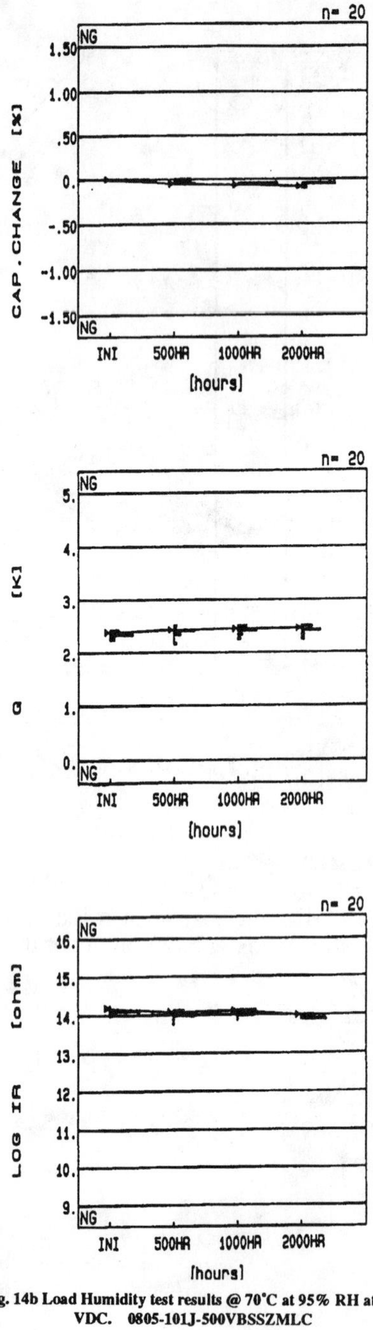

Fig. 14b Load Humidity test results @ 70°C at 95% RH at 500 VDC. 0805-101J-500VBSSZMLC

CONCLUSION

1. The non-reducing and low temperature fireable dielectric formulations consisting of BaO, SrO, SiO_2, and ZrO_2 were investigated.

2. We have developed the technical skill that is necessary to co-fire the above mentioned ceramics material with the copper electrodes in a slightly reducing atmosphere.

3. As a result of this investigation, we have developed the high frequency NPO MLC's whose high frequency characteristics is superior to the MLC's using palladium electrodes, and have reduced the metal cost of Hi-Q MLC's.

REFERENCES

1. Y. Yokotani, Proc. Jpn. Ceram. Soc. Annual Meeting, 171 (1986).

2. Y. Sakabe, "Recent Progress on Multilayer Ceramic Capacitors", Mat. Res. Soc., Vol. 10, 119 (1989).

3. G. Nishioka et al, Jpn. Pat. No. Hei-1-102808 (1989).

4. G. Nishioka et al, "Temperature Compensating Dielectric Ceramic Composition", U.S.P. 4816429 (1989).

5. H. C. Lin and W. R. Foster, "Studies in the System BaO-Al_2O_3-SiO_2: The Ternary System Sanbornite-Celsian-Silica", Amer. Ceram. Soc., 51[10]549-551 (1970).

PARAMETERS IMPORTANT FOR SURFACE MOUNT APPLICATIONS OF MULTILAYER CERAMIC CAPACITORS

by

Bharat S. Rawal
Kumar Krishnamani
John Maxwell

AVX Corporation
P.O. Box 867
Myrtle Beach, SC 29577

ABSTRACT

With increasing use of multilayer ceramic capacitors (MLCs) in surface mount technology (SMT), the understanding of the mechanical properties and thermal stress resistance parameters of MLCs is essential for zero defect soldering and sub ppm failure rates. In this paper, various aspects of SMT including zero defect design, placement consideration, soldering techniques, thermal stress resistance parameters and post solder handling are reviewed. Special emphasis is given to parameters responsible for thermal shock behavior of MLCs with review of the effect of overall component thickness, temperature gradients, and termination of MLCs.

INTRODUCTION

A detailed review of various aspects of surface mount manufacturing has been undertaken to understand various sources of defects which may arise in multilayer ceramic capacitors (MLCs) after assembly. In line process modifications, marginal components and rework are no longer acceptable in obtaining sub-ppm failure rates for surface mount assemblies. It is essential to understand the design considerations and possible sources of defects in these assemblies. It is also mandatory that zero defect soldering designs be used and that they be manufacturable. The next aspect of this subject is the understanding of placement of MLCs on boards and various considerations thereof to have zero defect placement before any reflow is carried out. Soldering techniques themselves are then considered to understand the needs of the mechanical and thermal properties required from MLCs. Post solder handling is reviewed as this may cause defects and result in early failures.

Various aspects of the thermal stresses of MLCs are reviewed and some experimental results are reported; emphasis is placed on, as an example, a detailed understanding of the termination of MLCs. The complexity of this part of the study clearly shows that various models need modifications to really understand the behavior of MLCs.

Various tests to review the reliability of MLCs on test boards are outlined and failure rates are estimated.

REVIEW OF SURFACE MOUNT MANUFACTURING

Zero Defect Design

Surface mount technology (SMT) for assembly of boards primarily uses reflow soldering techniques which include infra-red reflow and vapor phase, and wave soldering. In reflow soldering, solder is applied prior to placement of components. Wave soldering is an additive process because the solder fillet is formed as a result of the transfer of solder from a reservoir of molten solder to the chip and board . For reflow soldering techniques the parameters important for zero defect soldering are:

(i) *Solder Mass*: When the solder reflows, surface tension forces exerted by the molten solder are counteracted by the mass of the device and the moment arm created by the component length and adhesion of molten solder to the opposite end. Too much solder results in draw-bridging or a missing solder joint for MLCs, small outline transistors (SOTs) and small integrated circuits (ICs). Inadequate solder results in weak or missing solder joints for passive components or coplanarity problems with ICs. It is therefore recommended that the total equivalent wet laydown of solder paste should be 0.25 to 0.30mm and must take the board solder plating into account.

(ii) *Pad Design*: There is a profusion of recommended pad designs by component manufacturers, government agencies, industry associations and component users. Comprehensive details are available from The Institute for Interconnecting and Packaging Electronic Circuits, with details in IPC-SM-782[13]. It should be noted that this document contains compromises of many industry representatives and may not have an understanding of the impact of yields. Therefore, the best source for pad designs are users or vendors that have reliable, low defect solder joint histories. These designs should be tried and tested for manufacturability followed by thermal cycling and environmental testing.

(iii) *Trace-Pad Interactions*: Details of parameters for remodels need modifications to really understand the behavior of MLCs.

Various tests to review the reliability of MLCs on test boards are outlined and failure rates are estimated.

REVIEW OF SURFACE MOUNT MANUFACTURING

Zero Defect Design

Surface mount technology (SMT) for assembly of boards primarily uses reflow soldering techniques which include infra-red reflow, vapor phase, and wave soldering. In reflow soldering, solder is applied prior to placement of components. Wave soldering is an additive process because the solder fillet is formed as a result of the transfer of solder from a reservoir of molten solder to the chip and board . For reflow soldering techniques the parameters important for zero defect soldering are:

(i) *Solder Mass*: When the solder reflows, surface tension forces exerted by the molten solder are counteracted by the mass of the device and the moment arm created by the component length and adhesion of molten solder to the opposite end. Too much solder results in draw-bridging or a missing solder joint for MLCs, small outline transistors (SOTs) and small integrated circuits (ICs). Inadequate solder results in weak or missing solder joints for passive components or coplanarity problems with ICs. It is therefore recommended that the total equivalent wet laydown of solder paste should be 0.25 to 0.30mm and must take the board solder plating into account.

(ii) *Pad Design*: There is a profusion of recommended pad designs by component manufacturers, government agencies, industry associations and component users. Comprehensive details are available from The Institute for Interconnecting and Packaging Electronic Circuits, with details in IPC-SM-782[13]. It should be noted that this document contains compromises of many industry representatives and may not have an understanding of the impact of yields. Therefore, the best source for pad designs are users or vendors that have reliable, low defect solder joint histories. These designs should be tried and tested for manufacturability followed by thermal cycling and environmental testing.

(iii) *Trace-Pad Interactions*: Details of parameters for reflow soldering and trace-pad interactions are common industry practice[1,13]. Briefly, the designs should limit the numbers of traces entering a pad, the traces entering the pad should be symmetrical and there should be no vias or through holes in a pad. In addition, the ground planes should be isolated from components with necked down conductors, and via pads should be isolated from the component pads. Examples are shown in Figures 1(a) and (b).

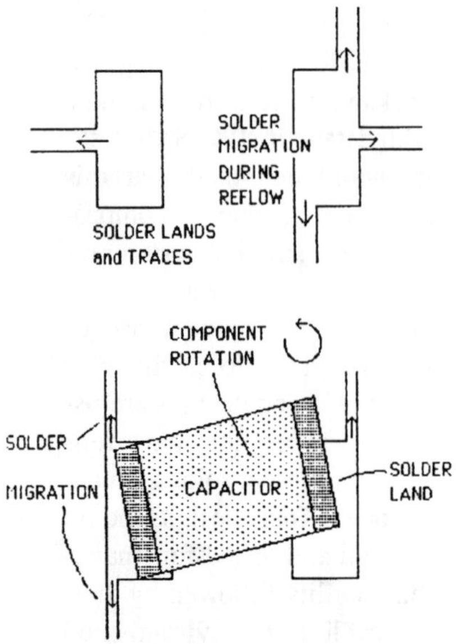

Figure 1(a). *Design Techniques to Avoid*

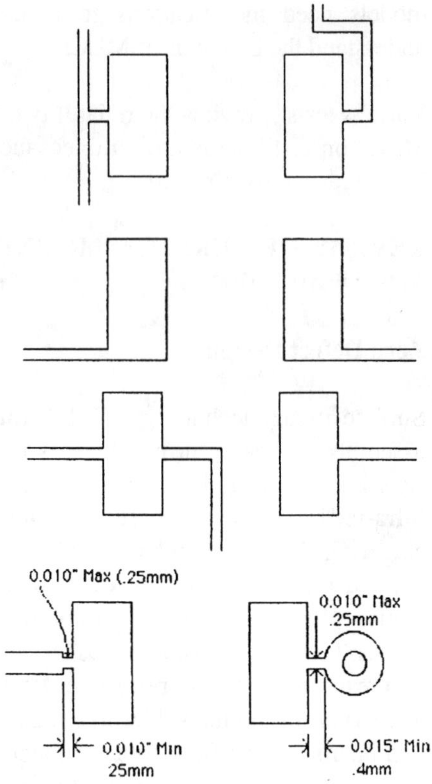

Figure 1(b). *Design Techniques That Work*

(iv) *Solder Mask Design*: Solder mask artwork should be oversized[1,13] by 10 mils in each dimension to allow for artwork mis-registration or slump of wet film to avoid contamination of pad surface; examples are shown in Figures 2(a) and (b). In addition, having no solder mask material between pads eliminates solder mask thickness

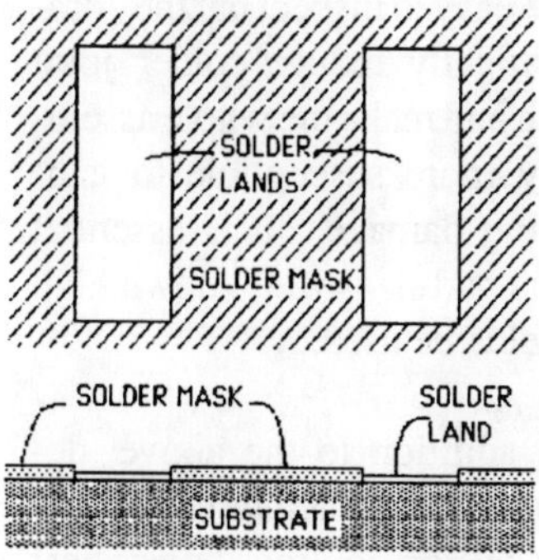

Figure 2(a). *Ideal Solder Mask*

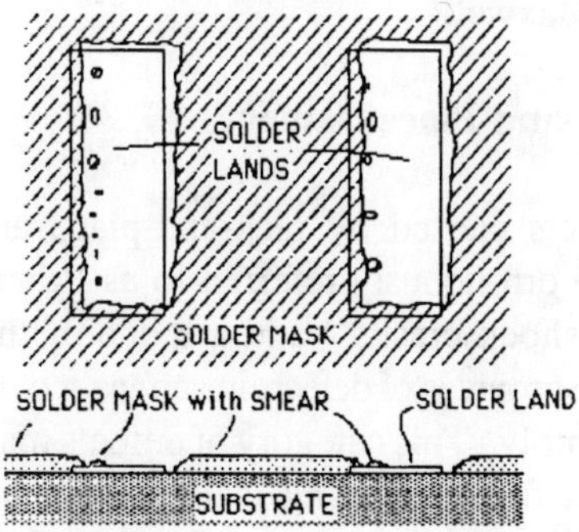

Figure 2(b). *Actual Solder Mask*

problems as shown in Figures 3(a) and (b).

(v) *Component Orientation*: These defects are a result of low mass component termination and differential heating. Each termination should enter

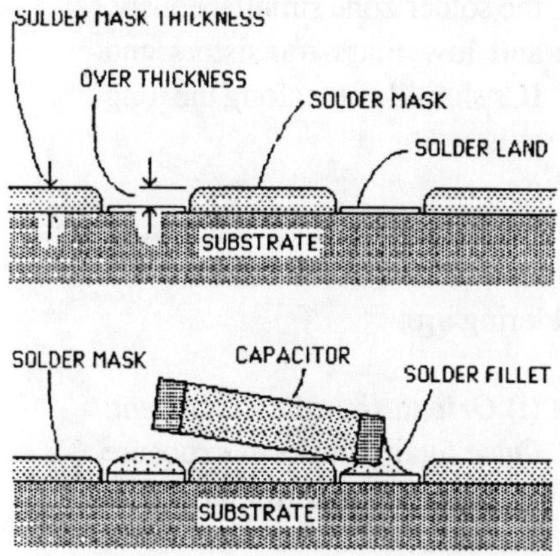

Figure 3(a). *Solder Mask Induced Drawbridge*

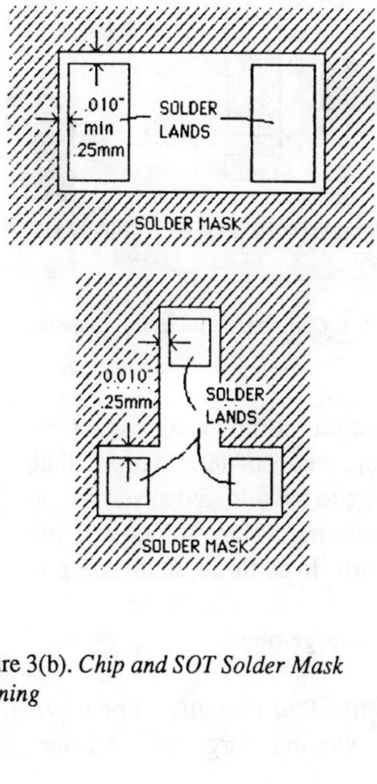

Figure 3(b). *Chip and SOT Solder Mask Opening*

the solder zone simultaneously, and low mass transistors and ICs should enter along the long axis.

For wave soldering, the parameters important for zero defect soldering are:

(i) *Orientation and Placement*: Once again, uniform exposure of component terminations to the solder bath is critical to achieve low defects and reliable solder joints as shown in Figure 4. In addition to orientation, components cannot be upstream of terminations that are to be soldered as starved or missing solder joints will result. In addition, small components cannot be in the shadow of large ones.

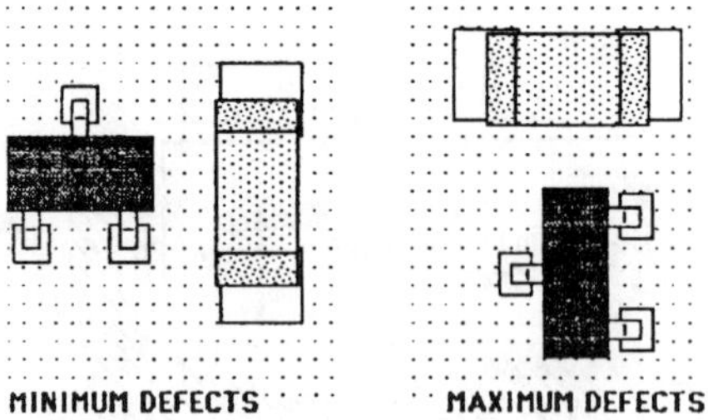

Figure 4. *Component Soldering Orientation*

(ii) *Pad Designs*: For wave soldering, large solder fillets are easy to inspect but they dramatically reduce solder joint life and make the whole assembly more susceptible to handling damage. It is essential that pad sizes with known high yields be used.

In addition to the above, designs should include considerations for tooling holes, test points, component separation, power/ground planes, and post solder assembly stresses. Details of these are outlined by Maxwell[1].

Pick-and-Place Damage

Defects caused by pick-and-place are often erroneously referred to as "thermal shock cracks" and are one of the largest sources of defects in surface mount assembly. This category of defects may result from centering jaws or vacuum pick-up bits and are briefly described below:

(i) *Vacuum Pick-Up-Bits*: This defect is visible as a circular or a half-moon shaped crushed area with ragged edges and is caused by excessive z-axis placement force. Another possible manifestation of the excessive placement force is where, for reflow soldering, the terminations are supported by solder paste allowing the unsupported component body to crack in the flexure mode. Pneumatic actuators may be used but component thickness variations, air pressure variations, solder thickness variations and board warpage should be monitored. In summary, z-axis placement force has to be monitored and controlled for zero placement defects.

(ii) *Centering Jaw Damage*: Even though machine vision is being increasingly used, centering jaws (or mechanical alignment) are used extensively to center and orient the components before placement. Typical pick-and-place machine jaws are 1 mm wide and 0.25-0.5mm deep, and this results in small contact areas between the jaw and the components resulting in cracked components as shown schematically in Figure 5. In addition, worn out jaws or misorientation of components result in near-point contacts and high concentration of stresses causing chipouts at the impact points and this is schematically shown in Figure 6.

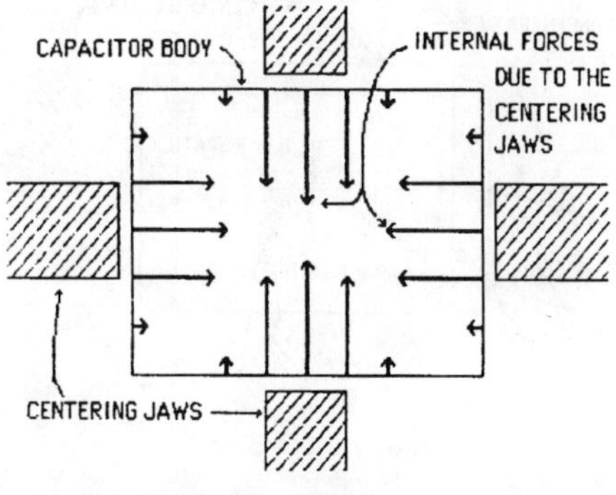

Top View of Capacitor Body, Centering Jaws and Force Concentration

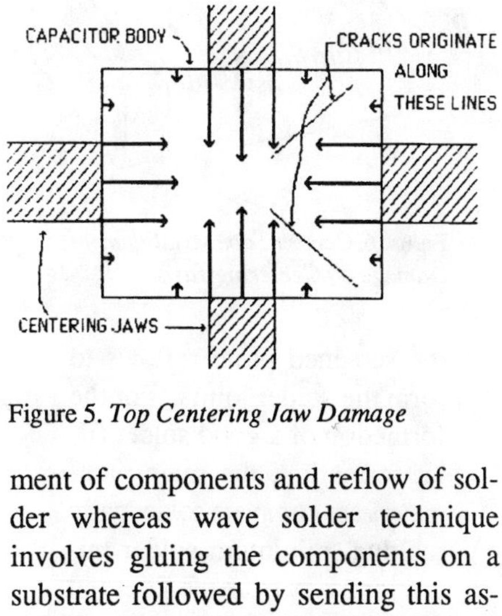

Figure 5. *Top Centering Jaw Damage*

Soldering Techniques

As described earlier, infra-red (and hot plate) and vapor phase reflow techniques involve screening of solder paste, placement of components and reflow of solder whereas wave solder technique involves gluing the components on a substrate followed by sending this assembly over a solder bath. Details of process profiling are covered in the literature[3,14] and are summarized below:

(i) *Infrared (IR) and Hot Plate Reflow*: For IR reflow, heat conducts from the substrate to component land patterns where

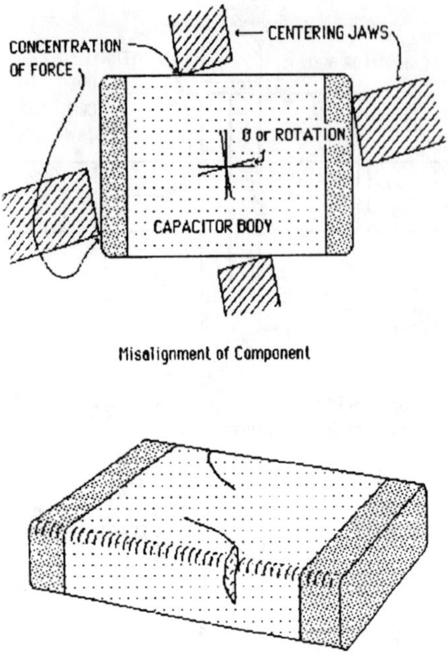

Figure 6. *Carrier Tape Misalignment Damage by Centering Jaws*

the screened paste reflows to form the solder joints. For the formation of a good solder fillet formation, the main parameters are complete solder paste melting, minimum solder migration away from the fillet and minimum component temperature exposure. The maximum rate of use or cool down of temperature in an IR reflow profile should be 4°C/sec. and a peak temperature of 215-219°C with 45-60 seconds above the melting point for Sn/Pb eutectic solders. Before the maximum temperature is reached, the boards are allowed to soak in the preheat zone at 100°C and at 150°C to activate the flux and to allow uniform heating of the board respectively. A recommended profile is shown in Figure 7.

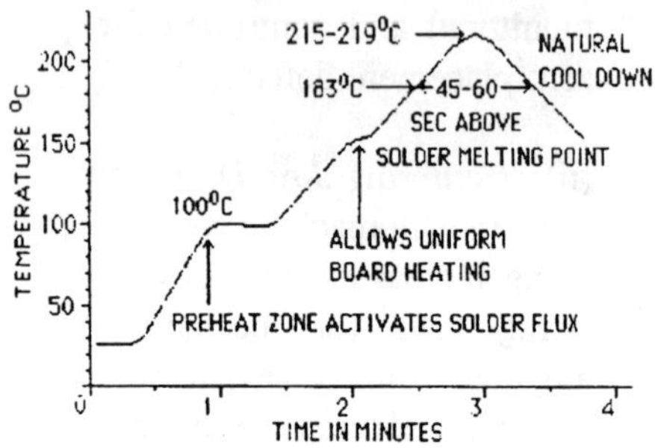

Figure 7. *IR Reflow Solder Profile*

(ii) *Vapor Phase Reflow (VPR) Soldering*: VPR uses surface heating from the latent heat of vaporization of the condensing vapor to reflow the solder paste. Preheat is suggested to avoid rapid rates of temperature in excess of 50°C/sec, to activate solder flux and to drive off the solvent volatiles. Natural cool down is also used to eliminate rapid changes in temperature. A preferred reflow profile is shown in Figure 8.

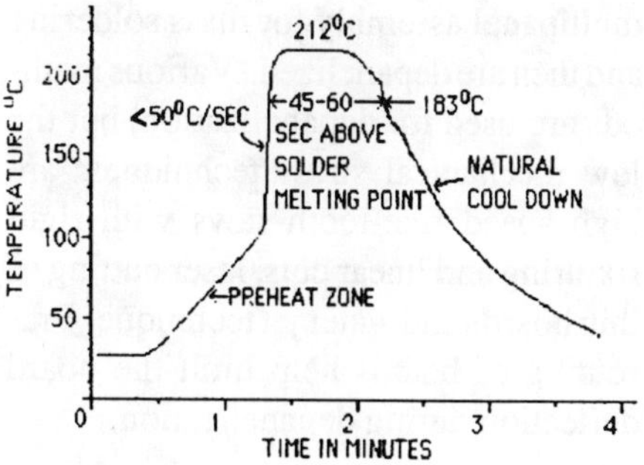

Figure 8. *Vapor Phase Reflow Solder Profile*

(iii) *Wave Soldering*: This process uses infrared preheat and liquid solder pots resulting in very rapid rates of heat transfer to the devices being soldered. This soldering technique has therefore the largest temperature gradients and is the most difficult to process. Minimum degradation of insulation resistance of MLCs as a result of thermal shock with a low temperature solder wave, high assembly bottom preheat temperatures and moderate belt speeds of 1.2 to 1.5 meters/minute are recommended. The actual solder wave temperature should be 232° +- 2°C for 60Sn/40Pb solder and the assembly bottom preheat should exceed 140°C with a dwell time of the components not to exceed 10 seconds; a recommended wave solder profile is shown in Figure 9.

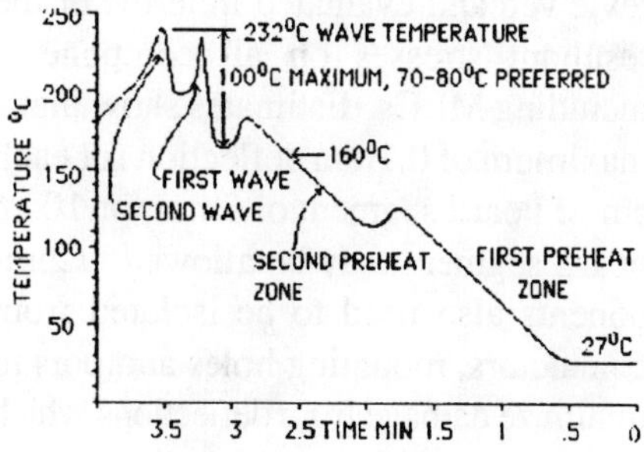

Figure 9. *Wave Soldering Profile for Minimum Thermal Shock*

The key to bottom side assembly preheat is that the top side of the assembly not be heated, and this can be accomplished by proper choice of preheating mechanisms. Area panel heaters offer the best choice as these heaters are close for direct preheating but furnish minimum convective heating[4].

In all three soldering techniques described above, temperature profiling is critical and this in itself is important for process control[4].

Post Solder Handling

Post solder handling includes testing, depanelization, installing connectors and putting in a chasis. These operations constitute another large source of defects in MLCs and is one of the least understood as far as specifications are concerned. It is essential that board deflections in any of these operations be reviewed and evaluated in terms of the resultant stresses on all components including MLCs. Estimates show that a maximum of 0.1mm deflection for each cm of board segment or 3mm for 10cm board segment may be allowed. Components also need to be isolated from connectors, mounting holes and pots to minimize damage from deflections which may occur during board handling; typical board warpage cracks are shown schematically in Figure 10.

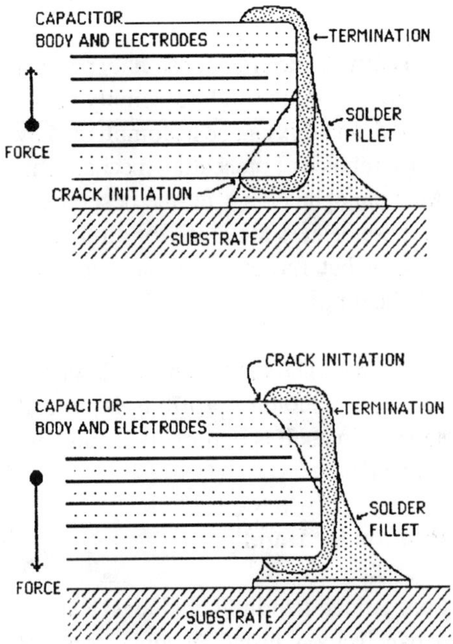

Figure 10. *Typical Board Warpage Cracks*

In many cases, boards are built on a multipanel assembly for mass soldering and then are depanelized. Various methods are used for depanelization, but the low mechanical stress techniques[2] are high speed fine-tooth saws with rigid fixturing and linear cuts, laser cutting of thin boards and water jet technique. Prerouting of boards help limit the board deflection during depanelization.

THERMAL AND MECHANICAL PROPERTIES OF MLCS

As clearly demonstrated earlier[5,6], the mechanical and thermal shock resistance parameters of MLCs are important for understanding the SMT applications of these devices. In particular, it was shown that thermal stresses resulting from rapid changes in temperature (for example, wave soldering of MLCs) can explain the thermal shock properties of these capacitors. These thermal stresses can be defined by:

$$\sigma_s = \frac{E\alpha}{(1-\mu)} \quad \frac{s\text{Ø}t^2}{(k/dc_p)} \qquad \text{..........(1)}$$

Where E is the elastic modulus, α is coefficient of linear thermal expansion, S is the shape factor, Ø is the rate of change of temperature, t is the overall thickness, μ is the Poisson's ratio and (k/dc_p) is the thermal diffusivity with k, d and c_p being the thermal conductivity, density and specific heat respectively. As equation (1) suggests, and as a number of analytical studies have shown[7,8], materials with lower E, lower α and lower μ along with higher thermal diffusivity are desirable. The above equation clearly suggests that lower rates of change of temperature and thinner geometrics are desirable. For this reason, proper preheat, for example, in wave solder as well as in IR and vapor phase soldering techniques is so critical. The thermal stresses by themselves are not sufficient for failure to occur; crack initiation and crack propogation, and the corresponding thermal stress resistance parameters R' and R" play an important role in the thermal shock behavior of chips[8]. Further evaluation of dielectric materials[9] show that parameters like thermal diffusivities show significant changes to effect thermal stresses and may be utilized to improve the thermal shock behavior of MLCs. The σ_s, R' and R" parameters are dependent on the material characteristics such as homogenity, porosity, and pre-existing flaws, and this makes the analysis very complex. Considering that the failure criterion of fracture occurs when the thermal stress reaches the fracture stress, it is possible to estimate the maximum temperature difference which a sample can withstand, and this temperature difference is directly proportional to R' and inversely proportional to half the sample thickness t and to the surface heat transfer coefficient h. Examples will be shown below to isolate some of the parameters mentioned above and show their effect on thermal shock behavior of MLCs.

EXPERIMENTAL PROCEDURE

Selected examples of experiments and their results are shown to isolate various parameters listed in equation (1) where thermal stresses depend on square of the overall thickness, the rates of change of temperature, and wetting characteristics of the termination. The complexity of this study is further shown by evaluating materials with varying K_{1c} and stresses in the termination itself.

Devices are tested by carrying out life tests and 85°C/85%RH testing of components mounted on substrates using wave solder technique.

Ceramic Dielectrics Evaluated

In an effort to isolate various parameters mentioned above, experiments were carried out with dielectric ceramic materials classified by EIA as X7R and Z5U.

(i) *Dependence of Fracture Toughness, K_{1c}*: MLCs were prepared with $BaTiO_3$ based X7R formulations referred to as ceramic A and ceramic B, and their K_{1c} values were determined using the Vicker's microindentation technique[10]. Samples were also prepared with Z5U ceramics C and D.

Table 1. Effect of K_{1c} on Thermal Shock

Chip Style	Capacitance Value (µf)	Ceramic Type	Chip Thickness (mm)	K_{1c} (MPa.m$^{1/2}$)	Thermal Shock*
1206	0.001	A	0.85	1.3	0/50
1206	0.001	B	0.85	0.9	37/50

(Number of Failures to Total Number Tested)*

(ii) *Thickness Dependence*: MLCs of varying thickness were prepared with ceramic B above to study the effect on σ_s.

Effect of Termination

It has been shown earlier[5] that MLCs with Pd/Ag termination and plated terminations show large differences in thermal shock due to the rate at which the heat transfer occurs across the termination. In this paper, other parameters which affect the termination are investigated:

(i) *Effect of Glass Frit in the Termination:* Composition of a termination used in MLCs (made from ceramic C) was altered with glass frit and its effect will be shown.

(ii) *Effect of Residual Stress in Termination:* Ni barrier plated in two different plating baths with varying pH was investigated, and the resultant stresses in the Ni film were systematically varied. These stresses are measured and their effects are shown.

Thermal Stress Evaluation Test Method

The wave solder test technique had been described in detail in an earlier publication[5] and this was used to evaluate the effect of thermal stresses. Briefly, it consists of mounting MLCs on FR4 boards using an epoxy, and soldering devices using no preheat at belt speeds of 3 meters/minute with 60Sn/40Pb solder at 260°+-5°C.

Life and 85°C/85% RH Testing

MLCs reflowed after wave soldering were tested at 125°C using twice rated bias voltage of 100 volts. In addition, in selected cases these devices were tested at 140°C and eight times rated voltage of 400 volts. Other groups of these devices were tested at 85°C/85% RH with a bias voltage of 100 volts.

RESULTS AND DISCUSSION

Effect of Fracture Toughness, K_{1c}

In Table 1, thermal shock results

are shown for 1206 0.001μf X7R chips made from ceramic A with K_{1c} value of 1.3 MPa.m$^{1/2}$ and from ceramic B with K_{1c} value of 0.9 MPa.m$^{1/2}$ respectively.

As the results in Table 1 clearly show, the susceptibility to thermal shock changes from no failures to MLCs showing about 75% failure rate. This clearly demonstrates that ceramic materials with higher K_{1c} are desirable.

Effect of Chip Thickness

In Table 2, thermal shock results are shown for Z5U ceramic D as a function of thickness. The table shows that (for a chip style) as the thickness increases, the number of failures increase rapidly and as the chip size increases, this increase is faster because σ_s also proportional to $\Delta\alpha$ (difference in coefficient of thermal expansion between the metal and ceramic).

Table 2. Effect of Chip Thickness on Thermal Shock

	Ceramic D Overall Chip Thickness (mm)		
	0.5	1.0	1.5
Chip Style	Thermal Shock Failures (%Failures)		
1206	0	0	3
1212	0	2	24
1812	0	11	41

Effect of Termination

(i) *Wetting Characteristics*: It was shown in an earlier work[5] that MLCs with plated terminations wet about a hundred times faster compared to MLCs with Pd/Ag or Ag terminations. This in turn demonstrated that the rate of heat transfer Ø can be varied to the extent that Ni plated components had approximately 100% failures whereas standard thick film terminations of Ag and Pd/Ag had no failures when subjected to the thermal shock testing; this result is shown in Table 3.

Table 3. Effect of Wetting Characteristics on Thermal Shock

1210 Chips Ceramic C 0.1 μF 1.55mm Thick		
Ag Termination	Ag with Plated Ni Termination	Ag withNi Treated @ 400°C,1hr
0/200*	199/200*	0/200*

*(*Number if Failures to Total Number Tested)*

To demonstrate that this was indeed a result of the Ni wetting characteristics, the Ni layer was oxidized while still maintaining excellent adhesion to the chip termination and the number was zero again. This result was obscured by the possibility that there are residual stresses in the Ni film.

(ii) *Residual Stresses in the Ni film:* MLCs made with ceramic B which had a low K_{1c} were plated in two different types of Ni baths, P and Q, with varying pH values. Change in the pH results in different residual stresses in the Ni film, and these stresses are measured by a gearless contractometer.

Samples were then tested for susceptibility to thermal shock and as the results show in Table 4, the number of thermal shock failures are independent of these stresses within the range of stresses measured. This result clearly demonstrates that the conclusions made earlier regarding the wetting characteristics and the rate of heat transfer are the dominating parameters.

(iii) *Glass Frit Content of Termination*: The thermal shock behavior of chips may be changed by increasing the amount of glass frit in the Ag termination. Table 5 shows the results of two ceramics having lower K_{1c} values purposely chosen to make chips with higher thermal shock failure rates so that the effect of glass frit may be evaluated.

Table 4 Effect of Residual Stresses in Ni Film on Thermal Shock

Group	Ni Bath Type	pH	Tensile Residual Stresses in Ni Film (MPa)	Thermal Shock Failures (%)
1	P	2.9	0	6.6
		4.5	13	5.4
		5.0	76	6.6
1	Q	3.0	328	3.8
		4.1	119	2.9
		5.5	77	3.3

Table 5. Effect of Glass Frit in Termination on Thermal Shock

Chip Type	Ceramic B 1206 0.1μF	Ceramic C 1825 1.0μF
Conc. of Glass Frit (Weight %)	Thermal Shock (Failures %)	
2.7	54	56
4.3	41	24
5.0	35	22

In both cases, as the amount of glass frit increases, the thermal shock failure rate drops; this is attributed to the lower thermal conductivity of the termination. This is deduced from the observation that the glass frit is distributed along the grain boundaries of the termination Ag. The behavior of the Ag termination is dominated by the glass frit which has a thermal conductivity of about 1 w/m°K compared to about 400 w/m°K for Ag. It should be noted that the porosity in all cases is about the same.

Life Test and 85°C 85% RH Results

Detailed experiments like the ones shown above can be carried out to optimize the K_{1c} values, the chip thickness, the wetting characteristics of the termination, the residual stress in the Ni film and the glass frit in the termination to build MLCs for surface mount assemblies exhibiting very low failure rates. This is demonstrated by mounting chips on experimental surface mount boards, subjecting them to wave solder processing and subjecting them to accelerated life testing at 100V, 125°C or highly accelerated life testing at 400V, 140°C followed by 85°C/85% RH. Results of some of these tests are reported in an earlier publication[6]. Models on conduction and failure mechanisms published earlier[11] show that voltage and temperature acceleration factors can be used to predict component failure rates at use conditions of, for example, 65°C, 25V (a conservative use condition; actual use voltages are usually 5 or 12 volts). In the tests listed above, 100 piece samples nominally show 0 or 1 failures and the failure may occur after 10 to 20 minutes of testing.

$$\frac{t_1}{t_2} = \left(\frac{V_1}{V_2}\right)^n \exp\left[\frac{Es}{K}\left(\frac{1}{T_1} - \frac{1}{T_2}\right)\right] \quad ...(2)$$

The time to failure t, temperature T, and voltage V may be estimated from an empirical relation arrived at by Prokopowicz and Vaskas[12] where Es is the pseudo-activation energy, K is the Boltzmann's constant and n is the voltage exponent. In our experiments, Es equals approximately 1.8-1.9 eV for various dielectrics, n equals approximately 3, and accelerated test voltage and temperature are 400 volts and 140°C, respectively. This equation (2) suggests that the failure occurring after only 1.0 minute under accelerated test conditions as outlined above will fail after 39 years in the use conditions mentioned earlier; in other words, the failure rate is sub-ppm. As noted above, actual failures occur after 10 to 20 minutes of testing, and therefore the failure rates are very low.

The various studies demonstrate the capability of the model where a few parts of this complex study can be used to improve the reliability of MLCs at the board level.

CONCLUSIONS

In this paper, various aspects of the surface mount applications of MLCs have been outlined. These consist of:

A. Zero Defect Soldering Designs

B. Pick-and-Place Considerations

C. Soldering Techniques

D. Post Solder Handling

E. Thermal and Mechanical Properties of MLCs

An understanding of the thermal and mehanical properties of MLCs along with key elements of the model where the thickness of MLCs, the fracture toughness of ceramics and the role of terminations are optimized and allow us to redesign these devices. With this redesign, it is shown that failure rates of MLCs in various surface mount assembly operations are at sub-ppm levels.

ACKNOWLEDGEMENT

We would like to thank those people in the Corporate Research and QC Laboratory who helped in compiling the experimental data. We also wish to thank Debbie Grissett for typing the manuscript.

REFERENCES

1. J. Maxwell, "Surface Mount Zero Defect Design Checklist", AVX Technical Information Series, 1989.

2. J. Maxwell, "Cracks: The Hidden Defect", Proceedings of the 38th Electronics Components Conference, Los Angeles, California, 1988, pp. 376-384.

3. J. Maxwell, "Surface Mount Soldering Technique and Thermal Shock in Multilayer Ceramic Capacitors", AVX Technical Information Series, 1987.

4. J. Maxwell, "Temperature Profiles: The Key for Proper Surface Mount Assembly Process Control", AVX Technical Information Series, 1987.

5. B. S. Rawal, R. Ladew, and R. Garcia, "Factors Responsible for Thermal Shock Behavior of Chip Capacitors", Proceedings of the 37th Electronic Components Conference, Boston, Massachusetts, 1987, pp. 145-156.

6. B. S. Rawal, M. Childs, A. Cooper, and B. McLaughlin, "Reliability of Multilayer Ceramic Capacitors After Thermal Shock", Proceedings of the 38th Electronic Components Conference, Los Angeles, California, 1988, pp. 371-375.

7. W. D. Kingery, "Factors Affecting

Thermal Shock Resistance of Ceramic Materials'', J. American Ceramic Society, 38(1) 3-15 1955.

8. D. P. H. Hasselman, ''Unified Theory of Thermal Shock Fracture Initiation and Crack Propagation of Brittle Ceramics'', J. American Ceramic Society, 52(11) 600-604 1969.

9. G. S. White, C. Nguyen, and B. Rawal, ''Young's Modulus and Thermal Diffusivity Measurements of Barium Titanate Based Dielectric Ceramics'', in Proceedings of Nondestructive Testing of High Performance Ceramics, 1987, pp. 371-379.

10. T. L. Baker and S. W. Freiman, ''Fracture Behavior in Ceramics Used in Multilayer Capacitors'', in Proceedings of the Third U.S.-Japan Seminar on Dielectric and Piezoelectric Materials, Toyoma, Japan, 1986.

11. B. S. Rawal, and N. H. Chan, ''Conduction and Failure Mechanisms in Barium Titanate Based Ceramics Under Highly Accelerated Conditions'', in Proceedings of the 34th Electronics Components Conference, New Orleans, 1984, pp. 184-188.

12. T. I. Prokopowicz and A. R. Vaskas, ''Research and Development, Intrinsic Reliability, Subminiature Ceramic Capacitors'', Final Report, ECOM-90705-F, 1969 NTIS AD-860444068, 177 pp.

13. IPC-SM-782, ''Surface Mount Land Patterns (Configurations and Design Rules)'', The Institute for Interconnecting and Packaging Electronic Circuits, 1987.

14. J. S. Baker, ''Reflow Soldering and Cleaning (Part 2)'', SMT/TAB News, August/September, 3 (5) 27-32, 1989.

SINTERING IN SILVER-GLASS ELECTRODE SYSTEMS

Takashi Yamaguchi and Hiroyuki Imai
Faculty of Science and Technology, Keio University, Yokohama, Japan 223

ABSTRACT

The densification and microstructure development in silver-glass electrode systems for thick film and multilayer applications were studied. The grain growth was promoted by the glass, and the extent of promotion was dependent upon the glass composition. Behavior of glasses was a function of the firing temperature relative to the softening point, but was not necessarily that of the solubility of silver. Location and morphology of the glass phase in the film were studied in correlation with the effects of firing conditions as well as parameters of materials such as volume fraction, composition and particle size of the glass.

1. INTRODUCTION

Silver-based electrodes are widely used in electronics for thick film and cofired applications, and a variety of characteristics are required. In order to achieve these requirements, the electrode systems should be developed based on microstructural considerations. However, it is felt that no systematic studies seems to have been established in silver-glass electrode systems. Some studies have been made on the sintering of bulk silver powder compact[1,2], but very little is reported on the effect of glass[3].

In view of these considerations, we studied the sintering behavior of silver-glass thick films with and without substrate materials and tried to understand the mechanism of microstructure development. Parameters studied are composition, particle size and content of glass, temperature, and substrate type.

Table 1. Typical Features of Glasses

	Composition(wt%)	softening point	Density
Glass 1	$PbO(90)-B_2O_3(10)$	320°C	6.7g/cm^3
Glass 2	$PbO(54)-B_2O_3(6)-SiO_2(40)$	480°C	4.7g/cm^3

2. EXPERIMENTAL

2.1 Materials

The silver powder used had an average particle size of 0.05μm by SEM. Glass powders with different compositions were classified into two fractions, larger and smaller than 2μm by water elutriation. They were designated as coarse and fine particles, respectively. Typical feature of the glass are listed in Table 1. $CaTiO_3$ ceramics and 96% alumina substrates, both commercial, were used.

2.2 Procedures

Silver-glass inks were prepared by intimate mixing of weighed silver and glass powders in a 10wt% ethylcellulose solution of butylcarbitol. The obtained inks were printed on a polyethylene film and dried at 80°C for 1 hour. The film obtained was removed from the base film. Thick films were fired isothermally between 600°C and 800°C for 10min. Inks were printed on appropriate substrate materials, dried and fired under the required time-temperature conditions. The film thickness was approximately 80μm after firing. The average grain size was obtained by the line intercept method using SEM's of free film surface. Extreme care was taken in polishing the section of thick films. Fired films were mounted in resin and polished with 3μm diamond paste with very small applied pressure. No etching was applied.

3. RESULT AND DISCUSSION

3.1 Effect of silver-glass interaction on the densification and grain growth

Figure 1 illustrates the densification behavior of silver

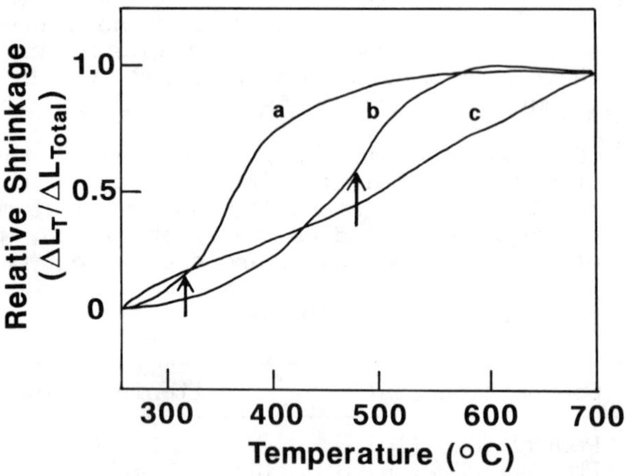

Fig.1 The densification behavior of silver and silver-glass films a) glass-1 13.4vol%, b) glass2 13.4vol%, c) Ag

and silver-glass films, in which the ordinate indicates the relative shrinkage between 250°C and 700°C. Obviously, addition of glass promotes the densification near the softening temperature of the respective glass, as shown by the arrows. The acceleration of densification is explained by the rearrangement of silver particles which is enhanced by the glass phase as a "liquid phase" in the liquid phase sintering. Consequently, the degree of promotion should depend on the glass-silver interaction, which is stronger in glass-1 as judged from the steeper densification curve near the softening temperature. This point of view was supported by the sessile drop experiment, in which pellets of glass-1 and 2 placed on a silver plate were heated at progressively increasing temperatures. At 200°C above the softening temperatures of the respective glass compositions, the glass-1-pellet was converted into a film with the contact angle of approximately 20 deg., whereas glass-2-pellet did not spread extensively with a contact angle of 50deg. Another factor to be considered is the degree of sintering relative to the glass softening. Thus, silver particles bonded firmly before glass2 softens, whereas silver particles are free to move when glass1 softens. This situation is important in microstructure development also, as will be mentioned later.

Figure 2 shows the effect of glass on the grain growth of silver. The grain growth in pure silver is extensive above 700°C, but addition of glass promotes the growth especially for glass1. The observed difference in the effect of glass composition on the grain growth has been interpreted as due to the difference in the solubility of silver in glass[4,5]. So, the apparent solubility of silver was determined by the diffusion couple method combined with EPMA. The results are shown in Fig.3. The apparent solubility is larger for glass-1. Close parallelism between the solubility and the grain growth behavior

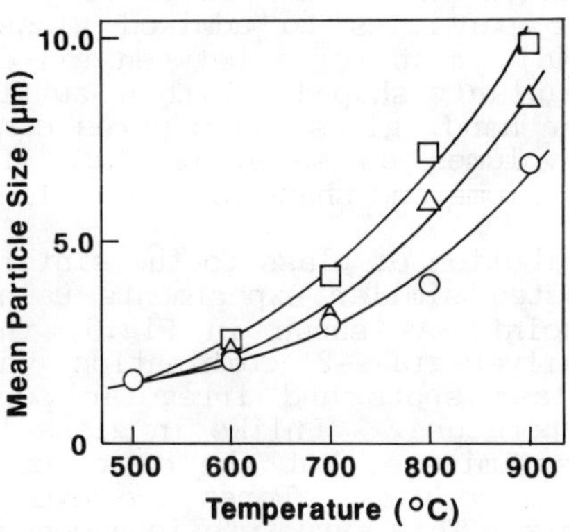

Fig.2 The effect of grain growth of films, (□)glass-1 13.4vol% (△)glass-2 13.4vol%, (○)Ag

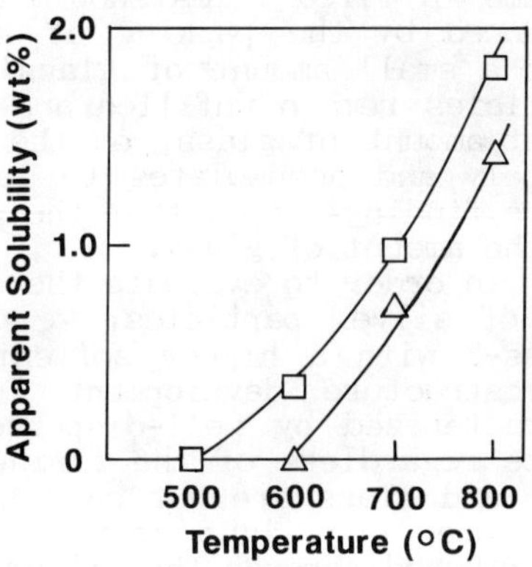

Fig.3 Apparent solubility of silver, (□)in glass-1, (△) in glass-2

should be noted. It must be mentioned that the diffusion coefficient, in addition to the solubility, is related to the growth behavior if the grain growth proceeds by the Ostwald ripening mechanism. Viscosity data of glasses indicated that glass-1 should have a diffusion coefficient larger than that in glass-2.

3.2 Microstructure development in silver-glass films and effects of materials parameters thereon.

Figure 4 illustrates typical microstructures of the polished cross sections of silver-glass 1 films of fine glass particles fired at different temperatures for 10min. In backscattered electron images(BEI) the glass phase is identified by the white areas in the gray matrix of silver. The dark areas are pores. Immediately after the shrinkage at 600°C, grain growth of silver was not extensive and glass phase was scattered in the matrix containing small pores(Fig.4a). At 700°C, however, glass phase became more discernible, and small pores are no longer observed. Firing at 800°C(Fig.4e) increased the number of pores appreciably. An important point is that some pores are partially filled with glass. The observed microstructure development in silver-glass films should be interpreted by referring to the silver-glass interaction, since good wetting of silver to glass and appreciable solubility of silver in glass were confirmed. Thus, the microstructure development depends on the rate of interparticle bonding of silver relative to that of softening of glass. Softening of glass enhances the rearrangement of silver particles, resulting in the extensive densification. The glass penetrates into the grain boundaries through capillary force, while leaving pores behind. Then, grain growth of silver grains squeezes and pushes back the glass phase to grain junctions, where the original glass particles were located in the unfired film. Thus, the number of pores partially filled with glass increased in 800°C-film(Fig. 4e).

The effect of glass content on the microstructure (Fig.4d through f) is probably explained by the above mechanism, the pore volume in fired films should depend on the volume ratio of pore produced by the packing of silver particles to admixed glass. With a small amount of glass(Fig.4d), most pores between silver particles remain unfilled and irregularly shaped. With a sufficient amount of glass, on the other hand, glass fills pores completely and accumulates to large volumes as shown is Fig. 4f. These findings imply that the pore volume and shape are controlled by the amount of glass.

In order to evaluate the contribution of glass to the sintering of silver particles, we conducted similar experiments using glass-2 with a higher softening point. As shown in Fig.5, the microstructure development for silver-glass-2 combination was characterized by well-dispersed glass spots and irregular pore shape regardless of the firing temperature. Unlike in glass-1, pore and glass are not easy to discriminate, but the dark spots are pores and light spots contain more glass. Pores are evenly distributed through the silver matrix. This structure is a consequence of extensive sintering to form a strong skeleton of silver grains before softening of glass takes place, so that the movement of glass is highly limited. This point of view is supported by the densification curve different from that of glass-1(Fig.1). Increase in firing temperature did not cause any substantial change in microstructure. Increase in glass content increased the pore volume, contrary to what was observed with glass-1. Pore and glass were not spheroidized even at higher temperature. These findings with glass-2 are probably explained by the poor mobility of glass in the stage in which silver densifies and also by the less extensive silver-glass interaction as compared with that for glass-1.

Comparison of Fig.4c and 5c will reveal the effect of glass particle size on the microstructure development. In glass-1, softening and infiltration of a glass particle into the grain

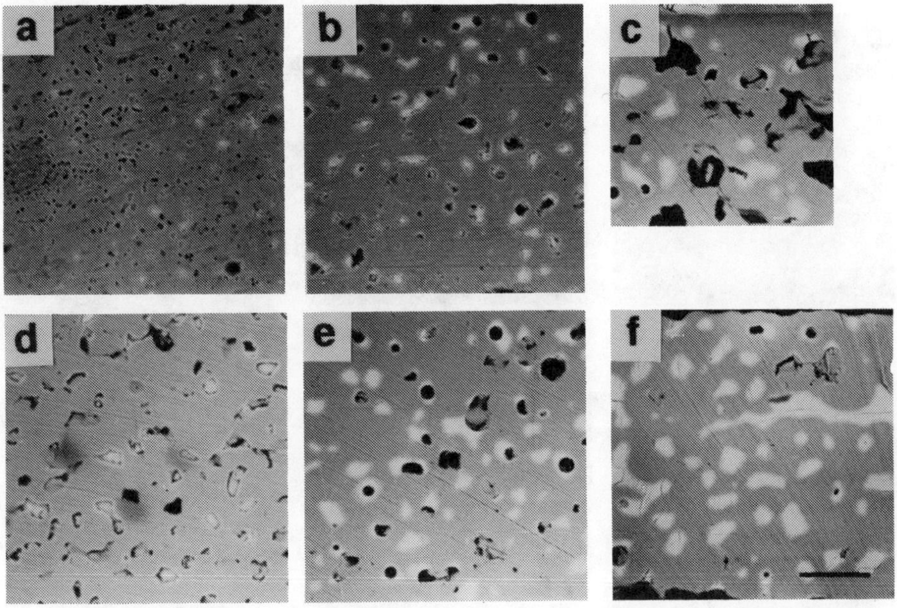

Fig. 4 typical microstructure of Ag-Glass films

a) glass-1(fine) 13.4vol% 600°C,10min
b) glass-1(fine) 13.4vol% 700°C,10min
c) glass-1(coarse) 13.4vol% 700°C,10min
d) glass-1(fine) 6.7vol% 800°C,10min
e) glass-1(fine) 13.4vol% 800°C,10min
f) glass-1(fine) 23.4vol% 800°C,10min

bar = 20 μm

boundaries leaves a pore with similar size to that of the original glass particle. Thus, an increase in glass particle size results in large pores as well as in inhomogeneity. Finer glass particles are recommended for the dense uniform microstructure.

3.3 The effect of substrate material on the microstructure development in silver-glass films.

In an attempt to extend our study to practical applications, sintering of silver-glass films on the substrate materials was studied. In the presence of a substrate material, microstructure development depends on the wettability, reactivity of glass with the substrate materials, and porosity of the substrate materials[6,7] which will play important roles in the sintering.

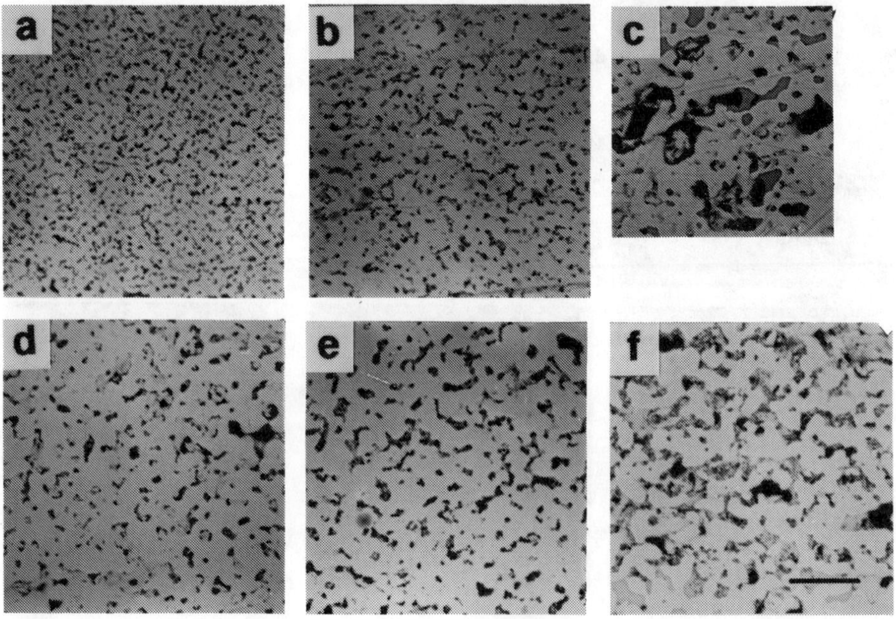

Fig. 5 typical microstructure of Ag-Glass films

a)	glass-2(fine)	13.4vol%	600°C,10min
b)	glass-2(fine)	13.4vol%	700°C,10min
c)	glass-2(coarse)	13.4vol%	700°C,10min
d)	glass-2(fine)	6.7vol%	800°C,10min
e)	glass-2(fine)	13.4vol%	800°C,10min
f)	glass-2(fine)	23.4vol%	800°C,10min

bar = 20 μm

Figure 6 illustrates typical microstructures obtained by firing the silver-glass thick films printed on a 96% alumina substrate and $CaTiO_3$ ceramics. The white layers are the thick film and the dark areas are substrate. Glass has penetrated into the substrate, resulting in the depletion of glass in the thick film. The microstructure is characterized by the large pores, which were not observed in the absence of the substrate. In 96% alumina, a glass layer was observed between the substrate and thick film. In $CaTiO_3$ ceramics, a reaction layer was observed, which was identified to be a solid solution of lead titanate and calcium titanate by XRD.

Glass-2 exhibited a different feature; glass did not penetrate into the substrate regardless of the composition. Therefore, no substantial change in microstructure except for the accumulation of glass at the film-substrate interface was observed even in the presence of substrate materials.

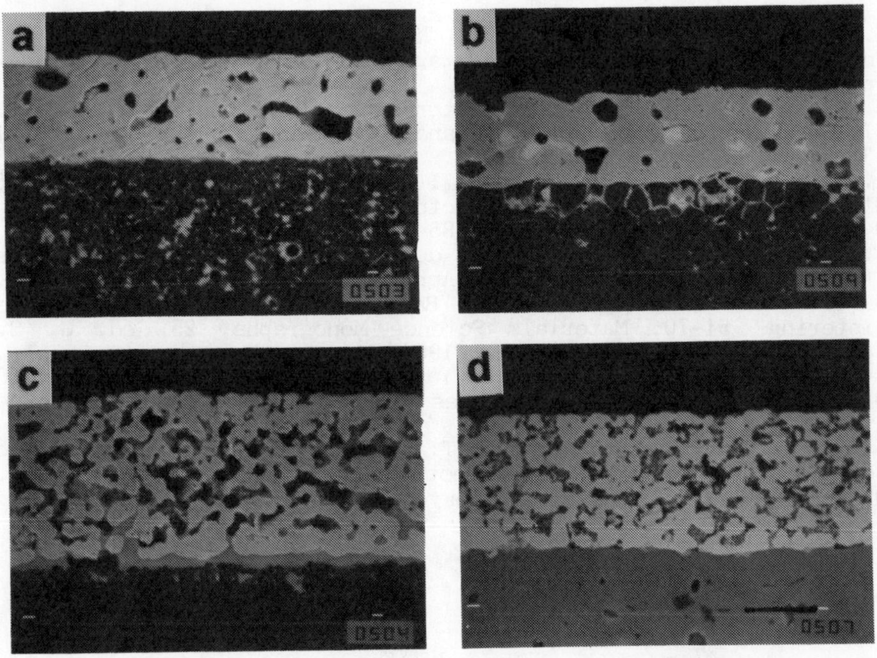

bar=20 μm

Fig. 6 The effect of substrates for microstructure (fired at 800°C for 10min.)
a) glass-1(fine) 13.4vol% printed on a 96%Al_2O_3
b) glass-1(fine) 13.4vol% printed on a $CaTiO_3$
c) glass-2(fine) 23.4vol% printed on a 96%Al_2O_3
d) glass-2(fine) 23.4vol% printed on a $CaTiO_3$

4. SUMMARY

1. Grain growth of silver in the films is promoted by the additional glass. The degree of promotion depends on the solubility of silver in glass and the viscosity of glass.
2. The densification of films is accelerated by the rearrangement of silver particles enhanced by the glass phase as a "liquid phase" in liquid phase sintering.
3. The microstructure of films is determined by the densification of silver particles near the softening point of glass. Finer glass particles lead to a dense uniform microstructure.
4. The glass-substrate interaction strongly affects the sintering of silver-glass thick films, and thick film formation should be done with reference to the substrate materials.

Acknowledgment: The authors express their appreciation to TDK electronics for the financial support.

Reference

1. S. Iwata and T. Sahashi "Sintering of Ultrafine Metal Powders. I. Coalescence Growth Stage of Au and Ag", J. Appl. Phys. Jpn., 19[6] 1039-1044 (1980).
2. S. Iwata and K. Hayakawa "Sintering of Ultrafine Metal Powders. II. Neck Growth Stage of Au, Ag and Cu" J. Appl. Phys. Jpn., 20[2] 335-340 (1981).
3. K. Yajima and T. Yamaguchi "A Feature on the Microstructure of Glass-Bonded Ag Thick-Film Conductors" IEEE Trans. Compon. Hybrids Manuf. Technol., CHMT-7[3] 281-285 (1984).
4. W. D. Kingery "Densification during Sintering in the Presence of a Liquid Phase I. Theory" J. Appl. Phys., 30[3] 301-305 (1959).
5. G. Petzow and W. A. Keysser "Basic Mechanisms of Liquid Phase Sintering" 51-70, Materials Science Monographs, 25, ed. G. S. Upadhyaya, Elsevier, Amsterdam (1984).
6. K. Harano, K. Yajima and T. Yamaguchi "Spreading of Hybrid IC Glass Frits on Alumina Substrates and Reaction Therewith" Yogyo-Kyoukai-shi 92[9] (1984).
7. T. E. Tayler, S. Arai and S.T. Joslin, "A Chemical Approach to Conduction Performance on Dielectirics and Alumina", Proc. Int. Microelec. Conf., 1988 Proc., Tokyo, 177-185(1988).

A COMPARISON OF AQUEOUS AND NON-AQUEOUS SLURRIES FOR TAPE-CASTING, AND DIMENSIONAL STABILITY IN GREEN TAPES

Paul Nahass, W.E. Rhine, R.L. Pober, H.K. Bowen,
Ceramics Processing Research Laboratory, M.I.T., Cambridge, MA;
W.L. Robbins, C.S. Draper Laboratory, Cambridge, MA.

ABSTRACT

Aqueous and non-aqueous formulations were developed for tape casting alumina for the purpose of comparison of the resulting green sheets. The aqueous system employed an acrylic emulsion binder while the organic slurry contained the same polymer dissolved in toluene. Green tapes from both systems had similar physical properties. Samples cast from the aqueous formulation exhibited lower linear shrinkage than those from the organic system after aging for 35 days under controlled-ambient conditions. Treatments intended to improve dimensional stability had little effect on aqueous-based green sheets but yielded solvent-based samples which experienced a dimensional change of ~0.15% over 35 days and no difference in shrinkage in the tape casting direction compared to the transverse direction. Solvent-based tapes had greater linear firing shrinkage than aqueous-based samples but lower shrinkage anisotropy on firing. A treatment which involved 24-hour exposure to saturated water vapor at 80°C yielded solvent-based green tapes which exhibited the lowest shrinkage and anisotropy after aging and after firing.

INTRODUCTION

Tape-cast ceramic green sheets for multilayer electronics applications are generally punched, screened, stacked, and laminated prior to firing. The punch step usually involves mechanical perforation of *vias* through each layer which are subsequently filled with conductor pastes to interconnect the circuitry between layers. Aside from exhibiting strength and flexibility sufficient for subsequent handling and processing steps, green sheets must exhibit excellent dimensional stability over time to ensure accurate interlayer registration upon lamination[1,2]. Even slight variations from design tolerances can lead to device rejection. As ceramic components become more densely loaded with circuits, the necessity for maintaining rigid device tolerances becomes of the utmost importance.

The objective of this investigation was to develop formulations for tape-casting alumina using water-based and organic solvent-based slurries and to determine the effect of changing solvent systems on slurry processing and green tape quality. Considerations involve selection of compatible materials to effect a valid comparison and evaluation of processing steps necessary for each system. Green tape comparisons with regard to physical and mechanical properties, dimensional stability upon aging, response to stabilization treatments, and firing shrinkage behavior will be addressed.

Evaluating the effect of changing any one component of a slurry formulation requires maintaining all other components as constant as possible, keeping in mind component interactions which may change an important property such as particle packing, slurry viscosity, or polymer morphology.

MATERIALS AND PROCESS SELECTION

AKP-30 alumina powder (Sumitomo Chemical USA, New York) was chosen for use in this study. AKP-30 showed rheological and dispersion properties superior to those of other alumina powders evaluated, including MPA-4 (Ceralox Corporation, Tucson, AZ). AKP-30 has an average particle diameter of 0.3 μm, surface area of 6.4 m^2/g, and purity of 99.99%.

Binders from two general classes were recommended by industrial contacts for use in water-based tape-casting systems: standard water-soluble polymers and stable polymer emulsions dispersed in water. Rhoplex B-60A (Rohm and Haas, Philadelphia, PA), a proprietary emulsion formulation of acrylic polymers, yielded the toughest films with the simplest preparation. The emulsion afforded very high loadings of polymer (nearly 50 wt% solids, as received) at viscosities *2 to 3 orders of magnitude* less than similar concentrations of polymer solutions. This property is especially desirable for tape casting; low solvent content keeps drying times down and can lower the large volume reduction (as high as 70%) encountered on drying, limiting drying cracks.

A potential drawback of using an emulsion for tape casting is film formation. Onions[3] describes the general behavior of these latex films. He indicates that emulsions filled with TiO_2 formed uniform films with little or no porosity if the particle concentration was lower than the *critical pigment concentration*, or CPC[4]: the morphology of the film may reflect the emulsion latex particle size. Particle size distribution of Rhoplex B-60A emulsion droplets and AKP-30 powder particles dispersed with Darvan®C (R.T. Vanderbilt, Norwalk, CT), an ammonium salt of poly(methacrylic acid), in water were measured using a Horiba CAPA 500 centrifugal analyzer. The mean diameter of the ceramic particles was ~2.5 times that of the latex, revealing the size of the defects which might be incurred above the CPC.

Chemically, little is known about Rhoplex products. Transmission Fourier Transform Infrared (FTIR) spectroscopy (IR-44, IBM Instruments) was performed on a dried film of B-60A . The spectrum resembled many acrylate and methacrylate polymers, especially those of poly(ethyl acrylate) and poly(methyl methacrylate). A mixture of these polymers could provide a glass transition temperature near that of B-60A (reported at about 15°C).

B-60A films were dried at 100°C for 24 h. Small amounts of these dried films did not dissolve in either ethanol or isopropanol, but readily dissolved in toluene and toluene/ethanol mixtures. Toluene/ethanol was considered acceptable as a solvent system which could dissolve the dried B-60A acrylic polymer as well as an effective water-soluble dispersant for AKP-30. Toluene alone is responsible for dissolution of B-60A solids in a toluene/ethanol system; similarly ethanol alone might be responsible for dispersant dissolution. A 50/50 mixture of toluene and ethanol (by weight) readily formed a 20 wt% polymer solution and also formed a 10wt% solution of p-HBA (para-hydroxybenzoic acid, Aldrich Chemical, Milwaukee, WI). Therefore this mixture was chosen as the solvent system to be compared with water for casting tapes.

Darvan®C was found to give stable dispersions of AKP-30 in water with powder settling densities near 50% of theoretical after 30 days. p-HBA also yielded stable dispersions with settling densities near 50% in water; however p-HBA seemed to break down the emulsion in Rhoplex B-60A and therefore could not be used. Results for p-HBA in toluene/ethanol were not as promising, but p-HBA in pure ethanol provided adequate settling densities for comparison. The maximum density for the poorer dispersant (pHBA) occurred at 2.0 wt% pHBA (by weight of powder) and was 43% of the theoretical density for alumina. The same settling density was achieved using 0.8% Darvan®C in water, hence these amounts were used in slurry formulations.

Poly(propylene glycol), or PPG (molecular weight 1200, Fluka Chemical Corp., Hauppage, NY) was used as plasticizer for both aqueous and organic systems. A solution of lecithin (capsules, CVS, Woonsocket, RI) in isopropanol was applied to tempered glass plates as a release agent for the dried green tapes. A fluorocarbon dry release was used for the solvent-based slurries (MS-136, Miller-Stephenson, Danbury, CT).

Cracking was common in many water-based tapes. Garino[5] found the principal factors responsible for cracking in ceramic slurries were binder content, surface tension, solids content, and film thickness. Since desired film thickness and binder content were fixed for this study, cracking was deterred by increasing solids content and reducing surface tension in the slurry. Slurries containing powder, dispersant, and solvent were centrifuged at 3500 rpm for >30 min to remove excess water not necessary for powder dispersion. Usually 60-70% of the water necessary for initial dispersion was decanted after centrifugation while flow properties were maintained.

Surface tensions of solvents and solvent-binder mixtures were measured to uncover slurry differences which might induce cracking in the aqueous system. A Rosano Surface Tensiometer (Biolar Corp., North Grafton, MA) which employs the Wilhelmy Plate principle of surface tension measurement was used. Pine oil was recommended as a surfactant, by the binder manufacturer, for Rhoplex acrylic latexes for use in pigment-filled films[6]. Significant reduction in surface tension of water was achieved with small additions of pine oil to the emulsion. Additions of nearly 1 wt% pine oil (too much according to Allyn) did not reduce the surface tension of water or water and binder to near that of the solvent system, but 0.1 wt% seemed to eliminate cracking in tapes.

The slurry formulations used for this study are presented in Table 1. To cast a tape, release agent was applied to a glass plate and allowed to dry. Dispersant was dissolved in the appropriate solvent, which was then added to the powder and stirred until fluid. The mixture was sonicated 3-7 min (until no visibly undispersed ceramic particles remained) using a 400-W Ultrasonic Disperser with a 0.5-in probe (Harvey Scientific, Buffalo, NY) set at 40 W. More solvent was added if the rheological properties were unsuitable for sonication. The mixture was added to the binder-plasticizer mixture (often a small amount of solvent was necessary for miscibility in the aqueous system) and ball-milled for at least 4 h to intimately mix the components. This step enabled tapes with low polymer concentrations to have high strength and toughness; without milling the same slurries formed brittle, cracked tapes. Ball-milling was carried out on a roller mill in porcelain jars filled to slurry height with alumina media. The slurry was filtered through a 400-mesh stainless steel screen to remove large particles and bubbles. The solvent-based slurries often required pressure for efficient filtration. Slurry viscosity was then measured using a Model RVT Synchro-lectric viscometer (Brookfield Inc., Stoughton, MA). Slurries were cast inside a glove box: under flowing air at 23°C and 50% relative humidity (RH) for aqueous system; in pure N_2 for solvent system. The casting head had two blades and was driven by magnetic field along rails (Anorail-5, Anorad Corp. Hauppage, NY) at ~2 cm/s. A plastic cover was placed over the cast to slow the drying process. Polymer films were made by manually casting mixtures of solvent and all organic components on glass.

Table 1. Slurry Formulations for Comparison

Component	Organic System (g)		Aqueous System (g)	
Powder	Al_2O_3, AKP-30	55	Al_2O_3, AKP-30	55
Solvent	Ethanol/Toluene[#]	85	DI Water	15
Dispersant	p-HBA	1.1	Darvan® C	0.7
Binder	Dried Rhoplex	10	Rhoplex B-60A	20
Plasticizer	PPG-1200	4	PPG-1200	4
Surfactant	None		Pine Oil	0.1
Solids Content (%)		10-15		44*
Slurry Viscosity (cp)		~5000		700
Surface Tension (dyne/cm²)		29		42

*40% of this was due to polymer solids in Rhoplex.
[#]Equimass mixture used.

After casting, slurries were allowed to dry at least 24 h with room conditions 23 ± 2°C and 50 ± 7% RH. Lower drying times enabled by controlled low-humidity atmospheres always caused aqueous-based tapes to crack severely on drying. Drying cracks were observed to occur parallel to the direction of the cast while cracks due to tape brittleness usually occurred normal to the casting direction. Tapes were removed from the glass plates with a wide razor blade and cut into two-inch squares, then punched by an automatic programmable punch (Model SK 5470; Schneider & Marquard, Newton, NJ, with model MD5-5 Berger-

Lahr controller) with 0.25 cm measurement holes, and subjected to appropriate storage conditions or treatments. Strips of tape and polymer films ~0.5 cm x ~6 cm were cut parallel and normal to the casting direction for directional property measurements.

Punched tape samples and accompanying strips for both aqueous and organic systems (in duplicate) were subjected to one of several treatments in order to assess dimensional change on aging. Sample treatments included: 1) a simple bake (2 h at 120°C in flowing air); 2) thermal cycling (-5°C for 30 min, then 120°C for 30 min, 4 times); 3) solvent exposure[7] (saturated toluene vapor for 1 h at room temperature, then *simple bake*); 4) humidity[8] (24 h at 100% RH, 80°C, then 24 h at 80°C in flowing air). Control samples were left untreated. All samples were then stored under controlled conditions, 23 ± 2°C and 50± 7% RH. Positions of punched holes were measured using an optical comparator with radial viewing template at 20x magnification. Hole positions were measured after punching, immediately after treatments, periodically during aging, and after firing.

Green tapes were sintered for 30 min at 1500°C in air. A porous alumina coverplate was placed on the sample during sintering to ensure flatness of the sample while allowing organics to escape. Fired densities were measured geometrically and by the Archimedes method with t-butanol as the solvent, with some at nearly 99%. Sintering temperature and time were not optimized.

Properties of green tapes were analyzed by several techniques. Weight loss was measured by thermal gravimetric analysis (TGA) on a 7 Series Thermal Analysis System (Perkin Elmer) at a constant heating rate of 10°C/min to 500°C. Thickness was measured mechanically using an Ono Sokki measurement device. Bulk and directional green densities were measured dimensionally. Bulk porosity was calculated from green density measurements and TGA results. Surface finish was analyzed over a 1 cm profile. Glass transition temperatures of tape strips and polymer films were measured by dynamic mechanical analysis on a Rheovibron Direct Reading Dynamic Viscoelastometer (Model DDV-II, Imass, Hingham, MA) over a temperature range of -120°C to 150°C.

RESULTS

Table 2 summarizes the physical properties of green tapes. All data in this study are from individual casts of solvent- and aqueous-based slurries. Overall, physical and compositional consistency was maintained between the aqueous-based and solvent-based tapes. TGA weight loss represents the amount of organics and other components which decompose at temperatures less than 500°C. Green tape thicknesses were averaged from 50 to 75 points over the entire cast (15 x 100 cm), excluding 5 cm from the beginning and end of the cast and 1 cm from the side edges. Standard deviations of thickness were low and green densities were satisfactorily high. The low porosity indicates powder concentrations in green tapes were probably below the CPC for each system, hence drastic morphological differences should not be present. Green tape samples did not dissolve in toluene or hexane.

Dimensional stability results for green tapes are shown in Figure 1;

Table 2. Physical properties of green tapes for comparison.

	Organic Slurry	Aqueous Slurry
TGA Weight Loss (%)	21.8	21.3
Avg. Thickness (μm)	239	274
Avg. Bulk Density (g/cm³)	2.43	2.46
Porosity (%)	0.8	0.6

estimated shrinkage error is 0.05% for all data. Untreated solvent-based samples exhibited anisotropic shrinkage on aging, with parallel shrinkage significantly greater than transverse shrinkage. During stabilization treatments, solvent-based tapes experienced up to 3% linear shrinkage, with all samples shrinking more in the casting direction. The baked samples showed significantly lower shrinkage during treatment than all other samples. After completion of treatment, all solvent-based tapes underwent significantly less shrinkage than untreated tapes and showed little or no anisotropy. Samples which underwent the bake shrunk the most after treatment and had the most anisotropy while samples which underwent the solvent and humidity treatments had less than 0.2% linear shrinkage after 35 days.

Untreated aqueous-based green tapes shrunk less than solvent-based tapes on aging. However, all aqueous-based tapes experienced significantly less shrinkage during stabilization treatments than solvent-based samples did. After treatment, little change was observed in the magnitude of shrinkage from that of untreated samples, though anisotropy was lowered. As in the solvent system, the humidity treatment caused the most dramatic difference in shrinkage behavior. Reproducibility was assessed by repeating the slurry preparation procedure to form a new batch of green tape. Tape samples were subjected to solvent and humidity treatments and again compared to untreated tape. All data agreed with that from the previous batch within the measurement error.

Green tapes from the two systems did exhibit some similar behavior. Samples within each slurry system showed wide variations in overall shrinkage (combining shrinkage during and after treatments) which were accompanied by

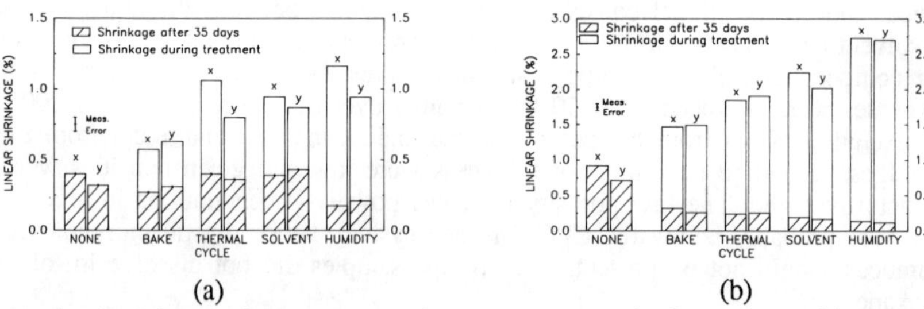

(a) (b)

Figure 1. Shrinkage of green tapes after treatments and after aging: (a) water-based, (b) solvent-based.

corresponding green density changes. No quantitative correlation could be made, however, due to the comparatively low accuracy of density measurement capability. Sample weights did change *during* treatments, but only untreated samples experienced weight changes greater than 0.1% on aging.

Firing Behavior

Directional linear shrinkage incurred during firing of samples is given in Table 3. Each shrinkage value was averaged from two samples, with four data points per sample. Several samples from different slurry batches were fired to check reproducibility of firing shrinkage. All green tapes from the solvent-based slurry had less shrinkage anisotropy on firing, in general. Stabilization treatments performed on solvent-based green tapes reduced the magnitude of firing shrinkage noticeably and consistently except for the humidity treatment, which caused more shrinkage reduction than the other treatments and eliminated anisotropy.

Fired densities were measured for all samples by geometric methods and were consistent for all samples. Archimedes measurements gave artificially high results for all samples. The discrepancy between the results from the measuring techniques was attributed to open porosity in films which is known to be present in ceramics below 90% of theoretical density. The same samples refired at 1600°C for 1 h sintered to between 99 and 100% of theoretical density (densities were equivalent using both measurement techniques), therefore samples fired to 1500°C for 30 min can be considered to have been underfired.

The aqueous-based samples shrunk less than the organic, solvent-based samples, in general, and with more shrinkage anisotropy. As in the solvent-based

Table 3. Linear firing shrinkage of green tape layers (%); all layers were aged 40 days prior to firing unless otherwise noted.

Solvent System:	Organic		Aqueous	
Direction:	parallel	transverse	parallel	transverse
No Treatment	19.0	18.7	16.5	17.0
	18.0[#]	17.8	16.5[#]	16.9
	17.6[+]	17.3		
Bake	18.6	18.4	16.1	16.7
Thermal Cycle	18.6	18.4	16.3	16.8
Solvent	18.6	18.4	16.2	16.6
Humidity	17.4	17.4	15.9	16.5
	17.4[#]	17.5		
Avg. Density (% theor.)*	87.2		86.5	
Std. Dev.	0.98		0.86	

[#]Aged 70 days. [+]Aged 100 days. *Measured geometrically.

system, treatments lowered the magnitude of firing shrinkage without altering anisotropic behavior. Samples exposed to the humidity treatment underwent the lowest firing shrinkage (as in the solvent-based system) but shrinkage anisotropy remained the same as for all other aqueous samples. Fired densities for aqueous, emulsion-based tapes were again consistent for all samples and were similar to those for solvent-based samples.

Reproducibility of firing shrinkage was investigated by firing samples which were prepared from different slurry batches. These samples had been aged longer before firing (70 or 100 days) than the two batches used for most of the data collection in this study (40 days). The magnitude of the shrinkage for untreated solvent-based samples decreased with aging time, x/y ratio remained constant. The magnitude of the firing shrinkage for the sample which had aged 100 days was nearly identical to that for the humidity-treated sample. The untreated, aqueous-based sample aged 70 days experienced the same shrinkage as tape from a different aqueous slurry batch which had been aged only 40 days. The solvent-based sample which underwent the humidity treatment and 70 days of aging had the same shrinkage (within measurement error) same as the humidity-treated sample which had aged only 40 days.

DISCUSSION

Every effort was made to keep systems as similar as possible in this study, in order to isolate green tape differences due to solvent and those caused by changes in binder states. Processing techniques and formulation compositions needed to be slightly altered for each system to produce samples which would be valid for comparison (see Materials and Process section). While the results presented are for only one cast for each system, they reflect numerous attempts, as both systems had major processing difficulties to overcome. The solvent-based slurry, very much more volatile and irritating to process, was relatively insensitive to perturbations; it produced strong, uniform green tapes nearly every time. The aqueous slurry, without dissolved polymer, was easy and not unpleasant to process but had a smaller tolerance to minor changes in drying conditions, casting composition, or film thickness; it produced crack-free, uniform green tapes only when all variables were controlled extremely well.

The concept of aging of polymers and polymer systems is the subject of an exhaustive work by Struik[9]. Amorphous polymers experience aging only below their glass-transition temperature where aging is defined by Struik as the process of approaching thermodynamic equilibrium. Glasses, for example, are known to undergo dimensional change on very long time-scales. Polymers in the rubbery state (above T_g) are able to achieve equilibrium very quickly (essentially faster than measurement can detect), and are therefore considered (by Struik) not to age. Semi-crystalline polymers and rubbery polymers filled with carbon black were found to age at temperatures above their T_g. The hypothesis for this phenomenon is that polymeric material near or attached to the surface of the particle or crystallite behaves as part of this rigid matrix while that material further away from particle surfaces behaves as amorphous polymer. The composite system cannot achieve equilibrium rapidly due to the lowered mobility of the attached polymer due to constraint by the rigid filler/attached-polymer matrix and the broadened glass transition of the amorphous phase. The model proposed by Struik was verified by correlations of aging with particle size of the filler.

The driving force then, for aging in filled polymer systems is internal stress in the system. Stresses are generated when polymer chains are constrained by the matrix, usually due to processing. Therefore, it can be seen that two factors influence aging in these systems: (1) the degree of stress in the system, or how far it is from thermodynamic equilibrium determines the magnitude of the conformational changes which might occur; and (2) the mobility of the bulk polymer phase in the matrix will influence the time-scale at which equilibrium is approached, or the rate of aging.

The purpose of the stabilization treatments was to determine their effect on limiting dimensional change in ceramic green sheets. All the treatments involved heating which accelerated the aging (hence shrinkage) process. The bakeout involved heating at 120°C, just above the boiling points of toluene (110.5°C) and water (100°C), and was intended to eliminate nearly all residual solvent that might cause dimensional change on evaporation over time. Thermal cycling provided more resistance to dimensional change than constant temperature treatment for the same total time. Four cycles of 30 min each at -5°C and 120°C were administered to green tapes. The total time at 120°C was equal to the bakeout time, 2 h. Cycling caused volume changes which allowed the system to rearrange and relax. Solvent-vapor saturation is believed to relieve residual stresses in the green tape incurred during constrained drying[7], by swelling the polymer and causing an increase in mobility due to higher free-volume in the swelled state. The humidity treatment was taken from an ASTM test for dimensional testing of plastics after simulated aging conditions[8]. It also causes swelling which increases mobility and causes shrinkage.

Higgins[10] found by infrared spectroscopy that poly(methyl methacrylate) chemisorbs on alumina surfaces, sometimes displacing other surface species. This phenomenon may have occurred in these systems, with soluble poly(alkyl acrylates and methacrylates) present. A ceramic-polymer interaction like this would contribute to property differences between systems on aging and firing. The presence of toluene (a good solvent for the polymer) in the solvent-based slurry might thermodynamically prevent some polymer from adsorbing, thereby allowing for more polymer mobility and hence more shrinkage over time. This issue is currently being investigated.

Another factor probably affecting shrinkage differences between the two systems and between x and y directions within the same green tape is the physical state of the binder in the slurries. Shear under the doctor-blade during tape-casting causes alignment or deformation in both polymer systems. Soluble polymer chains become *stretched* or aligned more in the casting direction and then become frozen in that state by drying; hence more shrinkage occurs in the casting direction. Dispersed latex particles in the emulsion may be dragged along by the doctor-blade, and then in order to form a continuous film, polymer chains must stretch somewhat in the casting direction as the water evaporates, causing anisotropic shrinkage in that system due to the film-formation mechanism. Shrinkage differences between the two slurry systems are most likely affected by the solvent-drying rate combined with the polymer conformation.

REFERENCES

1) B. Schwartz, "Review of Multilayer Ceramics for Microelectronic Packaging," *J. Phys. Chem. of Solids*, 45[10], p.1051, 1984.

2) R.R. Tummala, "Ceramic Packaging," Ch. 7 in Microelectronics Packaging Handbook, R.R. Tummala and E.J. Rymaszewski eds., Van Nostrand Reinhold, New York, 1989.

3) A. Onions, "Films from Water-Based Colloidal Dispersions," Manufacturing Chemist, March 1986, p.55; article continued April 1986, p.66.

4) G.P. Bierwagen, "The Critical Pigment Volume Concentration and Its Effects in Latex Coatings," in Organic Coatings Science and Technology, Vol. 5, G. Parfitt and A. Patsis, eds., Marcel Dekker, New York, 1983.

5) T.J. Garino, "Patterning, Drying, and Sintering of Particle Films on Rigid Substrates," PhD. Thesis, M.I.T. Dept. of Materials Science and Engineering, Cambridge, MA, May 1987.

6) G. Allyn, "Acrylic Ester Emulsions and Water-Soluble Resins," Ch. 1 in Characterization of Coatings: Physical Techniques, Vol. 1, Part 1, R.R. Myers and J.S. Long eds., Marcel Dekker, New York, 1976.

7) G.F. Hait and R.W. Nufer, "Process for the Elimination of Dimensional Changes in Ceramic Green Sheets," US Patent #3,953,562, April 27, 1976.

8) "Determination of Weight and Shape Changes of Plastics Under Accelerated Service Conditions", D756-78, Annual Book of ASTM Standards, Vol. 8.01, American Society for Testing of Materials, New York, 1983.

9) L.C.E. Struik, Physical Aging in Amorphous Polymers and Other Materials, Elsevier Scientific Publishing Company, New York, 1978.

10) R.J. Higgins, "The Chemistry of Carbon Retention During Non-Oxidative Binder Removal from Ceramic Greenware," PhD. Thesis, M.I.T. Dept. of Materials Science and Engineering, Cambridge, MA, January, 1990.

Mo-Mn METALLIZATION ON AlN SUBSTRATE

S. Nozaki, A. Kanda, S. Iio, Y. Nakayama, and N. Kato
NTK Technical Ceramics, Division of NGK Spark Plug Co. Ltd., Komaki, Aichi, Japan

ABSTRACT

Mo-Mn post fire metallization, which has been widely used on alumina substrates to make a electro-conductive layer for electronic devices, was applied to AlN substrates. Metallization on AlN substrate with adhesion strength of more than 20 MPa was obtained under firing conditions at 1543-1593K in a little wet H_2 and N_2 mixed gas with dew point at 243-258K, different from that for Al_2O_3 substrates. SiO_2 addition to metallization paste was effective for AlN material with no grain boundary phase which reacts with metallization paste. A glassy intermediate layer including nitrogen and oxygen was observed between metal layer and AlN substrate. The metallized AlN substrate passed our inspections for Al_2O_3 substrate such as baking test, die attach test, wire bondability and solder wettability.

1. INTRODUCTION

Nowadays, the most important demand for electronics industry is high speed and high density circuits. Therefore silicon chip size becomes much larger and larger power is supplied to silicon chips, and problem of heat dissipation from chips is becoming critical as a result even though alumina (Al_2O_3) ceramics, which have been currently used for package and substrate materials, have fairly good thermal and electrical properties. Recently, high thermal conductive ceramic materials such as BeO, SiC and AlN have been developed. Above all, AlN material is thought to be the most promising material, because thermal conductivity of AlN material is as high as BeO and thermal expansion coefficient is similar to silicon chip. In addition, AlN has electrical resistance, dielectric constant and mechanical strength comparable to Al_2O_3, and has no toxicity like BeO.

In order to utilize AlN substrates, it is necessary to metallize these substrate. In case of Al_2O_3 substrate, most common metallization is Mo-Mn post firing metallization technique. Adhesion strength of 20-100 MPa can be obtained for these Al_2O_3.[1-5]

Many attempts[6-11] have been made to make metallized layers on AlN substrate. However it seems to be difficult to obtain desired properties such as good adhesion strength, hermeticity and high reliability for applications in electronics such as heatsinks, hybrid ICs and packages.

One reason for these difficulties is that AlN material has no grain boundary phase, especially in case of high thermal conductive AlN material, and has little of oxygen. Therefore it hardly reacts with metallizing paste in contrast to Al_2O_3. Another reason is that mismatch of thermal expansion between AlN and metallized layer leads to residual stress. In our work, addition of SiO_2 into Mo-Mn paste was attempted because liquid phase exists at temperature higher than 1524K in SiO_2-MnO two component system[12], and effect of firing condition such as firing temperature and dew point of environmental gas on the adhesion strength, microstructure of intermediate layer and thermal conductivity was studied.

Table 1. Properties of AlN substrate.

Properties	Typical Value
Density (g/cm³)	3.25
Bending strength (MPa)	400-500
Thermal conductivity (W/m·K)	150
Thermal expansion coefficient (R.T.~ 673 K) (1/K)	4.5×10^{-6}
Volume resistivity (Ω·cm)	$> 10^{13}$
Dielectric constant at 1MHz	9.0
Dielectric loss at 1MHz	8×10^{-4}

2.EXPERIMENTAL PROCEDURE

The AlN substrates used in this study were prepared by well known pressureless sintering process with CaO addition and some properties of the obtained substrates are summarized in Table 1. These AlN substrates were cleaned by alkali and acid.

Metallization paste was prepared by the following process. Raw material powders shown in Table 2 were mixed by ball milling and after the addition of organic binder, it was mixed again. Viscosity of the paste was controlled by the amount of solvent for printing. The pattern was printed on the AlN substrate by screen printing and printed thickness was about 20um. The printed substrate was heated up to 523K and kept for 4 hours to remove the binder. The procedure of sample preparation is shown in Fig. 1.

Then the substrate was fired in tunnel-type metallizing furnace at various temperatures ranging from 1493 to 1643K in reducing atmosphere (N_2+H_2) with various dew point ranging from 233 to 323K. The duration at maximum temperature was 3 hours and total time for firing was 12 hours.

After metallization, electrolytic Ni plating was treated upon metallized layer.

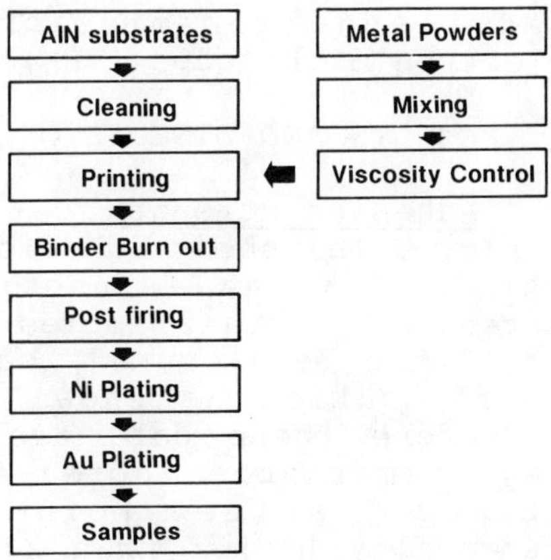

Fig. 1. The procedure of sample preparation.

Ni wire (0.6mm diameter) was soldered with 40%Pb-60%Sn solder on metallized pattern for adhesion strength measurement between AlN substrate and metallized layer, as shown in Fig. 2 schematically. Five samples on each firing condition were used for adhesion strength measurement.

Table 2. Metallization paste composition.

Materials	Content (%)	Specific surface area (m²/g)	Average particle size (μm)	Purity (%)
Mo	80	5.22	0.67	99.94
Mn	10	0.33	13	99
SiO_2	10	0.46	5	99.5

Microstructures of the AlN substrates with metallizing layers, especially the intermediate layers fabricated under various firing conditions were investigated by electron probe micro analysis (EPMA), X-ray diffractometry (XRD) and transmission electron microscope (TEM).

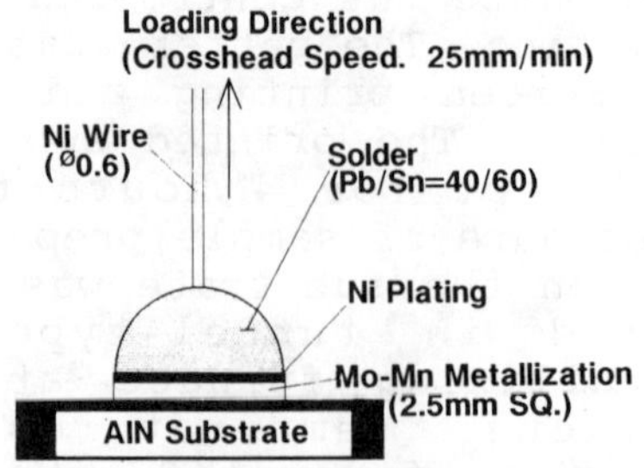

Fig.2. Method of adhesion strength measurement.

3. RESULTS AND DISCUSSION

Adhesion strength

Fig. 3 (a) shows adhesion strength for samples fired at 1543K as a function of dew point. Adhesion strength for the samples fired at dew point of 238K could not be measured, because of the incomplete metallization as shown in the following section. Therefore these data are not represented in Fig. 3 (a). Remarkable increase of adhesion strength was observed at dew points of higher than 243K, especially higher than 278K. However the scatter of strength also increased. Fig. 3 (b) shows the adhesion strength for the samples fired at various temperatures with dew point of 258K. It indicates that the adhesion strength had maximum value at 1593K.

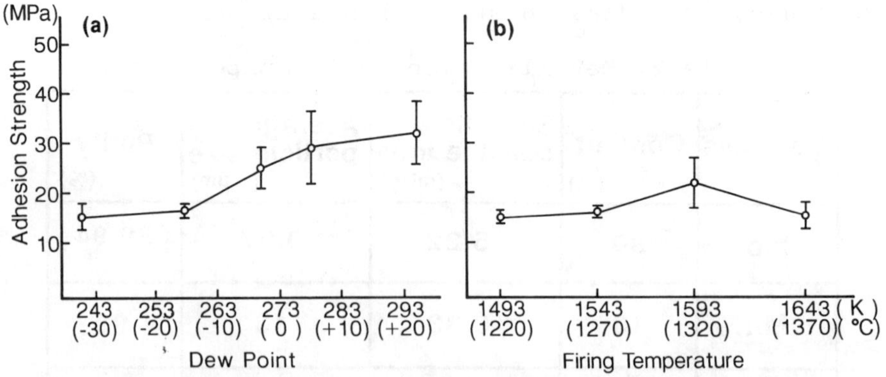

Fig. 3. Adhesion strength as a function of firing condition (a) fired at temperature of 1543K (b) fired with dew point 258K.

EPMA, TEM and XRD investigation

Fig. 4(a)-(d) shows the typical cross section photographs by SEM for the samples fired at various temperature with various dew points. A porous metallized layer was observed for the sample fired at 1543K with a dew point of 233K, shown in Fig. 4(a). The formation of low electro-conductive Mo_3Si was detected by XRD , resulting in low electrical conductivity of metallized layer. It was thought that AlN reacted with oxygen of SiO_2 in the metallizing

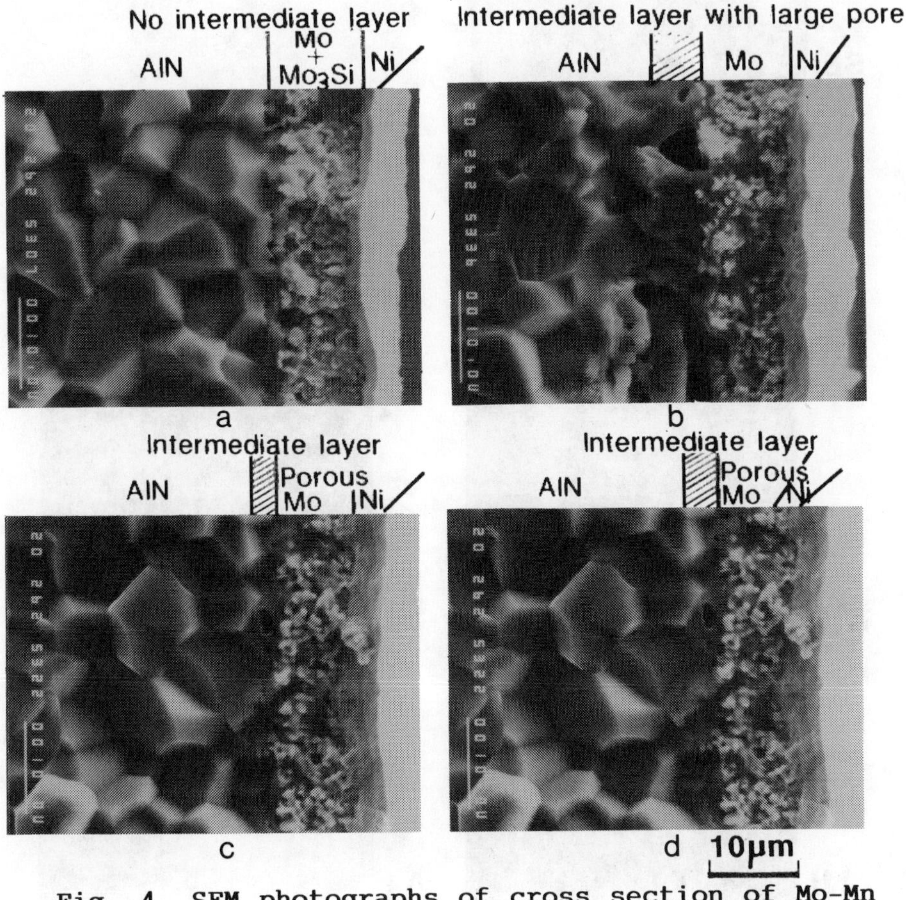

Fig. 4. SEM photographs of cross section of Mo-Mn metallization on AlN substrate: (a) dew point 233K, firing temperature 1543K. (b) dew point 278K, firing temperature 1543K. (c) dew point 243K, firing temperature 1493K. (d) dew point 243K, firing temperature 1643K.

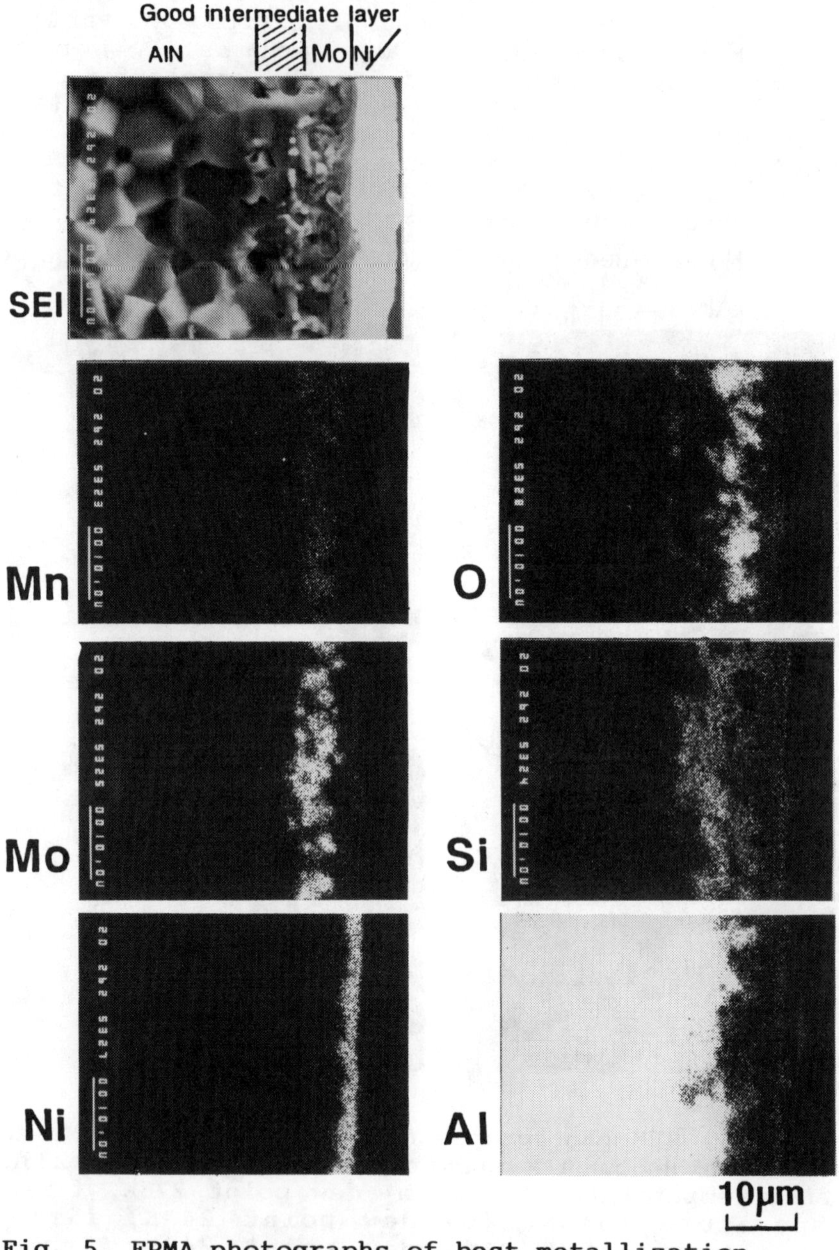

Fig. 5. EPMA photographs of best metallization condition: dew point 258K, firing temperature 1593K.

paste to form free Si, and consequently the following reaction occurred.

$$3Mo+Si \rightarrow Mo_3Si$$

Co-existence of Ni in this layer was also observed. It was thought that the Ni ion in electrolytic plating liquid flew into pores during the Ni plating process.

On the other hand, under the condition of higher dew point (278K) at the same firing temperature (1543K), oxygen was easily supplied from atmosphere, and many larger pores, which seemed to have been produced by the severe oxidation of AlN substrate accompanied by release of nitrogen gas, were observed in the intermediate layer, as shown in Fig. 4(b). It was thought that the adhesion strength was affected by these pores and was scattered as a result.

Fig. 4(c) shows the cross section for the low dew point (243K), and low firing temperature (1493K). It shows that the reaction between metallizing paste and AlN substrate did not occur; an intermediate layer cannot be found. On the other hand, when the firing temperature was high, a dense metallized layer was not formed because of the Mo volatilization as shown in Fig. 4(d). When the firing temperature was 1593K and the dew point was 258K, the best firing was carried out, the EPMA photographs being shown in Fig. 5. The dense intermediate layer of 5-6um thickness connected the metallized layer with AlN substrate.

TEM observation with EPMA analysis was carried out to clarify the microstructure of the intermediate layer. Co-existence of Al, Mn, Si, O and N elements was confirmed, being shown in Fig. 6.

It was thought that the intermediate layer was vitreous and consisted of some nitrogen glass.

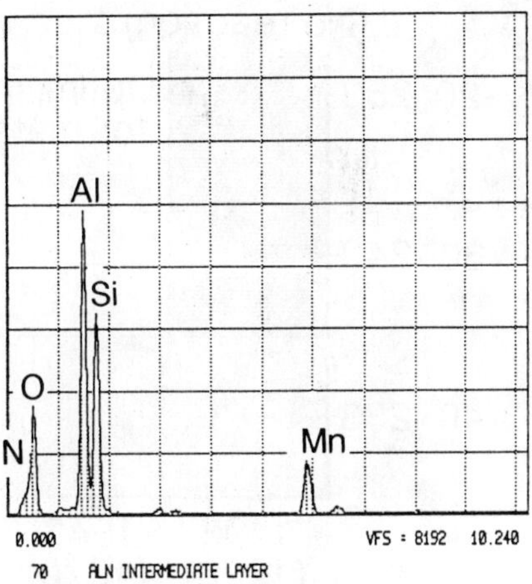

Fig. 6. Peak profile of intermediate layer by EPMA.

However further investigation will be necessary to fully understand the microstructure of it.

Baking test

The reliability of plating such as stability against thermal exposure is important for metallized AlN substrate. The baking test (723K for 5min) was conducted to evaluate this factor. Swelling of plating was observed in the sample fired at 1643K with dew point at 243K.

The above mensioned results are summerized in Fig.7

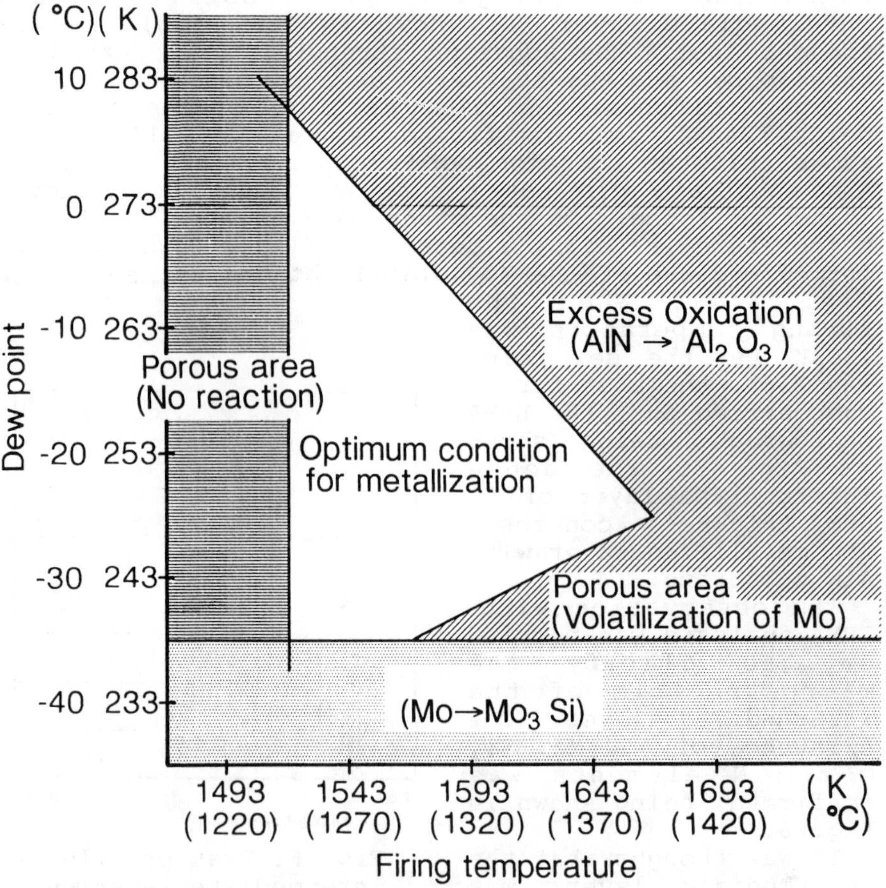

Fig. 7. Characteristics of metallized layer under various firing condition.

4. PROPERTIES OF THE METALLIZED AlN SUBSTRATE

The effect of the metallization on the thermal conductivity of AlN substrate was examined using a laser flash measuring equipment. It was found that the thermal conductivity of AlN substrate was not affected by the layer of metallization, and the value of thermal conductivity similar to that of non-metallized AlN substrate was obtained under any metallizing condition.

The metallized AlN substrate was tested under our inspection procedures for Al_2O_3 metallized substrate such as baking test (723K for 5min.), die attach test and solder wettability. The result were all satisfactory for use. Some properties of the Mo-Mn metallized AlN substrate are summarized in Table 3.

Table 3. Properties of metallized AlN substrate

Properties	Specification
Adhesion Strength (pull off) (MPa)	> 20
Sheet Resistance (mΩ/sq.)	40
Thermal Conductivity (W/m·K)	150
Baking Test 723K - 5min.	OK
Die Attach Test	OK
Wire Bondability	OK
Solder Wettability	Excellent

*Mo-Mn metallization with Ni/Au plating

5. CONCLUSION

1) Metallization on AlN substrate with adhesion strength of 10-30 MPa was obtained under firing conditions at 1543-1593K in a little wet H_2 and N_2 mixed gas with dew point at 243-258K, different from that for Al_2O_3 substrates.
2) SiO_2 addition to metallization paste was effective for AlN material with no grain boundary phase which reacts with metallization paste.
3) Glassy intermediate layer including nitrogen and oxygen was observed between metal layer and AlN substrate.
4) The metallized AlN substrate passed our inspections for Al_2O_3 substrate such as baking test, die attach test, wire bondability and solder wettability.

References

[1]H. Pulfrich, "Ceramic to Metal Seals", U.S.Pat. Nos. 2 163 407 and 2 163 410, June (1939).

[2]H. J. Nolte and R. F. Spurch, "Metallizing and Ceramic Sealing with Manganese", U.S.Pat. Nos. 2 667 432 and 2 667 427, Jan. 24, (1954).

[3]A. J. Chick and L. J. Speck,"Fabrication of Metal-to-Ceramic Seals", U.S.Pat. No.2 708 787, May (1955).

[4]H. Takashio,"Alumina Ceramic-to-Metal Seals by the 'Mo-Mn Process'", Yogyo-kyokai-Shi, **79** [9] 330-39 (1971). [in Japanese]

[5]J. R. Floyd,"Effect of Composition and Crystal Size of Alumina Ceramics on Metal-to-Ceramic Bond Strength", Am. Ceram. Soc. Bull., **47** [2] 65-70 (1963).

[6]N. Iwase, K. Anzai, K. Shinozaki, O. Hirai, T. D. Thanh, and Y. Sugiura, "Thick Film and Direct Bond Copper Forming Technologies for Aluminum Nitride Substrate", IEEE Trans. on CHMT, Vol.CHMT-8 (2), 253-8 (1985).

[7]Y. Kurokawa, K. Utsumi, H. Takamizawa, T. Kamata, and S. Noguchi, "AlN Substrates with High Thermal Conductivity", Proc. of the 1st IEEE CHMT Sympo. 15-22 (1984).

[8]M. Takahashi, K. Yamada, T. Oukouchi, K. Kanai, and K. Kurihara, "Development of Mo Metallization Process of AlN Ceramic Surface", The Transactions of the Institute of Electronics, Information and Communication Engineers, Vol.J-72-C, 1-9(1989). [in Japanese]

[9]H. Hamaguchi, Y. Kurokawa, and Y. Shimada,"AlN Multilayer Wiring Substrate", Proc. of 2nd MES.'87 185-188 (1987).[in Japanese]

[10]H. Yamamoto, Y. Kurokawa, Y. Shimada, and T. Kamata, "Metallization Development and Application for Aluminum Nitride Substrates", Proc. of 39th ECC 23-28 (1989).

[11]H. Asai, F. Ueno, N. Iwase, H. Sato, N. Mizuno T.Kimura, K. Endo, T. Takahashi, and Y. Sugiura, "Titanium Nitride-Molybdenum Metallizing Method for Aluminum Nitride", Proc. of 2nd IEMT Sympo. 221-224 (1989).

[12]"Phase diagrams for Ceramists ", Edited and published by the American Ceramic Soc., Fig.101, p65, 1st edition, Columbus, Ohio, (1964)

PROCESSING MULTILAYER CERAMICS WITH INTERNAL COPPER CONDUCTORS

Ian Burn and William C. Porter

E.I. du Pont de Nemours & Co., Inc.
Electronics Department
Wilmington, DE 19880-0334

ABSTRACT

The processing of multilayer ceramic capacitors (MLC's) and multilayer ceramic interconnects (MLI's) with internal copper conductors is described. Ceramic tape was used with screen-printed copper conductors and both the tape and the copper paste were made using acrylic binders. The dielectrics were formulated using reduction-resistant oxides which sintered with good hermeticity at 1050 to 1065 ^{o}C. The multilayer structures were fired in a nitrogen atmosphere with additions of carbon dioxide and hydrogen. This gas mixture produced a controlled oxygen partial-pressure of below 1 part per billion (10^{-9} atm.) during the sintering process, protecting the copper from oxidation while providing for efficient burn-out of the organics in structures with twenty five or more conductor layers. Good electrical performance of the devices, including reliability under humidity-bias conditions, was verified in tests on MLC's.

INTRODUCTION

In many respects the processing of multilayer ceramic capacitors and multilayer interconnects are similar. The main difference, apart from size, is that the conductors are interconnected with vias in multilayer substrates (Fig.1), whereas connections to the internal electrodes of multilayer capacitors are usually made with end terminations (Fig.2), although vias can be used [1]. If tape technology is employed, as in the present work, the multilayers are formed by screen-printing the conductors on to pieces of the tape, followed by lamination of the layers. The individual multilayers are then fired to produce a dense hermetic structure, at a temperature which depends on the melting point of the metal used for the conductors. Base metals, such as W, Mo, Ni, Cu or Pb can be used for the conductors but special procedures must be followed. Except for Pb, which has been used in capacitors by injecting the molten metal into cavities in the fired ceramic, the multilayer must be fired in an atmosphere with low oxygen content, to prevent oxidation of the conductors. In the case of Cu, the ceramic must have a low sintering temperature because of the relatively low melting point of copper (1083 °C). Nevertheless, copper is the ideal conductor for MLC's and MLI's based on its conductivity relative to other metals (Table 1), and because of its low cost, small skin effect, and its better resistance to metal migration compared with silver.

Table 1. Conductor Properties

Metal	Melting Point (C)	μΩ.cm at 20 C
Ag	960	1.6
Cu	1083	1.7
Au	1063	2.4
Mo	2625	5.2
W	3380	5.4
Ni*	1452	7.8
Pt	1774	10.6
Pd	1550	10.8
Pb	327	22

(* magnetic - has large skin effect)

While early attempts to use copper as the internal electrodes in multilayer capacitors [2,3] generally resulted in some compromise in electrical properties, the feasibility of developing high-performance Cu compatible dielectrics has been demonstrated by continued work for capacitor applications [4-8], and for substrates [9-11]. However, details of the processing of multilayers with Cu conductors are not readily available, so this topic is covered here.

ATMOSPHERE CONTROL

Copper / Copper Oxide Equilibrium

To avoid oxidation of the copper conductors when the multilayer is fired, the oxygen level in the furnace must be less than that for equilibrium between copper and copper oxide. The equilibrium conditions for the reaction

$$1/2\,(O_2) + 2\langle Cu \rangle = \langle Cu_2O \rangle$$

are determined by the Gibbs free energy ΔG. The reaction constant, K_p, is given by

$$\Delta G = - RT \ln K_p = - 4.576 \, T \log K_p$$

where R is the gas constant and T is temperature (in ^oK). The oxygen partial-pressure, $p(O_2)$, in equilibrium with copper and copper oxide at temperature T is given (in atmospheres) by

$$\log p(O_2) = 2 \log K_p$$

The Gibbs free energy (in cal) can be obtained [12] in the form

$$\Delta G = A + BT \log T + CT$$

where A = 40,500, B = 3.92 and C = -29.5.

The equilibrium oxygen partial pressure (in torr) can also be obtained from the following equation [13]

$$\log p(O_2) = -17{,}400 / T + 10.18$$

(760 torr = 1 atmosphere = 1.01 bars = 1.01 x 10^5 Pa).

Oxygen pressures calculated for temperatures between 750 ^oC and 1050 ^oC are given (in atmospheres) below:

Table 2. Oxygen Partial Pressures for Copper / Copper Oxide Equilibrium

Temperature	$p(O_2)$ (ref.12)	$p(O_2)$ (ref.13)
750	2.0. 10^{-10}	2.7. 10^{-10}
850	6.4. 10^{-9}	8.0. 10^{-9}
950	1.2. 10^{-7}	1.4. 10^{-7}
1050	1.4. 10^{-6}	1.5. 10^{-6}

Even though there is some slight disagreement in the data, it can be seen that the oxygen partial pressure should be less than 1.5 ppm in order to avoid oxidation of the copper at 1050 ^oC, and less than 0.3 ppb at 750 ^oC. It turns out that, for firing copper in contact with ceramic dielectrics, the oxygen partial-pressure must be appreciably less than that for copper / copper oxide equilibrium, to avoid excessive interaction with the dielectric.

Protective Gases

Nitrogen gas containing about 1 ppm of oxygen as impurity is often used for processing microcircuits with copper conductors at 900 - 950 ^{o}C. However, in order to achieve the appropriate oxygen levels to protect the copper from oxidation, oxygen must be removed from the nitrogen by burn-out of organics and by oxidation of the inside of the furnace (belt and muffle). The final oxygen pressure is unknown and is uncontrollable, leading to problems with reproducibility when part size and/or belt loading are varied.

Various gas mixtures can be used to give a buffered oxygen partial-pressure. In this case the system maintains an equilibrium oxygen pressure, even if disturbed by removal of oxygen, as in binder burn-out, or by the addition of oxygen, as when air intrudes. The most commonly used mixtures are:

- water vapor + hydrogen [9]
- carbon dioxide + carbon monoxide [14]
- carbon dioxide + hydrogen [15]

In each of these systems, nitrogen can be used to reduce the amount of "active" gas needed to maintain a particular flow rate. The carbon dioxide + hydrogen system was chosen for the present work because, in calculating the mixtures needed, it proved to be the most convenient and the safest: it was not necessary to store and pipe carbon monoxide, and only non-explosive levels of hydrogen were required. The oxygen partial-pressure was regulated by adjusting the ratio of gases in the mixture, according to the gas reaction

$$CO_2 + H_2 = CO + H_2O$$

The reaction constant K_3 is given by

$$K_3 = \frac{p(CO).p(H_2O)}{p(CO_2).p(H_2)} = K_2 / K_1$$

where values of K_1 and K_2 can be obtained from thermodynamic data for CO/CO_2 and H_2/H_2O equilibria, respectively. For example, values of K_1, K_2 and K_3 at various temperatures are given in Table 3.

The ratio of carbon dioxide to hydrogen needed is given [16] by

$$p(CO_2) / p(H_2) = x\,(1 + x) / (x + 1/K_3)$$

where x is the ratio of CO_2/CO which would give the required oxygen partial-pressure. For example, at 1050 ^{o}C the ratios for various oxygen partial-pressures are given in Table 4.

Table 3. Reaction Constants for $CO_2 + H_2$ Equilibrium

Temperature (C)	K_1	K_2	K_3
650	$2.75. 10^{11}$	$1.35. 10^{11}$	0.49
750	$7.59. 10^{9}$	$5.75. 10^{9}$	0.76
850	$3.98. 10^{8}$	$4.37. 10^{8}$	1.10
950	$3.35. 10^{7}$	$4.90. 10^{7}$	1.46
1050	$4.11. 10^{6}$	$7.59. 10^{6}$	1.85

Table 4. Calculated CO_2 / H_2 Ratios for 1050 °C

Oxygen Pressure (atmospheres)	CO_2 / CO	CO_2 / H_2
10^{-8}	411	411
10^{-9}	130	131
10^{-10}	41.1	41.6
10^{-11}	13.0	13.4

It can be seen that the required ratios for carbon dioxide with hydrogen are very similar to those for carbon dioxide with carbon monoxide. Based on earlier work [2,3], it was expected that oxygen partial-pressures between 10^{-9} and 10^{-10} atm. would be required, so carbon dioxide/ hydrogen ratios of about 50/1 were used.

The composition of the furnace atmosphere can be calculated from the ratio of the input gases. For example, with a 41.6 input ratio of carbon dioxide and hydrogen (which produces an equilibrium oxygen pressure of 10^{-10} atmospheres at 1050 °C), the furnace atmosphere would contain, theoretically, 0.9533 atm. of carbon dioxide, 0.0232 atm. of carbon monoxide, 0.0232 atm. of water vapor, and 0.0003 atm. of hydrogen, as well as the 10^{-10} atm. of oxygen, if there was no dilution with nitrogen. The predicted oxygen pressure is not changed significantly by diluting the active mixture with nitrogen, since the partial-pressure depends on the ratio of carbon dioxide and carbon monoxide (and hydrogen and water) in the furnace. However, excessive dilution will diminish the ability of the system to maintain the required oxygen partial pressure during binder removal and/or to handle small air leaks.

The Cladan Furnace

A Cladan 304/305 kiln was used[1] . The main features of this bottom-loading batch furnace (Figs. 3,4) are:

- A suitable size for R&D but large enough for pilot, or small scale production
- Programmable profiles
- Gas inlets and flow meters provided
- Computer-controlled gas valves
- Low thermal mass insulation

In spite of using a buffered gas mixture, it was not possible to achieve a low oxygen pressure in the furnace in its as-received state because the insulation was not adequately sealed. Therefore, thin sheets of stainless steel were riveted on the top and bottom of the furnace, and a cover plate was attached over the lead connections to the heating elements. Also, a fiber-glass gasket was used to seal the elevated base, and the seams in the stainless steel shell were caulked with silicone cement. When these modifications were completed, the furnace would maintain a small positive Magnehelic pressure (of about one half inch of water) with a total gas flow of about 70 standard cubic feet per hour. A change was made to the gas piping to allow the "active" gas to flow through one meter and plant nitrogen to flow through the other. These gases could then be switched on or off independently by the computer.

A flow rate of 6 liters/min of bottled carbon dioxide was selected as a starting point. A bottled mixture of 4% hydrogen in nitrogen was blended at a rate of 3 liters/min with the carbon dioxide to produce a CO_2/H_2 ratio of 50/1. These gases were then blended with plant nitrogen ($\approx$1 ppm oxygen) in the ratio of about one quarter "active" mixture to three quarters nitrogen.

The In-Situ Oxygen Sensor

In order to verify that the desired oxygen partial-pressure was being obtained by the gas mixture, a zirconia oxygen sensor was constructed and positioned inside the alumina exhaust tube in the center of the furnace (Fig.5). The sensor, shown schematically in Figure 6, was made from a yttria-stabilized zirconia tube (closed end)[2] with platinum contacts on the inside and outside of the tip. Oxygen was fed to the inside of the tip through an alumina tube, to act as a reference gas. The sensor registered an output voltage (E) related to the temperature and the difference in oxygen partial-pressure between the inside and outside of the zirconia tube, given by

---

[1] Integrated Ceramic Technology Inc., San Marcos, CA.

[2] Zircoa Products, Corning Glass Works, Solon, OH

$$E = \frac{RT}{4F} \ln [p(O_2)_1 / p(O_2)_2]$$

$$= -4.96.\ 10^{-5}\ T \log p(O_2)$$

where T is in ^oK, the Faraday constant (F) is $9.6487.10^4$, and the gas constant (R) is 8.3143 (in MKS). A high impedance voltmeter is needed to measure the output, to avoid drawing current. A calibration curve can be drawn for various temperatures, as shown in Figure 7, with the oxygen pressures for Cu/Cu_2O equilibrium superimposed. With the CO_2/H_2 ratio set at 50/1, as discussed above, the sensor reading was close to 0.6V, as predicted, clearly above the the value for copper/copper oxide equilibrium.

In order to verify reproducible conditions from run to run, the readings from the sensor, and from a thermocouple placed adjacent to its tip, were recorded via a data acquisition board to an IBM (AT) computer. Software written in ASYST [3] was used to access and display the data. Typical profiles for the temperature and oxygen partial-pressure used for firing MLC's are shown in Figure 8. A slower initial ramp rate (approx. 6 ^oC per minute) was used when firing MLI's, to avoid cracking and to ensure complete removal of binder. The change in oxygen partial-pressure with temperature in Figure 8 is a consequence of the temperature dependence of the reaction between the carbon dioxide and hydrogen in the gas mixture, except near the end of the cycle when the CO_2/H_2 mixture was switched off. Stable sensor readings were obtained at temperatures above about 350 ^oC.

MATERIALS AND PROPERTIES

Cu Compatible Dielectrics

One of the main requirements for the ceramic dielectric used in the multilayer is that it densify to a hermetic structure at a temperature $\leq$ 1065 ^oC. Another important requirement of the ceramic is that it will tolerate firing in atmospheres of low oxygen content without deterioration of its dielectric properties. Choice of dielectric depends on the dielectric constant and temperature stability needed. Dielectrics based on SiO_2 (K= 4) are suitable for MLI's because a low dielectric constant is desirable. Also, silica is stable in atmospheres of low oxygen content, so good electrical characteristics can be obtained if the sintering temperature of the silica is lowered using reduction-resistant glasses, or fluxes [e.g.17]. This approach was used in the present work, with good results both for MLI's, and for low capacitance MLC's.

---

[3] Macmillan Software Co., NY, NY.

For high capacitance MLC's, new Cu-compatible compositions based on barium titanate [5] resulted in dielectric constants as high as 10,000. Dielectrics based on magnesium titanate [8] gave lower values of K (≈20) but were suitable for applications requiring COG temperature stability (+/- 30 ppm per °C from -55 to 125 °C).

All of the dielectrics were processed into tape by conventional (doctor-blade) processing using acrylic binder (Du Pont 5200) with butyl benzyl pthalate as plasticizer. Tape for MLC's was cast to give a fired thickness close to 25 microns. Tape for MLI's was about 100 microns fired.

Copper Conductors

The copper conductor was prepared using copper powder, with a particle size close to 1 micron, in a butyl methacrylate binder (Elvacite® 2044, Du Pont) dissolved in beta terpineol [11]. Acrylic binders usually have significantly lower carbon residues compared with those with a cellulose base when fired in atmospheres of low oxygen content [18], so they offer an advantage, particularly for thick multilayers with many conductor layers into which access by the gas mixture in the furnace can be difficult. The beta terpineol was not a solvent for the acrylic binder in the tape, so interaction between the conductor composition and the tape was minimized, allowing good uniform deposits to be obtained by screen-printing. No bonding agents were needed in the Cu paste. Fired conductor thickness was ≤ 2.5 microns. While capacitor performance at high frequencies (> 100 MHz) was found to be significantly enhanced by using copper instead of Pd or Pd-Ag alloy [19], there appeared to be no benefit in using conductors thicker than 2.5 microns.

Multilayers

Examples of multilayers made by the procedures outlined above are shown in Figures 9 and 10. Figure 9 shows a cross-section of a leaded MLC with 15 electrode layers and Figure 10 shows a multilayer with 26 electrodes. The electrodes are smooth and uniform, with no evidence of delamination, or significant dissolution of the copper into the dielectric.

Reliability of the multilayers was assessed by measuring leakage current stability during accelerated testing at 125 °C with 200 V applied (24 hours), for MLC's with 25 micron layer thickness. Ability to survive humidity/bias testing at 85 °C and 85 % relative humidity with 50 V applied (for 1000 hours) was also checked. In fact, MLC's with a variety of dielectrics with K of 4, 20, 2200, or 9000 were made which would survive these tests.

SUMMARY

A method has been described for processing ceramic multilayers with internal copper conductors. The method applies to both multilayer capacitors and multilayer interconnects. Acrylic binders were used in both the ceramic tape and the copper conductors to minimize carbon retention during binder burn-out. Binder removal was also facilitated by using a mixture of carbon dioxide and hydrogen, diluted with nitrogen, and by firing at temperatures in the range 1050 - 1065 ^{o}C. The oxygen partial-pressure during firing was controlled in the range of 0.1 to 1 ppb by using an appropriate ratio of carbon dioxide to hydrogen, and was monitored by an in-situ oxygen sensor. This procedure coupled with the use of new low-firing reduction-resistant dielectrics with either low K, for multilayer interconnects, or high K, for multilayer capacitors, enabled reliable multilayers to be processed with good compatibility with internal copper conductors.

REFERENCES

1. W.L. Robbins, "Viad Chip Capacitor and Method for Making Same", U.S Patent 4,864,465 (1989).
2. I. Burn, "Monolithic Base-Metal Glass-Ceramic Capacitor", U.S. Patent 4,101,952 (1978).
3. I. Burn, "High Q Monolithic Capacitor with Glass-Magnesium Titanate Body", U.S. Patent 4,308,570 (1981).
4. J. Kato, Y. Yokotani, H. Kagata, and H. Niwa, "Multilayer Ceramic Capacitors with Copper Electrodes", Jap. J. Appl. Phys. **26** (Suppl. 26-2) 90 (1987).
5. I. Burn, "High K Dielectric Composition for Use in Multilayer Ceramic Capacitors having Copper Internal Electrodes", U.S. Patent 4,855,266 (1989).
6. Y. Sakabe, "Recent Progress on Multilayer Ceramic Capacitors", Proc. MRS Int. Mtg. on Adv. Mats., **10** 119 (1989).
7. M. Madai, Y. Sakabe and J.P. Canner, "Multilayer Ceramic Capacitors with Copper Electrodes (Abs.)", Bull. Am. Ceram. Soc., **68** (9) 1671 (1989).
8. I. Burn, "Low-Firing Dielectric Composition", U.S. Patent 4,845,062 (1989).
9. L.W. Herron, R.N. Master and R.R. Tummala, "Method for Making Multilayer Glass-Ceramic Structures Having Internal Distribution of Copper-Based Conductors", U.S. Patent 4,234,367 (1980).
10. N. Kamehara, "Method for Producing Multilayer Glass-Ceramic Structure with Copper-Based Conductors Therein", U.S. Patent 4,504,339 (1985).
11. I. Burn, "Method for Making a Ceramic Multilayer Structure Having Internal Copper Conductors", U.S. Patent 4,766,027 (1988).
12. O. Kubaschewski, E.L. Evans and C.B. Alcock, "Metallurgical Thermochemistry" 5th Ed., Pergammon Press, p.421 (1979).

13. J. Dellacherie and D. Balesdent, "The Electrical Conductivity of Cuprous Oxide", J. Phys. Chem. Solids **37** 539 (1976).
14. I. Burn and G.H. Maher, "High Resistivity $BaTiO_3$ Ceramics Sintered in CO-CO_2 Atmospheres", J. Mats. Sc., **10** 633 (1975).
15. R.J. Leonard, "Controlled Atmosphere Firing of Periclase-Chrome Refractories", Bull. Am. Ceram. Soc. **54** (7) 662 (1975).
16. A Maun and E.F. Osborn, "Phase Equilibria Among Oxides in Steelmaking", Addison-Wesley, p.49 (1965).
17. I. Burn, "Low Dielectric Constant Compositions", U.S. patent 4,879,261 (1989).
18. R.J. Higgins, "The Chemistry of Carbon Retention During Non-Oxidative Binder Removal from Ceramic Greenware", Ceram. Proc. Res. Lab. Rep. (MIT) #101, p.47 (1989).
19. I. Burn and W.C. Porter, "MLC's with Copper Electrodes for High Frequency Applications", To be presented at the 40th Elect. Comp. and Tech. Conf., Las Vegas, (May 1990).

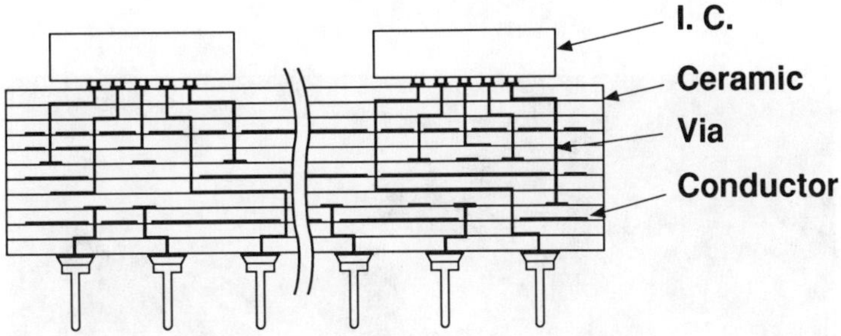
I. C.
Ceramic
Via
Conductor

Fig. 1

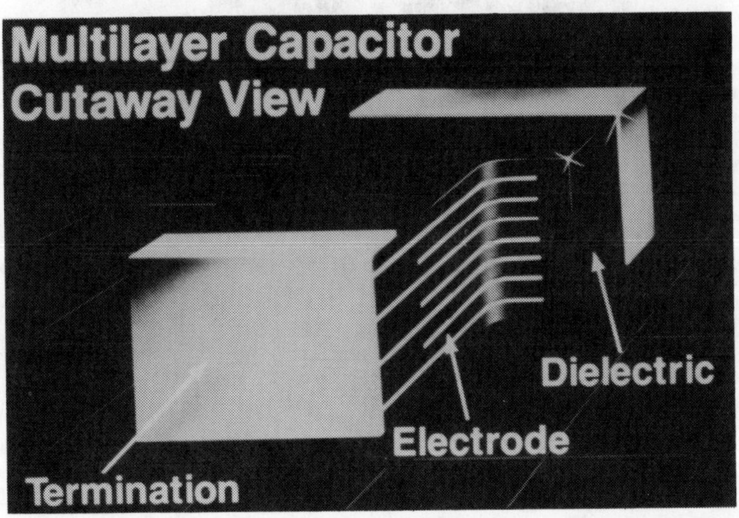
Multilayer Capacitor
Cutaway View
Dielectric
Electrode
Termination

Fig. 2

Fig. 3

Fig. 4

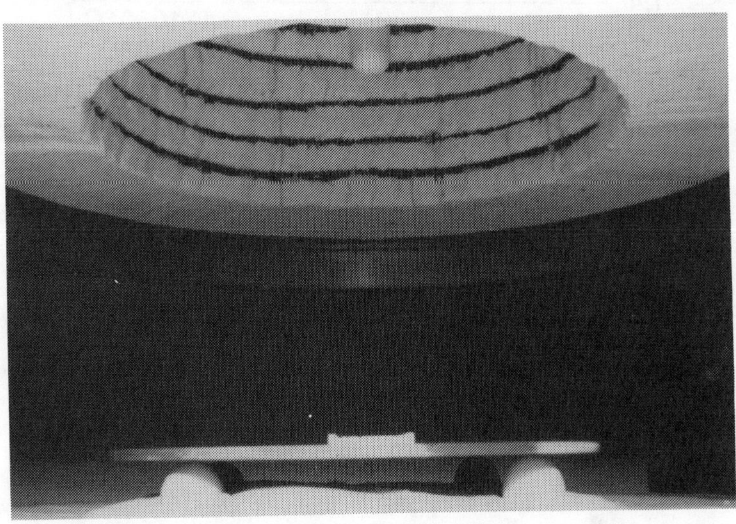

Fig. 5

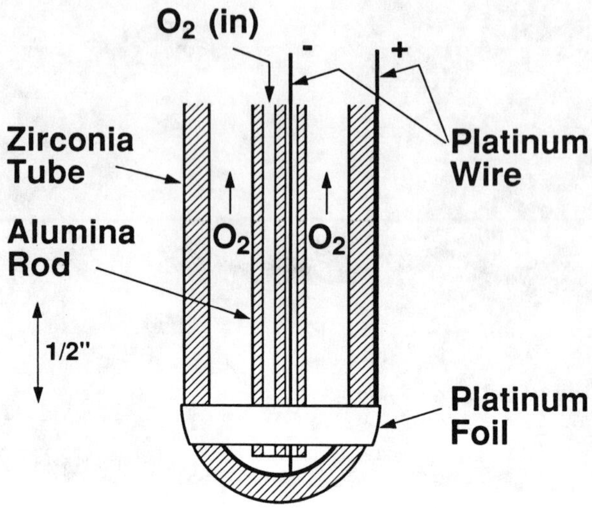

O2 (in)
-
+
Zirconia Tube
Platinum Wire
Alumina Rod
O2
O2
1/2"
Platinum Foil

Fig. 6

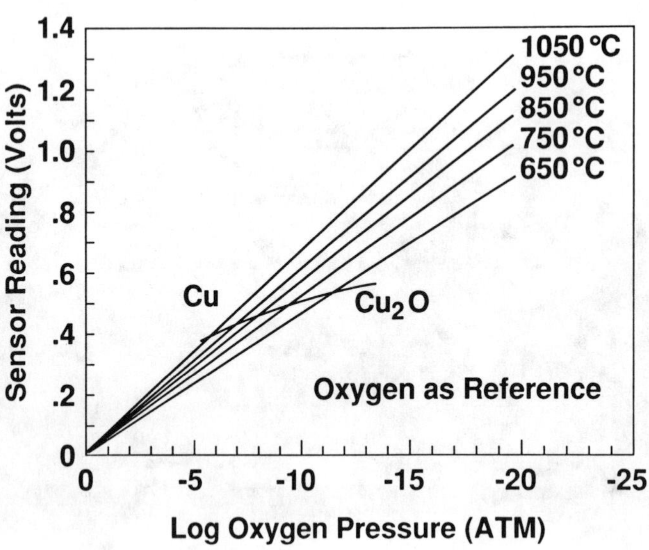

1.4
1.2
1.0
.8
.6
.4
.2
0
Sensor Reading (Volts)
1050 °C
950 °C
850 °C
750 °C
650 °C
Cu
Cu2O
Oxygen as Reference
0
-5
-10
-15
-20
-25
Log Oxygen Pressure (ATM)

Fig. 7

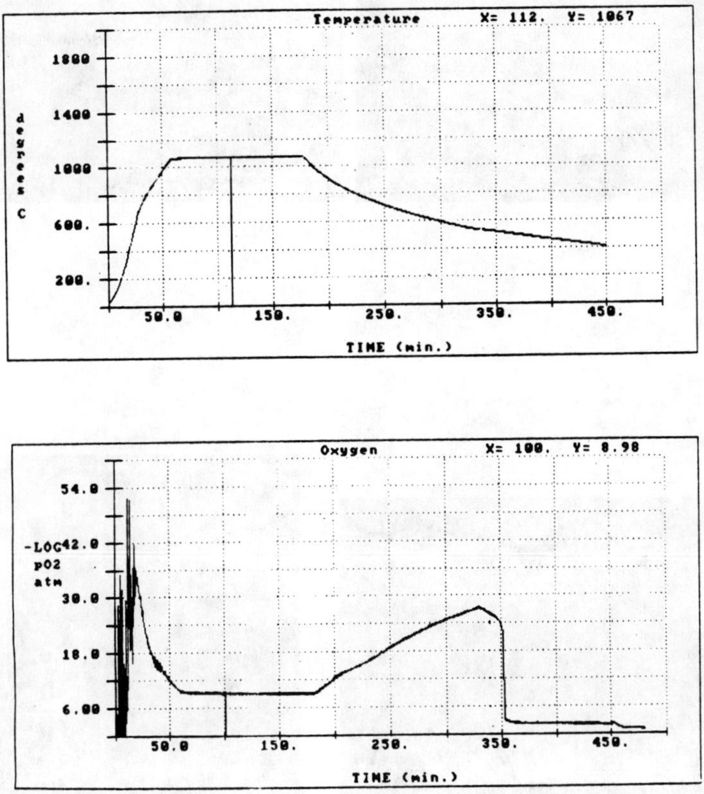
Temperature
X= 112. Y= 1067
1800
1400
1000
600.
200.
degrees C
50.0
150.
250.
350.
450.
TIME (min.)
Oxygen
X= 100. Y= 8.98
54.0
42.0
30.0
18.0
6.00
-LOG pO2 atm
50.0
150.
250.
350.
450.
TIME (min.)

Fig. 8

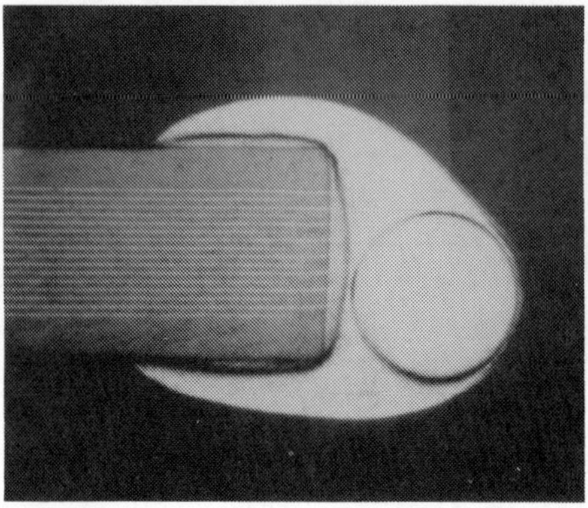

Fig. 9a

Fig. 9b

Fig. 10a

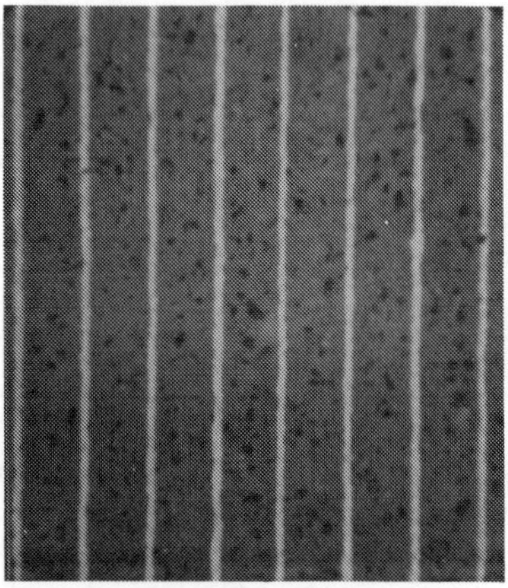

Fig. 10b

A LOW TEMPERATURE COFIRED MULTILAYER CERAMIC L-C FILTER WITH COPPER CONDUCTORS

Harufumi Mandai, Kikuo Wakino, Hisatake Okamura
Murata Mfg. Co., Ltd.
2-26-10 Tenjin
Nagaokakyo-shi
Kyoto, 617 Japan

James P. Canner
Murata Erie N.A., Inc.
1900 West College Ave.
State College, PA 16801

ABSTRACT

A low temperature fireable ceramic (BSSZ) consisting of BaO, SrO, SiO_2, ZrO_2, can be fired with copper electrodes at 990 ~ 1000°C in a slightly reducing atmosphere. It can be used for high frequency hybrid circuits because of the low resistance conductor and the low dielectric loss. It can be used to make compact microelectronic devices combining capacitors and inductors within the same unit. It has a stable NPO TCC (Thermal Coefficient of Capacitance) and high reliability.

A Chip Multilayer Ceramic (CMC) L-C filter (made of BSSZ) is a very small surface mount device (SMD) for high frequency (200 - 1000MHz) application in personal communication devices. Its unique structure allows short lead lengths between L and C elements in the filter in a package size comparable to an integrated circuit DIP.

INTRODUCTION

Due to the recent trends toward miniaturization, high density, high speed, high reliability, and low cost, a low temperature cofireable ceramic was developed. A previous paper describes the use of BSSZ as a multilayer high frequency capacitor. This paper describes the use of a new low temperature cofireable ceramic (BSSZ) with copper conductors in applications for L-C filter. For L-C filters, the most important points are miniaturization, SMD- type, high frequency use, and low cost. The CMC - L-C filter has all of these properties.

BSSZ DIELECTRIC

Composition:

The BSSZ dielectric consists of BaO, SrO, SiO_2, ZrO_2, and small amounts of additives. The relationship between electrical properties and the main composition (BaO, SrO, SiO_2, and ZrO_2) are as follows:

1. Dielectric Constant: The Dielectric Constant increases from 8 to 15 as the amounts of (Ba, Sr) O and ZrO_2 increase.
2. Dissipation Factor: The Quality Factor (Q) increases from 400 to 2000 as the amounts of SiO_2 decrease.
3. TCC: The TCC decreases from P180 to NPO with increasing amounts of (Ba, Sr) O.

Additives - TiO_2:

The relationship between the TCC and the amounts of TiO_2 is shown in Fig. 1. By adding TiO_2, the TCC can be adjusted from +100ppm to -200ppm to obtain NPO characteristics while maintaining desired firing characteristics and electrical parameters.

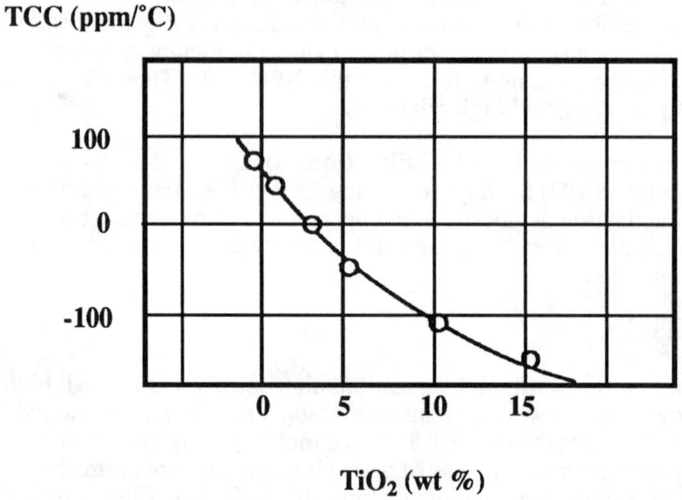

Fig. 1 The TCC change with the amounts of TiO_2.

Microstructure:

X-ray diffraction patterns of the calcined powder (fired at 825°C x 2 hours in air) and the sintered ceramics (fired at 1000°C x 2 hours in $N_2/H_2/H_2O$ atmosphere) are shown in Fig. 2.

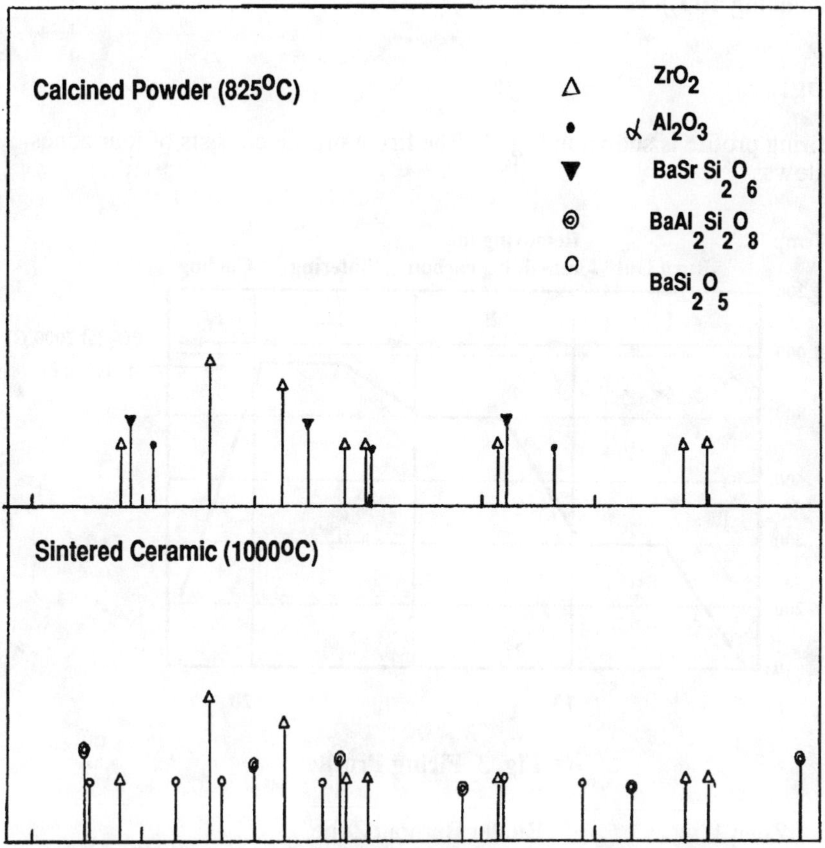

Fig. 2 X-ray diffraction patterns of the calcined powders (a) and the sintered ceramics (b).

Three crystalline phases (ZrO_2, α - Al_2O_3 and $BaSrSi_2O_6$) are observed in the calcined powder. On the other hand, in the sintered ceramics, two new crystalline phases ($BaAl_2Si_2O_8$ (celsian) and $BaSi_2O_5$) appear and one crystalline phase (ZrO_2) still remains and two crystalline phases (α- Al_2O_3 and $BaSrSi_2O_5$) disappeared.

From the SEM photograph of the cross section of the sintered ceramics, at least three kinds of phases are observed as follows:

Phase I Zr - Rich Phase (presumed to be ZrO_2)
Phase II Al - Rich Phase (presumed to be $BaAl_2Si_2O_8$).
Phase III Ti - Rich Phase (presumed to be glass phase including TiO_2)

Firing:

The firing profile is shown in Fig. 3. The firing profile consists of four zones as follows:

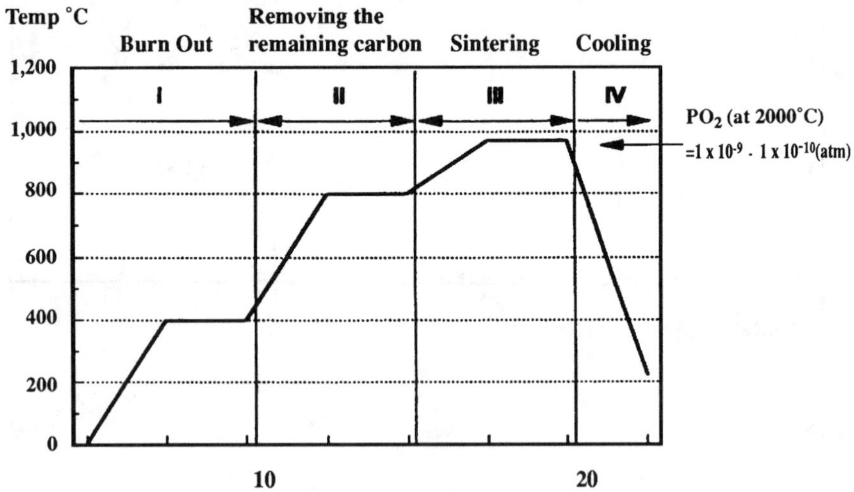

Fig. 3 Firing Profile

Zone 1	Binder Burnout Zone
Zone 2	Removal of the remaining Carbon Zone
Zone 3	Sintering Zone
Zone 4	Cooling Zone

The firing atmosphere is a slightly reducing atmosphere using $N_2/H_2/H_2O$ gas system. The partial pressure of oxygen (PO_2) in the firing profile must be strictly controlled because the copper conductors must be maintained in the metallic state and carbon must be removed as CO and CO_2. The relationship between firing temperature and the remaining carbon content is shown in Table 1. After sintering at the temperature of 1000°C, the residual carbon falls below 10ppm.

Table 1 Relationship between firing temperature and remaining carbon content	
Temp (°C)	**Residual Carbon(wt%)**
Before firing	5%
450°C	0.4%
900°C	120 ppm
1000°C	10 ppm

The variation of temperature at the sintering temperature must be small because the TCC is sensitive to the sintering temperature. At the lower temperature, the TCC becomes positive. On the other hand, at the higher temperature, the TCC becomes negative.

Properties:

The typical BSSZ substrate properties are shown in Table 2.

Table 2 Typical BSSZ Substrate Properties	
Dielectric Constant (@1MHZ)	15
Dissipation Factor (@1MHz)	5×10^{-4}
Temperature Coefficientof Capacitance (TCC) (-55 +125°C)	±30~-30 ppm/°C
Log Resistivity	$>1 \times 10^{14} \Omega \cdot cm$
Dielectric Strength(kV/mm)	$>50^{kv}$/mm
Density	3.6 g/cm^3
Flexural Strength	2300 kg/cm^2
Thermal Expansion Coefficient (25-600°C)	6.8×10^{-6}/°C

CMC - L-C FILTER

L-C Filter Design:

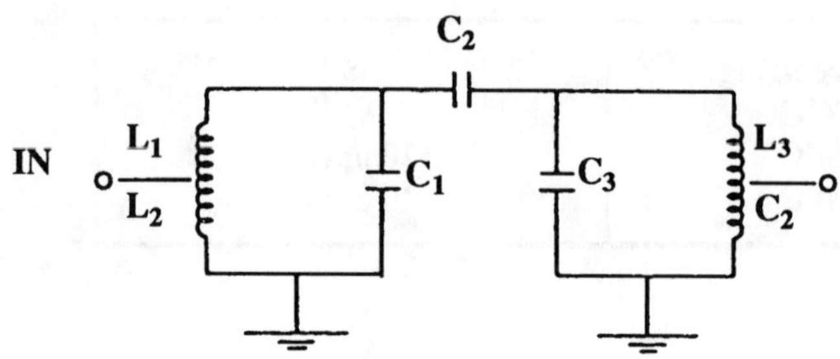

Fig. 4 Electrical Equivalent Circuit for Band Pass Filter (BPF 254 MHz)

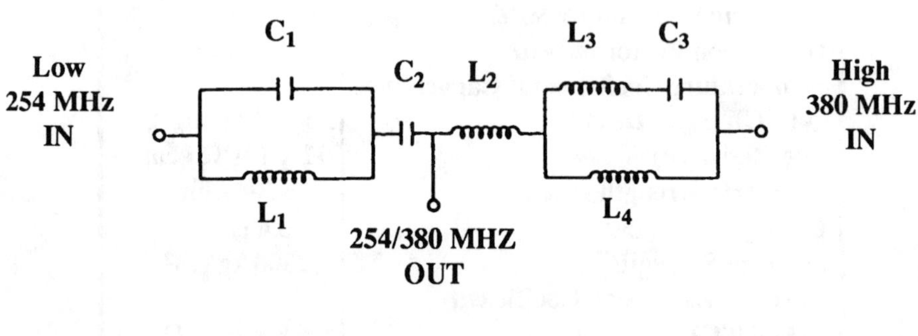

Fig. 5 Electrical Equivalent Circuit for Duplexer (DPX 254-380 MHz)

The electrical equivalent circuits for band pass filter (BPF) and duplexer (DPX) are shown in Fig. 4 and Fig. 5. The center frequency (fo) of BPF and DPX are formulated according to the relationship shown below:

BPF

$$fo = \frac{1}{2\pi\sqrt{C_1 (L_1 + L_2)}}$$

$C_3 = C_1$

$L_3 = L_1$

$L_4 = L_2$

C - - Capacitor
L - - Inductor

DPX

$$fo = \frac{\sqrt{L_2 + L_3}}{2\pi\sqrt{C_3 (L_2 L_3 + L_2 L_4 + L_3 L_4)}} \quad \text{(380MHz)}$$

$$fo = \frac{1}{2\pi\sqrt{L_1 (C_1 + C_2)}} \quad \text{(254MHz)}$$

Insertion Loss (I. L) is influenced by dissipation factor of the ceramic, R (Resistance) of circuit and circuit design. The other fundamental characteristics (Zin, Zout (Impedance), ATT (attenuation), pass bandwidth, etc., are determined by C (capacitors) and L (inductors). So the L-C filter design is very important.

Structure:

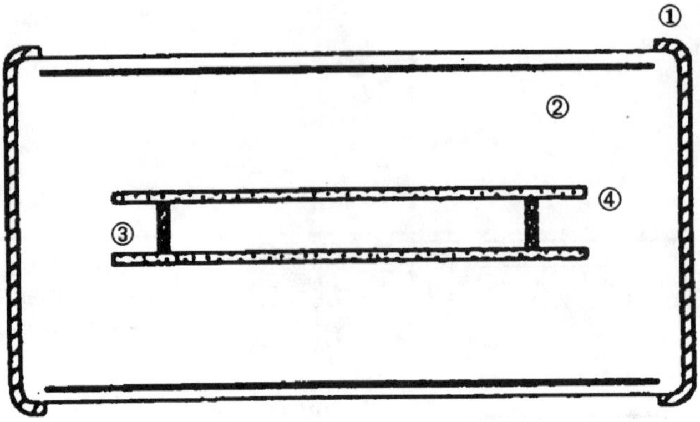

Fig. 6 Cross Section Diagram of CMC-L-C Filter

The cross section diagram is shown in Fig. 6. The shield conductors are formed near the surface. The inner conductors for capacitors (C) and inductors (L) are formed in several layers and are connected via holes. The inductors consist of spiral patterns of one or two turns. Inductance is typically 20 to 30nH. The capacitors are made by stacking square patterns to obtain capacitance of 20 to 30pF. The filter circuits are then constructed by interconnecting capacitor and inductor elements using via holes.

The inner conductors are located near the center layer. This location is important because of the following two reasons. One is the stray capacitance between the shield conductor and the inner conductors, which makes the L-C filter design very difficult. The other one is the Q of the inductors, which is decreased if the distance between the shield conductor and the inner conductor is too short. A photograph of BPF for cordless telephones and hand-held (cellular) telephones is shown in Fig. 7. Both of these are the SMD type - BPF and very small.

Their dimensions are the following:

BPF for Cordless Telephone (254MHz) -8mm x 5mm x $2mm^t$
BPF for Hand-held Telephone (810MHz)-5.7mm x 5mm x $2mm^t$

Fig. 7 CMC-L-C Filter (BPF)

Fabrication:

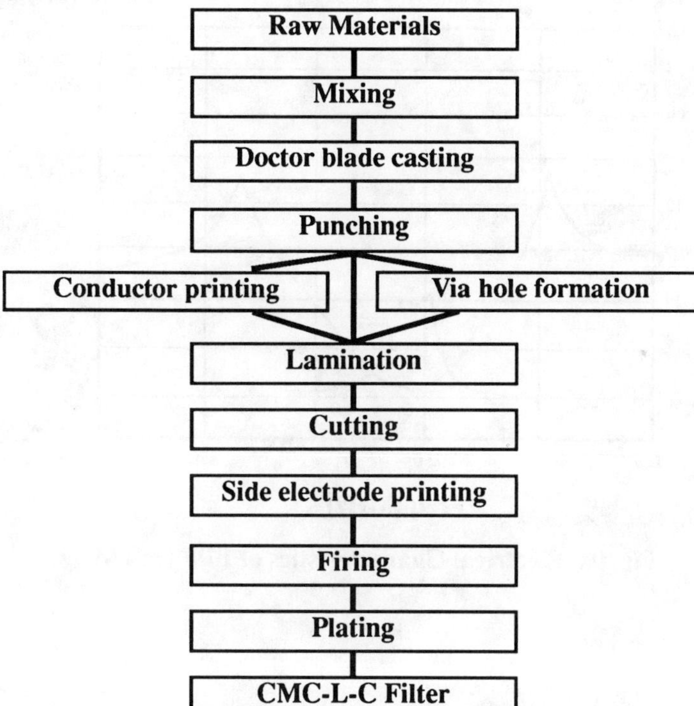

Fig. 8 Fabrication Flow Chart for CMC-L-C Filter

The fabrication process chart is shown in Fig. 8. Raw material (calcined powder) was mixed by ballmilling. Slurry, consisting of powder and organic vehicles, was cast by doctor blade method. The green sheets were 40 ~ 200μm in thickness. Via holes (0.2mm in diameter) were formed by using an numerical control drill machine. A copper thick film paste was printed on these sheets and filled into the via holes. The printed sheets were then laminated together. The laminated sheets were cut into chips. Then the paste was printed on the edges of the chips. The chips were fired at 990 to 1000°C in a slightly reducing atmosphere ($N_2/H_2/H_2O$). Finally, the side conductors were plated by Ni and Sn metal or solder dipped.

Electrical Properties:

Electrical characteristics of BPF (254MHz) are shown in Fig. 9a. The comparison between simulated data and measured data of BPF (254MHz) is shown in Fig. 9b. In the attenuation curve, simulated data is identical with measured data.

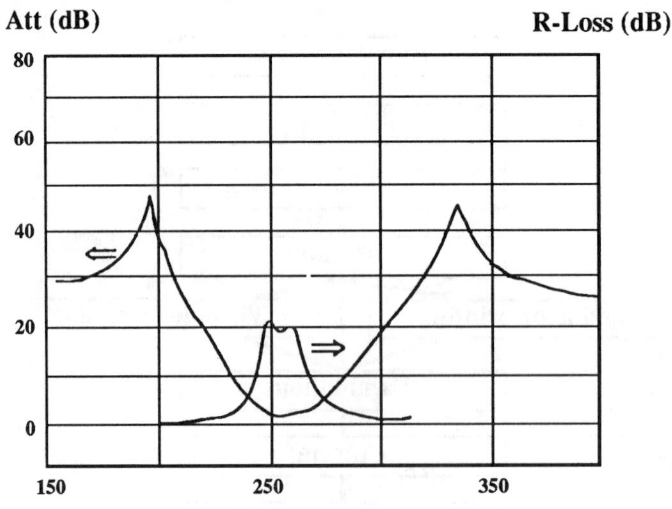

Fig. 9a Electrical Characteristics of BPF (254 MHz)

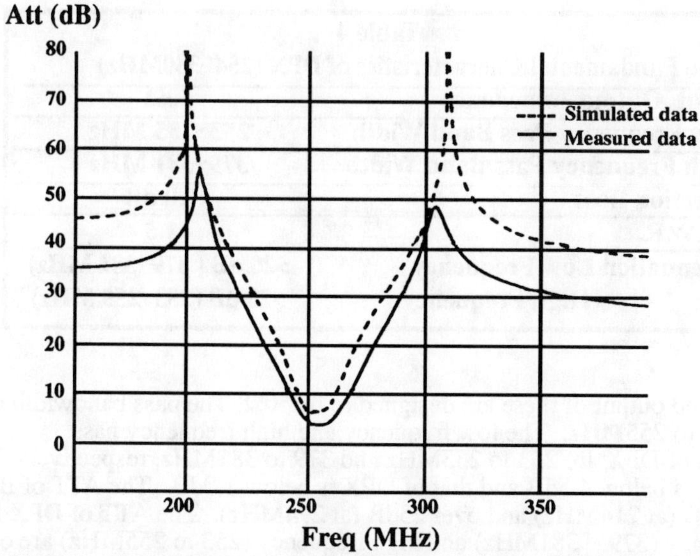

Fig. 9b The Comparison between Simulated Data and Measured Data in Electrical Characteristics of BPF (254 MHz)

The fundamental characteristics of BPF (254MHz) and DPX (254 - 380MHz) are shown in Table 3 and Table 4.

Table 3 The Fundamental Characteristics of BPF (254 MHz)	
Input, Output Impedance	**50Ω**
Pass Band Width	**253 ~255 MHz**
Insertion Loss	**<4.5 dB**
Pass Band Ripple	**<1.0 dB**
Attenuation	**>27 dB (at 214 MHz)** **>20 dB (253 ~ 255 MHz)**

Table 4 The Fundamental Characteristics of DPX (254~380MHz)	
Input, Output Impedance	50Ω
Low Frequency Pass Band Width	253~255 MHz
High Frequency Pass Band Width	379~381 MHz
Insertion Loss	<1.0 dB
V.S.W.R.	<1.5
Attenuation Low Frequency High Frequency	>20 dB (379~381 MHz) >20 dB (253~255 MHz)

The input and output of these are designed to be 50Ω. The pass bandwidth of BPF is 253 to 255MHz. The low frequency and high frequency pass bandwidths of DPX are 253 to 255MHz and 379 to 381MHz, respectively. The I.L. of BPF is below 4.5dB and that of DPX is below 1.0dB. The ATT of BPF is over 27dB (at 214MHz) and over 25dB (at 294MHz). The ATT of DPX in low frequency (379 ~ 381MHz) and high frequency (253 to 255MHz) are over 20dB. The change of fo and I.L. by temperature is shown in Table 5. The change of fo is below 0.6MHz in the temperature range (-25 to +85°C) and I.L. is below 1dB in the temperature range (+25 to +85°C). Both of the changes are small because the change of fo depends mainly on the TCC of BSSZ.

Table 5 The Change of the Center Frequency (fo) and Insertion Loss (I.L.) by Temperature	
Center Frequency (fo) (-25 ~+85°C)	<0.6 MHZ [<30ppm/°C]
Insertion Loss	< +1 dB

Reliability:

After temperature cycle test (-55°C to +125°C 100 cycles), high temperature load test (85°C @ 500 hours), cold temperature load test (-25°C @ 500 hours), and humidity load test (40°C, 95% RH @ 500 hours), the electrical properties of BPF show no change. These test results show that the CMC - L-C filter has good reliability.

Applications of CMC - L-C Filter:

As stated above, the CMC - L-C filter has good properties as a filter at high frequency. Its applications for equipment are shown in Table 5. It will be used for cordless telephones, hand-held (cellular) telephones, pagers, BS tuners (I.F.), and CATV's (I.F.), etc. They are used in the frequency range between 200MHz and 1GHz. Considering the trend of electronic devices to use high frequencies, many more applications of CMC - L-C filter will develop in the near future.

Table 5
Application of CMC-L-C Filter

Cordless Telephone	**254 MHz** **380 MHz**
Handheld Telephone	**800 MHz ~ 1GHz**
Pager	**400 ~ 500 MHZ (Europe)**
B.S. Tuner	**400 ~ 500 MHz**
CATV (IF)	**610 MHZ**

In order to make CMC - L-C filters smaller with better filter properties, we have developed a new low temperature cofireable ceramic that has dielectric constant greater than 15, NPO TCC, and dissipation factor of less than .005.

CONCLUSION

The new low temperature cofireable ceramic (BSSZ) with copper conductor has the following properties:

1. Good Electrical Properties in High Frequencies
2. NPO TCC
3. High Reliability

One of the applications of BSSZ is as an L-C filter. The L-C filter (CMC - L-C filter) made with BSSZ can be used as a surface mount device because of its small size. It is excellent at high frequencies due to its integrated structure and shielding and low electrode loss. It is intrinsically reliable due to the low electrode resistance and integrated structure. Its cost advantage is realized through the use of copper electrodes, the compact size, and the surface mounting capability.

REFERENCES

1. S. Tosaka et al, proceeding IEEE CHMT Symposium, 29 (1984).
2. S. Nishigaki et al, ISHM, 429 (1986).
3. A. L. Eustice et al, IMC Proceedings, 49 (1986).
4. H. Mandai et al, IMC Proceedings, 61 (1986).
5. T. Nishimura et al, IMC Proceedings, 249 (1986).
6. H. C. Bhedwar et al, IMC Proceedings, 89 (1988).
7. N. Yoshida et al, IMC Proceedings, 98 (1988).
8. H. Saeki et al, IMC proceedings, 352 (1986).

HIGH-PERFORMANCE PACKAGING USING LASER LITHOGRAPHY*

Lionel M. Levinson, C.W. Eichelberger, R.J. Wojnarowski, and R.O. Carlson

GE Corporate Research and Development
Schenectady, New York 12301

ABSTRACT

High-Density Interconnects (HDI) is a novel hybrid packaging approach that uses polymer layer overlays laminated over bare chips mounted on a substrate. The overlays are laser-patterned with copper to connect the chips and I/O. The advantages of the HDI approach can be summarized as follows: the overlay layer makes the entire chip area available for interconnect lines; the interconnect has very high density and 2-mil pitch has been demonstrated; via and line formation are under computer control and no patterning mask is used. The HDI process is ideal for prototype or moderate volume production of high-performance circuits.

INTRODUCTION

The ceramic packaging engineer is typically concerned with the fabrication of enclosures for single integrated circuit chips. The electronic system designer attempts to place these enclosures in packages as closely together as possible in order to increase the chip density and decrease the system size. The standard interconnection vehicle, the printed wiring board, is normally offered with a 0.100-inch plated through hole and a wireability of about 40 lines/inch/layer. Many interconnect layers are possible, but 2- to 8-layer boards are most common.

Modern VLSI silicon processing technology has produced dramatic improvements in chip functionality and speed. However, in many situations, the increase in system performance available from advanced VLSI silicon chips is not realized because electronic packaging and assembly techniques as commonly practiced today negate much of the potential performance gain. This arises because the decrease in chip-to-chip interconnection has not kept pace with the rapid decrease in transistor-to-transistor interconnect wiring.

To properly utilize the available VLSI chip performance in the system as a whole, the chips have to be placed much closer to each other than allowed by typical single-

*Partially supported by the U.S. Air Force Weapons Laboratory

chip package and printed wiring board techniques. It is also necessary to use minimum dielectric constant materials for the interconnect insulator support network and provide for sufficient wireability to properly interconnect the high-pin-count integrated circuits produced by VLSI technology.

Multichip modules provide a technique to shrink interchip wiring. A variety of approaches are possible[(1)] and can provide dramatic improvement in the system packaging density and performance. We describe here a specific multichip packaging approach based on an adaptive laser lithography.

CONCEPT OVERVIEW

High-Density Interconnects (HDI) is a unique, novel, hybrid approach for the interconnection of high-performance integrated circuit (IC) chips in a dense configuration. The essence of the concept can be summarized as follows: bare IC chips are mounted face up on a substrate; a thin, flexible polymer film is laminated over the top of the IC chips; laser drilling is used to form vias in the polymer over the chip bonding pads; copper interconnects are defined on the polymer film to run connections from chip-to-chip and to the hybrid substrate input/output pads.

The HDI process has a number of advantages over more conventional high-density hybrid approaches. Some of these are as follows:

- The overlay layer makes the entire chip area available for interconnect lines.
- The interconnect has very high density; 2-mil pitch has been demonstrated.
- The bonds between the chip pads and interconnect are by thin film metallization; thus there are no wire bonds or solder joints on the hybrid.
- Via and line formation are under computer control, and no patterning masks are used. Chip misalignment is accommodated by computer-adaptive writing, i.e., the computer modifies the interconnect pattern to adjust for variations in chip placement.
- Copper is the conductor metallization.
- The chips can be placed to be almost touching, thereby giving maximum packing density. Slight gaps between chips provide strain relief.
- The interconnect technology accommodates any chip size and can accommodate mixed chip technologies.
- The process is ideal for prototype or moderate volume production because it is flexible and quickly implemented.
- The overlay can be removed and replaced repeatedly without chip damage.

- The overlay can be applied to a single chip in a socketable, single chip package for testing and burn-in, and then removed.
- The bare chips are mounted directly on the substrate. This facilitates heat removal.

FABRICATION

Figures 1, 2, and 3 show the basic assembly of the HDI structure, which consists of three layers: substrate, frame, and patterned overlay. To date, most of the frames and substrates have been alumina, because of its ready availability. We have, however, demonstrated the compatibility of our process with other substrate and frame materials such as silicon and Kovar® alloy.

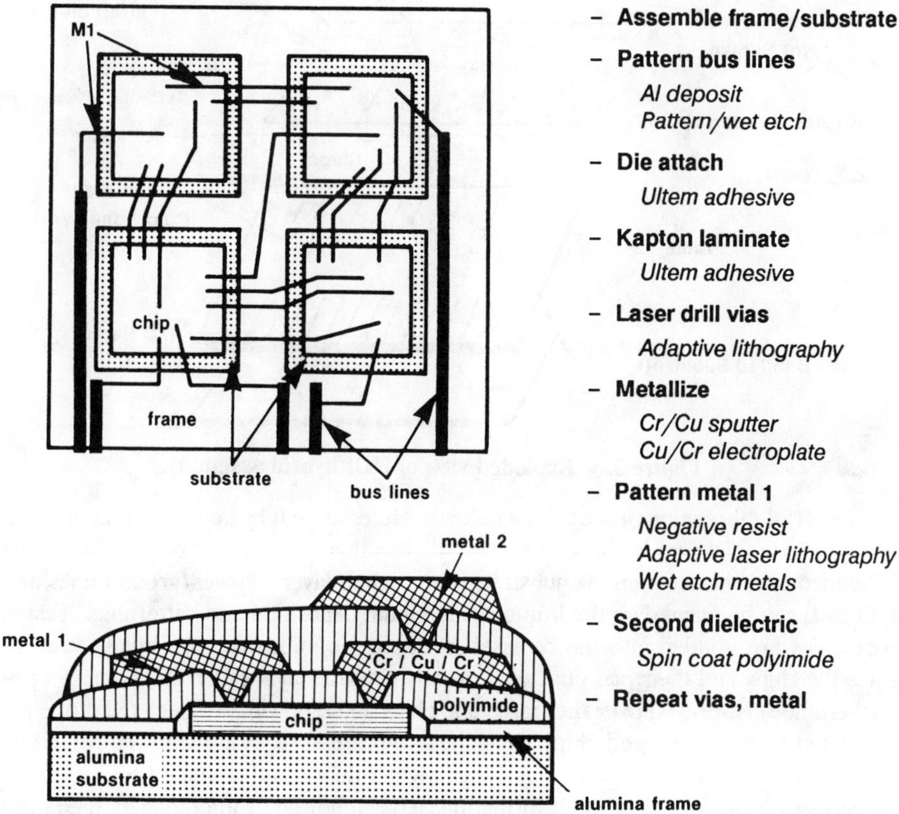

Figure 1. HDI process.

® Kovar is a registered trademark of Carpenter Technology Corp.

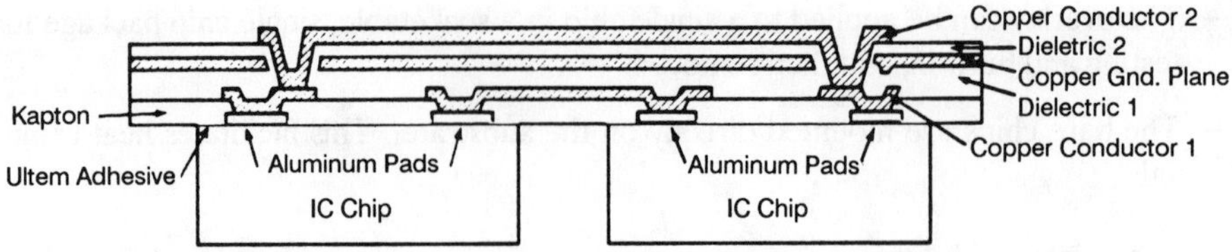

Figure 2. Schematic cross section of a two-signal-layer HDI structure with a ground plane separating the two layers.

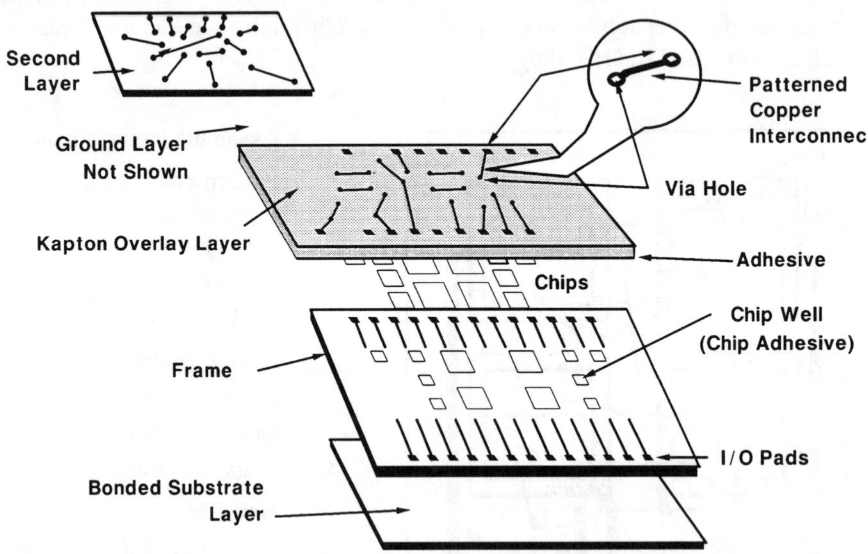

Figure 3. Exploded view of HDI hybrid assembly.

The HDI fabrication process is as follows: Holes are cut by laser in a thin alumina frame, whose thickness is chosen to be a little less than the thinnest chips. The frame is bonded to a thicker alumina substrate with a glass layer. Power/ground lines and I/O pads can be formed on the frame by aluminum deposition and patterning. Tested bare chips are bonded into the recesses in the frame/substrate. An image scan locates the chips and their pad positions. A polymer glue layer (such as Ultem®, a polyetherimide) is sprayed over the frame and chips and a polyimide sheet (Kapton®) is laminated to the frame and chips under heat and pressure. Vias are then drilled by the laser over the chip pad positions. Any polymer debris and residual film can be removed by a plasma etch. The via formation step is followed by interconnect metalliza-

® Ultem is a registered trademark of General Electric Company

® Kapton is a registered trademark of E.I. DuPont de Nemours & Co., Inc.

tion. Chromium followed by copper is sputtered over the whole hybrid surface (including the drilled via holes) and the metallization built up by copper plating followed by a chromium overcoat. Figure 4 shows two vias of different size and taper over an aluminum bonding pad. Via taper can be adjusted by varying the laser scan conditions. Via bottoms down to 10-μm are achievable with high reliability. The metallization exhibits excellent step coverage and continuity in the via hole.

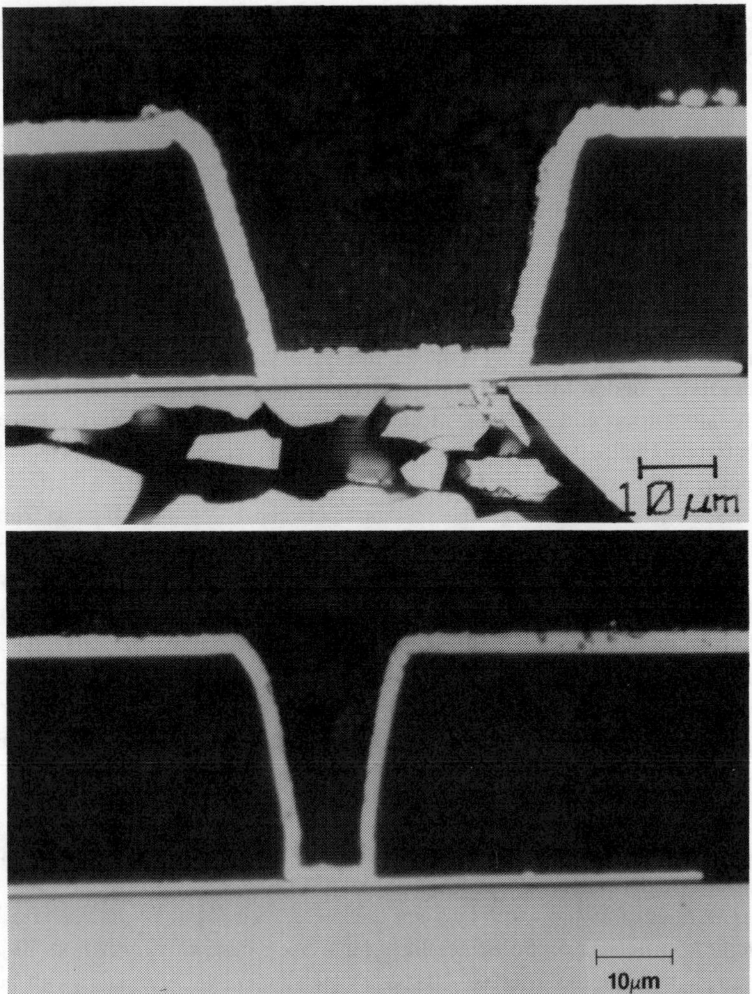

Figure 4. Via hole and metallization for two via shapes. Vias are drilled in 25-μm Kapton film (overlay layer). Metallization is plated Cu on sputtered Cu/Cr.

The signal metallization is defined by exposing a negative resist using the laser beam. Metal etching is used to remove the unwanted Cu/Cr. Chip misalignment is accommodated by computer-adaptive writing. The ideal chip pad locations and interconnect network are resident in software. Identification of the actual positions of two chip pads in a slightly misaligned chip serves to define the real position of all pads in a particular chip. The interconnect configuration is then slightly modified (shifted) to adapt to the actual chip pad positions.

The next signal layer is built up following a process sequence similar to that of the first layer. The polymide dielectric is spun over the first metal layer. Via holes are opened with the laser, and metal deposition and patterning are carried out as before. Power and ground lines (if not already on the frame, or if more are desired) can be incorporated into the signal layers, or, for better impedance control, separate power and ground planes can be formed between, on top, or beneath the signal layers.

RESULTS

Test substrates involving discrete chip functions, ring oscillators, and long chain interconnects through vias, have been fabricated to demonstrate line density, speed performance, signal fidelity, overlay removal and replacement, chip removal and replacement, durability under MIL-STD-883 environmental and mechanical tests, and survivability under total and transient radiation dosage. As an example of one of these test substrates, Figure 5 shows two views of the Pad Array Test Substrate. The four chips shown in the top view are 0.678×0.678 cm in size and fit into a frame cavity of 1.389×1.389 cm. Thus the average spacings between chips and chips to frame wall are only about 100 μm, dramatic proof of the close spacing of the chips possible with this overlay interconnect technique. The closeup in Figure 5 shows 100-μm-wide by 5-μm-thick copper lines interconnecting short 23-μm by 1-μm-thick aluminum lines formed on an oxidized silicon chip. The copper lines are on two signal layers. A close look shows there are two sets of vias contacting the vertical copper lines, one set to connect from aluminum to first-level copper and a second set to connect first-level copper to second-level copper. A cross section of a two-level staggered via interconnect is shown in Figure 6.

The chain resistance of the daisy weave pattern is very useful in demonstrating consistent near-100% yield in via connections and as a tool for proving interconnect durability through various stress tests. This and other test substrates were formed with copper lines of 25-μm width by 5-μm thickness, with 25- to 50-μm spacing, leading to a pitch of 50 to 75 μm. Still finer lines have been fabricated, down to 10-μm line and spacing, over limited lengths. Vias are typically 25 by 25 μm (they can be laser-drilled to be square, rectangular, or circular), but can be made as small as 8 by 8 μm. Table 1 gives a listing of typical HDI parameters.

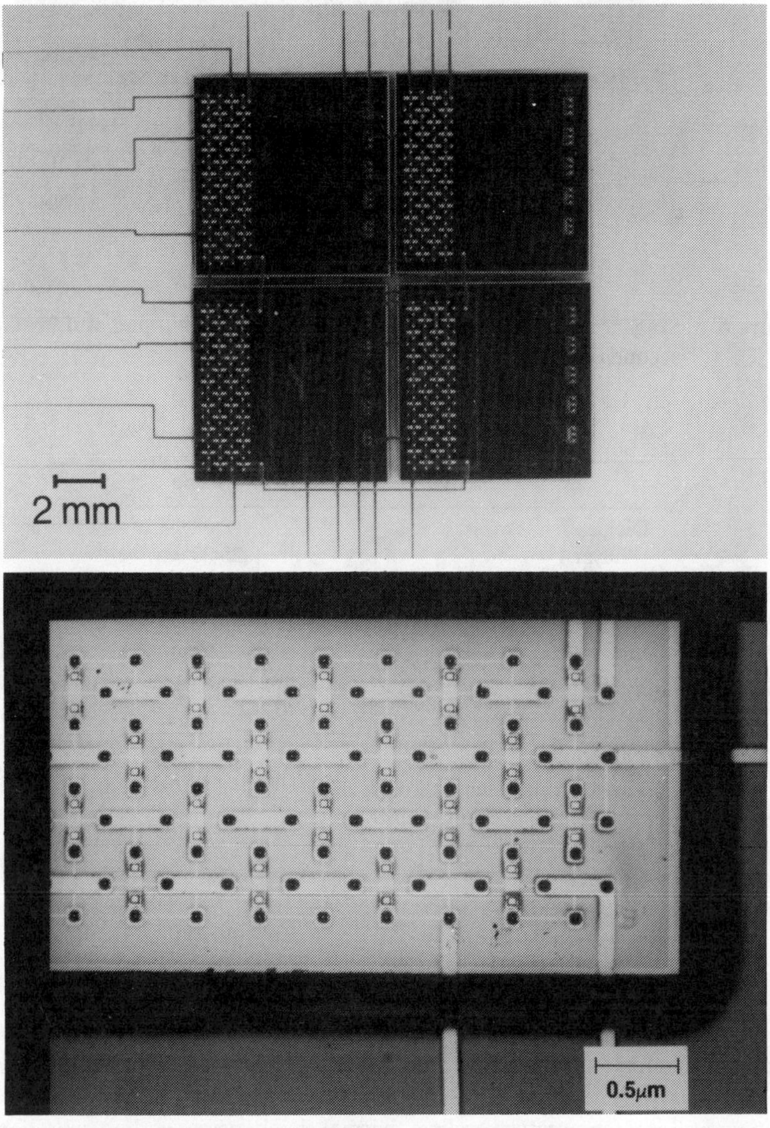

Figure 5. Photo of pad array test circuit at two different magnifications. Note the high quality of the metallization on the curved polyimide film bridges between the chips. Laser adaptive lithography (see text) accommodates the evident misalignment of chips.

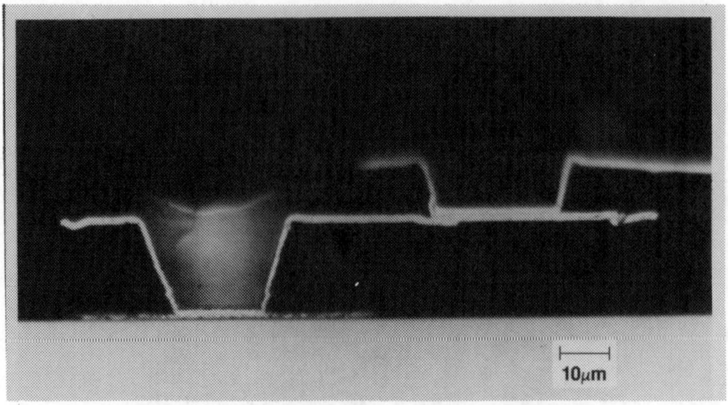

Figure 6. Staggered vias showing interconnect between chip pad and first and second signal interconnect layers.

Table 1. HDI Process Parameters

Property	Value
Dielectric thickness	
Layer 1	25-30 μm
Layers 2, 3, 4,	10-30 μm
Line width	25 μm
Line spacing	25 μm
Via bottom	10-30 μm
Conductor thickness	2-5 μm
Chip-to-chip spacing	≳100 μm
Thermal conductance through chip, substrate, and chip-to-substrate bonding layer	~1 °C/W
Die shear	>30 lb for 1 cm^2 chip
Dielectric properties	
Dielectric withstand	>1 kV
Dielectric constant	about 3
Interconnect capacitance	0.7-1 pF/cm

Any viable interconnect scheme must survive an appropriate set of environmental stresses. We have examined HDI interconnect behavior under temperature bakes, burn-in under bias, temperature cycling, thermal shock, power cycling, vibration, drop shock, and centrifuge. Populated substrates were visually examined and electrically tested after stress as appropriate. Table 2 summarizes some of the tests undertaken in HDI substrates. No failures were found that could be ascribed to the interconnect

Table 2. Stress Tests for HDI Substrates

Test Type	Description	Comments
Stabilization bake	150 °C, 48 h	No failures/changes
Temperature cycling	−55 °C to +150 °C	No failures/changes to 1000 cycles
Burn-in	125°C, 168 h with bias	No failures/changes from HDI process
Thermal shock	−65 °C to +150 °C liquid to liquid	No failures/changes to 1000 cycles
Thermal shock	Liquid nitrogen to hot plate at 150 °C	125 cycles, no failures/changes
Power cycling	Self heating, 20 to 150 °C	1000 cycles, no failures/changes
Variable frequency vibration	20 to 2000 Hz, 20 G level or 0.06 in. peak-to-peak	No failures/changes
Constant acceleration	7000 G, centrifuge	No failures/changes
Mechanical shock	5 drops to 1500 G level	No failures/changes if Al_2O_3 substrate reinforced
Polymer outgassing	Baked at 100 °C for 24 h after sealing in hermetic package	<1500 ppm moisture worst case, 500 ppm typical

system. The few failures observed invariably derived from extraneous sources, such as IC chip failure or mechanical weakness in the Al_2O_3 substrate.

In order to evaluate the speed capability of the HDI approach we have also fabricated CMOS ring oscillator circuits with various lengths of interconnect between the oscillator stages. A ring oscillator circuit is shown in Figure 7. The operating delay per stage was measured and compared with calculated values using both RC and transmission line models as appropriate. Good agreement was obtained between experiment and calculations. A 7-stage oscillator, with each stage purposely separated from the preceding stage by 3.7 cm of interconnect, oscillated at 40 MHz. This is to be compared with an oscillation frequency of 220 MHz for a similar ring oscillator with close to zero stage-to-stage spacing. The delay per stage of 1.8 ns corresponds well with the calculated value.

In addition to various test circuits, a four-chip hybrid module was fabricated. This module is comprised of four identical 256-point fast Fourier transform chips. These VLSI chips have 148 I/Os each; 352 I/Os are connected internally and 240 are brought out to the module periphery. Figure 8 is a photograph of the completed module.

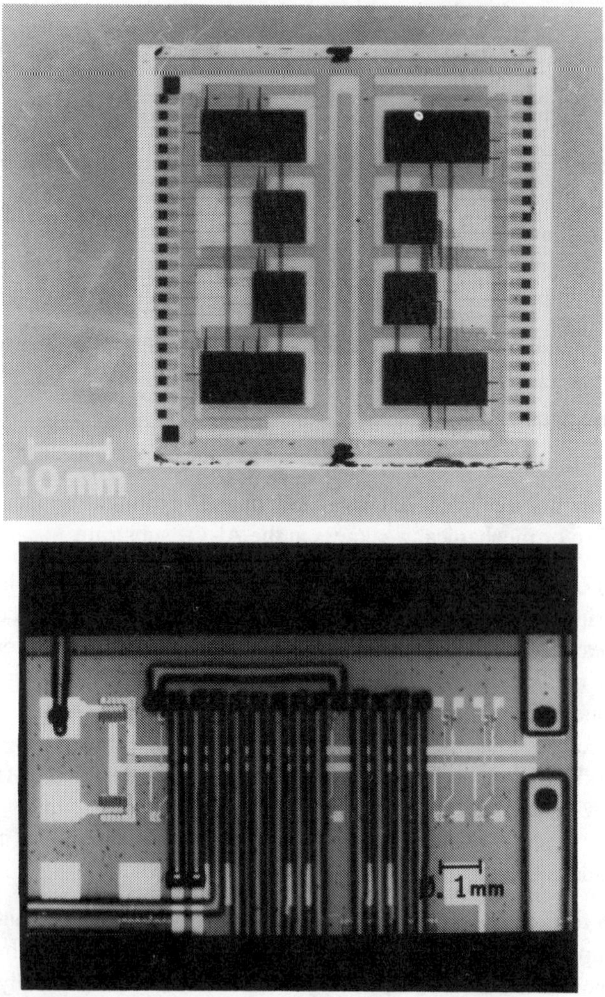

Figure 7. Photograph of ring oscillator circuit. The closeup shows interconnect to inverters using 25-μm pads.

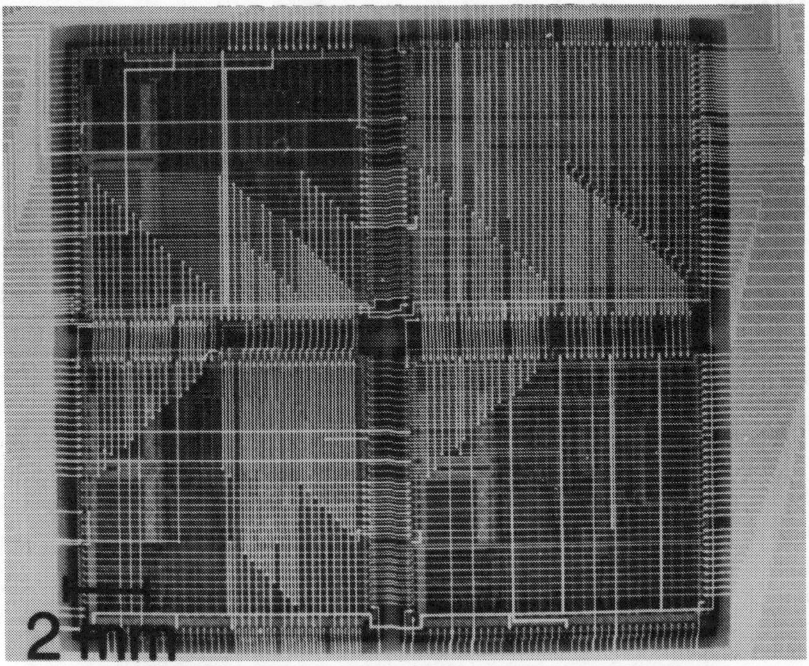

Figure 8. Photo of HDI interconnected 1024-point FFT signal processor. The four VLSI chips being interconnected have 148 I/Os each. The circuit is 1024-point FFT signal processor composed of four 256-point FFT chips. 352 I/Os are connected internally; another 240 are brought out to the module periphery.

A major issue that must be addressed for any workable hybrid interconnect scheme relates to repair and rework. The HDI approach addresses rework by the demonstrated ability to remove the overlay layer and any faulty chip after test of the completed module. If the fault detected is minor, such as shorts or opens in the interconnect, or if a change in the interconnect pattern is desired, the overlay can be removed and a new overlay formed over the existing bare chips. If, in addition, one or more of the chips have been determined to be faulty, the overlay can be removed, followed by removal of the individual faulty chips. New chips are then bonded to the substrate and a new overlay formed. A set of experiments has confirmed that the HDI approach has extensive rework and repair capability. Details will be presented in a forthcoming publication.

SUMMARY

High Density Interconnect (HDI) is a unique, novel, hybrid approach that uses polymer layer overlays laminated over bare chips mounted on a substrate. The overlays are laser-patterned with copper to connect the chips and I/O. The HDI approach has a number of advantages, including the following:

- The overlay layer makes the entire chip area available for interconnect lines;
- The interconnect has very high density: 2-mil pitch has been demonstrated;
- Via and line formation are under computer control and no masks are required;
- The overlay can be removed and replaced without chip damage;
- Chips are mounted directly on the substrate for good heat dissipation.

Initial tests have indicated that the process is manufacturable, reliable, and repairable. A pilot line capability is currently operational in a class 100 clean room at GE Corporate Research and Development and is fabricating various memory and logic modules on a routine basis. We believe that HDI will prove ideal for prototype or moderate volume production of high performance circuits.

REFERENCE

1. C.A. Neugebauer, R.O. Carlson, R.A. Fillion, and T.R. Haller, "High Performance Interconnections Between VLSI Chips," *Solid State Technology,* Vol. 31, No. 6, p. 93, June 1988.

CERAMIC PHOTOLITHOGRAPHY

W.J. Nebe
E.I. DuPont Co.
Wilmington Delaware

ABSTRACT

Photochemistry, polymerization and ceramic technology have been combined to yield a photoformable thick film paste which has been used to generate small diameter vias and fine line conductor lines.

This photoformable thick film is screen printed onto a substrate, dried and exposed with ultraviolet light, followed by processing in solvent or aqueous developer. The via size and conductor line width are superior to that which can be achieved with screen printing.

CERAMIC PHOTOLITHOGRAPHY

Light energy is a critical ingredient in some of the most important natural and modern technological processes. Among these are photosynthesis, vision, photography, xerography, television and photopolymerization.

The combination of photochemistry and polymerization science has resulted in the development of photopolymerization technology. The additional combination of photopolymerization with optics permits generation of polymeric material in specific regions of a composition which is being imaged. With the aid of photomasks, lenses, lasers, etc., photopolymerization can be used to cause physical changes in specific regions of monomer/polymer mixtures, allowing generation of an image. The ability to photolytically selectively change materials, particulary to cause irradiated material to become insoluble, is the basis of this present application and the means of photogeneration of images containing ceramic materials. This combination of photopolymerization chemistry with thick film ceramic technology is used to identify fine line circuit lines in conductors and small diameter vias in dielectrics which can be used in conjunction with microchip packaging and printed circuit board manufacture.

PHOTOCHEMISTRY

The science of photochemistry entails the molecular absorbtion of electromagnetic radiation, of wavelengths ranging from approximately 2000A to 10,000A, and the subsequent reaction/decomposition of the resulting excited molecule. The excited molecule can emit radiation from its singlet state, lose its energy in the form of heat, cross over to its triplet state, or, from its triplet state, lose energy in the form of heat or light. The excited singlet or triplet can break a chemical bond (photolysis) or react with another molecule. They can also transfer their energy to another molecule (sensitization). These are some of the basic reaction pathways followed by a photoactivated molecule (Fig 1,2).

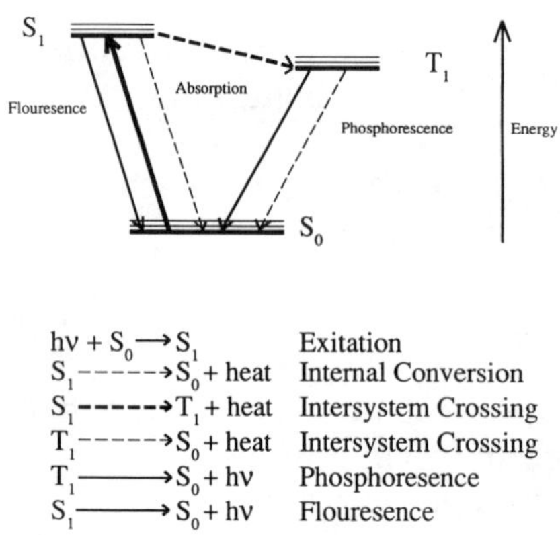

$h\nu + S_0 \longrightarrow S_1$	Exitation
$S_1 \dashrightarrow S_0 + \text{heat}$	Internal Conversion
$S_1 \dashrightarrow T_1 + \text{heat}$	Intersystem Crossing
$T_1 \dashrightarrow S_0 + \text{heat}$	Intersystem Crossing
$T_1 \longrightarrow S_0 + h\nu$	Phosphoresence
$S_1 \longrightarrow S_0 + h\nu$	Flouresence

Fig. 1. Exited State Photochemistry

Photochemistry has some unique features in that the amount of energy put into a molecule can be controlled by selection of the desired wavelength of light. Additionally, because light is the activating reagent in a photochemical reaction, the material to be exposed, in many cases will have excellent shelf life, as the activating agent, namely light, is absent from the reaction until the sample is irradiated.

The application of this technology, in which an exited

molecule breaks into radicals, one of which reacts with a monomer, to start a chain polymerization, is the basis of photopolymer chemistry.

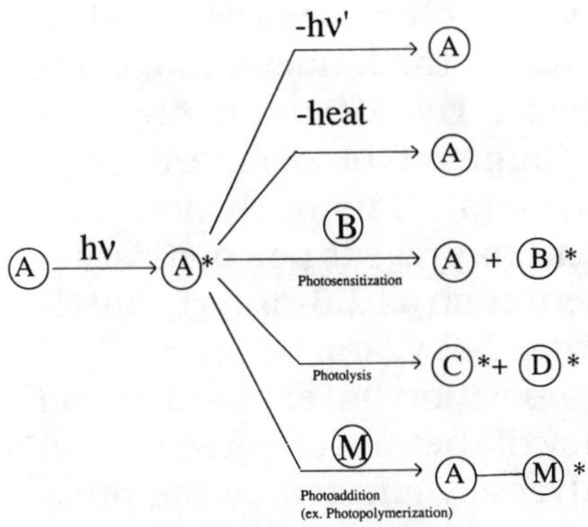

Fig.2. Photochemical Reaction and Excitation Pathway

Sensitization	$S \xrightarrow{h\nu} S^* \xrightarrow{I} S + I^*$
Initiation	$I \xrightarrow{h\nu} I^*$
Radical Formation	$I^* + RH \longrightarrow IH\bullet + R\bullet$
Chain Propagation	$R\bullet + nM \longrightarrow R(M)\bullet_n$
Termination	$R(M)\bullet_n + R(M)\bullet_m \longrightarrow R(M)\bullet_{n+m}R$
Inhibition	$R(M)\bullet_n + O_2 \longrightarrow R(M)\bullet_n OO\bullet$

Fig.3. Mechanism of Photopolymerization

POLYMERIZATION

A mechanism of photopolymerization Fig.3, involves the light induced activation of a sensitizer material, such as Michler's ketone, to its singlet state, followed by crossover to its triplet excited state, followed by energy transfer to an initiator. The activated initiator generally splits into two radicals, one of which reacts with the monomer, causing polymerization to occur. Acrylates and methacrylates Fig.4, are commonly the monomers of choice because of their high polymerization rate, low cost, and low toxicity. Additionally acrylates and meth-

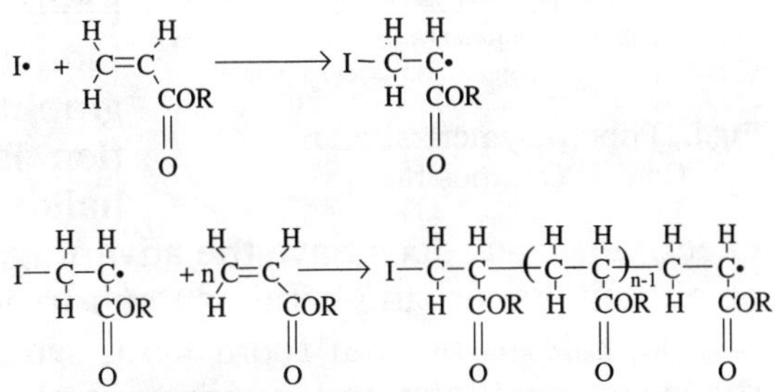

Fig.4. Mechanism of Radical Polymerization of Acrylates

acrylates readily and cleanly degrade under low oxygen firing conditions when the photopolymer composition is used in a thick film formulation.

Photopolymerizable compositions also contain preformed polymers, known as binders, which are used to give the unexposed composition higher viscosity and other desired physical characteristics. In the exposed material, the binders, in combination with the polymerized monomer give desirable physical properties such as insolubility. Common binders are polyacrylates and polystyrene copolymers. Photopolymerizable compositions also contain stabilizers to prevent polymerization before exposure. Common stabilizers such as 2,6-di-tert.-butyl-4-methylphenol, give shelf life of from 1-2 years.

The photopolymerizable composition is exposed using ultraviolet light Fig.5. The efficiency of the rate of a free radical polymerization is markedly and adversely effected by the presence of oxygen. This is due to the capability of oxygen to act as an excellent radical trap, reacting with sensitizer, initiator, monomer and polymer radicals, giving the stable, and therefore unreactive, peroxy radical. The peroxy radical, because it is unreactive, essentially stops the polymerization. Only when the oxygen has been consumed by reaction with the sensitizer, initiator, etc. radicals does the polymerization proceed, uninhibited, Fig.6. Because of this induction period the photographic speed of polymerization is slower than a silver halide imaging system. This effect, however, does have the advantage of allowing processing of photopolymerizing compositions in yellow light of intensity comparable to normal room light, which permits handling of these compositions under near normal working conditions. The composition is developed using an organic solvent or basic

•Light Sensitive Layer

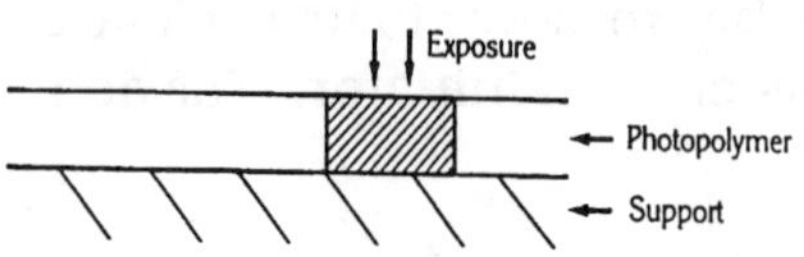

•Conversion of monomer yields physical change.
•Physical property differances between exposed and unexposed regions allow development.

Fig.5. Photopolymerization of Coated Composition

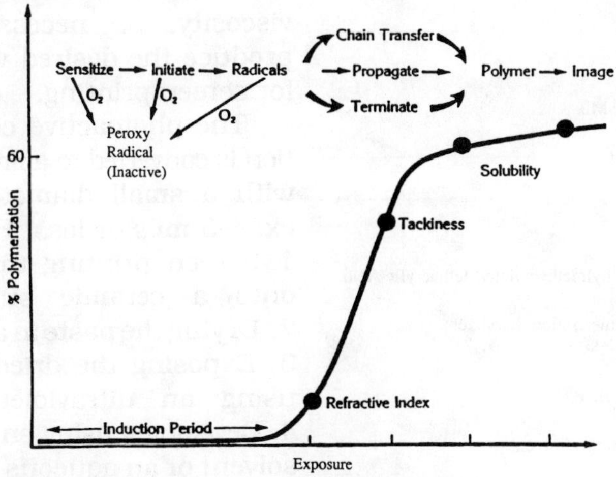

Fig.6. Physical Changes as Determined by Extent of Photopolymerization

aqueous solution, depending on the formulation of the imaged material. The exposed formulation, as determined by its composition, may also be developed using changes in tackiness, adhesion diffusion, refractive index etc. properties, Fig.7.

Physical Property Change	Readout
Soluability	Washout
Adhesion	Delamination
Softening Point	Thermal Transfer
Permeability	Modulated Diffusion
Tackiness	Toning
Refactive Index	Refraction

Fig.7. Development Method vs. Physical Property Change

PHOTOFORMED CERAMIC - PHOTOFORMED DIELECTRIC VIAS

Photopolymer technology has been combined with thick film technology to yield a photoformable thick film dielectric paste. The present work extends the resolution capabilites previously identified by Jolley et. al.[1]. The dielectric composition Fig.8, is composed of alumina, glass frit and the organic portion, which is a solution of polyacrylate binder, monomer, initiator, stabilizer and solvent to dissolve the polyacrylate. The solvent is

Dielectric Powder Mix
- Al_2O_3
- Blue Pigment
- Glass Frit

Vehicle
- Binder:
  - Polymethylmethacrylate/Methacrylic Acid
- Photoinitiator
- Radical Polymerization Stabilizer
- Solvent

Monomer - Triacrylate

Viscosity Adjustment - Solvent

Developer - Organic or Aqueous Solvent

Fig.8. Photoformable Dielectric Formulation

also used to modify the paste viscosity, as necessary, to produce the desired viscosity for screen printing.

The photoactive composition is converted to a fired paste with a small diameter vias, ex. 4-5 mils or less, Fig.9, by 1. Screen printing the paste onto a ceramic substrate, 2. Drying the paste in an oven, 3. Exposing the dried paste, using an ultraviolet lamp, 4. Developing using an organic solvent or an aqueous sodium carbonate solution in a spray developer[2]. The developed parts have open areas which correspond to areas of the photomask which have not permitted passage of light during the exposure (negative working image). The exposed and developed dielectric is fired through an air atmosphere furnace or a nitrogen atmosphere furnace, depending on the metal being used in conjunction with the dielectric.

This composition yields exposed and developed vias of approximately 4-5 mil in diameter for a fired dielectric thickness of 20 microns.

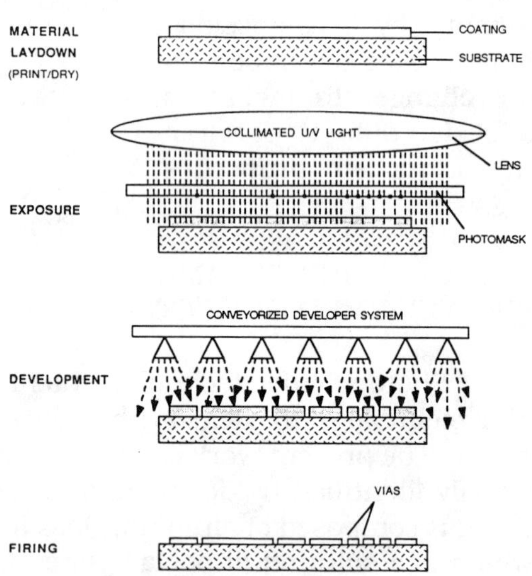

Fig.9. Processing Scheme For Photoformation of Vias[2]

Exposure times are in the range of 3-6 seconds (70 millijoules/cm^2) when a 1000 watt mercury vapor light source is used at a distance of two feet. Exposures are run in a nitrogen environment, as oxygen, as described above, will retard a free radical polymerization. The exposed parts are developed for approximately 30 seconds using a spray developer. The difference between a screen printed via and a photogenerated via is shown in Fig. 10. The screen printed via shows sloping side walls due to the flow of paste after the via is printed. This paste flow prevents generation of small vias, as the flow tends to close the bottom of the via hole. Because of this effect, conventional pastes must be formulated with minimum flow characteristics. This, however, results in printed pastes which contain pin holes and other imperfections. Development of the photoformed via, however, is carried out on dried paste where paste flow is not a concern, therefore the paste can be formulated to flow (level) when printed, giving a smooth pinhole free surface.

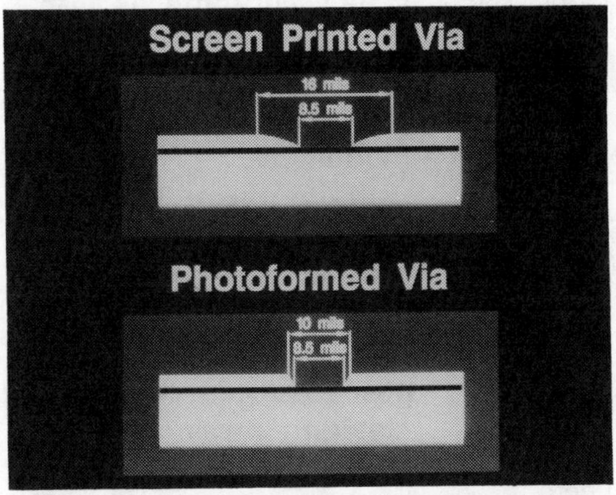

Fig. 10. Screen Printed vs. Photoformed Vias

The factors governing the ultimate resolution character of the photoformable dielectric paste are reflection and refraction of the imaging light when passing through the paste during exposure, thickness of the composition, and developing parameters.

These considerations and experimental results suggest that generation of a 4-5 mil via by means of screen printing is extremely difficult, whereas photogeneration of 4-5 mil vias is readily accomplished by this method.

PHOTOFORMED CONDUCTOR

This photoformable technology has also been combined

with conductor materials such as gold and copper to form very fine conductor lines. These compositions consist of the conductor metal, a glass frit, and the organic vehicle. The photoactive vehicle contains, as with the dielectric, a polyacrylate binder, a photoinitator, stabilizer and a monomer.

The paste is processed Fig. 11, in the same manner as the dielectric. It is screen printed onto a substrate, dried and exposed through a photomask, to ultraviolet light to polymerize all areas corresponding to a circuit line (negative working imaging). The substrate containing the polymerized and nonpolymerized material is developed with an organic solvent or an aqueous sodium carbonate solution using a spray developer. The developed parts are fired in either air for gold or nitrogen for copper.

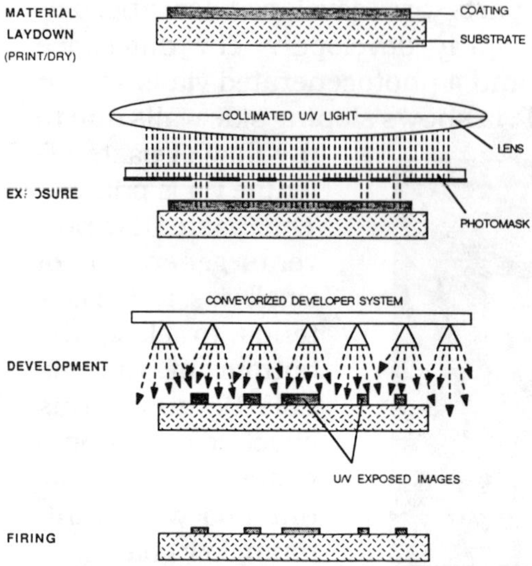

Fig. 11. Processing Scheme for Photoformation of Conductor Lines[2]

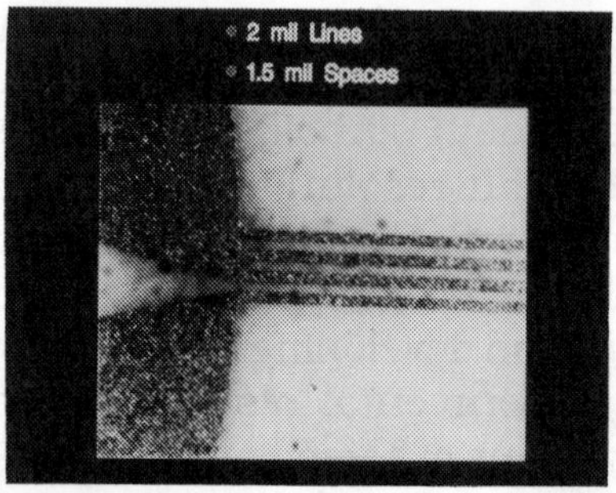

Fig. 12. Fired Conductor With 2 mil Lines and 1.5 mil Spaces

This procedure gives a fired conductor line of 1-2 mils width at a thickness of between 9-11 microns, Fig. 12.

PHOTOFORMED MULTILAYER

The photoformable dielectric and conductor technologies have been combined to prepare a multilayer assembly Fig 13.

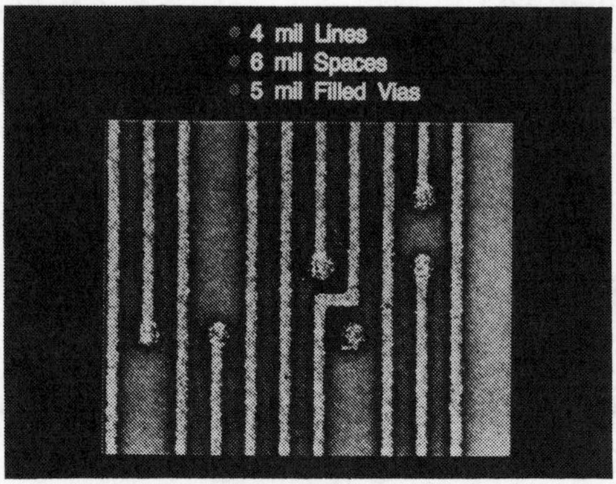

Fig. 13. Multilayer with 4 mil Lines and 5 mil Vias

CONCLUSION

Aqueous and solvent processable photoactive dielectric and conductor thick film pastes have been prepared and used to prepare 4-5 mil vias and 1-2 mil conductor lines. The following general and specific characteristics for the dielectric and conductor pastes are listed in table I and II.

TABLE I
PHOTOFORMABLE PASTE CHARACTERISTICS

- High resolution pattern created by photoimaging
- Fine conductor lines and spaces
- Small via diameter dielectric
- Smooth surface free of pinholes and mesh marks
- Can be processed in nontoxic aqueous developer

TABLE II
PHOTOFORMED LINES AND VIA CHARACTERISTICS

• Line width/space	2/5 ± 1/2 mils
• Via diameter	4 ± 1/2 mil
• Dielectric thickness	2 ± 1/2 mil
• Conductor thickness	10 microns ± 1
• Via to via distance	5 ± 1/2 mil

REFERENCE

1. Jolly, J.E., Weaver, R.V., Wells, J.H., Felten, J.J., Roe, D.W., SOLID STATE TECHNOLOGY, MAY 1974, page 1.

2. Hagan, N.C., Henderson,J.F., Nebe, W.J., Osborne, J.J., INTERNATIONAL JOURNAL FOR HYBRID MICROELECTRONICS, DECEMBER 1989, page 175.

ACKNOWLEDGEMENT

The author gratefully aknowledges the assistance and contributions of J.J. Osborne, N.C. Hagan, J.F. Henderson, M.S. Critzer and J.K. Holzapfel.

IN SITU PREPARATION OF $BaTiO_3$ THIN FILM BY HYDROTHERMAL-ELECTROCHEMICAL SYNTHESIS

Masahiro Yoshimura, Seung Eul Yoo, Motoo Hayashi and Nobuo Ishizawa

Research Laboratory of Engineering Materials, Tokyo Institute of Technology, 4259 Nagatsuta, Midori, Yokohama 227, Japan

ABSTRACT

The thin films of $BaTiO_3$ have been prepared in situ on Ti metal substrates in high temperature-high pressure Ba^{2+} solutions by a new technique of hydrothermal-electrochemical method, which conducts the electrochemical reaction under the hydrothermal conditions. Thin films of polycrystalline cubic $BaTiO_3$ have been formed at temperatures above 100°C in 0.2 ~ 0.5N-$Ba(OH)_2$ solutions by applying the current densities of 10 ~ 100 mA/cm^2 for 10 ~ 80 min. The obtained films had the thickness of about 0.1 ~ 0.3 μm and had smooth and homogeneous surfaces without any defects of visible pores or cracks. These films are paraelectric with the dielectric constant of ≈300.

Key-words: $BaTiO_3$, Thin film, Hydrothermal, Electrochemical, Capacitors

INTRODUCTION

Barium titanate is of interest because of its usefulness in electric devices. For example, it is widely used as a capacitor material because of its ferroelectric properties. Therefore various attempts have been made for the preparation of $BaTiO_3$ film. As the vapor deposition techniques, vacuum evaporation[1-3], evaporation by electron bombardment[4] and sputtering[5-8] from $BaTiO_3$ targets etc. have been reported to prepare $BaTiO_3$ thin films. On the other hand, hydrolysis of organometallic compounds[9,10], spray coating, dip coating and/or sol-gel methods[11,12] have been also widely used as chemical processings. However, these methods generally require calcining at appropriate high-temperatures to obtain well crystallized $BaTiO_3$ films as well as those with good electrical properties. These heating processes often bring about the reaction of the film with the substrates and/or atmospheres, and then result in incorporation of impurities, generation of pin-holes, cracking and/or peeling of the films. Particularly, the organometallic processings have some difficulty to form dense films, because the decomposition of organometallic compounds would give porous ceramics consisted of fine grains. Therefore the in situ formation of the crystalline $BaTiO_3$ film has been desired for the purpose of excluding these disadvantages.

We have reported previously that $BaTiO_3$ crystalline powders could be prepared by a new technique, hydrothermal anodic oxidation[13,14], where the anodically oxidized TiO_2 film on Ti metal reacts with Ba components in hydrothermal $Ba(NO_3)_2$ solutions at high temperatures under pressures when the film suffers an electrical breakdown by applying excess potentials. Recently, another new technique of hydrothermal-electrochemical synthesis has brought about a successful preparation of the $BaTiO_3$ thin film[15], which has been crystallized in situ on the Ti metal plate by preventing both the formation and the breakdown of TiO_2 film. The present study deals with the formation condition of these $BaTiO_3$ films.

EXPERIMENTAL

A titanium metal plate(working electrode) was connected to the anode by a platinum wire, then it was set in an autoclave together with a platinum plate(counter electrode) as the cathode as shown in Fig.1. The distance between the working electrode and the counter electrode was 3.0 cm. A direct electric current was applied in the electrolytic solution at elevated temperatures up to 200°C under the saturated vapor pressures ranging between 0.1 and 1.8 MPa. The treatment was performed in $Ba(OH)_2$ electrolytic solutions. The hydrothermal-electrochemical treatment was carried out galvanostatically with the applied current density between 10 and 100 mA/cm^2 using a galvanostat (Kikusui electronics Co., Type PAD 1K-0.2L) in an electrolytic autoclave(NRD Co., Type AAC-2030). A teflon beaker was set for preventing the autoclave wall from corrosion by contacting electrolytic solution.

The thin films formed on the Ti metal plate after the hydrothermal-electrochemical treatment were examined by X-ray diffractometry and scanning electron microscopy(SEM). The capacitance of the film was also measured using an LCR meter (YHP Co., Type 4284A) by depositing Ag electrode of 2 mmϕ in diameter.

RESULTS and DISCUSSION

Figure 2 shows X-ray diffraction patterns of the surfaces of $BaTiO_3$ films produced after hydrothermal-electrochemical treatments of titanium metal plates in 0.5N-$Ba(OH)_2$ solution. Polycrystalline $BaTiO_3$ as a cubic or pseudo-cubic phase could be detected on the Ti metal plate treated at temperatures above 100°C. It should be noted that no TiO_2 phases could be detected in the products above 100°C. On the contrary, TiO_2(Anatase) was only the product at low temperatures as a room temperature. The TiO_2 formation was similarly observed by the anodic oxidation of Ti metal plates at low temperatures in $Ba(NO_3)_2$ solutions reported in our previous studies[13,14]. These results clearly indicate that the reactions between Ba component and Ti component to form $BaTiO_3$ film require the high temperatures above 100°C at lowest during the hydrothermal-electrochemical treatment of Ti metal plates in $Ba(OH)_2$ solutions.

Under conditions at temperatures above 100°C with the current density of 10-100 mA/cm^2 in 0.5N-$Ba(OH)_2$ solutions, continuous thin films showing multiple interference colors, yellow, blue, green, purple, etc., were formed. The scanning electron micrograph of the thin film obtained in 0.5N-$Ba(OH)_2$ solution at 200°C with an applied current density of 10 mA/cm^2 for 60 min is shown in Fig.3. This $BaTiO_3$ film had smooth and homogeneous surface without any defects of visible pores or cracks.

Figure 4 illustrates the formation diagram of the film products under various conditions in $Ba(OH)_2$ solutions. In high concentration regions of $Ba(OH)_2$ solutions, $BaTiO_3$ film formed above 100°C, whereas anatase was only produced even at temperatures of 100°-150°C in lower concentration of solutions such as below 0.1N under the conditions studied. As shown in this diagram, $BaTiO_3$ film has not been formed at lower temperatures, nor in low-concentration barium solutions. These results apparently indicate that the production of $BaTiO_3$ is dependent on the treatment temperature and/or solution concentration.

Figure 5 shows typical cell voltage-time(V-t) curves, where the voltage between the titanium anode as working electrode and the platinum cathode as counter electrode was measured, during hydrothermal-electrochemical treatments of titanium metal plate in $Ba(OH)_2$ and $Ba(NO_3)_2$ solutions. In the $Ba(OH)_2$ solution, the cell voltage shows nearly constant during the treatment. In this case, thin film of polycrystalline $BaTiO_3$ has been produced directly on Ti metal plate, and any TiO_2 phases such as anatase and rutile have not been detected as shown in Fig.2. On the other hand, the cell voltage-time relation in the $Ba(NO_3)_2$ solution showed very different phenomenon from that in $Ba(OH)_2$ solution; that is, the cell voltage in $Ba(NO_3)_2$ increased with time up to 40 to 50V and then showed vigorous fluctuation. In these conditions, powders and thick films consisted of aggregates of $BaTiO_3$ powders with submicron grains have been produced from Ti metal plates after treatments[13,14].

The considerable fluctuation of cell voltage during the formation of $BaTiO_3$ powders in $Ba(NO_3)_2$ solutions indicated that $BaTiO_3$ powders might be formed when surface-oxidized TiO_2 films, which had been grown anodically to a certain thickness, were decomposed by an electrical breakdown accompanied by sparking[13,14]. Thus, the hydrothermal anodic oxidation technique requires the formation of the TiO_2 film oxidized anodically on the Ti metal plate for the preparation of $BaTiO_3$ powders in $Ba(NO_3)_2$ solutions. The X-ray diffaraction and SEM observation[13,14] supported the formation of TiO_2 film before the formation of $BaTiO_3$.

On the other hand, the present technique to form $BaTiO_3$ thin film seems not to need to form apparent TiO_2 films. This does not exclude the formation of amorphous TiO_2 thin films, but the formation mechanism of $BaTiO_3$ thin film in $Ba(OH)_2$ solution is probably different from that of $BaTiO_3$ powders which results from the breakdown of surface-oxidized TiO_2 film in $Ba(NO_3)_2$ solution. We suggest a following possible mechanism for the formation of $BaTiO_3$ thin film in the present technique:

$$Ti^{o} = Ti^{4+} + 4e^{-} \quad \text{-----} \quad (1)$$

$$Ti^{4+} + 6OH^{-} = Ti(OH)_6^{2-} \quad \text{-----} \quad (2)$$

$$Ba^{2+} + Ti(OH)_6^{2-} = BaTiO_3 + 3H_2O \quad \text{-----} \quad (3)$$

In the initial stage of reaction, Ti metal ionizes as Ti^{4+} metal ions by electrochemical reaction of Ti metal plates in hydrothermal $Ba(OH)_2$ solutions according to the elementary anodic reaction of (1). Ti^{4+} metal ions react with OH^- ions to form such hydroxyl titanium complex as $Ti(OH)_6^{2-}$ on the Ti metal surface. It is supposed that these intermediate products as $Ti(OH)_6^{2-}$ react with Ba^{2+} ion in the $Ba(OH)_2$ solution to form a continuous $BaTiO_3$ film layer directly on Ti metal plate. These reactions, containing dissolution-precipitation reactions with a three-step process, might proceed in the hydrothermal $Ba(OH)_2$ solution near the anode to precipitate crystalline $BaTiO_3$ on Ti metal plate of the anode. Since electrochemical and hydrothermal reactions proceed simultaneously during the reaction, the present technique was named as hydrothermal-electrochemical synthesis[15].

The thicknesses of the $BaTiO_3$ films prepared at 200°C in 0.5N-$Ba(OH)_2$ solutions with the applied current density about 15 mA/cm^2 increased with the treatment time as shown in Fig.6. It reached to about 300 nm after 60 min treatment. These values have been calculated from the weight gain after treatments assuming that the reacted titanium was all converted to $BaTiO_3$ without dissolution, and that the density of $BaTiO_3$ was 6.02 g/cm^3.

The capacitance measured for the film formed at 150°C by the applied current density of 13 mA/cm^2 for 80 min was 70 nF with $\tan\delta \approx 15\%$. The value corresponds to the dielectric constant of ≈300 for 0.1 μm of film thickness estimated from the weight gain of the film. These values appear to be ordinary for the paraelectric films with cubic or pseudo-cubic $BaTiO_3$ phase.

We have also succeeded in preparing $BaTiO_3$ thin films from the sputtered titanium on the organic film substrates as well as glass substrates by this hydrothermal-electrochemical method[16].

NOTE ADDED IN PROOF.

Just recently, the authors observed the existence of very thin layer of TiO_2(amorphous or anatase) between the $BaTiO_3$ film and Ti metal substrate.[16] Furthermore, the formation of $BaTiO_3$ thin films with the thickness of 50 nm or thinner has been observed even without electrical currents under similar conditions to the present work. These results suggest following things: $BaTiO_3$ might be formed through such ionic reactions as indicated by equations (1) ~ (3) but it might grow due to diffusion processes. Electrical currents might affect rather on the growth process.

REFERENCES

1) C.Feldman, "Formation of Thin Films of $BaTiO_3$ by Evaporation," Rev.Sci.Instrum., 26[5]463-66(1955).
2) E.K.Muller, B.J.Nicholson and G.L'E.Turner, "The Epitaxial Vapor Deposition of Perovskite Materials," J.Electrochem.Soc., 110[9]969-73(1963).
3) Y.Murayama, R.Ueda, T.Suzuki and F.Tokugawa, "Formation Process of $BaTiO_3$ Thin Film Evaporated in Vacuo,"

Oyo Buturi(Japan), 33[7]456-60(1964).
4) A.E.Feuersanger, A.K.Hangenlocher and A.L.Solomon, "Preparation and Properties of Thin Barium Titanate Films," J.Electrochem.Soc., 111[12]1387-91(1964).
5) Y.Shintani and O.Tada, "Preparation of Thin $BaTiO_3$ Films by dc Diode Sputtering," J.Appl.Phys., 41[6]2376-80(1970).
6) Y.Shintani and O.Tada,"$BaTiO_3$ Films by RF Sputtering," Denshi-Tsuushin-Gakkai-Ronbunshi, 56-C[4]205-12(1973).
7) H.Igarashi and M.Yuasa, "PTC Effects of $BaTiO_3$ Thin Film," Electronic Ceramics(Japan), 18[9]28-33(1987).
8) V.S.Dharmadhikari and W.W.Grannemann,"AES Study on the Chemical Composition of Ferroelectric $BaTiO_3$ thin Films RF Sputter-deposited on Silicon,"J.Vac.Sci.Technol. A,1[2]483-85(1983).
9) J.Fukushima, K.Kodaira and T.Matsushita, "Preparation of $BaTiO_3$ Films by Hydrolysis of Organometallic Compounds," Am.Ceram.Soc.Bull., 55[12]1064-65(1976).
10) M.I.Yanovskaya, N.Ya.Turova, E.P.Turevskaya, A.V.Novoselova, Yu.N.Venevtsev, S.I.Sagitov and E.M.Soboleva, "Preparation of Ferroelectric Barium Titanate Films from Solutions of Metal Alcoholates," Inorg.Mat.,17[2] 221-24(1981).
11) S.Sakka, K.Kamiya, K.Makita and Y.Yamamoto, "Formation of Sheets and Coating Films from Alkoxide Solutions," J.Non-Cryst.Solids, 63[2]223-35(1984).
12) T.Yoko, K.Kamiya and S.Sakka, "Preparation of $SrTiO_3$ and $BaTiO_3$ Films by the Sol-Gel Method and Their Photoelectrochemical Properties," Res.Rep.Fac. Mie.Univ., Vol. 12, pp.49-64(1987).
13) S.E.Yoo, M.Yoshimura and S.Sōmiya, "Preparation of $BaTiO_3$ and $LiNbO_3$ Powders by Hydrothermal Anodic Oxidation," Sintering '87, Proc. 4th Intern. Symp. on Sci. and Tech. of Sintering,Tokyo, 1987(Elsevier Appl.Sci., London, New York and Tokyo, 1988) vol.1, pp.108-13.
14) S.E.Yoo, M.Yoshimura and S.Sōmiya, "Direct Preparation of $BaTiO_3$ Powders from titanium Metal by Anodic Oxidation under Hydrothermal Conditions," J.Mater.Sci.Lett. 8[5]530-32(1989).
15) M.Yoshimura, S.E.Yoo, M.Hayashi and N.Ishizawa, "Preparation of $BaTiO_3$ Thin Film by Hydrothermal Electrochemical Method,"Jap.J.Appl.Phys., 28[11]L2007-09(1989).
16) M.Hayashi, N.Ishizawa, S.E.Yoo and M.Yoshimura, "Preparation of Barium Titanate Thin Film on Titanium-Deposited Glass Substrate by Hydrothermal-Electrochemical Method," Jpn.J.Ceram.Soc.(Seramikkusu Ronbunshi), submitted.

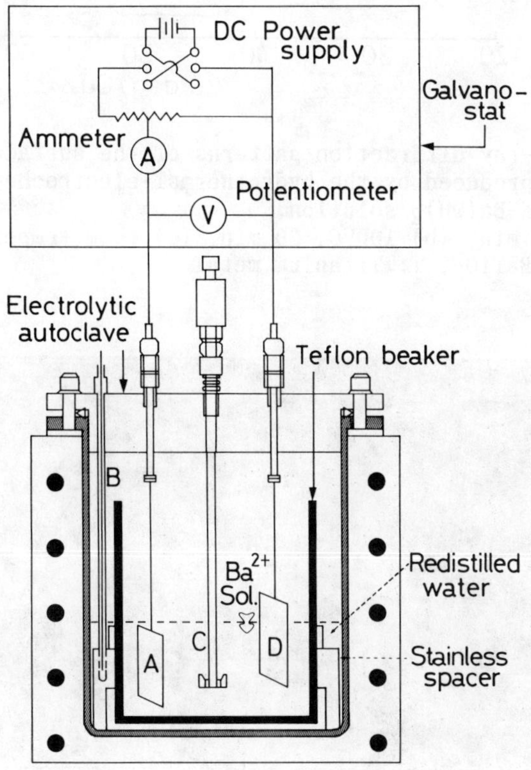

Figure 1. Schematic illustration of the electrochemical cell and circuit arrangements for hydrothermal-electrochemical method. (A) Pt plate, cathode, (B) Thermocouple, (C) Stirrer, (D) Ti plate, anode

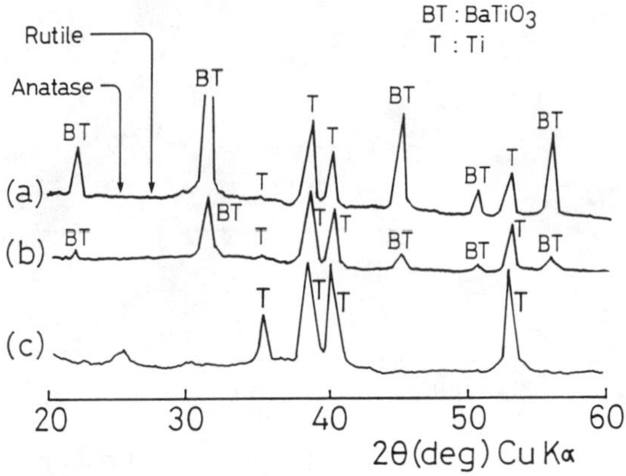

Figure 2. X-ray diffraction patterns of the surfaces of $BaTiO_3$ films produced by the hydrothermal-electrochemical method in 0.5N-$Ba(OH)_2$ solution.
(a) 200°C, 30 min, (b) 100°C, 30 min, (c) room temperature, 60 min, BT: $BaTiO_3$, T: Titanium metal

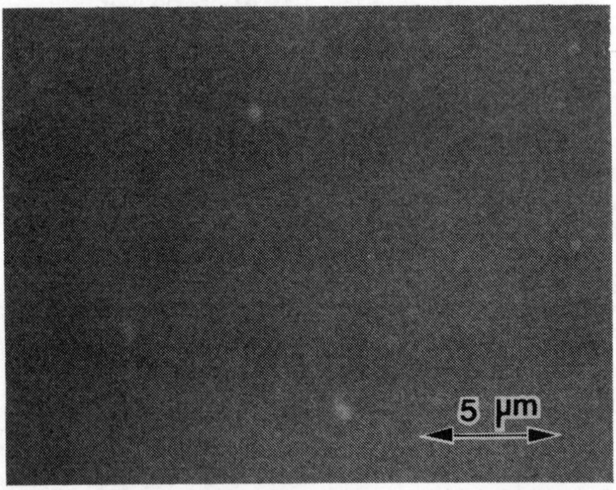

Figure 3. Scanning electron micrograph of the surface of $BaTiO_3$ thin film prepared on Ti plate at 200°C in 0.5N-$Ba(OH)_2$ solution with the current density of 10 mA/cm^2 for 60 min.

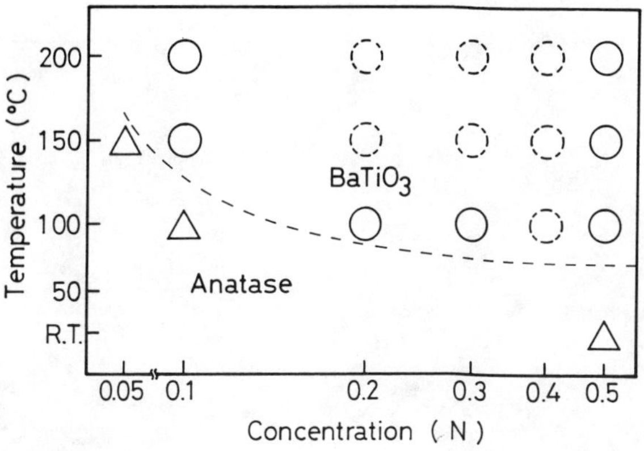

Figure 4. Formation diagram of the film products by hydrothermal-electrochemical method under various conditions in $Ba(OH)_2$ solutions.

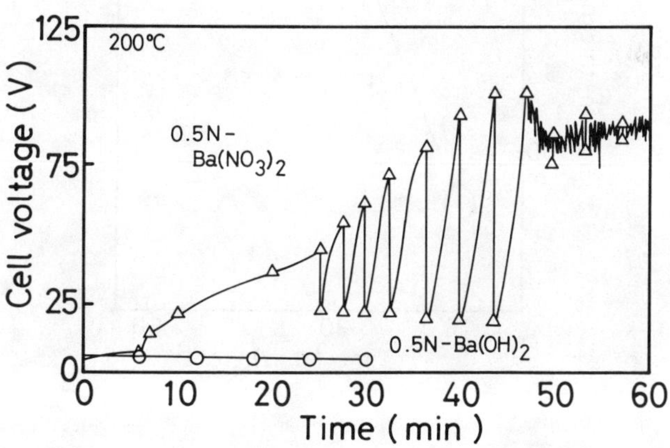

Figure 5. Typical cell voltage-time(V-t) curves during hydrothermal-electrochemical treatments of Ti plate at 200^oC in Ba^{2+} solutions.

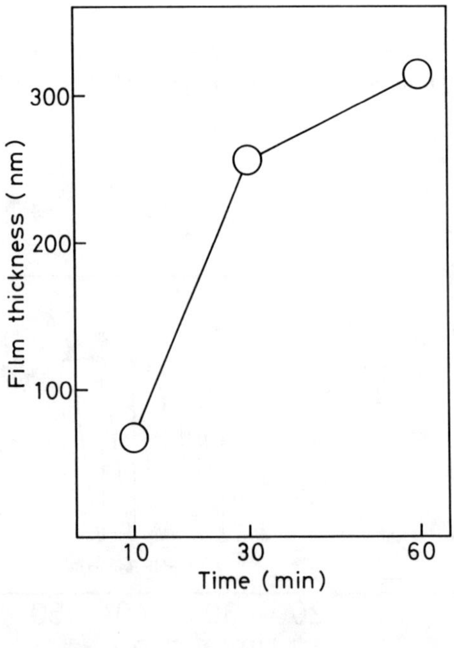

Figure 6. Relation between the treatment time and the thicknesses of $BaTiO_3$ films prepared at 200°C in 0.5N-$Ba(OH)_2$ solutions with the applied current density about 15 mA/cm^2.

EFFECT OF DEPOSITION CONDITIONS ON DEPOSITION MECHANISM AND SURFACE MORPHOLOGY OF TiO_2 THIN FILMS DEPOSITED BY CHEMICAL VAPOR DEPOSITION

Cheol Seong Hwang and Hyeong Joon Kim

Dept. Inorganic Materials Engineering, Seoul National University

Seoul, Korea

Abstract

TiO_2 thin films were deposited on Si and Al_2O_3 substrates by chemical vapor deposition using $Ti(OC_2H_5)_4$ as raw material. The deposition mechanism was infered from the thickness variation as a function of deposition temperature, flow velocity and reactant partial pressure. The morphology of deposited thin film was quite different according to the deposition temperature region. This difference in morphology could be explained theoretically.

1. Introduction.

The thin films deposition process, such as sputtering and chemical vapor deposition (CVD), has been widely used in the semiconductor device fabrication process. The deposition process was also introduced to ceramic processing and has played an important role in the deposition of wear-resistant and heat-insulating ceramic films. Recently, research for thin or thick films of electronic ceramics is very active since electronic parts become to be smaller-size, higher-density and lower-weight. The deposition process of electronic ceramics is more sophisticated and difficult than that of the common ceramic thin films since thin films of electronic ceramics keep high purity, uniform suface morphology and better surface characteristics. The deposition mechanism, especially in CVD process, should be determined relating with the deposition condition if we want to control surface morphology and surface characteristics of thin films. But the deposition mechanism of a CVD process is very sophisticated and the study of it is quite difficult due to so many experimental variables and the interrelationships between them. However the deposition process and mechanism in some simple atmospheric thermal CVD has analyzed theoretically and experimentally.

The rapid developement in semiconductor devices and sensor materials require new material synthesis process. Recently the MOCVD process becomes very attractive and versatile process in semiconductor and various electronic ceramic fields. MOCVD provides low deposition temperature, high deposition speed and high purity deposition process. And so the understanding of deposition mechanism and effects of deposition condition on surface morphology of deposited thin film in MOCVD is essential in furture developement.

TiO_2 ceramic has been mainly used as a base material of dielectric capacitor due to its high dielectric constant, but is presently used as a catalytic material as well as an oxide semiconductor because of its photo-electronic effect and nonstoichiometry. Especially it attracts strong interest in the application of automobile exhaust oxygen sensor since it exhibits the resistive change properties in variant oxygen atmosphere.[1]

In the experiment, we deposited TiO_2 thin films by MOCVD and determined the deposition mechanism by measuring the deposition rate as a function of deposition condition. And then we related the deposition mechanism to surface morphology of deposited thin films and also explained this relation using the surface perturbation theory.

2. Experimental procedure.

The schematic diagram of CVD system is shown in Fig.1. Most of system is the same as the common CVD system, but using the evaporator [6] for a source vapor is different. Since no gas sources for Ti are available, liquid sources such as $TiCl_4$ and $Ti(OC_2H_5)_4$ should be used in CVD system. Most of investigators for TiO_2 thin films have used TiCl $_4$, but we adapted $Ti(OC_2H_5)_4$ as a source vapor of Ti and O because of lower deposition temperature and high deposition rate. The reactant vapor in the evaporator, which was kept in a constant-temperature bath, was carried into the reactor by a carrier gas of high-purity Ar. The reactor wall and the inlet pipe were wraped by heating tape and kept at 150° C to avoid the condensation of reactant vapor during flowing.

Two kinds of substrates were used to deposit TiO_2 thin films: one was Si wafer, of which a surface had thin SiO_2 layer of 85 nm thick and the other was high-purity(99.9%) Al_2O_3 substrate of 0.6 mm thick. Both of them were ultrasonically cleaned in TCE, aceton and ethanol by turns, and then they were located in the susceptor[7].

Prior to deposition, the reactor was purged for 1 hr by Ar flowing through a bypass-line. During purging process, the susceptor was heated up to and held at a depositon temperature until thermal and mass transfer equilibrium in the reactor was reached. After a deposition was completed, 10 slm of Ar was flowed through the reactor to get rid of residual gas and by-product gas in the reactor and also the susceptor was cooled down to room temperature in the reactor.

The thickness of deposited thin films was determined by two methods. One is the stylus method and the other is SEM observation of the fractured cross-section of thin films. In the case of the stylus method, a part of thin film was etched with HF-solution to make a step, and a height of step was measured by a stylus surface profiler(Tencor α-step 200). Surface morphology of thin films was observed by SEM(Akashi ISI-30) after Au coating.

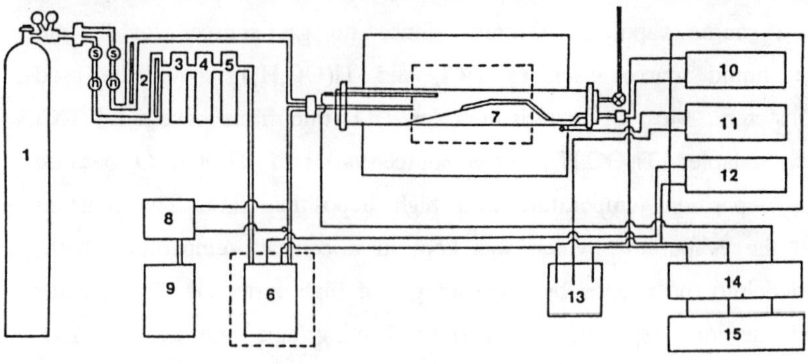

Fig.1 Schematic disgram of APCVD system built in house for growing TiO_2 thin films. (1)Ar gas cylinder (2)flowmeters (3)conc. sulferic acid (4)$CaCl_2$ powder (5)silica gel (6)evaporator (7)susceptor (8)thermostat (9) AC power supply (10)substrate temperature controller (11)wall temperature controller (12)evaporator temperature controller (13)thermocouple cold junction (14)switching unit (15)digital multimeter.

3. Results and discussion.

3.1 Determination of deposition mechanism.

Depositon mechanism during CVD process is governed by thermodynamic reaction of vapor phases, reactor shape and deposition condition such as deposition temperature, deposition time, gas flow velocity and total pressure. Since all these parameters are closely related one another, it is so difficult to determine deposition mechanism. But it is well known that deposition rate is generally determined by two deposition mechanisms in atmospheric-pressure CVD. One is the surface chemical reaction controlled mechanism and the other is the mass transfer reaction mechanism.[2-4] In the former the overall deposition rate is controlled by chemical reaction on substrate surface and has exponential dependence of temperature since chemical reaction and adsorption and desorption of vapor molecules or atoms are thermally activated processes. In this case apparent activation energy can be obtained from a slope in ln(deposition rate) vs 1/T plot. In the latter deposition rate is sensitive not to temperature but to carrier gas velocity because gas velocity affects the thickness of boundary layer. The deposition rate in mass transfer controlled region is expressed as Eq.(1), which is derived from the simple boundary layer theory considering experimental condition.[5]

$$V = \frac{C_T(P_r/P_T)[v_1/(v_1+v_2)]\rho^{1/2}D_G(v_1+v_2)^{1/2}}{2(L\mu A)^{1/2}N_1} \quad \text{--------} \quad (1)$$

where V is deposition rate, C_T molar density of total gas, P_r partial pressure of reactant gas, P_T total pressure, ρ density of total gas, D_G diffusion coefficient of gas phase, L the thickness of boundary layer, μ viscosity of gas phase, N_1 molar density of film, v_1 carrier gas velocity, v_2 diluent gas velocity and A is the cross sectional area of reactor. Since each parameter has linear or parabolic dependence of temperature, the deposition rate has dependence of $T^{1.5\text{-}2.0}$. Therefore a slope of a plot of ln v vs 1/T looks almost flat.

Fig.2 shows the variation of deposition rate as a function of temperature for different partial pressure of $Ti(OC_2H_5)_4$. From this figure, we can find that as the partial pressure of reactant gas increases, the transition temperature from surface chemical reaction controlled mechanism to mass transfer controlled mechanism increases and the apparent activation energy in the surface reaction controlled range is 6.74 Kcal/mole. The former result is reasonable since higher partial pressure of reactant vapor makes deposition rate less sensitive to mass transfer step during deposition. The latter is very similar result obtained by Yokozawa et. al.[6] They used tetraisopropylorthotitanate instead of ethyltitante and obtained 6.5 Kcal/mole and 35.6 Kcal/mole of activation energy with 5 - 60% oxygen and without oxygen, respectively. But they did not explain the reason for discrepancy in the activation energy between oxygen-involved and oxygen-exclusive reactions. According to their result, oxygen seemed to be envolved in this reaction during process. Anyhow, it is not obvious how to introduce oxygen into the reactor.

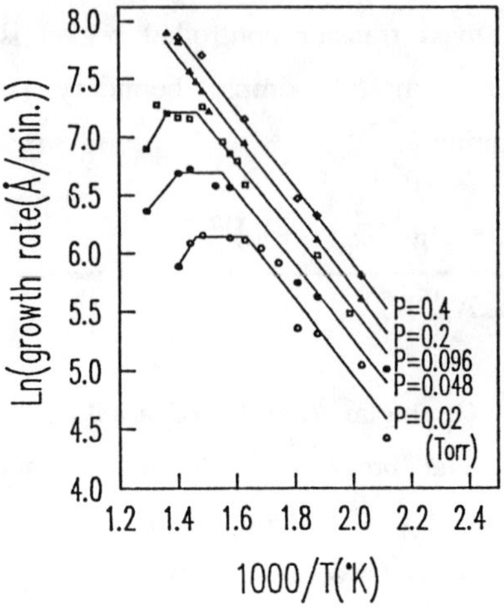

Fig.2 Change of ln(growth rate) as a function of temperature(1/T) under different reactant gas partial pressure, total flow rate of Ar gas was fixed to 755 SCCM.

The deposition rate calculated from Eq.(1) and that experimently obtained are 134 nm/min. and 129nm/min., respectively. In this case, the reactant vapor pressure, v_1 and v_2 were 0.48 Torr, 151 sccm and 604 sccm, respectively. To make a calculation of Eq.(1), we assumed that the total pressure of reactor, the thickness of boundary layer and diffusion coefficient of gas were 1 atm, 10 cm and 0.1 cm^2/sec, respectively. These results show that the calculated value is in good agreement with the value experimentaly obtained. This implies that in the mass transfer controlled region the deposition rate can be easily estimated by calculation of a simple equation based on the simple boundary theory. This kind of estimation was also adapted to the growth of Si epitaxial layer using SiH_4 and H_2 gases by Everstein et. al.[7] and was proved to be good.

In the surface chemical reaction controlled-deposition condition, the adsorption mechanism is able to be deduced from the relationship between deposition rate and partial pressure of reactant vapor.[8-9] Fig.3 shows a linear relation of deposition rate with partial pressure of reactant vapor and this implies that the adsorption mechanism is Rideal-Elley mechanism, which indicates that ethyltitanate vapor is adsorbed in a molecular form on a substrate and then decomposed into TiO_2, C_2H_4 and H_2O as shown in Eq. (2).

$$Ti(OC_2H_5)_4\ (g) \rightarrow Ti(OC_2H_5)_4\ (ad)\ :\ \text{adsorption}$$

$$Ti(OC_2H_5)_4\ (ad) \rightarrow TiO_2(s) + 4C_2H_4(ad) + 2H_2O(ad) \quad \text{-------} \quad (2)$$

$$:\ \text{heterogeneous reaction}$$

$$4C_2H_4(ad) + H_2O(ad) \rightarrow 4C_2H_4(g) + 2H_2O(g)\ :\ \text{desorption}$$

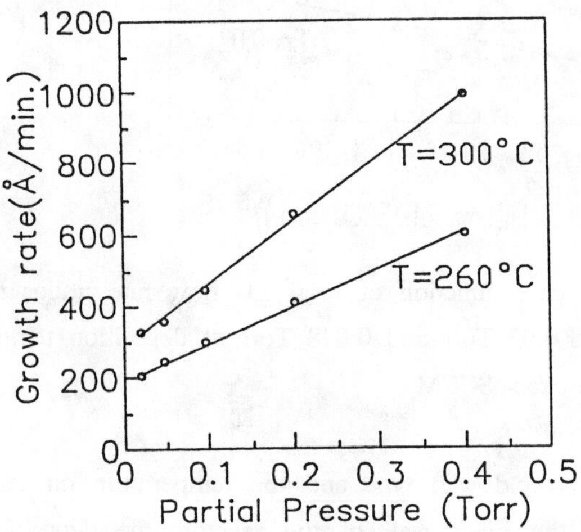

Fig.3 Deposition rate as a function of reactant gas partial pressure at deposition temperature of 260°C and 300°C. (total Ar gas flow rate : 755 SCCM).

As mentioned before, a slope of linear plot of ln(deposition rate) vs. ln(flow rate) is 1/2 in the mass transfer controlled deposition condition. This relation is observed in this experiment as shown in Fig. 4. In the case of low partial pressure, 0.02 Torr, of reactant vapor, a slope of line is 1/2, but for high partial pressure of 0.048 Torr a slope of line increases from 1/2 to higher value at a flow rate of around 755 sccm. This behavior is due to the change from the mass transfer controlled condition to the surface reaction controlled condition according to high flow rate of carrier gas.

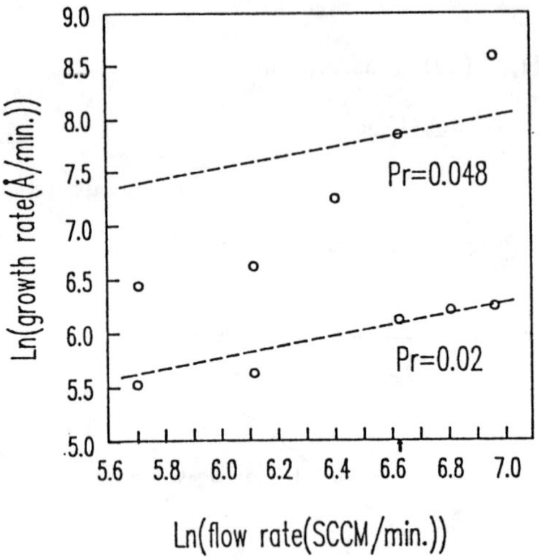

Fig.4 Deposition rate as a function of total gas flow rate under reactant gas partial pressure of 0.02 Torr and 0.048 Torr at deposition temperature 440°C. arrow indicate 755 SCCM

Since the reactor is cold wall type and the temperature on sample is higher than that of the inner wall of the reactor, the deposition rate reduces at the low flow rate condition regardless of reaction pressure. In this case the heat convection occurs from the center to the wall of the reactor due to the temperature gradient. The heat convection makes the buoyancy force on the surface, which reduces mass transport from gas stream to the substrate and results in the low deposition rate at the low flow rate condition. As the flow rate increases, the gas streaming force parallel to substrate increases and move away the buoyancy force from the surface of substrate.

There are two causes to reduce deposition rate at high temperature region. One is the reactant gas deficiency,[10-11] which occurs because the reactant gas is easily decomposed and mostly deposited at the fore edge part of substrate. The other is homogeneous reaction.[12]

3-2 Surface morphology of deposited thin films.

The surface morphologies obtained in the experiment are at most divided into two groups. One is obtained from mass transfer controlled-deposition condition and the other is obtained from surface chemical reaction controlled-deposition condition. The used substrates were Si wafers with wet thermal oxidized SiO_2 thin films of 850A thinkness, Rough surface Al_2O_3 ceramic sheet (surface roughness of about 500 nm) and mirror polished Al_2O_3 ceramic sheets (surface roughness of about 10 nm). Fig.5 is the SEM micrograph of fractured interface of deposited TiO_2 thin film on SiO_2/Si substrate, which was deposited in the mass transfer controlled region and of which thickness is about 2 μm.

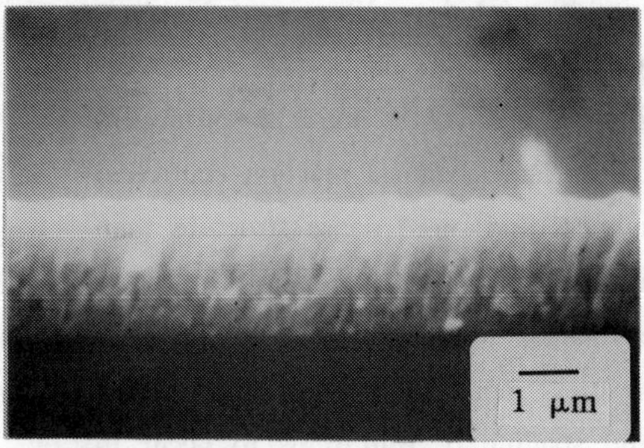

Fig.5 SEM micrograph of cross section of smooth surface film deposited on SiO_2/Si substrate. (Deposition condition: 440°C,0.096Torr, 634SCCM,10min.)

Fig.6 is the SEM micrograph of fractured interface of deposited TiO_2 thin film of approximate thickness of 4 μm which was deposited under the same condition but the deposition time was twice that of the sample in Fig.5. We can know from these results that in the initial stage of deposition time the thin film was deposited very dense and smooth, but as the deposition time increased the surface morphology became rougher and rougher and finally an enhanced columnar structure.

Fig.6 SEM micrograph of overview of rough surface film deposited on SiO_2/Si substrate. (Deposition condition: 400°C,0.096Torr, 755SCCM,20min.)

Though in the same mass transfer controlled region, the thin film morphology on rough Al_2O_3 substrate showes columnar structure even in the initial deposition time region such as Fig.7. It is obvious from Fig.7 (b) and (c) that the surface roughness increases, and the morphology has more obvious colummar structure as the deposition time and the thickness of thin film increases. The deposition time of Fig.7(b) sample was 20 min, and that of Fig.7(c) was 30 min, compared to 10 min of Fig.7(a). The distance between columns shown from SEM micrograph of surface in Fig.7(a) is about 0.3 μm, but that of Fig.7(c) is about 2~3μm. On the contrary, the films deposited in the surface chemical reaction controlled region has much smaller surface roughness and more dense structure than those deposited in mass transfer controlled region, as are shown in Fig.8, and Fig.9. As we can see in these figures, the deposited film has very dense structure, and thus the internal stress increases as the film thickness increased. This leads the cracking and spalling of thin films to occur in the case of SiO_2/Si substrate due to the large difference in thermal expansion coefficient between the deposited TiO_2 thin film and substrate. And so we could not obtain the thin film that exceeds thinkness of 1 μm on the SiO_2/Si substrate. Therefore we could use only polished Al_2O_3 substrate in the surface chemical reaction controlled region to get the

Fig.7(a) SEM micrograph of cross section of rough surface film deposited on Al_2O_3 substrate. (Deposition condition: 440°C,0.096Torr, 755SCCM,10min.)

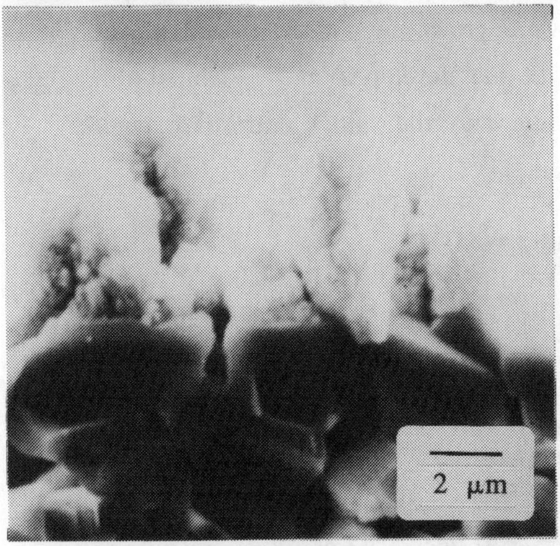

Fig.7(b) SEM micrograph of cross section of rough surface film deposited on Al_2O_3 substrate. (Deposition condition: 440°C,0.096Torr, 755SCCM,20min.)

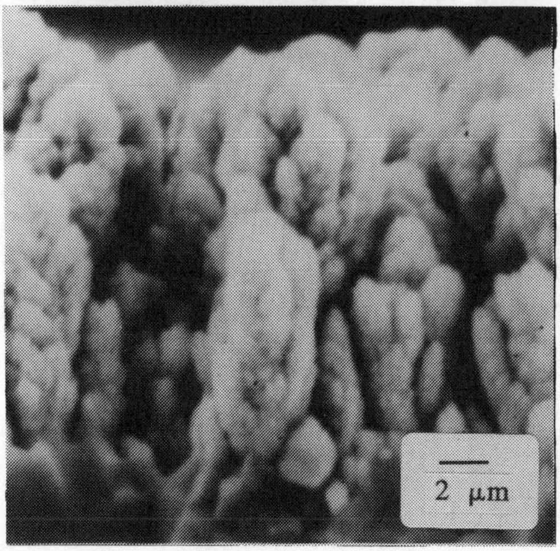

Fig.7(c) SEM micrograph of cross section of smooth surface film deposited on Al_2O_3 substrate. (Deposition condition: 440°C,0.096Torr, 755SCCM,30min.)

surface morphology of more than 1 μm thinkness sample. The estimated thinkness of Fig.8 sample is about 1.3 μm, which was deposited for 3 hours, and that of Fig.9 sample is about 2 μm, which was deposited for 5 hours. We can see from the above two figures that the surface morphology of samples, which were deposited in the surface chemical reaction controlled region, is very smooth relative to that of the samples which were deposited in mass transfer controlled region. And we can also see that the surface morphology of the above samples did not change as the deposition time increased. Especially we can see by a close examination of Fig.9 that the films were grown by the developed nuclei, planar agglomeration and coalescence of the developed nuclei, which were formed on the whole surface uniformly, and the initial uniformity could be retained by this growth mechanism.

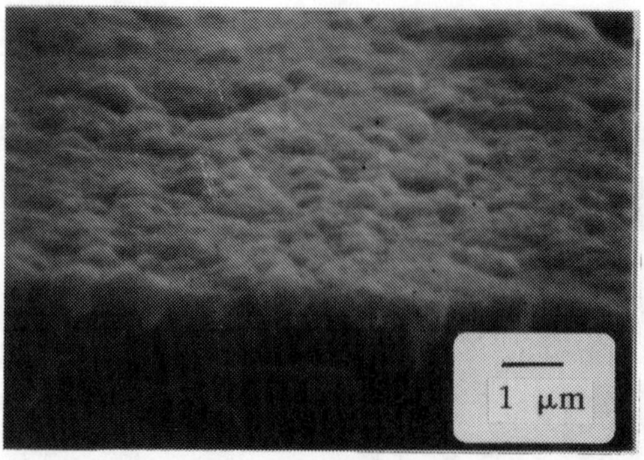

Fig.8 SEM micrograph of surface morphpology of deposited film on smooth Al_2O_3 substrate. (Deposition condition: 300°C,0.096Torr, 755SCCM,3hours, estimated thickness:1.3μm.)

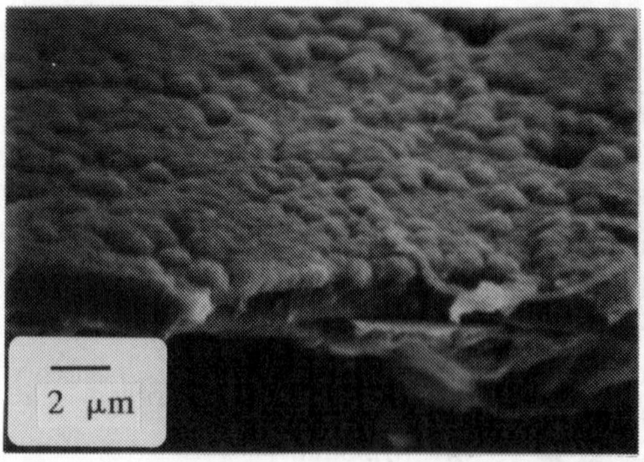

Fig.9 SEM micrograph of surface morphpology of deposited film on smooth Al_2O_3 substrate, showing island type nuclei. (Deposition condition: 300°C,0.096Torr, 755SCCM,5hours, estimated thickness:2μm.)

This growth mechanism is very different from the growth mechanism of the films which were grown in the mass transfer controlled region. In the mass transfer controlled region the columns formed in the initial stage of deposition processes grew independantly and eventually resulted in the very rough surface morphology eventually.

All the above results can be well explained by the surface perturbation theory of C.H.J. van den Brekel[13]. According to his theory about the surface morphology in CVD, in the deposion process after the initial period of nucleation had elapsed, and when the perturbation of surface morphology has occured, the surface morphology can be described by the Fouriers sine series solution such as eq.(3).

$$z(\omega,t) = \Sigma\ E(\omega,t)\ \sin(\omega x) \quad \text{--------} \quad (3)$$

Where E(ω.t) is the amplitude for given ω and time t. ω is the spatial frequency of the surface hormonic, t is time, and x is the spatial distance on the surface. Here, we should notice that the E(ω.t) was given as a function of ω, and t, espeially of ω such as eq.(4).

$$E(\omega,t) = E(0)\ \exp[f(\omega)t] \quad \text{--------} \quad (4)$$

where E(0) is the value of E at time t=0 for some given ω.

From eq.(4), we can see that for a given ω, the amplitude of surface harmonic of that ω is exponentially decreased with time if the f(ω) is negative, but it increases exponentially with time if the f(ω) is positive. It is needless to say that the extent of amplitude increase or decrease with time depends on the absolute value of f(ω) for the given ω.

If f(ω) is negative, the surface harmonic-surface perturbation of morphology of that ω can not grow even more and disappear. It is possible that we regard the f(ω) contains all physical elements which affect the deposition process. Therefore the consideration on f(ω) enables for us to analyze the effect of various adapted deposition parameters on deposition rate. The f(ω) could be written as following eq.(5).

$$f(\omega)=V_mDG_c\omega\frac{2G_c-2C_e\Gamma\omega^2+m(G+G')}{2[1+(\omega\delta)/Sh]G_c+m(G+G')} \quad \text{-------} \quad (5)$$

Where Vm is atomic volume of solid phase, Ce is the equilibrium concentration at a planar interface, Gc is the concentration gradient in gas phase, Γ is capillity constant, G and G' is thermal gradients in gas and solid phase and Sh is Sherwood number equals to Kδ/D. Where K is surface chemical reaction constant. δ is boundary layer thinkness, D is diffusivity. And the factor m in eq.(5) is given as following eq.(6)

$$m = Ce(\Delta H/RT_o^2) + (Cs - Ce)(\Delta E/RT_o^2) \quad \text{------} \quad (6)$$

Where Cs is interface concentration, T_o is the unperturbated interface temperature, and ΔH is reaction enthalpy, ΔE is the apparent activation energy. Note that the Sherwood number Sh is determined by the deposition conditions, that is as the deposition condition approached to the perfect surface chemical reaction controlled region, the Sh number approached to zero. But as the deposition condition approached to the perfect mass transport controlled region, the Sh number has very larger positive value than 1. The examination of f(ω) variation versus log ω at various values of Sh show that f(ω) has a constant positive value for all ω in case of much smaller value of Sh than 1 and it has some maximum value at some particular ω value in case of larger value of Sh than 1. For very large value of ω region the f(ω) has largely negative value irrespective of ω. Since in this ω region the surface had very large specific area and the thin film system has excessively lage surface energy, it must be energetically unstable. This experimental result can be expressed in terms of eq.(5) by the large negative value of $2Ce\Gamma\omega^2$ term and f(ω) to have large negative value, when ω is large. For better understanding we show the variation of f(ω) as a function of log ω in Fig.10 for hot wall case from reference 13.

The figure like Fig.10 for the system in the experiment could not be obtained owing to the deficiency of thermodynamic and kinetic data, and the system was cold wall type but the temperature gradient in the adapted gas flow range was so small that we could apply this result in Fig.10 to our experiment. And so we can not know the absolute value f(ω) but it is sure that the variation trend of f(ω) with respect to log ω itself is smiliar to that of Fig.10.

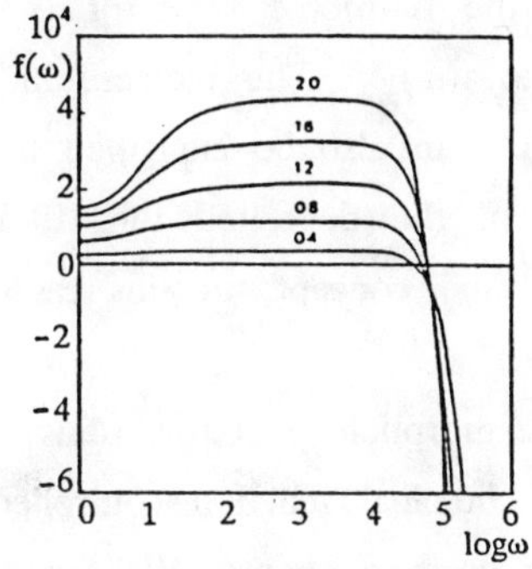

Fig.10 Stablity function for deposition of poly crystalline silicon with S_h as parameter.[13]

However the relaxation time τ defined as equation (7) also affects the change in morphology.

$$\tau = 1/f(\omega)_{max} \qquad \text{-------------} \qquad (7)$$

So it is obvious that even in the case of large positive f(ω), the initial surface morphology could be maintained in some initial short time $t<\tau$ region. And it is obvious that τ has different values for different Sh number.

The variation of surface morphology with time in the mass transfer controlled region (Sh >> 1) can be well explained from above theory as follows. In many surface harmonic waves which constitute the Fourier sine series solution for perturbed surface morphology, the amplitude of the wave of ω which corresponds to the maximum f(ω) or waves near the maximum f(ω), grows more rapidly with time than the other surface harmonic waves. So as the deposition time elasped, the surface morphology should show more clearly the surface harmonic wave of maximum f(ω). The appearance of surface roughness beyond the relaxation time irrespective of initial substrate roughness, that is rough Al_2O_3 substrate or very smooth SiO_2/Si substrate, can be explaind by the previous theory. The surface morphology of Fig.7(c) shows the harmonic wave of ω which is more closer to the ω corresponding to $f(\omega)_{max}$. The increase in inter columnar distance shown in Fig.7(a) to (c) can also be explained by above theory. The initial surface smoothness of the deposited thin film in Fig.5 can be explaind by the relaxation time concept; in tihs case the relaxation time has not been elapsed.

Also, the constancy of surface morphology of thin films with time which were deposited in the surface chemical reaction controlled region could be explained very well from the previous theory. We can see from Fig.10 the fact that the f(ω) for all pertinant ω range in surface chemical reaction controlled case (Sh << 1) is constant. And so the initial small surface roughness of nucleation stage can be maintained during the whole deposition process. Therefore, it is possible for us to get very smooth surface thin film in the surface chemical reaction controlled region while the rough surface thin film are obtained in the mass transfer controlled region. Fig.11 shows the complex film morphology which contain both smooth and rough structure regions on polished Al_2O_3 substrate. From this figure we could see that for given sample the surface morphology could be changed with the subsequent change in deposition conditions. The difference in morphology of deposited thin films was obvious under the two different deposition conditions.

The thickness change of the deposited thin films occured by heat treatment can give us the information about the internal density of the deposited films. The thin film which was deposited in the mass transfer controlled region and undergoes heat treatment of 1100°C in air atmosphere for 24 hours shows the decrease in thickness of about 30%. From this result we can find that the thin film deposited in mass transfer controlled condition was rather porous.

And it is also shown from Fig.12(a), Fig.12(b) and Fig.13 that after heat treatmen the morphological change of thin film deposited in mass transfer controlled region is very prominent.

surface

mass transfer controlled condition

surface reaction controlled condition

(Deposition conditions step1:300°C,0.2Torr,1hour, step2:440°C,0.096Torr,30min, equal flow rate of 755 SCCM)

Fig.11 SEM micrograph of surface morphpology of deposited film on smooth Al_2O_3 substrate, showing two step deposition.

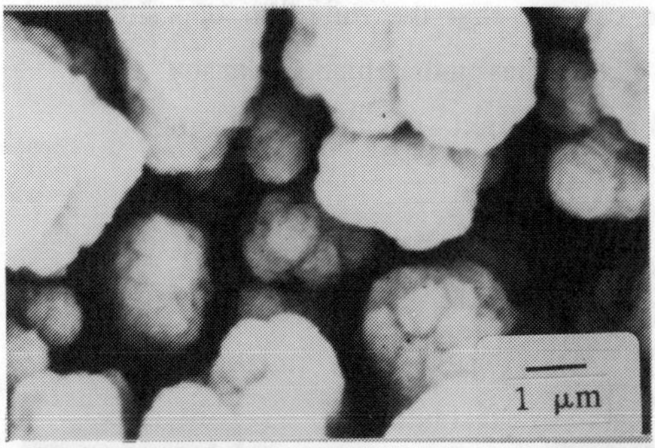

Fig.12(a) Surface morphology of as-grown film on Al_2O_3 substrate under diffusion controlled condition.

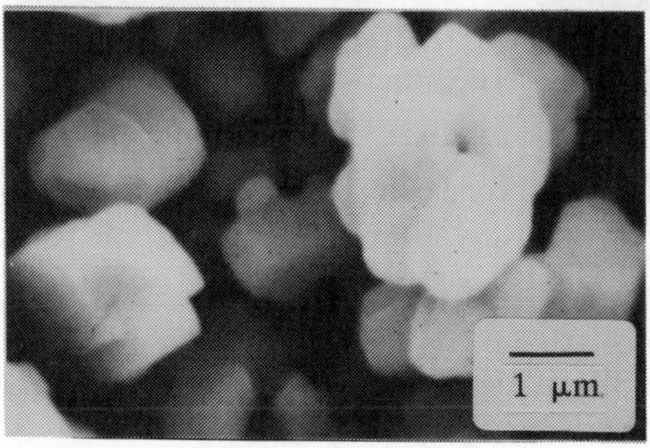

Fig.12(b)Surface morphology of heat-treated film on Al_2O_3 substrate. (Heat treatment condition: 1100°C,24hours,in air)

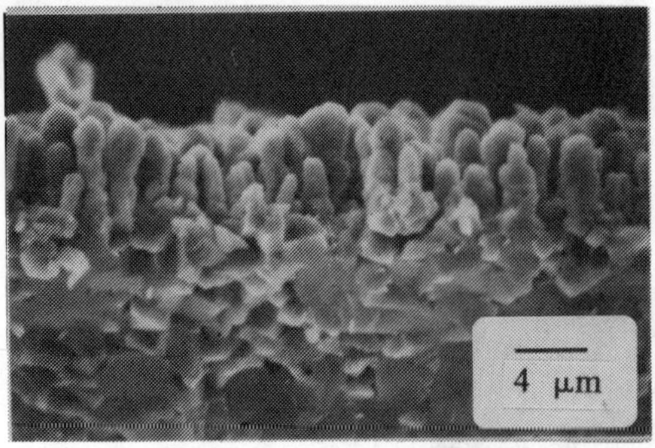

TiO_2 thin film

Alumina substrate

Fig.13 SEM micrograph of fractured interface morphology of fig.12(b) sample.

But from the comparison of Fig.8 and Fig.14, we can see the fact that the thickness decrease and morphological change could not be detected for the thin film deposited in surface chemical reaction controlled region after the heat treatment. Therfore we can see that the thin film deposited in the chemical reaction controlled region has more higher density and lower defect concentration than that deposited in the mass transfer controlled region.

Fig.14 SEM micrograph of fractured interface morphology of heat-treated TiO_2 thin film on Alumina substrate (Heat treatment condition: 1100°C,24hours,in air)

4. Conclusion.

TiO_2 thin film was deposited on silicon and Al_2O_3 substrate using ethyltitanate as raw material by the cold wall horizontal tupular type reactor AP CVD system. To determine the deposition mechanism the deposition rate was measured as a function of deposition temperature, gas flow rate, and reactant partial pressure. The estimated apparent activation energy in the surface chemical reaction controlled region was 6.74Kcal/mole. The deposition rate in the mass transfer controlled region when reactant partial pressure was lower than 0.2 Torr could be calculated by the simple boundary layer theory and, the result was in good agreement with the experimental results. And the adsorption mechanism was proved to be Rideal-Elley mechanism since the deposition rate vs. reactant partial pressure hold linear dependance. If the flow rate were lower than 755 SCCM. The deposition rate was smaller than the expected value by simple boundary layer theory. The causes of this phenomena were the relatively strong free convection effect and the depletion effect of reactant gas at these lower flow rate region.

The surface morphology of deposited thin films can be largely divided into two groups; one is obtained from the mass transfer controlled region, and the other is obtained from in the surface chemical reaction controlled region. The former has typical columnar structure and thus it's surface roughness is relatively high, and the roughness also increases as the deposition time increases. But the latter is relatively very smooth and does not show the morphological variation as the deposition time elapses. These results can be well explained by the surface perturbation theory suggested by C.H.J. van den Brekel. It is found in the experiment that the surface morphology of thin films could be controlled by the change of deposition conditions. The deposited thin films exhibited different heat treatment effects according to the deposition condition. And from these results it is revealed that the density of thin films deposited in mass transfer controlled region has lower value than that of thin films deposited in the surface chemical controlled region.

5. References

1. T. Takeuchi, Proc. of the 2nd inter. meeting on chemical sensors, BORDEAUX, 1986, PL-6 pp 69-77.
2. W. Kern and V.S. Ban, Thin film processes (eds.Vosen and Kern) Academic Press, pp. 267-270, 1978.
3. A.S. Groove, physics and Technology of Semiconductor Devices John Willey Son's, pp. 11-13, 1967.
4. J.M. Blocher, Jr., "Chemical Vapor Deposition" in Deposition Technologies for Films and Coatings, eds. Bunshah et.al., Noyes Publications, pp. 348-352, 1982.
5. A.S. Groove, physics and Technology of Semiconductor Devices John Willey and Son's, pp. 14-21, 1967.
6. M. Yokozawa, H. Iwasa, I. Teramoto, Jpn. J. Appl. Phys. 7, 96, (1968).
7. F.C. Everstein, J. Electrochem. Soc. 117, (7), 925-931, (1970).
8. M.E. Jones and D.W. Shaw, "Growth from Vapor" in "Treatise on Solid State Chemistry Vol.5 Change of State", (ed. by N.B. Hannay), Plenum Press, (1982).
9. C. Cobianu and Oavelescu, J. Electrochem. Soc., 130, 9, (1988).
10. M. Levin and A.C. Adams, J. Electrochem. Soc., 129, 588, (1982).
11. C. Adams and C.D. Capio, J. Electrochem. Soc., 126, 1043, (1979).
12. K.J. Sladek, J. Electrochem. Soc., 118, 4, 654, (1971).
13. C.H.J. van den Brekel, "Growth and Morphology in CVD process" in the Proceedings of the 8th international conference on CVD, eds. J.M. Blocher,Jr., G.E. Vuillard and G. Wahl, pp 116, (1981).

NANO-INDENTATION HARDNESS OF THIN FILMS

M. Abouelleil, L. Conopask, W. Nighan, and W. Roman
United Technologies Research Center
East Hartford, Connecticut 06108

D. Price
United Technologies Corporation, Pratt & Whitney Group
East Hartford, Connecticut 06108

ABSTRACT

Thin wear-resistant films can significantly improve the life of aerospace components. A key characterization parameter is film hardness. Accurate hardness measurements of ≤ micron thick films require special techniques. An ultra-sensitive microprobe "Nanoindenter" indentation depth 0.01–1 μm was used to study the hardness of TiN, TiB_2, and DLC films. Measurements revealed two significant phenomena: (1) An indentation size effect (ISE) manifested by an increase in measured hardness and elastic modulus with decreasing indentation depth, and (2) A film-substrate effect (FSE) manifested by the measured hardness being less than the true film hardness due to interaction with the substrate. Both effects influence the determination of intrinsic film hardness and must be accounted for. Models permitting correction for both ISE and FSE effects, and hence determination of intrinsic film hardness, are discussed.

1. INTRODUCTION

High wear resistance is an essential property for determining the reliability and service life of mechanical components. This is particularly important for aerospace applications because of two reasons: (1) They are often subjected to severe frictional loads, high operating temperatures, and/or erosion damage, and (2) Their failure can be catastrophic. To protect and enhance the service life of such components it is often necessary to coat them with a hard, thin film of wear resistant material, such as titanium nitride, zirconium oxide, or diamond-like-carbon (DLC). The mechanical properties of such films depend on the conditions of film deposition and are not necessarily the same as those of similar bulk materials. Thus, it is necessary to evaluate the mechanical properties of such films by carrying out tests on the films themselves. Such measurements can also provide a method for studying the effect of different deposition parameters on the mechanical properties of fabricated films.

Measuring of the mechanical properties (e.g. hardness, elastic modulus, fracture toughness, etc.) of micron and submicron films is very difficult, if not impossible, using conventional mechanical testing machines. However, a recently developed mechanical properties microprobe, the "Nanoindenter" [1], provides a tool for obtaining measurements of both the hardness and the modulus of elasticity from indentations as shallow as 20 nanometers. We utilized a Nanoindenter to study submicron indentations of thin films and bulk materials. While the analysis of our measurements work is general, some of the details pertain to the Nanoindenter specifically. Hence, a brief description of the Nanoindenter is in order.

2. THE NANOINDENTER

The Nanoindenter is a submicron hardness indentation tester, incorporating a triangular pyramid diamond indenter having the same projected area to depth ratio as the Vickers square pyramid indenter. However, the mechanism of load application and indentation area measurement is different from that of the Vickers indenter, in such a way as to permit submicron indentation measurements, as well as determination of both plastic and elastic properties.

Figure 1 shows a schematic of the Nanoindenter mechanism. A nearly stepless load is applied through an electromagnetic coil to the indenter head. The load range is 0 to 100 mN with a 0.003 mN resolution. The indenter displacement is measured by a glass capacitor having a depth measurement resolution of 0.5 nm. Load and displacement data are digitized and fed to a computer which controls the voltage applied to the electromagnetic coil, and processes the data required to develop a continuous load displacement curve for a complete cycle of loading and unloading. The computer also controls the movements of a microscope, magnification to 1500x, and a video camera for observation and accurate positioning of the indentation site. Figure 2 is a photograph of the Nanoindenter illustrating its main components.

To better illustrate the working principles of the Nanoindenter, Figure 3 shows a complete cycle of loading and unloading plotted against indentation depth. During the loading cycle, the load is gradually increased from zero to a maximum load L, at which time the indenter penetrates the sample to the maximum depth d_t. The corresponding

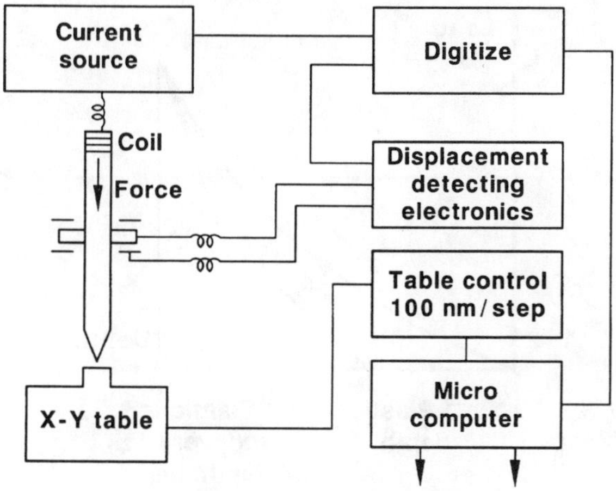

Fig. 1 Diagram of the Nanoindenter

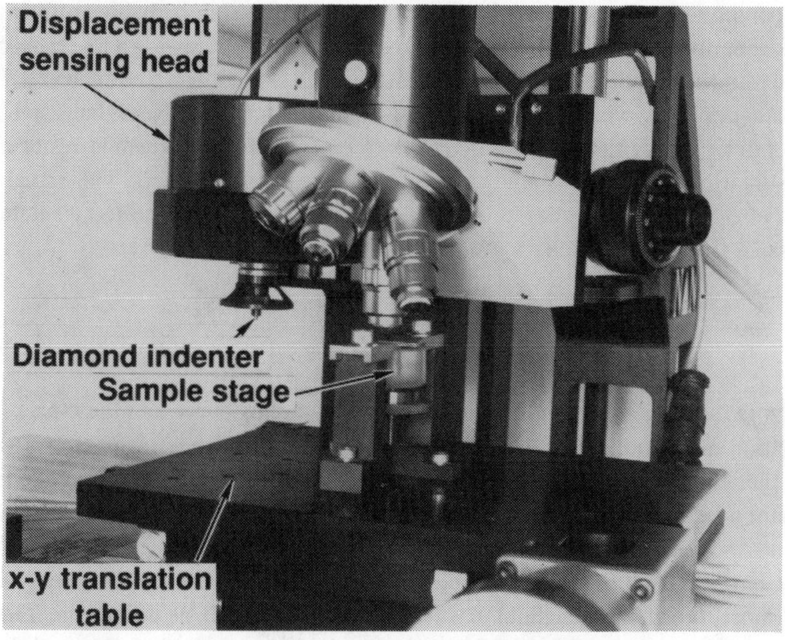

Fig. 2 Photograph of the Nanoindenter

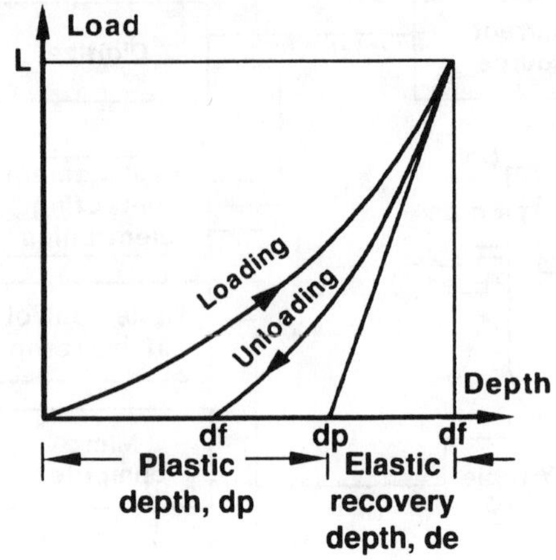

Fig. 3 Indentation Cycles

deformation of the indenter surface at this point is comprised of both plastic and elastic components. During the initial portion of unloading, a nearly linear unloading curve is generally observed [2], indicating an elastic regime. Drawing a line tangent to the unloading curve at the point of maximum load and extending it to zero load defines the plastic depth d_p, which is utilized to calculate the hardness. Note that use of the final depth d_f, rather than the plastic depth to estimate hardness would result in an over–estimation of hardness due to changes in the indent shape. The remaining portion of d_t is the elastic recovery depth $d_e = d_t - d_p$, which is used to calculate the stiffness, S, from the relation:

$$S = L/d_e = \tan\psi \tag{1}$$

where ψ is the angle between the tangent to the unloading curve and the x–axis. This information can be utilized to calculate the elastic modulus of the specimen. It is worth noting that the surface hardness and the modulus of elasticity are among the most important mechanical properties required for the characterization of thin films. This is because the wear resistance of surfaces is linearly related to the surface hardness [3], so that hardness data provide the best quantitative measure of wear resistance. Furthermore, the yield strength of elastic/plastic materials [4], as well as the fracture toughness of brittle materials [5], depend on both the hardness and the modulus of elasticity.

3. HARDNESS MEASUREMENTS

The hardness determined by the Nanoindenter is defined as the maximum load divided by the projected area. Figure 4 shows a schematic of the Nanoindenter diamond tip. The projected area, A, can be expressed in terms of the length of the side of the triangular indent, D, or the indent depth, d, and the indenter angle, Θ, as:

$$A = (3)^{1/2}D^2/4 = (27)^{1/2} \tan^2\Theta \ d^2 = 24.5 \ d^2 \tag{2}$$

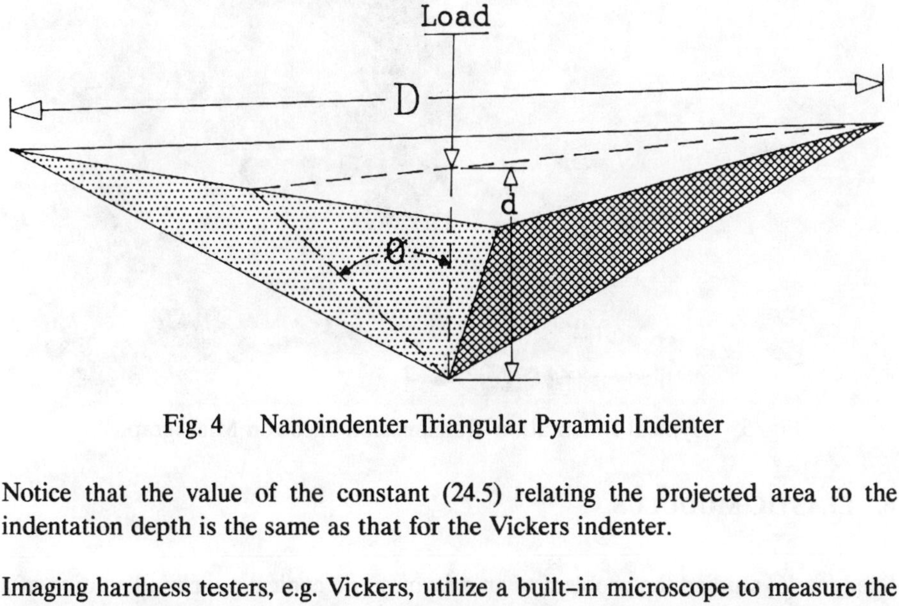

Fig. 4 Nanoindenter Triangular Pyramid Indenter

Notice that the value of the constant (24.5) relating the projected area to the indentation depth is the same as that for the Vickers indenter.

Imaging hardness testers, e.g. Vickers, utilize a built-in microscope to measure the indent size, which is then used to calculate the area. The Nanoindenter uses an ultra-sensitive depth sensing capacitor to measure the indentation depth, d, as small as 10 nm, which is then used to calculate the area using a relation similar to Eq. 1. It should be noted that Eq. 1 assumes a perfect, sharp indenter tip. Practical indenters may have a rounded tip (typical tip radius = 1.5 nm) as well as a small error in the angle Θ. Thus, it is important to obtain an area function which reflects the actual geometry of the indenter. This is accomplished by making indentations at different depths, imaging them by electron microscopy, and calculating the actual area-depth relation. To calculate the area function, we utilized transmission electron microscopy

and a carbon replica technique have been used to image indentations in aluminum, at depths in the range 50–1000 nm. Figure 5 shows a typical TEM replica photograph of a indentation in aluminum. Figure 6 shows a graph of the calibrated area function (solid line), and the ideal indenter function (dashed line). Notice that at shallow indentations the actual area is as large as twice the ideal indenter area.

Fig. 5 Typical Indentation Transmission Electron Micrograph

4. ELASTIC MODULUS

Sneddon [6] studied the problem of an axisymmetric conical punch indentation of a semi-infinite elastic solid. Loubet [7] applied Sneddon's solution to the elastic recovery during unloading of the indenter. He showed that the measured elastic modulus, E, is related to the stiffness and the plastic depth by the expression:

$$E = kS/d_p \tag{3}$$

where k is a numerical constant whose value depends on the shape of the indenter (k = 0.179 for the Nanoindenter).

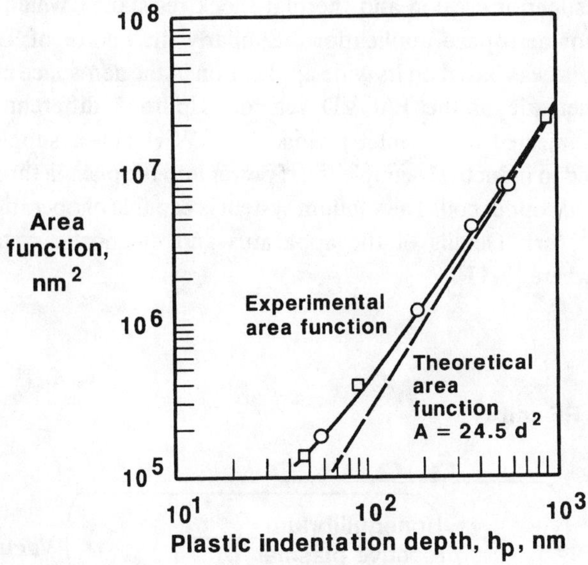

Fig. 6 Area Function Calculation

The calculated value of E is utilized together with the known value of the elastic modulus E_i of the indenter to determine the elastic modulus E_m of the indented material from the relation:

$$1/E = [1 - \nu_m^2]/E_m - [1 - \nu_i^2]/E_i \tag{4}$$

The Poisson's ratios ν_i of the indenter, and ν_m of the indented material need not to be known with great precision in order to obtain a good estimate of E_m. This is because both ν_i and ν_m are in the order of 1/3 and hence $\nu^2 << 1$.

5. EXPERIMENTAL

Three ceramic thin film specimens, TiB_2, DLC, and TiN, and three bulk specimens, TiB_2, M50 steel,* and soda lime glass, were tested. Two of the thin films, TiB_2 and DLC, were fabricated at United Technologies Research Center by plasma assisted chemical

---

* Composition of M50 steel: 0.8% C, 0.25% Mn, 0.25% Si, 4% Cr, 0.1% Ni, 4.5% Mo, 1% V and 89.1% Fe.

vapor deposition "PACVD" [8]. Our interest in TiB_2 and DLC coatings was based on their expected superior erosion and thermal shock resistance, which are important requirements for aerospace applications. Similarly the choice of Ti–6Al–4V as a substrate material was based on its wide application in the aerospace industry. Figure 7 shows a schematic of the PACVD reactor. Up to 5 different gases can be simultaneously injected at the inlet region. A 5 kW rf power supply operating at 13.5 MHz is used to inductively couple the rf power into the plasma through a multiple turn water cooled copper coil. The vacuum system is capable of operating at pressures as low as 10^{-8} Torr. Details of the apparatus and diagnostic system have been published elsewhere [9–11].

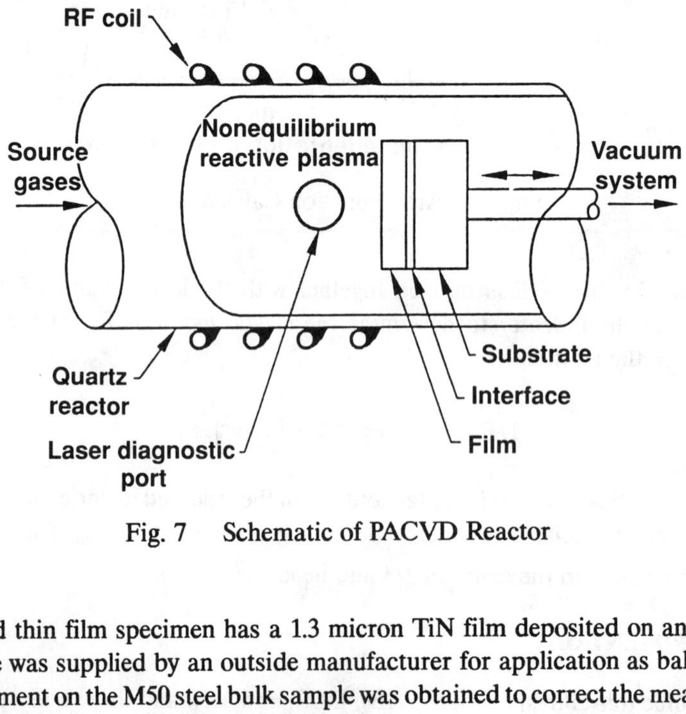

Fig. 7 Schematic of PACVD Reactor

The third thin film specimen has a 1.3 micron TiN film deposited on an M50 steel substrate was supplied by an outside manufacturer for application as ball bearings. Measurement on the M50 steel bulk sample was obtained to correct the measured TiN film hardness for the effect of the softer substrate.

Finally, measurement of the soda lime glass bulk specimen was obtained to better understand the indentation size effect, "ISE", since most silicate glasses, at least on a macroscopic scale, are nearly ideal elastic solids [12].

The surfaces of all specimens were mechanically polished and carefully examined at 1500x to ensure flat, scratch-free surfaces before indentation. The only exception was the glass specimens which were indented as received. Nanoindentation tests were then carried out by making indentations in the depth range of 10–600 nm. The hardness and elastic modulus values were calculated as the average and standard deviation of five measurements at the same depth.

Because of the submicron sizes of the indentations and the very small thickness of the TiN film it is to be expected that some of our measurements may include contributions from both the indentation size effect "ISE" and the film/substrate effect "FSE". A brief discussion of these two effects is therefore presented below.

6. INDENTATION SIZE EFFECT

When surfaces are tested by pyramidal indenters at indentation depths > 10 microns, the hardness is almost independent of indentation depth. On the other hand, it has been recognized for some time that at shallower depths the measured hardness increases with decreasing indentation depth [13]. Several factors have been proposed to account for this "indentation size effect":

1. Strain-hardening of the surface layer due to polishing or other cold working processes.

2. Formation of a very thin oxide or other chemical film on the indented surface [14].

3. The presence of residual stresses resulting from the fabrication process in thin films [15].

4. Variation of elastic recovery, during unloading, with the indentation size. Such effects have been proposed to explain the observed change in the ratio of the long to short diagonals of a Knoop impression from that of the actual Knoop indenter [15, 16].

In general there is enough experimental evidence to substantiate the ISE, however, there is no mechanistic model to account for it. The general approach to deal with the ISE phenomenon is to use Meyer's emperical law [17]:

$$H = 1.854\ ad^{(n-2)} \tag{5}$$

where the ISE index n and the constant,a , are characteristics of the material being tested. When $n = 2$ there is no ISE; for $n < 2$ the hardness increases with decreasing indentation depth, for n > 2 the hardness decreases with decreasing indentation depth. Most materials have n values in the range 1.6–1.9 [13].

7. FILM/SUBSTRATE EFFECT

When indentation hardness tests are conducted on thin films the measurements reflect contributions from both film and substrate, since both deform under indenter load. Two models have been advanced to account for the FSE. The Johnson and Hogmark [18] model used the relative sizes of the load-supporting areas to relate the composite hardness H_c to both the film hardness H_f and the substrate hardness H_s as follows:

$$H_c = H_f (A_f/A) + H_s (A_s/A) \tag{6}$$

where A_f, A_s, and A are the film, substrate, and total area supporting the indenter load, respectively. Application of this model fails when the depth of indentation was less than the film thickness [19].

Burnett and Rickerby [19] proposed a model based on the volume law of mixtures to address the FSE effect. This model related H_c to H_f and H_s as follows:

$$H_c = H_f (V_f/V)^3 + \chi H_s (V_s/V) \tag{7}$$

where V_f, V_s, and V are film, substrate, and total deformed volumes, respectively. The parameter χ is an interference parameter that allows for modification of the deformed volume of the softer material. The numerical value of χ is chosen empirically to provide a best fit to the experimental results. However χ was found to be a function of the ratio of the quotient of the hardness and elastic modulus of both the film and the substrate [20] Burnett applied this formula for soft films on a hard substrates [21], and hard film on soft substrates [20].

It is easy to recognize that equation 6 represents a force balance, while equation 7 represents an energy balance; thus, both have reasonable physical bases. It is very difficult, however, to measure the deformed areas and volumes. However, the ratios in equation 6 and 7 can be related, by approximate relations, to measurable quantities, e.g. t/d, the indenter angle Θ, and the ratio of the elastic modulus to the hardness of both film and substrate [19, 21].

8. RESULTS AND DISCUSSIONS

I. TiB_2 and DLC

Figures 8a and 8b show typical load–depth curves for TiB_2 and DLC films respectively. The thickness of both films was about 20 microns. The unloading curves show moderate levels of elastic recovery, as evidenced by the slope of the tangent to the initial part of the unloading curve. The hardness–depth relations for TiB2 and DLC are shown in Figures 9a and 9b, respectively. The measured hardness at 20 nm is higher than that at 175 nm by a factor of 2 in the case of TiB2 and 3 in the case of DLC, Figures 10a and 10b show the elastic moduli of the two films as calculated from the indentation measurements. The elastic moduli, E, of both films increase with decreasing indentation depth, such that the ratio of E at 20 nm to E at 175 nm is nearly 1.8 in the case of TiB_2 and 1.9 for DLC. The systematic increase in the elastic modulus reported here is interesting because, unlike the ISE, it has not been reported in the literature. Furthermore, it has more important implications as regards the physics of the indentation mechanism. This is because unlike H, which may vary with residual stress, E is considered to be a material property.

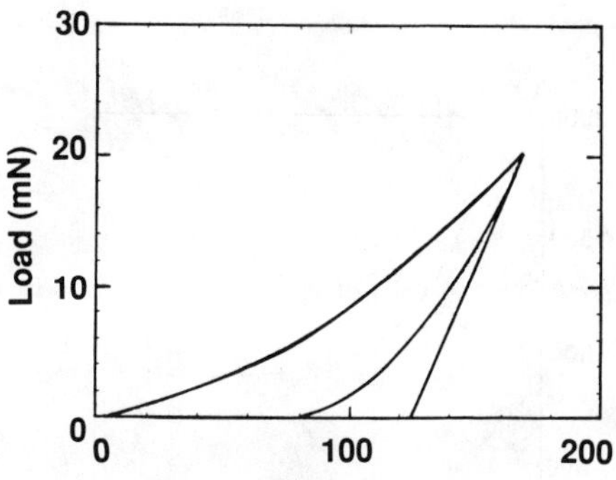

Fig. 8a Typical Load-Displacement Curve of TiB_2 Film

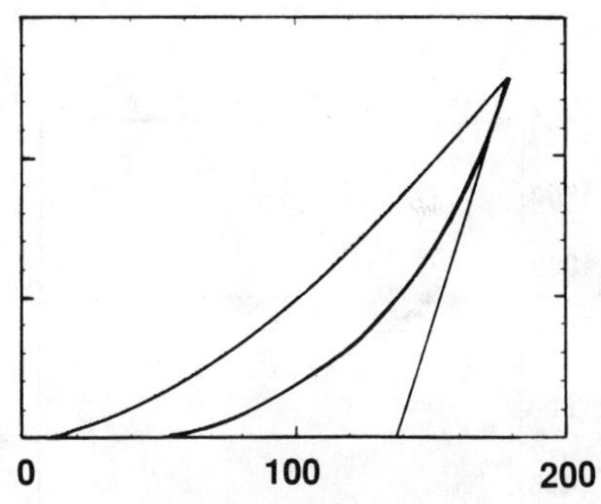

Fig. 8b Typical Load-Displacement Curve of DLC Film

It is proposed that the change in either H or E in the TiB_2 or DLC measurements is due to a substrate contribution, since the ratio of the film thickness to indentation t/d,

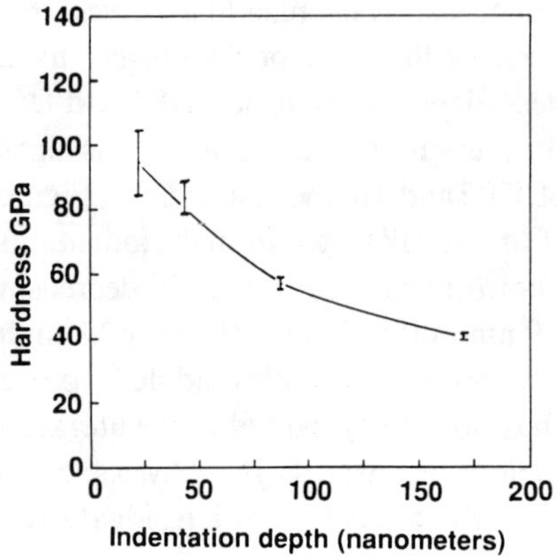

Fig. 9a Hardness of TiB_2 Film

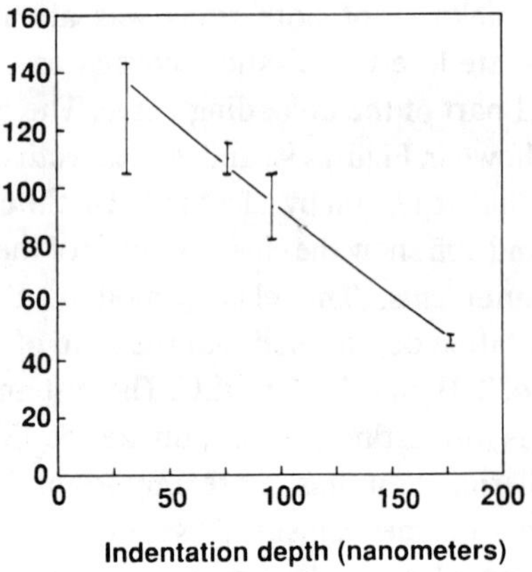

Fig. 9b Hardness of DLC Coating

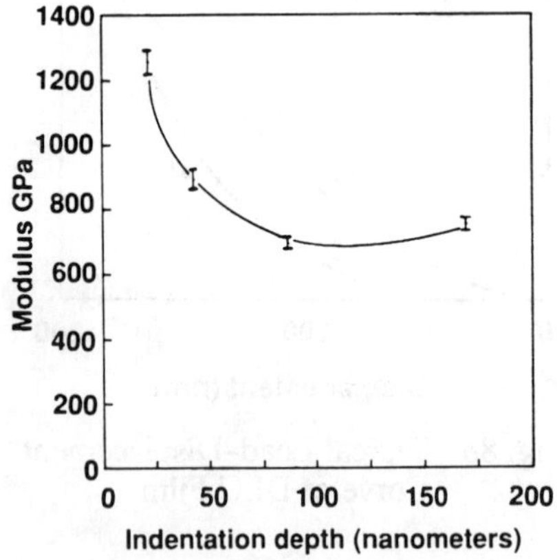

Fig. 10a Modulus of Elasticity of TiB_2 Film

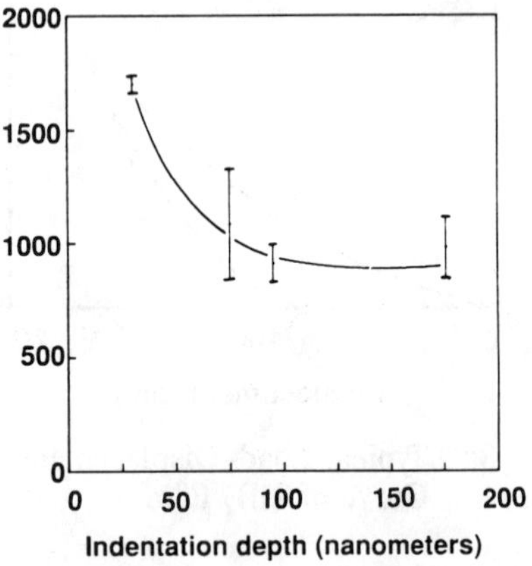

Fig. 10b Modulus of Elasticity of DLC Coating

depth in these measurement is more than 100. This was further tested by measuring both H and E on a bulk TiB_2 sample. Comparison of measurements of film and bulk hardness and modulus of elasticity is shown in Figures 11a and 11b respectively. The close agreement for both H and E between film and bulk specimens confirms that the observed changes in H and E are related to an ISE and not to an FSE.

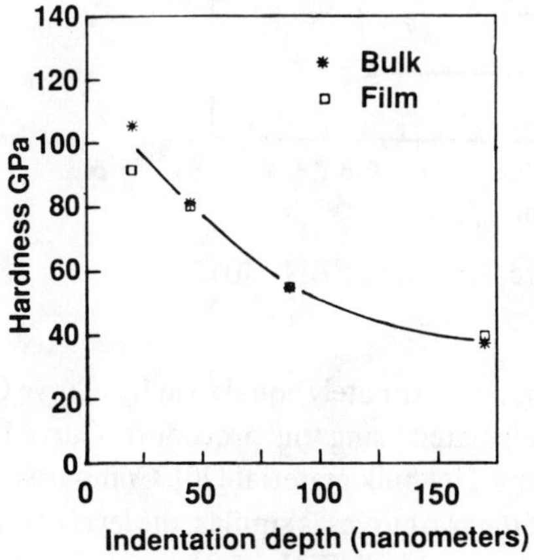

Fig. 11a Comparison of Hardness of Bulk and film TiB_2

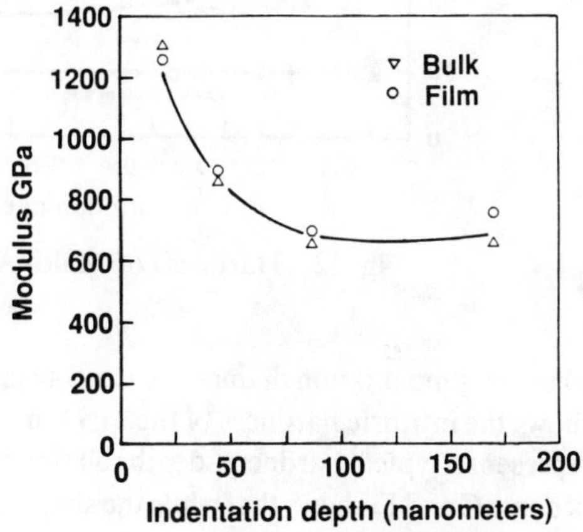

Fig. 11b Comparison of Modulus of Elasticity of Bulk and Film TiB_2

II. TiN Results

The load depth curves for TiN are quite similar to those of TiB2, and hence are not shown here. Since the thickness of the TiN film was only 1.3 microns, the results were expected to show a FSE effect, i.e. the measured hardness is expected to be that of a TiN/M50 steel composite. To account for the FSE we measured the hardness of the M50 steel substrate, Hs. The results are represented by curve A in Figure 12. Curve B, Figure 12 shows the measured film/substrate composite hardness, H. The intrinsic film hardness was calculated from curves A and B using Equation 7 With $\chi = 1.2$. The size of the deformed zone, b, was estimated from the spherical cavity model [22] as:

$$b/D = c(E/H) \cot^3 \Theta \tag{9}$$

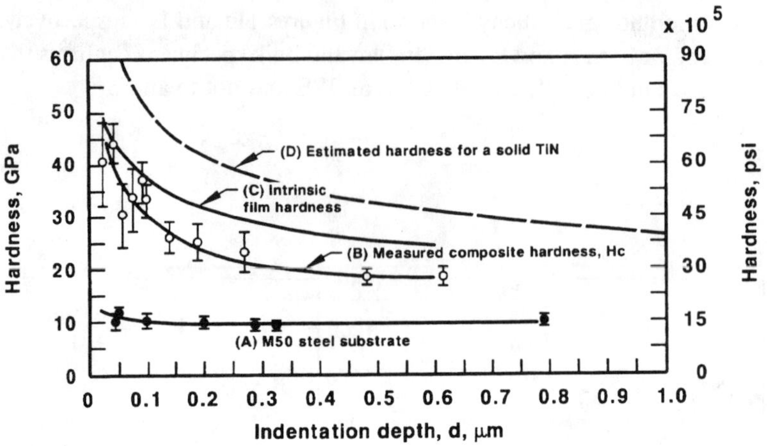

Fig. 12 Hardness of Multi–Arc TiN Film, SKN 71042

where D is indentation diameter, c is a constant approximately equal to unity. Curve C shows the intrinsic hardness of the TiN film calculated using this procedure. Curve D represent a typical hardness–depth relation for a TiN bulk material [20]. Comparison of curve C and D shows that while the shape of the two curves is similar, the level of the hardness af the TiN film is lower than that of typical bulk TiN.

III. Soda Lime Glass Results

Figure 13 shows a typical load displacement curve for a soda lime glass specimen. The unloading curve reveals more elastic recovery than in the case of the ceramic films discussed above. However there is a considerable inelastic component of deformation remaining after load removal. Figure 14 shows a TEM micrograph of a 400 nm deep indentation in soda lime glass. The micrograph reveals several features of interest:

1) The radial and lateral cracks, characteristic of brittle fracture [5], are not observed in the indentations. This observation was confirmed by examining several indentations made at depths in the range 20–1000 nm, where neither radial nor lateral cracks could be observed. Measurements conducted by Marshall and Oliver [23] on glass fibers also did not show any radial or lateral cracks. This is probably because the nanoindentation size range is less than the threshold indent size required for crack initiation [5].

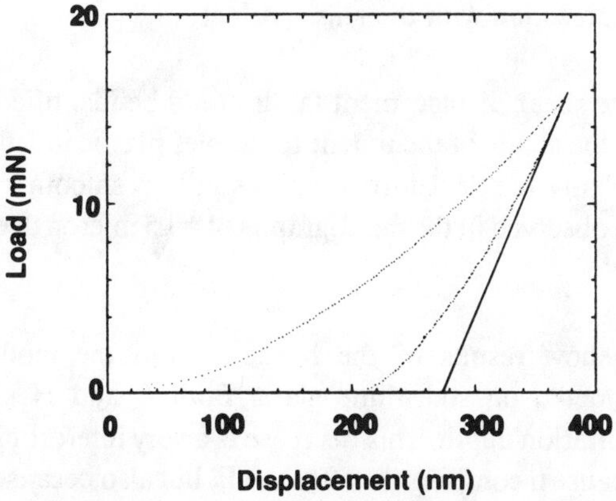

Fig. 13 Typical Load–Displacement Curve of Sodalime Glass

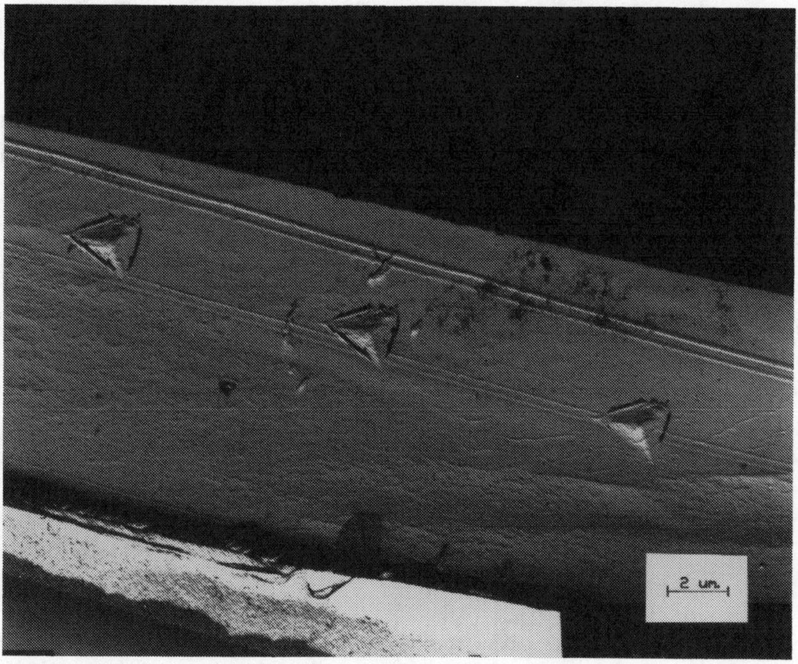

Fig. 14 Nanoindents on Glass

2) There is only one fault system (appears as a shadowed step) near the top of the indent at the intersection with the glass surface.

3) No other discrete shear displacement faults could be identified on the deformed surfaces. In fact the shape of the indent resembles plastic indentation of a ductile material rather than a brittle deformation. A similarly smooth indented surface on glass can also be observed in the micrographs of a 1.5 micron deep indent shown in Ref. 23.

Figure 15 and 16 show results of the hardness and the modulus of elasticity measurements conducted on soda lime glass. Both E and H show considerable dependence on indentation depth. This decrease is a very interesting manifestation of the ISE, not only because it confirms the effect on E, but also because of several unique features of glass, such as: (1) Glass is an isotropic amorphous material showing no effect of direction or grain size, (2) Glass is already an oxide, therefore, no hard oxide films are involved in the ISE, and (3) Glass is a brittle material such that no dislocation or work hardening effects are expected to influence the measurements. Nevertheless, comparison of results at 500 nm and 20 nm reveals an increase in H by a factor of 2.5, and in E by a factor of 1.5. This may be due to changes in the elastic recovery as suggested by Jindal and Quinto [15]. However, it is proposed that a more general model for explaining the ISE is needed. Such a model should consider the results on both ductile and brittle material and consider the probability of a brittle/ductile transition of the deformation process at micron and submicron indentation depths.

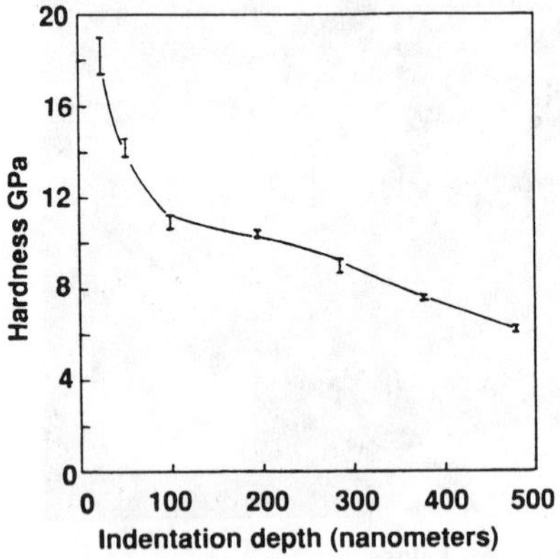

Fig. 15 Hardness of Sodalime Glass

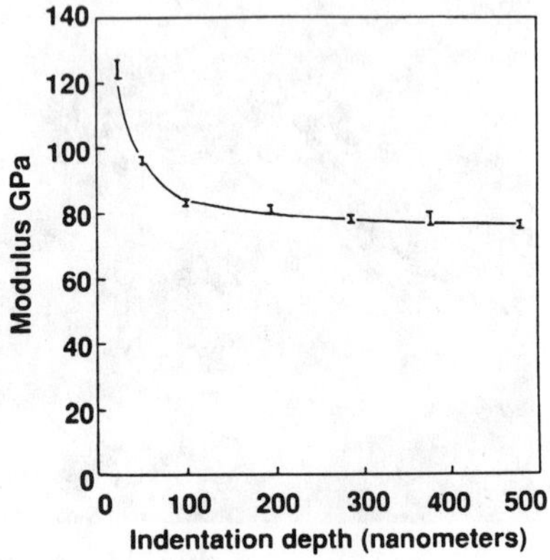

Fig. 16 Modulus of Elasticity of Sodalime Glass

9. CONCLUSIONS

Indentation measurements in the submicron range reveal that:

1. The ISE is a general phenomenon which applies to both ductile and brittle materials, as well as crystalline and amorphous materials.

2. The ISE affects both the hardness and the elastic modulus.

3. The deformation system of material during indentation at the submicron level exhibits a considerable non–elastic component.

4. Glass indentation at the submicron level reveal indent surface features similar to those of ductile materials.

REFERENCES

1. W. C. Oliver, "Progress in the Development of a Mechanical Properties Microprobe," MRS Bulletin, Sept./Oct, 15–19 (1986).

2. M. F. Doerner and W. D. Nix, "A Method for Interpreting the Data from Depth-Sensing Indentation Instrument," J. Mat. Res., 1 [4] 601–9 (1986).

3. D. Bucley, Surface Effects in Adhesion, Friction, Wear and Lubrication, pp. 472–475, Elsevier, New York, 1981.

4. S. S. Chiang, D. B. Marshall, and A. G. Evans, "The Response of Solids to Elastic/Plastic Indentation," Int. J. Mech. Sc., 53, pp. 289–311 (1982).

5. D. B. Marshall, and B. Lawn, "Indentation of Brittle Materials," pp. 26–46 in Microindentation Techniques in Materials Science and Engineering, Edited P. J. Blau and B. R. Lawn, ASTM STP 889, Philadelphia, 1985.

6. J. N. Sneddon, "The Relation between Load and Penetration in the asymmetric Boussinesq Problem for a Punch of Arbitrary Profile," Int. J. Eng. Sci., Vol. 3, pp. 47–57 (1965).

7. J. L. Loubet, J. M. Georges, O. Marchesini, and G. Meille, "Vickers Indentation Curves of Magnesium Oxide (MgO)," Trans. ASME, Vol. 106, pp. 43–48 (1984).

8. W. Roman, “Laser/Optical Diagnostic Techniques for Plasma Material Processing,” Symposium K Proceedings of 1987 MRS Symposium, Anaheim, California, April, 1987.

9. W. Roman, J. Stufflebeam, F. Otter, “Progress in Diagnostics of PaCVD System for depositing TiB2 Hard Face Coatings,” Proceedings of 8th International Symposium of Plasma Chemistry, Tokyo, Japan, Sept. 1987.

10. W. Roman, J. Stufflebeam, A. Eckbreth, C. Ultee, “CARS Diagnostics of Plasma Species in PACVD Process for Hard Face Coating Proceedings of International Symposium on combustion and Plasma Synthesis of High Temperature Materials,” San Francisco, CA, Oct. 23–26, 1986.

11. W. Roman, M. Collett, S. Hay, and A. Eckbreth, “CARS Diagnostics and Analysis of Species in Diamond Depositing Process,” Proceedings of Electrochemical Society 1st International Symposium on Diamond, Los Angeles, CA, May 7–12, 1989.

12. N. P. Bansal and R. H. Doremus, “Handbook of Glass Properties,” pp. 306–309, Academic press, Inc., New York, 1986.

13. P. M. Sargent, “Use of Indentation Size Effect on Microhardness for Materials Characterization,” pp. 160–174, in Microindentation Techniques in Materials Science and Engineering, Edited P. J. Blau and B. R. Lawn, ASTM STP 889, Philadelphia, 1985.

14. J. B. Pathica and D. Tabor, “Contact Characterized Metal Surfaces at Very Low Loads,” Surface Science, Vol. 89, pp. 182–190 (1979).

15. P. C. Jindal and D. T. Quinto, “Load Dependence of Microhardness of Hard Coatings,” Surface Coatings Tech., 36, 683 pp. 694 (1988).

16. L. E. Samuels, “Microindentation in Metals,” pp. 5–25 in Microindentation Techniques in Materials Science and Engineering, Edited P. J. Blau and B. R. Lawn, ASTM STP 889, Philadelphia, 1985.

17. W. C. Oliver, R. Hutchings, and J. B. Pethica, ” Measurement of Hardness at Indentation Depths as Low as 20 Nanometers,” pp. 90–108 in Microindentation Techniques in Materials Science and Engineering, Edited P. J. Blau and B. R. Lawn, ASTM STP 889, Philadelphis, 1985.

18. B. Jonsson and S. Hogmark, "Hardness Measurements of Thin Films, " Thin Solid Films, 114, pp. 257–269 (1984).

19. B. Burnett and D. S. Rickerby, "The Mechanical Properties of Wear Resistant Coatings I, " Thin Solid Film, 148, pp. 41–50 (1987).

20. B. Burnett and D. S. Rickerby, "The Mechanical Properties of Wear Resistant Coatings II, " Thin Solid Films, 148, pp. 51–65 (1987).

21. P. J. Burnett and T. F. Page, "Surface Softening by Ion Implantation," J. Mat. Sci., Vol 19, pp. 845–860 (1984).

22. B. R. Lawn, A. G. Evans, and D. B. Marshall, "Elastic/Plastic Indentation Damage in Ceramics:The Median/Radial Crack system," J. Am. Ceram. Soc., 63 [9–10] pp. 574–580 (1980).

23. D. B. Marshal and W. C. Oliver, "Measurement of Interfacial Mechanical Properties in Fiber-Reinforced Ceramic Composite," J. Am. Ceram. Soc., 70 [8] pp. 542–548 (1987).

SOL-GEL PROCESSING AND CHARACTERIZATION OF FERROELECTRIC PZT THIN-FILMS

A. R. Modak and S. K. Dey
Chemical, Bio and Materials Engineering and
Center for Solid State Electronics Research
Arizona State University, Tempe, Arizona 85287, U.S.A.

ABSTRACT

This paper illustrates the reproducible and reliable synthesis of sol-gel precursors and the fabrication of ferroelectric PZT thin-films. Thin-film microstructure and chemical analyses coupled with low and high field electrical properties of these nonlinear capacitors are presented and discussed.

INTRODUCTION

One dimensional inorganic ferroelectrics, particularly the oxygen octahedra class (e.g. $BaTiO_3$, $Pb_{1-x}La_x (Zr_{1-y} Ti_y)_{1-x/4} O_3$ or PLZT, $Bi_4Ti_3O_{12}$, $LiNbO_3$, and $Sr_xBa_{1-x}Nb_2O_6$ or SBN) have received considerable attention due to their multifunctional nature. In bulk form, these materials have high permittivities[1], large electromechanical coupling coefficients[2], exhibit pyroelectric behavior, and electro-optic effects[3,4]. Potential applications in thin-film form include decoupling capacitors[5], optical shutters and modulators, optical displays and memories[3,4], piezoelectric transducers[1,2], ferroelectric gate FETs[6], pyroelectric detectors[4] and more recently, ferroelectric memories[7-13] and components for artificial neural networks[14,15].

The advantages of ferroelectric $PbZrO_3$ - $PbTiO_3$ or PZT thin-films integrated onto semiconductors for bistable memory applications include a) high bit density b) nonvolatile low voltage operation over a wide temperature range, c) random access read/write cycles, d) short access and cycle times, e) low power requirements and f) less of an effect by nuclear and electromagnetic radiation(i.e., radiation hardness) than semiconductor or magnetic memories[7].

A number of processing techniques have been attempted for the reproducible fabrication of Pb-based perovskite thin-films including flash[16] and e-beam evaporation[17], rf diode[18] and magnetron[19] sputtering, ion beam deposition[20], and chemical vapor deposition[21]. The use of these vacuum techniques has been limited due to difficulties associated with compositional control, nonuniformity and cracking. In addition, relatively low values of dielectric permittivities and spontaneous polarization in thin-films compared to bulk materials were reported.

Recently, a reproducible sol-gel fabrication route of ferroelectric Pb-based perovskite thin-films with acceptable static and dynamic properties has been demonstrated[7]. Critical processing parameters were outlined and the control of polymeric structure and microstructure evolution was reported[7]. The potential compatibility of sol-gel processing of PZT thin-films with GaAs JFET planar technology now offers fabrication opportunities for highly complex nonvolatile programmable random access memory(NVPRAM) circuits that are radiation hard.

This paper outlines the synthesis of precursors and the fabrication of thin-films. Recent results of both physical and electrical characterization of PZT thin-films with varying Zr/Ti ratios are presented. The rationale for the difficulties encountered in interpreting the thickness and time dependences of the dynamic switching characteristics are outlined. The issues of the long term stability and switching characteristics of these ferroelectric thin-films are addressed.

EXPERIMENTAL PROCEDURE

Ti and Zr methoxyethoxides were reacted with anhydrous lead acetate in methoxyethanol under dry nitrogen using standard Schlenk technique[7] to synthesize the precursor complex:

$$Pb(Ac)_2 + xZr(OR)_4 + (1\text{-}x)Ti(OR)_4 \Rightarrow Pb(Zr_xTi_{1\text{-}x})O_2\,(OR)_2 + 2RAc$$

$Ac = CH_3\,COO$ $\quad$ $R = CH_3OCH_2CH_2$

Details of the procedure were reported previously[7]. The completion of the reaction was confirmed by distillate density measurements and gas chromatography. The precursor complex was then hydrolysed in the presence of a basic catalyst (NH_4OH) and aged to form a stable polymeric solution. This stable polymeric solution, which is a partial condensation product composed of metal-oxygen-metal linkages, was spin cast on Pt, Si, Pt passivated Si and GaAs substrates at 2000 rpm. Films were typically fabricated with multiple depositions accompanied by heating(300-400^{0}C) between layers. The films were typically fired at temperatures between 500-700^{0}C in air for crystallization and sintering.

Microstructural, physical and chemical characteristics of the thin-films were studied using a number of techniques including Rutherford Backscattering(RBS), Auger Electron Spectroscopy(AES), X-ray Diffraction(XRD) and Scanning Electron Microscopy (SEM).

Gold or platinum top electrodes (1 mm diameter) were deposited using a laboratory scale sputter coater and the electrical properties of these ferroelectric capacitor units were evaluated. Low and high field electrical characterization was carried out using the HP 4140B picoammeter/dc voltage source, HP multifrequency LCR meter, HP 54503A digitizing oscilloscope, Wavetek signal generator and other standard laboratory equipment. Hysteresis properties were evaluated by connecting the sample on a typical Sawyer Tower circuit (Figure 1). The samples were fatigued at a frequency of 25 kHz and hysteresis parameters were measured from time to time at different frequencies.

RESULTS AND DISCUSSION

Sol-gel Processing and Physical Characterization

It was observed that the topology and compactness of the thin-film structure depends on the nature of the polymeric solution and the spin-on conditions. Dense amorphous films form at 400^0 C and the crystalline perovskite structure develops within 10 minutes at temperatures ranging from 500-550^0 C depending on the Zr/Ti ratio. Increase in the Zr content increased the crystallization temperature. The physico-chemical issues related to the reproducible fabrication of the thin-films have been reported elsewhere[7]. Typically thicknesses of 800-900 A^0/ layer were obtained with a 0.5 molar solution spin cast at 2000 rpm.

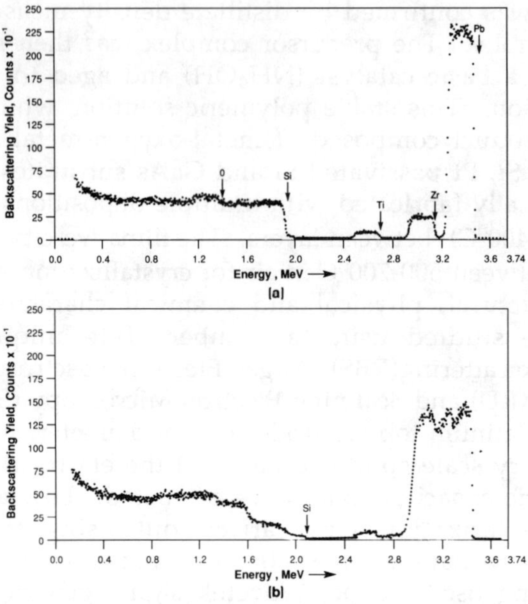

Figure 1 : 3.77 MeV He^+ RBS spectrum of (a) amorphous PZT(52/48) thin-film on Silicon, heatreated at 350^0C for 15 minutes, (b) crystalline PZT(52/48) thin-film on Silicon, heatreated at 700^0C for 30 minutes.

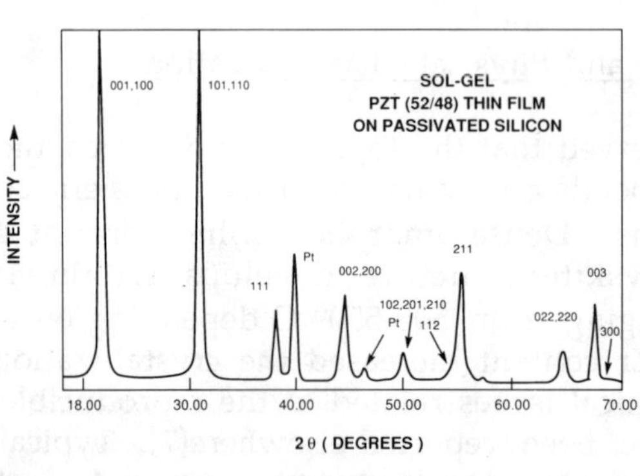

Figure 2 : X-Ray diffraction pattern for a PZT(52/48) thin-film on passivated Silicon, sintered at 550^0C for 10 minutes.

RBS was used to analyze the chemical composition of thin-films. Figure 1(a) illustrates a RBS spectrum of an amorphous PZT(52/48) thin-film on Si. The stoichiometries of thin-films obtained by analysis of the RBS spectra were in concurrence with precursor batch calculations. Also the thicknesses of the films as estimated from the RBS peak width analyses were in agreement with measurements from crossectional SEM and ellipsometry. Figure 1(b) illustrates the RBS spectrum of a crystalline film of the same composition on Si. The observed shift in the Si energy threshold from ~1.92 MeV in amorphous films to ~2.13 MeV in crystalline films clearly reveals the diffusion of Si almost upto the surface of the thin-film. Contamination of PZT thin-films with Si was responsible for retarding crystallization, and the effect would become more pronounced with decreasing film thickness. Films on Pt behaved similar to bulk PZT samples.

Figure 2 illustrates a well defined X-ray diffraction pattern for a PZT(52/48) thin-film on passivated Si. Note the crystallization of the perovskite phase at only 550^0 C in 10 minutes.

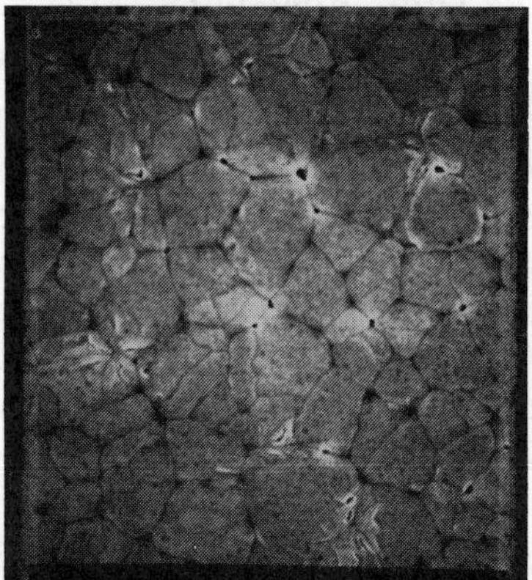

Figure 3 : SEM microstructure of PZT (52/48) thin-film on passivated Silicon, sintered at 700^0C for 30 minutes.

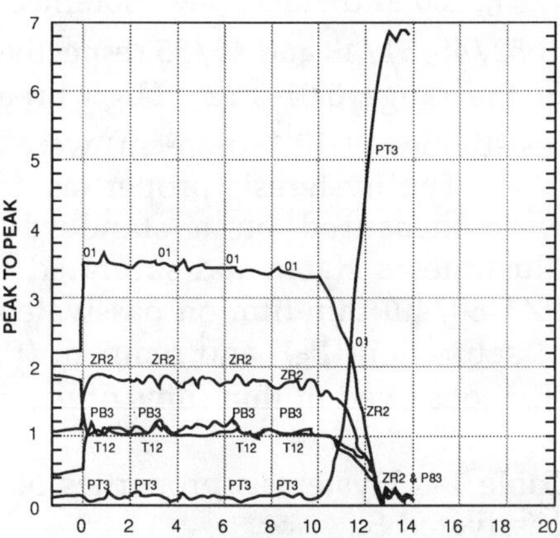

Figure 4: AES depth profile for Pb, Zr,Ti, and O in a PZT(51/49) thin-film on passivated Silicon., fired at 700^0C for 30 minutes.

Ideal microstructures with nearly theoretical densities developed for the sol-gel derived thin-films. Figure 3 is a SEM micrograph of the PZT(52/48) thin-film heat treated for 1/2 hr at 700^0 C. Note, the average grain size is in the same range as the thickness of the film. This indicates that the thin-film is single grain thick but can still support a high electric field. The observed intergranular porosity is totally eliminated by rapid thermal annealing at higher temperatures.

Figure 4 presents an AES depth profile of a PZT(51/49) thin-film on platinum passivated Si, heat treated at 700^0 C for 1 hr. The constant peak to peak intensity for the Pb, Ti, Zr and O signals indicates the chemical homogeneity or very uniform PZT concentration through the entire thickness of the film. The data also reveals a sharp Pt transition at the film substrate interface indicating very little Pt diffusion into the PZT film during heat treatment.

Electrical Characterization

Polycrystalline PZT thin-films on Pt, passivated Si and GaAs exhibited low field(100V/cm) relative permittivities (ϵ_r) comparable to bulk and were thickness (0.1-1.0 μm) independant. ϵ_r values of 1250,1200 and 1000 were obtained for PZT thin-films with Zr/Ti ratios of 52/48, 51/49 and 45/55 respectively. Corresponding tanδ values were in the range 0.01-0.02. High breakdown strengths (>10^6 V/cm) and resistivities (~ 10^{12} ohm-cm) were typical.

The hysteresis properties of the thin-film ferroelectric capacitor were measured on a standard Sawyer Tower circuit . Figure 5 illustrates a stable and saturated 60 Hz hysteresis loop obtained for a PZT(60/40) thin-film on passivated Si. Table 1 summarizes the values of saturation (P_s) and remnant (P_r) polarization and Coercive Fields (E_c) observed in thin-films of varying Zr/Ti ratios.

Table 1 : Hysteresis properties of Sol-Gel PZT thin-films(0.45 μm) on passivated Si.

COMPOSITION	P_r(μC/cm^2)	P_s(μC/cm^2)	E_c(kV/cm)
$Pb(Zr_{0.45}Ti_{0.55})O_3$	30-32	17-18	65-70
$Pb(Zr_{0..52}Ti_{0.48})O_3$	30-35	18-20	45-50
$Pb(Zr_{0.60}Ti_{0.40})O_3$	25-28	17-18	65-70

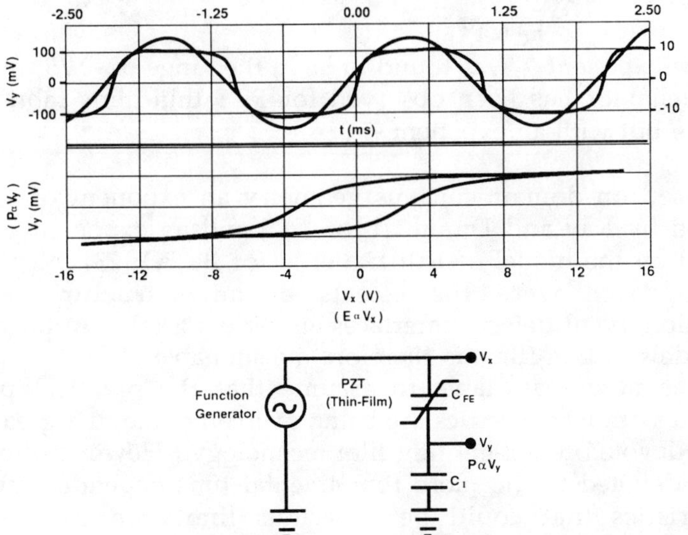

Figure 5 : (a) Characteristic ferroelectric hysteresis loop of a PZT(60/40) thin-film(0.45μm) on passivated Silicon, (b) Sawyer Tower Circuit diagram.

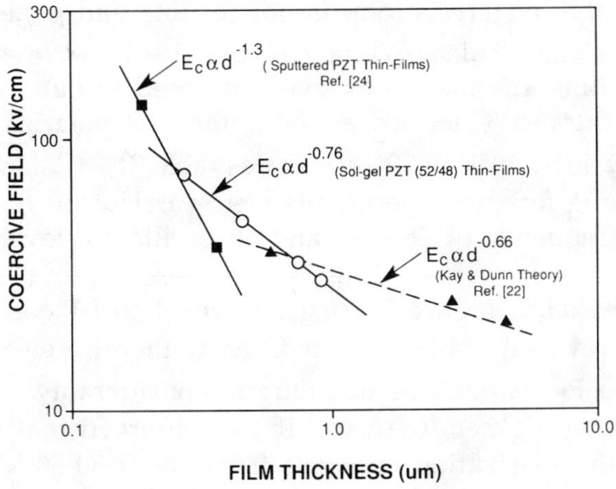

Figure 6 : Coercive Field (Ec) vs film thickness for Sol-Gel PZT(52/48) thin-film on passivated Silicon. Also reported for comparison is the thickness dependence of Coercive field for sputtered films[12], and the behavior predicted by the Kay and Dunn theory for TGS[22].

The thickness dependence of E_C illustrated in Figure 6 obeys a power law of the form:

$$E_C \alpha t^{\alpha}$$

where the exponent α was found to be in the range of -0.75 to -0.76. A similar behavior has been observed for PZT thin-films fabricated by sputtering but with an exponent -1.3.

Based on domain nucleus geometry an exponent of -0.66 was calculated by Kay and Dunn[22] for TGS and an exponent of -1 was predicted by the model of Fatuzzo et al for $BaTiO_3$[23]. Neglected in their approach were the effects of microstructure, chemical composition, point defects, interfaces and stress levels. Applicability of such models to thin-films is therefore questionable.

It is now reasonable to assume that the practical problems associated with ferroelectrics are being overcome due to a great deal of research devoted to sol-gel thin-film technology. However more effort must be devoted to the more fundamental time-dependent switching characteristics that could be a serious limitation to the use of ferroelectrics in digital designs. A major source of concern is the switching fatigue problem in that the amount of available signal depends on the number of polarization reversal cycles. Although the ferroelectric polarization can be reversed rapidly enough for most applications, the hysteresis loop is not reliable and generally fatigues with time (Dey and Zuleeg [7] have obtained ~ 10^{12} cycles of switching in PZT thin-films, the maximum ever disclosed to date). A thin-film with high endurance (defined as the number of polarization reversals for P_r to degrade by 50%) of 10^9- 10^{15} is desirable[7].

Switching fatigue experiments were carried out on thin-films at a switching frequency of 25 kHz and an oscillation level of $E > 3\ E_C$. Periodically the hysteresis loop parameters were measured at a frequency of 1kHz. Figure 7 illustrates results of P_r as a function of polarization reversals obtained on films with different Zr/Ti ratios. Note that the P_r values did not fatigue considerably even after 10^9 cycles indicating high endurances. It was observed that the E_C values increased with polarization reversals from about 50-60 kV/cm to 150-200kV/cm similar to a trend observed by Taylor[4] on fatigued single crystal $BaTiO_3$. Figure 7 also indicates a fatigue behavior with a positive curvature and approaching steady state values. This is contrary to the logarithmic dependence reported for bulk materials[4] and a negative curvature observed for sputtered films[24] . Interestingly, before approaching a steady state value, the fatiguing rate appears to be the lowest for morphotropic (52/48) and highest for tetragonal(45/55) with an intermediate rate for rhombohedral (60/40) phase compositions.

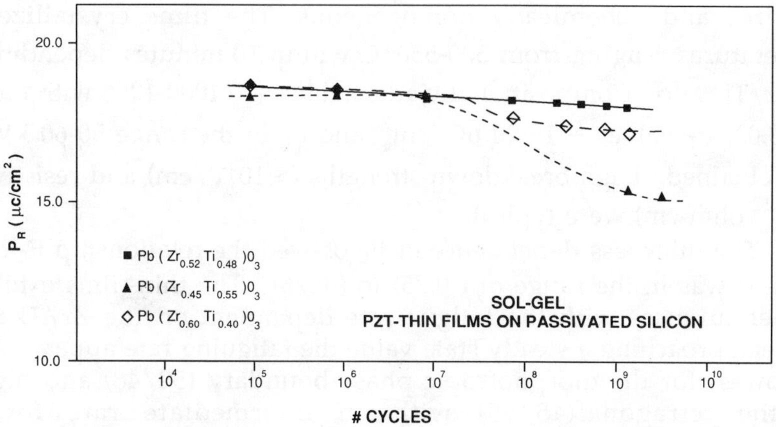

Figure 7 : P_r vs Polarization reversal cycles for Sol-Gel PZT thin-films (0.45-0.5μm) of different Zr/Ti ratios on passivated Silicon.

Proposed degradation models for the dynamic switching properties in bulk PZT are based on ferroelectric domain stabilization due to either PZT -electrode interfaces[25], impurity vacancy pairs[26], space charge effects[24], microcracking[27] and grain boundaries[28]. To date, there is no direct evidence to support any of these models. The validity of these models and their applicability to thin-films have raised questions that remain open. For instance the perturbation of polarization due to the film constrained to the substrate, and the stress effects as a result of constrained crystallites and piezoelectric effect will influence both, electrical and mechanical boundary conditions in the polycrystalline matrix. Undoubtedly these will reflect on the various dynamic parameters such as switching current and time, coercive field and domain wall mobility and their time dependences. Consideration of the above influences coupled with more advanced characterization will unambiguously elucidate the operative degradation mechanisms of fatigue in thin-films.

SUMMARY AND CONCLUSIONS

In this age of integration, Sol-gel processing is a relatively inexpensive, simple and reliable route for the fabrication of multicomponent ferroelectric thin-films on Pt, Si, passivated Si and GaAs.

The PZT thin-films fabricated by sol-gel processing were crackfree and chemically homogeneous. The films crystallized at temperatures ranging from $500\text{-}550^0$ C within 10 minutes depending on the Zr/Ti ratio. Compositionally adjustable $\epsilon_r \sim 1000\text{-}1250$ with $\tan\delta$ of 0.01-0.02, P_r values ~ 18-20 $\mu C/cm^2$ and E_c in the range 50-60 kV/cm were obtained. High breakdown strengths (> 10^6V/cm) and resistivities (~ 10^{12} ohm-cm) were typical.

The thickness dependence of E_c obeyed the relationship $E_c \; \alpha \; (t)^\alpha$ where α was in the range of (-0.75) to (-0.76). The thin-films exhibited high endurances with the fatigue rate dependent on the Zr/Ti ratio. Before approaching a steady state value the fatiguing rate appears to be the lowest for the morphotropic phase boundary (52/48) and highest for the tetragonal(45/55) with an intermediate rate for the rhombohedral(60/40) compositions. The relative influences of the intrinsic and extrinsic parameters on the electrical and mechanical boundary conditions are currently unknown in PZT thin-films. Consequently an unambiguous interpretation of the time and thickness dependences of the dynamic switching parameters is difficult. Applicability of bulk models to thin-films is also questionable at the present time.

ACKNOWLEDGEMENTS

This work was performed in the Advanced Ceramics Processing Laboratory and the Center for Solid State Electronics Research at Arizona State University supported by the Physics Division of the Army Research Office (Contract No. DAA-L03-89-K-0104) and McDonnell Douglas Corporation (Contract No. 89614221-H). We thank S. Bernacki of Raytheon for providing passivated Si substrates and AES data. We also thank Prof. J. Leavitt and Prof. L. McIntyre of the Physics Department of the University of Arizona for RBS analyses of the thin-films. Technical discussions with R. Zuleeg (of McDonnell

Douglas Corporation) and P. Brody, J. Benedetto, and J. McGarrity (all of Harry Diamond Laboratories) are appreciated.

REFERENCES

[1] B.Jaffe, W.R. Cook, and H. Jaffe, Piezoelectric Ceramics, Academic, New York, 135-183, 1970.
[2] K.Suzuki and M. Nishikawa," Poling Dependence of the Surface Wave Freq. -Generation on PZT", Jap. J. Appl. Phys.,**13** , 240-243, 1974.
[3] C.E. Land, P.D. Thatcher, and G.H. Haertling, "Electrooptic Ceramics", in Appl. Solid State Sci., Vol. 4, R. Wolfe, Academic, New York , 137-233, 1974.
[4] J.C. Burfoot and G.W. Taylor, Polar Dielectrics and their Applications, Berkeley and Los Angeles: Univ. Calif., 1979.
[5] S.K. Dey, K.D. Budd, and D.A. Payne,"Characteristics of Chemically-Derived Lead-Perovskite Dielectric Layers for High Speed Switching VLSI Chips", Electronics Packaging and Materials Science, MRS, Palo Alto, California, April 1986.
[6] S. Wu, " A New Ferroelectric Memory Device Metal-Ferroelectric Semiconductor Transistor", IEEE Trans. on Electron Devices, **ED-21** [8], 499-504,(1974).
[7] S. K. Dey and R. Zuleeg, " Processing and Parameters of Sol-gel PZT Thin-Films For GaAs Memory Applications", Ferroelectrics, **109**, 1990(In press).
[8] S. K. Dey and R. Zuleeg, " Integrated Sol-gel PZT Thin-Films on Pt, Si, and GaAs For Non-Volatile Memory Applications", Proceedings of the 7th International Meeting on Ferroelectricity, Aug. 28- Sept 1, 1989.
[9] S.K. Dey and D.A. Payne, " Reliability of Ferroelectric Thin-films," Krysalis Microelectronics Reports (1986-1987).
[10] S.K. Dey , K.D. Budd, and D.A. Payne, "Thin-film Ferroelectrics of PZT by Sol-gel Processing," IEEE Trans. on Ultrasonics, Ferroelectrics, and Frequency Control, **35** [1],80-81 (1988).
[11] S.S. Eaton, D.B. Butler, M.Perris, D. Wilson, and H. McMeillie, " A Ferroelectric Non-volatile Memory", Proc. Intl. Solid State Ckts. Conf., San Francisco, CA, Feb 1988.
[12] J.F. Scott et al, " Switching Kinetics of Lead Zirconate Titanate Submicron Thin-film Memories", J. Appl. Phy., **64** [2],787-792, 1988.
[13] Joint Govt.-University-Industry Meeting on Radiation Hard and Nonvolatile Ferroelectric RAM technology: United Technologies Microelectronics Center, Colorado Springs, Sept. 13-14, 1988.

[14] L.T. Clark, S. K. Dey and R.O. Grondin," Ferroelectric Connections for IC Neural Network," Proceedings of the 1st IEE International Conference on Artificial Neural Networks, Savoy Place, London, October 1989.
[15] S. K. Dey, L.T. Clark, and R.O. Grondin, " Ferroelectric Thin-Films in IC Neural Networks," Presented at the Symposium on Application of Ceramics in Modern Society, 1st Ceramic Science and Technology Congress, Anaheim, California, November 1989.
[16] L. S. Phillips," Ferroelectric films by flash evaporation of PZT," Electron. Comp., **12**, 523-526, 1971.
[17] M. Oikawa and K. Toda," Preparation of Pb(Zr,Ti)O3 thin-films by an electron beam evaporation technique," Appl. Phy. Letters, **29**, 491-492, 1976.
[18] A. Okada," Electrical Properties of lead zirconate- lead titanate thin films and their composition analysis by Auger Electron Spectroscopy," J. Appl. Phys., **49**, 4494-4499, 1978.
[19] S. B. Krupanidhi, N. Maffei, M. Sayer, and K. El-Assal, " rf planar magnetron sputtering and characterization of ferroelectric Pb(Zr,Ti)O3 films," J. Appl. Phys., **54**, 6601-6609, 1983.
[20] R. N. Castellano and L. G. Feinstein,"Ion beam deposition of ferroelectric lead zirconate titanate," J. Appl. Phys., **50**, 4406-4411, 1971.
[21] M. Kojima et al," Chemical Vapor Deposition of PbTiO3 thin-film," Proc. of the 4th Mtg. on Ferroelectric Materials and their Applications, Kyoto, May 1983.
[22] H.F. Kay and J.W. Dunn," Thickness Dependence of the Nucleation Field of Triglycine Sulphate," Phil. Mag., 7, 2027 (1962).
[23] E. Fatuzzo and W.J. Merz," Surface Layer in BaTiO3 Single Crystals", J. Appl. Phys., **32**, 9, Sept. 1961.
[24] J.F. Scott , Private Communication.
[25] G.W. Taylor," Electrical Properties of Niobium-Doped Ferroelectric Pb(Zr,Sn,Ti)O3 Ceramics,"ibid, **38** [12], 4697-4706 (1967).
[26] G.H. Jonker, " Nature of Aging in Ferroelectric Ceramics," J. Amer. Ceram. Soc.,**55** [1], 57-58 (1972).
[27] W.R. Salanek," Some Fatiguing Effects in 8/65/35 PLZT Fine-grained Ferroelectric Ceramic," ibid, **55** [4], 97-101 (1972).
[28] W.C. Stewart and L.S. Cosentino," Some Optical and Electrical Switching Characteristics of Lead Zirconate Titanate Ferroelectric Ceramics," Ferroelectrics, **1**, 149-167 (1970).

BARIUM TITANATE THIN FILMS BY CONVENTIONAL AND RAPID THERMAL ANNEALING.

J. R. HANRAHAN, E. SANCHEZ, AND J. J. SANTIAGO
Moore School of Electrical Engineering, University of Pennsylvania, 200 S. 33rd St. Phila., PA 19104

D. H. BERRY AND Q. JIANG
Dept. of Chemistry, University of Pennsylvania, 34th and Spruce st., Phila., PA 19104

ABSTRACT

Barium titanate ($BaTiO_3$) films have been deposited on platinum and silicon substrates via spin casting metallo-organic compounds and annealed in a conventional oven to 600^{0}C. This method called metallo-organic decomposition (MOD), produces $BaTiO_3$ films with the tetragonal structure and provides for excellent control of film stoichiometry(1). Variable thicknesses (0.1 - 5μm) were achieved by spin casting multiple layers of the films. The films were then annealed in a conventional furnace and/or rapid thermal annealing (RTA) apparatus, to temperatures as low as 600^{0}C. The films on both platinum and silicon were uniform and showed no cracks by scanning electron microscopy (SEM). The films show ferroelectric behavior as evidenced by a polarization vs. electric field hysteresis and a peak in the dielectric permittivity around 120^{0}C.

INTRODUCTION

In the MOD process, $BaTiO_3$ films are produced by spin casting carefully mixed near stoichiometric metallo-organic precursors using a photoresist spinner. The substrates used were platinum and low resistivity (.002 - .05Ωcm) silicon. Various film thicknesses (0.1 - 5μm) were produced by changing spin speeds and by multiple spin casting. Annealing was done by conventional furnace or RTA.

Experimental Details

The metallo-organic (MO) compounds used in the formation of $BaTiO_3$ films were similar to those reported by Shaikh and Vest(2). The precursors were barium neodecanoate and titanium dimethoxy dineodecanoate. Silicon substrates were chosen for their potential in electronic application. Two different silicon substrates were used: n-type (100) and (111) single crystal wafers with a resistivity of .002 to .05 Ωcm.

To produce non-cracking films the precursor was spin cast at 5000 rpm for 30 seconds. After spinning the films, thermal decomposition was carried out in a tube furnace. The films were heated from room temperature to 600^{0}C at 2^{0}C/min., held there for 4 hours, and cooled to room temperature at the same rate they were heated. It was necessary to make multiple film depositions to build up film thickness.

RESULTS AND DISCUSSION

In the early stages of film production cracking was an extremely difficult problem, especially on silicon substrates. It was found that the slower heating rate helped in reducing the crack size, but could not eliminate them completely. An experiment was performed to determine if the film thickness had an effect on the fracture pattern and it was found that spinning a thin enough film eliminated the cracks altogether. It was found that once the first layer was formed at 0.1μm subsequent layers could be deposited with no cracking at all.

The film's grain size was also determined by SEM. The film has five layers which were spun to build film thickness as previously described. The micrograph reveals grain size of about 0.3μm. Compositional analysis was performed on the $BaTiO_3$ films with a PHI 600 scanning auger multiprobe (SAM) as shown in figure 1. To correct deviations from stoichiometry barium titanate films were grown on platinum substrates and Auger depth profiling was then used as a tool to determine the actual film stoichiometry. The proportions of the two precursors were adjusted in the mixture until the proper stoichiometry was achieved.

The XRD patterns of $BaTiO_3$ films on Si(100), Si(111), and platinum are shown in figure 2. The sharpness of the $BaTiO_3$ peaks, as well as the d-spacings, suggest the formation of crystalline $BaTiO_3$ at 600^0C.

Polarization vs. electric field measurements were made with a modified Sawyer-Tower circuit for examining the film's hysteresis effect. The values of Ps, Pr, and Ec deduced from it are roughly:

$$Ps = 0.339\ \mu C/cm^2, \quad Pr = 0.100\ \mu C/cm^2, \quad Ec = 80\ V/cm$$

Capacitance versus temperature measurements were then made with a model ESI 252 impedance meter at 1000Hz. The sample was poled by applying 15 volts (E field of $5x10^4$V/cm) while cooling from 150^0C to room temperature. Then the temperature was varied at a rate of 1^oC/min., while the capacitance values were read from the impedance meter. Figure 3 is a plot of the relative dielectric constant (κ) versus the temperature (T) for a 5μm thick film deposited on a Si (111) substrate with a capacitor area of $7.07mm^2$. The graph shows a room temperature value of κ of about 15 and a peaked value of about 56.

The poor ferroelectric behavior of these films can be attributed to the existence of amorphous material between the grain boundaries and to the small size of the grains. Films annealed for different times at 600^oC in a conventional oven are compared in Figures 4(a,b). The film annealed for 4 hours shows the presence of uncrystallized material . Annealing for 16 hours eliminates the amorphous material and produces a formation that looks more like a ceramic.

Figure 4(b) also shows that these films' grain size is very small, of the order of 0.01 μm. It has been established(3,4) that the ferroelectric behavior of $BaTiO_3$ depends on grain size and that grain sizes of less than 0.1 micron show very little ferroelectricity, if any. Xu and Coworkers increased the grain size of their films by annealing their $BaTiO_3$ (also produced by MOD in a conventional oven) at temperatures as high as 1200 °C for a Pt substrate. Under those conditions they obtained films with a grain size of 0.2 μm but the films cracked under thermal stresses. They also produced films at lower temperatures and on different substrates having smaller grain sizes(4).

As an alternative method for achieving the high annealing temperatures required for producing bigger grains but without the cracking of the films, we produced films by rapid thermal annealing (RTA) the organometallic precursors instead of using conventional over annealing. The RTA can achieve temperatures as high as 1000°C in about 30 s. Two important consequences of this feature are the enormous increase in processing speed of each layer and the favoring of the formation of fast kinetic phases over slow ones.

Each layer of the films produced using RTA was spin casted as described above on top of preceding ones and RTA at 650°C for 30 s. Figure 5(a) shows the x-ray diffraction pattern for 10 layers (the thickness measurements are under way) each annealed to 650°C. Figure 5(b) shows the same sample after RTA the 10 layers to 950°C in an atmospheres of 10 mTorr of O_2. The postannealed sample shows a large increase on the $BaTiO_3$ peaks with a decrease in bandwidth, indicating both enhanced crystallization and increased grain size. These films remained with a homogeneous morphology (crackless) even at processing temperatures of 1000°C.

Figure 6 is a SEM micrograph of a film postannealed at 950°C in 10 mTorr of O_2. The micrograph shows the uniform morphology of the film and reveals grains of an average grain size of ~ 0.3μm larger than the grain sizes reported for the same temperature by Xu et al. This difference can be explained by assuming a diffusion mechanism for the grain growth.

Samples prepared under different atmospheres provide evidence in this direction. XRD patterns for samples prepared in the RTA at 650°C under a pressure of 10^{-3} Torr (Figure 8(a)) show no formation of $BaTiO_3$. In turn, when the annealing is done under 1 atmosphere of O_2 $BaTiO_3$ is formed (Figure 7(b)). Furthermore, Auger spectroscopy of samples in low vacuum (Figure 8(b)) shows a large peak at 280 eV corresponding to carbon in the films. Samples prepared in air or oxygen show only a small amount of carbon contamination. Auger depth profiles also show that the films prepared in oxygen have the smallest percent of carbon of all films produced.

This evidence suggests that when there is enough supply of oxygen (pure or in air), the formation of CO and CO_2 that then leaves the surface does not compete with the formation of $BaTiO_3$ at the grain boundaries that then diffuses to the grain. The presence of oxygen may be an important factor in the formation of bigger grain sizes for a given annealing temperature.

At this point dielectric constant versus temperature measurements for the films produced with the RTA are under way in order to determine their ferroelectric behavior. Details of the role of oxygen in the formation of $BaTiO_3$ using MOD and in grain size is also being explored.

CONCLUSIONS

Crack free thin ferroelectric films have been produced on silicon and platinum substrates via Metal-Organic Decomposition of successive layers. MOD permits excellent control of the film stoichiometry by simply modifying the organic precursors. As evidence by Auger Spectra, the films possess only minor amounts of carbon contamination.

The $BaTiO_3$ films produced by annealing in a conventional oven for 4 hours at a temperature of 600°C show ferroelectric behavior, but it is very small. This is attributed mainly to small grain size, although evidence is presented of the existence of an amorphous material surrounding the grains.

$BaTiO_3$ films were also produced using a Rapid Thermal Annealer apparatus. Depositing multiple layers via MOD at 650°C with the RTA and doing a final anneal at 950°C in a 10 mTorr of oxygen dramatically increases the $BaTiO_3$ XRD peaks and reduces their linewidth. SEM micrographs show the formation of grains with larger grain sizes than the ones reported by other workers at the same annealing temperature.

Auger spectroscopy and XRD of the RTA produced films suggest that the absence of oxygen gas during processing is an important factor hindering the formation of the $BaTiO_3$ films. Also the films processed in an oxygen environment show the least amount of carbon contamination of the studied films.

REFERENCES

1. R. Vest and J. Xu, Proc. Sixth IEEE Intl. Symp. Appl. Ferroelec., Bethlehem, PA, pp. 374 (1986).
2. A.S. Shaikh and G.M. Vest, J.Am. Ceram. Soc., **69**,683 (1986).
3. G. Arlt, D. Hennings, G. DeWith, J. Appl. Phys. **58**(4), 1619-1625 (1985).
4. J.J. Xu, A.S. Shaikh, R.W. Vest, IEEE Trans. UFFC **36**(3), 307-312 (1989).

ACKNOWLEDGMENTS

The authors wish to thank Bruce Rothman for his performance of Auger analysis and Eric Rainer for his help with the electrical measurements. This work was supported in part by NSF-MRL under grant no. DMR-8216718.

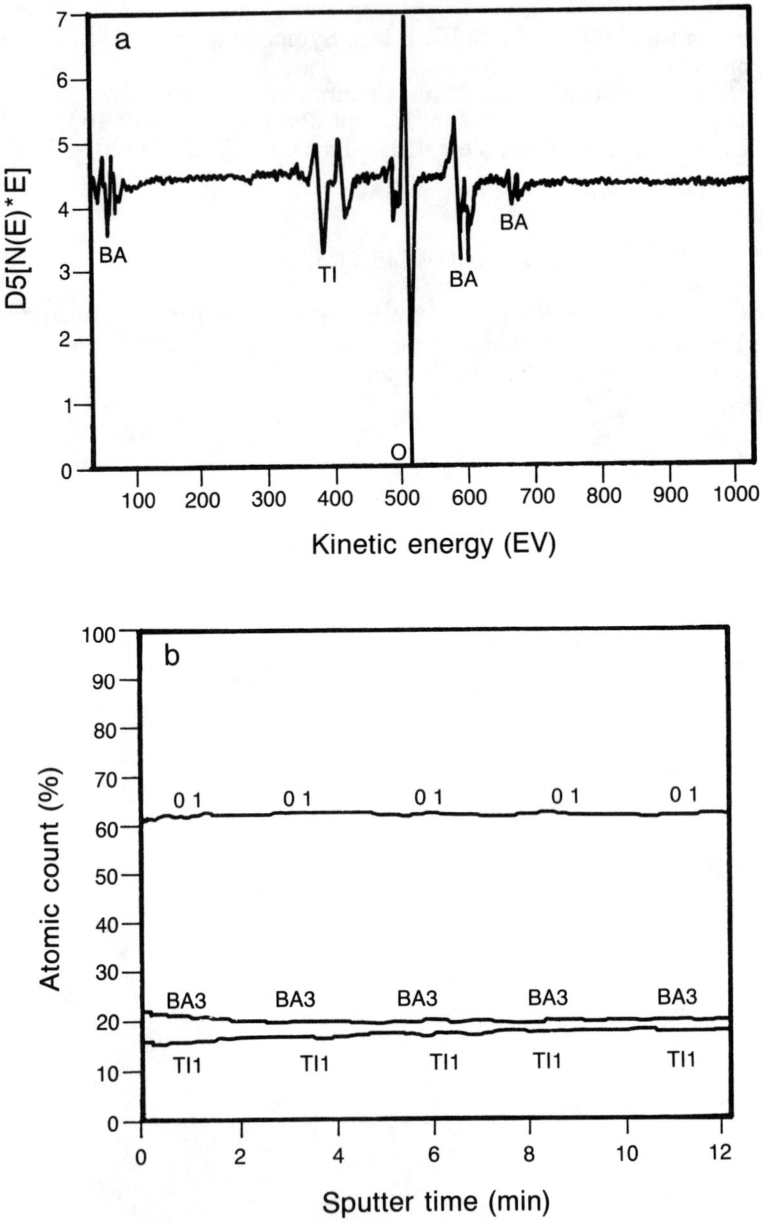

Figure 1a. AES spectum of a thin film deposited on platinum and annealed in a conventional oven to 600 °C for 4 hours.

1b. AES depth profile of the same sample.

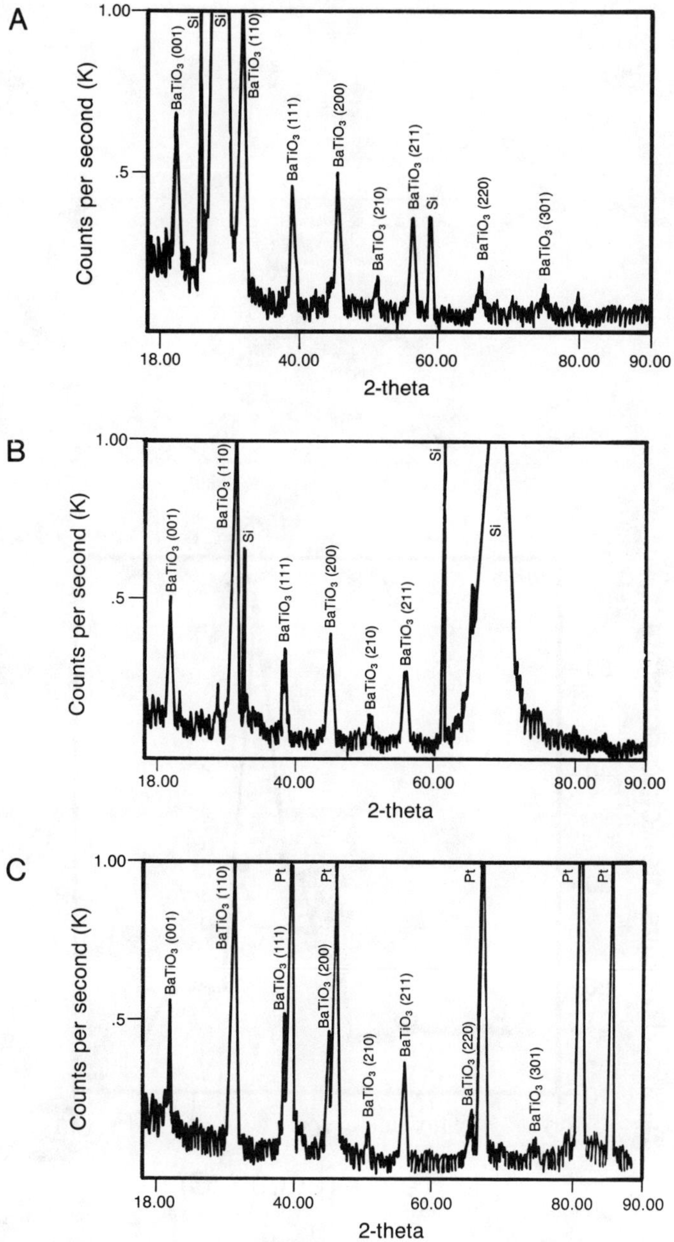

Figure 2. XRD for five coats of $BaTiO_3$ produced by MOD at 600°C for 4 hours on a. Si(111) b. Si(100) c. Pt

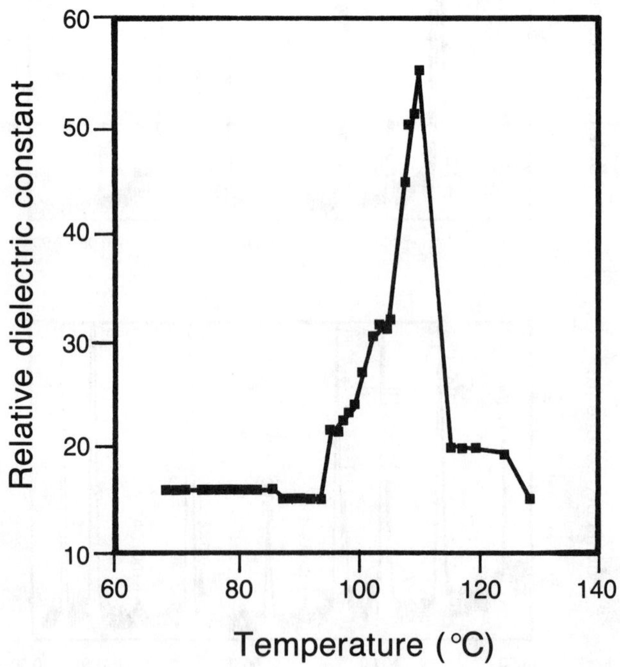

Figure 3. Relative dielectric constant as a function of temperature for a $BaTiO_3$ film produced at 600°C for 4 hours.

A

B

Figure 4. Micrographs of a sample produced on Si(111) by annealing at 600°C for a. 4 hours b. 16 hours.

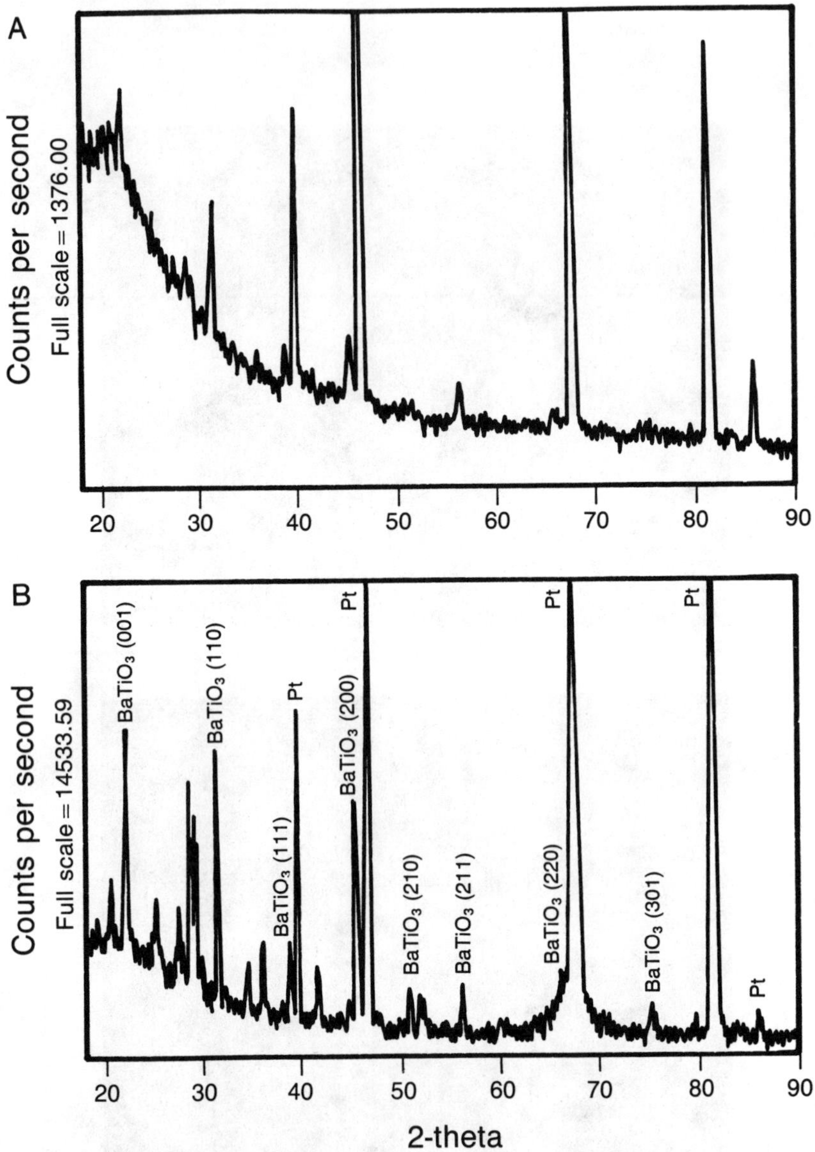

Figure 5a. XRD patterns of a sample consisting of 10 layers rapid thermal annealed in one atmosphere of O_2 to 650°C for 30 seconds each.

5b. Same sample as above but rapid thermal annealed in 10 mTorr of O_2 to 950°C for 15 seconds.

Figure 6. Micrograph showing the grains for the sample of figure 5b.

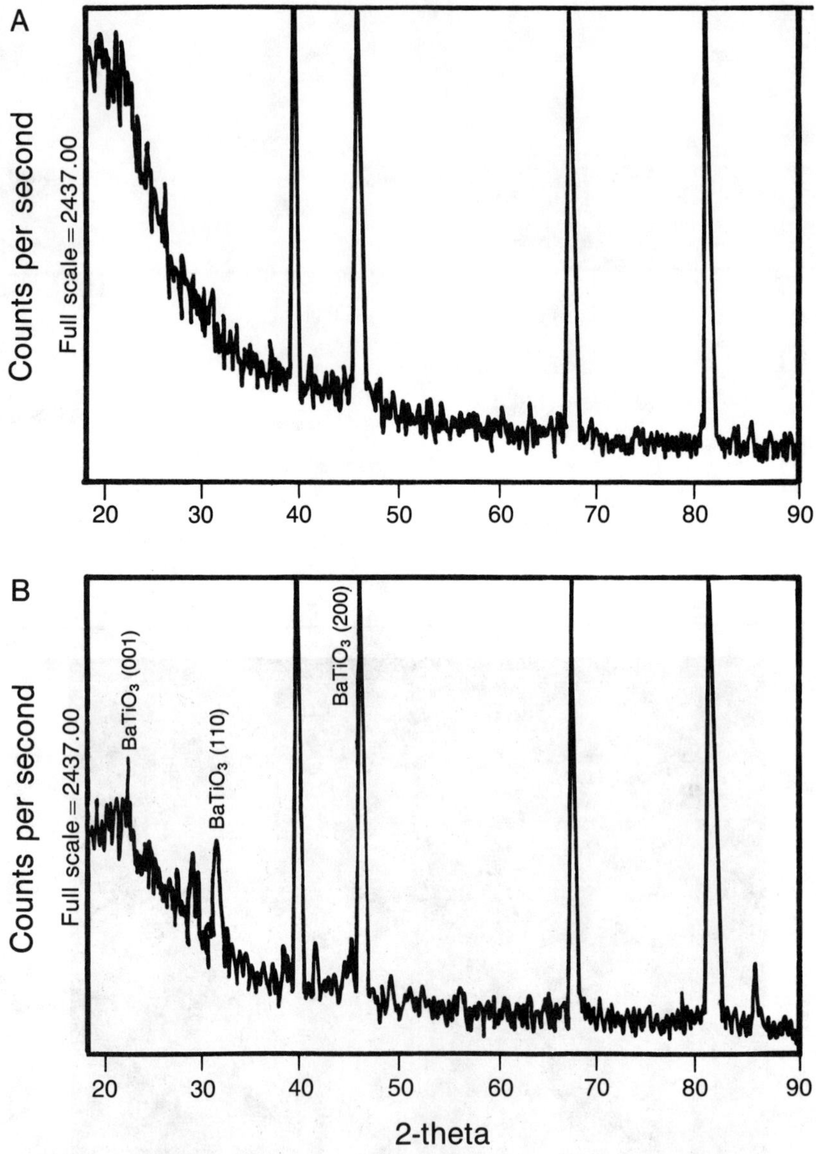

Figure 7. XRD patterns for 5 coats rapid thermal annealled to 650 ^oC for 30 seconds each
a. in a 10^{-3} Torr vacuum b. in a 10 mTorr of O_2 atmosphere.

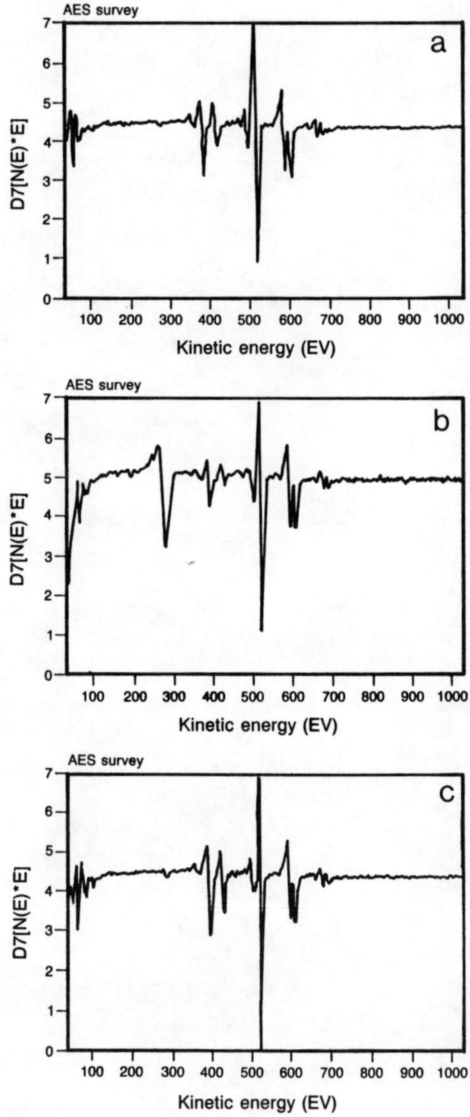

Figure 8. AES surveys for a layer deposited on Pt and rapid thermal annealed to 650°C for 30 seconds
a. in 1 atmosphere of air b. in 10^{-3} Torr c. in O_2.

MICROSTRUCTURE-PROPERTY-PROCESSING RELATIONSHIPS IN MICROELECTRONIC THIN FILMS

Dunbar P. Birnie, III and D. R. Uhlmann

Department of Materials Science and Engineering
Arizona Materials Laboratory
4715 E. Ft. Lowell Rd.
University of Arizona
Tucson, AZ 85712

ABSTRACT

Various processes that are used for producing thin films of ceramic materials are described. Major emphasis is placed on the details of the microstructures which develop and their relation to processing conditions. To help focus and limit the discussion, sol-gel chemically derived films are compared with films derived from physical vapor deposition (as from sputtering or evaporation). It is concluded that present understanding of thin film microstructures is relatively limited. It is suggested that developing the abundant opportunities in microelectronic thin films will require signficant further efforts at understanding this aspect of ceramic materials.

INTRODUCTION

Thin film applications for ceramic materials are increasing rapidly. These include materials for electronic, optical, and mechanical applications, such as capacitors, optical waveguides, and abrasion resistant coatings([1-5]). As with much product and device development, many thin film coating technologies are property driven; the desired property defines the composition of choice. Then the film processing effort proceeds, trying different deposition techiques and often varying several processing parameters to optimize final film properties. When several variables must be optimized, factorial analysis has been successful at improving final properties. Since the analysis of film microstructure is often a difficult problem, few attempts are usually made to correlate properties and processing with the resultant microstructures. Since, however, it is this "microstructure-property-processing" triangle that forms the basic paradigm for materials science, we shall in the present paper emphasize the microstructural aspects of this triad.

The above scenario of thin film development stems partly from the difficulty of obtaining quantitative information about microstructures in thin films. It should also be noted, however, that materials science is a relatively small field. Most development of thin films for specific applications is carried out by scientists and engineers with training in other fields who are less concerned with the understanding and control of microstructural features.

The present review is based on the belief that ceramists have a broad base of understanding about microstructures which should be tapped by industry when approaching new applications for ceramic thin film materials. This expertise comes from extensive experience with how processing conditions define microstructures and how microstructures affect final properties. It is here suggested that there are broad areas of technological and scientific challenge in understanding micostructures in thin films, and that with improved understanding of these features will come significant improvement in the properties of thin films and expanded ranges of their applications.

THIN FILM DEPOSITION

There is a great abundance of methods for making thin films. These include sputtering, evaporation, CVD, sol-gel, MBE, ion-assisted-deposition, and electrolytic deposition, to name a few. The present review will focus on two techniques: sol-gel and physical vapor deposition (including sputtering and evaporation and their variants). The comparison is therefore focused on techniques which typically yield polycrystalline films.

This section on film deposition is divided into three sections. The first two deal directly with the chosen deposition techniques. For each technique, attention will be directed to characteristics which affect the resultant microstructures. The third section is devoted to post deposition treatments that may be applied to either type of deposited film.

Sol-Gel Chemical Deposition: The sol-gel process is a solution-based chemical process for fabricating thin films. Solutions of metal alkoxides or metal salts are used as precursor solutions. The solutions initially have discrete metal ions or oligomers, each coordinated with organic groups or solvent ligands. These solutions are reacted - as by effecting hydrolysis and condensation reactions - to produce metal-oxygen-metal linkages. For polyvalent cations, an interconnected network begins to form. As the interconnectedness rises, the solution becomes more viscous and eventually gels (assuming that precipitation does not intervene).

For metal alkoxide solutions this bonding and linking process can be illustrated by two separate reactions. The first process is hydrolysis; an individual water molecule reacts with a metal alkoxide group yielding an alcohol molecule and a metal hydroxide:

Hydrolysis: M-OR + H_2O → M-OH + R-OH

Note that the metal is shown here with only one organic group. For many cations of interest, the coordination would be significantly higher, depending on their valence. The second important reaction is condensation; two metal hydroxyl groups (or an hydroxyl and an alkoxy group) react to leave a bridging metal-oxygen-metal bond:

Condensation: M-OH + M-OH → M-O-M + H_2O
or: M-OH + M-OR → M-O-M + R-OH

This bridging forms the basis for an interconnected, more rigid, structure. Higher coordination of organic groups around the metal atom in solution also ensures that the gel will be more interconnected and cross-linked, similar to glass structure itself. It can also be useful to consider this as a polymerization process. As condensation proceeds the metal-oxygen links become stranded (linear) or interconnected depending on the number of metal-oxygen-metal bonds that form.

In detail, the sol-gel process is quite complicated. Several variables must be carefully controlled ([6-9]). Important factors include: 1) choice of solvent; 2) choice of precursor molecules; 3) concentration of precursors and other reactants (including water); 4) chelating species and other processing aids; 5) temperature; 6) hydrolysis behavior (depends on both 1 and 2); 7) polymerization behavior (also depends on both 1 and 2); 8) post-gelation treatments; 9) removal of reacted species and solvent molecules (i.e., drying); 10) densification and/or crystallization behavior. Each of these variables can have a profound influence on the resultant microstructure and therefore on the final properties of the film.

The microstructures of sol-gel derived films can be quite homogeneous, both physically and chemically. They are often initially amorphous, but can give very fine grain polycrystalline structures after firing. Often the grain size in polycrystalline films is significantly smaller than the film thickness, ensuring a substantially isotropic (in-plane) and homogeneous film. However, the above described process variables can have a profound influence on the level of homogeneity, the structure, and finally, the properties of the resultant films. Three possible microstructural features which can be influenced by the sol-gel chemistry are chemical homgeneity, level of porosity, and film anisotropy.

Chemical homogeneity problems are most commonly encountered when attempting to make multicomponent ceramic films. The precursor molecules each have their own rates of hydrolysis and condensation. These rates depend on the central metal species as well as the coordinating chemical groups. When two or more cations are desired in the final film, then the chemical reaction behavior of all species must be well understood/controlled. If there is a significant disparity in reaction rates, then the initially homogeneous solution will spontaneously segregate chemically as the faster-reacting species hydrolyzes and polymerizes primarily with other molecules of the same species. This leaves the slower reacting species dissolved and substantially unreacted. Several approaches have been adopted to mitigate such problems, including partial pre-hydrolysis of the slower reacting species before mixing the two metal species together and then reacting the system as a whole ([10]). Another complicating factor in multicomponent systems is that a single solvent must be chosen which is compatible with all desired components. This becomes increasingly difficult as desired compositions become more complex.

The shape and size of porosity in the dried gel and fired film can be a result of local initial variations in density during the cross-linking and condensation process. These variations in density and porosity may remain even beyond the drying stage. Further influence on the porosity may come through the choice of precursor group and solvent (due to size or reactivity) as well as drying conditions.

Orientation effects and film anisotropy can result from a variety of considerations. The simplest is that during drying the film is constrained by a rigid substrate material. All shrinkage must therefore occur normal to the film surface; otherwise cracking and degradation occurs. Anisotropy associated with the film densification has indeed been observed([11]). Uniaxial shrinkage can cause preferential alignment of polymeric molecules into the plane of the film. It could also deform the shapes of pores along the direction of shrinkage; spherical shapes become ellipses or even almost disks, depending on the extent of shrinkage. Some of these shape changes may be mitigated by capillary effects driven by the pore surface tension.

Another potential source of orientation effects is possible anisotropy due to the film preparation technique. Sol-gel films may be applied in a variety of different techniques. They may be sprayed, dipped, spun-on, rolled, etc. Each of these techniques has its unique signature of fluid flow and shear phenomena acting during deposition. As an illustration of the alignment effects during film preparation, we will briefly examine the mechanics of dip coating a sol-gel film.

A substrate is dipped into a precursor solution and withdrawn at a certain velocity. A thin film of the liquid adheres to the surface as the substrate is removed. Several processes are occurring simultaneously. Solvent is evaporating. Atmospheric water is being incorporated when the film is exposed to the environment, causing hydrolysis and condensation reactions. This increases the cross-linking and polymerization of the film. Also, depending on the rate of substrate removal from the solution, there is unreacted fluid flow downward back into the container, especially in the miniscus region close to the liquid surface. Significant control of the structure of the gel film can be achieved simply by controlling the atmosphere around the dipping station; the vapor pressure of solvent influences the solvent evaporation rate and the water vapor pressure influences the hydrolysis and condensation reactions.

These dipping variables also influence bonding, which can lead to anisotropy in the resultant films. Evaporation of the solvent and introduction of the water vapor (or other gasseous species) into the film both occur at the film-vapor interface. Hence, the top-surface material has most accessibility to reactive species; one then expects differences in structure and properties which correlate with depth. Such variations with depth have been observed; but no clear assignment of their cause has been made ([12]). If the surrounding atmosphere is adjusted to keep the solvent and water exchanges slow, then the film structure is expected to be more uniform throughout the depth.

Another mechanism of alignment in the film can be related to the fluid flow aspect of film formation. This flow field defines a shear which acts in the dipping direction. This shear occurs simultaneously with some hydrolysis and evaporation, and should act to orient polymeric species in the coating solution. Alignment associated with the dipping direction has been observed for very thin (about 10 nm) sol-gel films ([13]), but the extent of this occurrence is unknown in thicker films (yet is expected to be smaller).

The above mechanisms are all operating during film deposition. They are primarily influenced by chemical reaction variables which are important in forming the interconnected structure. The as-coated and reacted film structure is an amorphous gel; it can also be regarded as a cross-linked inorganic polymer. After removal of solvent, the films have typically only about half their theoretical final density. They are then subjected to a heat treatment which causes densification. This densification is unidirectional (as is the drying process), and is typically carried out in an oxidizing atmosphere to help react and remove any organic species that remain in the film (without the oxygen, the organic species often decompose leaving carbonaceous residues in the films). The densification process may or may not be accompanied by crystallization. The densified crack-free films are limited to about one half micron in thickness: Thicker coatings do not densify in only one direction and cracking at the top surface results because of three dimensional shrinkage near the surface.

The sol-gel process has received considerable attention for making a wide variety of ceramic thin films. Sol-gel has several advantages over conventional processing routes, including lower temperature processing, higher purity starting materials, the ability to coat complex shapes (including insides of tubes), and fine grain, homogeneous microstructures.

Physical Deposition: This section will consider sputtering as a method for physically depositing thin films. Many of the concepts have, however, significant commonality with other related deposition techniques such as evaporation.

Sputtering is a vacuum deposition technique where an energetic gas phase impinges on the surface of the chosen target material, bumping atoms into the gas phase. These atoms travel through the gas phase and condense onto the chosen substrate material. Some critical parameters in the sputtering deposition process are ([14,15]): 1) target composition; 2) level of vacuum; 3) chamber gas phase composition; 4) sputtering power; 5) substrate choice; 6) substrate temperature; 7) substrate bias; 8) substrate orientation; and 9) target and substrate motions. When the target is some ionic material then the power source must usually be RF rather than DC.

As discussed above for sol-gel derived thin films, the process parameters can have a profound impact on the resultant microstructures and properties. Again, these process parameters influence chemical homogeneity, anisotropy, and porosity.

The starting composition of the target is seldom faithfully incorporated into the deposited film. This is partly due to different sputtering yields from the target - especially associated with elements having widely different atomic numbers or with differences between cations and anions. It is also partly due to different sticking coefficients onto the substrate or subsequent layers of deposited material. Variations in sputtering power, gas phase composition, and electrical bias applied to the substrate all influence the final film composition. In general, these are varied and optimized through some experimental grid of preparation conditions while measuring the quality and composition of the resultant films. Once the conditions have been optimized, films can be fabricated with relatively good reliably and reproducibility. However, the time invested in optimizing the deposition conditions for the desired composition may be substantial.

After deposition, physically deposited films often have a columnar microstructure([16-21]). This microstructure is strongly influenced by the angle of deposition and the temperature of the substrate. The basic principle is that whenever surface perturbations stick up above the plane of the surface, there is a "shadowing" effect on the growth of further layers of atoms. This shadowing effect causes columns to grow which are inclined with respect to the substrate, but which do not point directly at the incoming stream of atoms. The angle of column tilting, β, is related to the substrate tilting angle, α, by the well known tangent rule ([22-25]):

$$2 \tan \beta = \tan \alpha$$

For small angles, the columns approximately bisect the angle of deposition and the surface normal.

This columnar microstructure is also influenced by the ability of atoms to rearrange once they have come into contact with the surface. For most oxide materials, which usually have high melting points, the surface mobility is rather low and the columnar structure from the shadowing effect is rigidly maintained in the as-deposited film structure. For this reason, deposition is often carried out onto heated substrates. This heating of the substrate permits more rearrangement of the atoms before the next layer of atoms can build up.

As-deposited, sputtered films are also far from fully dense. The sputtering process commonly produces films which have perhaps only 70% of their theoretical density. Hence, there is a significant volume of porosity incorporated into the structure. As might be expected, when the solid material is arranged in somewhat dense columns of material, there are regions between the columns characterized by a strong deficit of material - i.e., regions of oriented porosity. As the tilting angle of the columns is increased the shadowing by the columns becomes more pronounced and the volume of porosity associated with the intercolumnar voids becomes greater. It should be emphasized that the material in the columns is not necessarily crystalline. With ceramic materials deposited well below their melting temperature, there is a strong tendancy to deposit amorphous material.

Another feature of the columnarity associated with sputtering deposition is an increase in the size of the columns with increasing film thickness[26-28]. This is true for both obliquely and vertically deposited films. It may be associated with preferred crystal growth directions if the columns are crystalline, and also with some extent of adatom mobility after deposition, although full column diameter motion is not required for column expansion. This evolution of column shape, even for vertical deposition, has recently been predicted with mathematical modeling [29]. Even when columns are strictly columnar and directed vertically with respect to the film surface, there is still inherent anisotropy in the film; properties normal to the surface will differ from properties in the plane of the film.

Post-Deposition Treatment: Both sol-gel and physical deposition techiques can produce films of high quality. However, there are often further treatments applied to the as-deposited films. These post-deposition treatments fall into three categories: further heat treatments, mechanical treatments, and chemical treatments.

The subsequent heat treatments can come in a variety of different forms; they can be applied slowly using furnace treatments or they can be done more or less rapidly using thermal imaging or even direct application of laser light for selective heat treatment. These heat treatments are often aimed at either further densifying the structure or at causing an amorphous film to crystallize. The densification processes have been modeled in the case of an amorphous film [30-35]; similar treatments for crystalline starting material can be imagined to be related to the conventional bulk analysis [36-39].

The heat treatments may or may not include actually melting the material in the films. Without melting, the densification occurs through normal solid-state or viscous sintering processes. Here however, the densification may proceed much more rapidly than with conventional bulk materials. This is due to the significantly smaller grain size/pore size often obtained in thin films (either amorphous or crystalline); this yields a large capillary driving force that causes densification at lower temperature than normally possible.

So far the discussion has centered on formation of polycrystalline or amorphous thin film oxide materials. However, there is increasing interest in obtaining single crystal films of various microelectronic materials. Post-deposition treatments can be aimed at this goal, as when melting does occur during the post-deposition heat treatments. Much effort has been directed at producing single crystal semiconductor materials using this technique [40-42], however, the unidirectional crystallization of oxide films is still relatively unexplored as yet.

Mechanical post-deposition treatments are more rarely employed. One technique which is being investigated for optical waveguide applications involves physical embossing of sol-gel films [43-45]. The impressed template deforms the gel-structure film into a particular shape (e.g., a grating or lens). This is later densified to retain the physically impressed features.

Chemical modification treatments are also possible. These include wet chemical and gaseous additions. These treatments can be used to incorporate more material into the pores; and, the incorporated material can even be of a different composition. This is more frequently attempted with sol-gel films than with sputtered films. One example of an atmospheric treatment is nitridation in a nitrogen atmosphere([46]); which is performed during heat treatment. Unless the as-deposited film is metallic to start with, it is usually difficult to effect complete transformation to a nitride phase; some oxygen is usually still present and an oxy-nitride phase results. Further, densification of the nitrided material is more difficult than with the corresponding oxide. Also, annealing treatments have been performed using nearby oxide sources with incorporation also occuring through the vapor phase([47]). Finally, modification using UV light has been observed([48]). It is thought that this facilitates bond rearrangement and helps densify the structure.

In summary, the sol-gel and the physical deposition treatments are quite different in both processing and in their resultant microstructures. However, many common features of film microstructures arise from both techniques. Films are often either amorphous or very fine grained polycrystalline (whether columnar or not). Films often have significantly lower density than their bulk parent materials, and often have some anisotropy associated with their deposition-related structural artifacts. The following section will discuss these micostructural features in more detail.

MICROSTRUCTURES OF FILMS

The previous section has briefly outlined two different techniques for preparing thin films of ceramic materials for microelectronic applications. The present section will go into further depth about specific features of these microstructures and their influence on film properties. Topics which will be addressed specifically are: 1) grain size; 2) porosity; 3) chemistry; 4) phase assembly; 5) orientation; 6) substrate-film interactions; and 7) multilayer films. As discussed above, many of these features have wide commonality between the two varied deposition techniques. Also, as discussed below, many of these features are inextricably interrelated (i.e. grain size and porosity).

Grain Size: As prepared films have a characteristic feature size that can be labeled "grain size". This is true even if the as-deposited film is amorphous. For sol-gel films, this grain size may be associated with the scale of local density fluctuations arising from the hydrolysis and condensation reactions occurring during film preparation. These density differences can be retained and are perhaps even reinforced during densification and removal of organics from the film during post-deposition processing. For physically deposited films, this grain size is more likely associated with the diameter of the columnar artifacts of deposition. The grain size is typically larger as the thickness of the vapor deposited film increases. If surface mobility of species is possible, the material may in fact be crystalline. In this case, the grain size has real crystallographic meaning and can be determined using standard X-ray diffraction techniques ([49]). Often these crystalline films also have columnarity; such films generally have tall, narrow columns, yielding a smaller grain size in the plane of the film than normal to the film. Only at high substrate temperatures will atom rearrangement be large enough to create coarse polycrystalline films. Significant atom rearrangement can also be facilitated by using an external ion bombardment. Impinging atoms of higher energy are known to help rearrangement by imparting kinetic energy to other atoms that are present. (This is the basis for ion-assisted-deposition, IAD).

The grain size is very important to further processing aimed at densifying films. Whether for viscous sintering of an amorphous material or for solid-state sintering of crystalline material, it is the solid-vapor interfacial curvature that drives densification. The grain size usually determines this curvature through simple geometric relationships; this is evidenced by the success of various sintering models which base their kinetic equations on the grain size rather than the pore size.

The kinetics of atom motion are usually dramatically higher in amorphous materials than in crystalline materials. Therefore whenever fully dense thin films are desired it is advantageous to densify before crystallization. The densification and the crystallization processes can be viewed as competing processes: if densification is faster, then a dense amorphous film can be achieved and perhaps later crystallized into a dense polycrystalline material; but if crystallization is faster, then densification can become too difficult and the final film may be polycrystalline but porous. It is necessary to carefully control deposition conditions such that processing routes exist for achieving high quality dense films. In any case, the competition between densification and crystallization can be described numerically; and windows of processing opportunity can be identified ([50]).

For physically deposited films where the columnar structure coarsens with thickness, the grain size will also be a function of the depth within the film. This gradient in grain size can lead to differential densification because the deeper layers have a larger driving force. However, it is more likely that the lower layers simply coarsen more rapidly until their size is in line with the upper layers.

In detail, the structure of the grains and pores will be governed by the capillary forces of the grain boundaries (if any) and interfaces, as is found in bulk polycrystalline materials ([51-55]). This has a strong influence on densification. If the dihedral angle of the pores is large then there will be a driving force for densification. However, if the dihedral angle is small then pores will grow because the solid-vapor surface curvature is in the wrong direction. This is another reason to promote densification before crystallization; it is anticipated that contacts between amorphous columns will be full amorphous bridges defining free-surfaces with net negative curvatures. Grain boundaries are only defined between crystalline grains where there is a discontinuity in the structure that has a particular excess energy associated with it (the grain boundary energy).

Porosity: Porosity is a ubiquitous feature of films deposited by wide varieties of techniques. The porosity can be discrete or interconnected. For sol-gel derived films, both shapes of porosity are known to exist. For physically deposited films the pores are more likely to be interconnected; the pore will be very strongly associated with the intercolumnar regions in the structure. In extreme cases these intercolumnar porous regions will be "pinholes".

The volume fraction of pores in thin films is usually quite considerable. This fact articulates the importance for post-deposition treatments to either fill pores or densify the film structure. On the other hand, these porous structures may have their own niche of application and can be used to advantage.

The size and shape of pores is very important for the film properties, besides being important for densification of the film structure. The vertical pores between columns are of some interest. Classically, tubular shapes are unstable to Rayleigh instabilities; pore channels will decay into a linear array of spherical pores. But, when the dihedral angle is small the pore channel may be completely stable to pinching off ([56,57]). However, if the tubular pore is relatively short (i.e., having a diameter near the thickness of the film), then surface tensions at the substrate interface or top surface may interfere. If the capillary actions of these top and bottom surfaces are not favorable, then the vertical pore could be stable or even grow([58-60]). This could provide a source of pinhole formation in thin films.

The capillary forces associated with pore surfaces may be controllable to some extent. This can be affected by changing the gas phase atmosphere or by other chemical post-treatments (surface tensions are reduced by adsorption of chemical species). Although this could reduce the driving force for densification by lowering the surface tension, it would directly raise the dihedral angle, possibly even changing the direction of curvature and "activating" densification. To our knowledge, none of these capillary effects have been tested in thin film microstructures.

It should be emphasized that the small scale of microstructure in thin films is no reason to abandon efforts to understand it better. In contrast, our rather advanced understanding of bulk microstructures could give broad insight into thin film microstructure and its development during deposition and processing. The validity of capillary forces on a small scale is emphasized by the relatively good representation of crystal nucleation phenomena ([61]): often the sizes of critical nuclei are in the range of tens of Angstroms or less.

Chemistry: An analysis of the chemistry of thin films must start with the overall composition. Ceramic materials usually consist of more than one component: at least one cation and (usually) oxygen; and many compositions of great technological interest contain several cations (e.g., $BaTiO_3$, $LiNbO_3$, $YBa_2Cu_3O_x$, etc). The careful attainment of specific compositions will be critical to achieving the desired properties in the final film product. This compositional specification includes 1) cation ratio, 2) cation-to-anion ratio (i.e. oxidation state), and 3) level of retained organic species or deposition gases. Further, to understand fully the chemistry of the system, it is important to know the level of homogeneity of these three factors.

The bulk composition is quite strongly tied to the deposition variables described in the previous section. Multicomponent compositions are particularly difficult to obtain exactly, using any thin film deposition technique. Often the processing variables must be repeatedly adjusted before the optimum conditions are reached for deposition of the desired chemistry.

One important aspect of the chemistry of thin film microstructures is their chemical interaction with vapor phases. This interaction is very important at all stages of film processing: deposition, post-treatment, and final use. The interaction of the vapor phase during deposition was emphasized in the discussion of the sol-gel technique above. Here, evaporation into the vapor phase controls drying, initial shrinkage, and overall film formation. Also, the interaction with airborne water molecules can be critical for hydrolysis and condensation reactions which help form the gel structure before firing. Later, as discussed in the section on porosity, atmospheric interactions can cause changes in surface tensions and thereby influence densification. Finally, interactions with atmospheric species may occur even after the film is fully processed. One of the most strongly interacting species is water[62-68]: some thin film devices can even be used as moisture sensors [69].

The vapor phase is also important for films derived from physical deposition processes. The chemical makeup of the sputtering gas is important. Some concentration of the "inert" sputtering gas is incorporated into the thin film. Also, reactive species (such as oxygen or nitrogen) can be included in the sputtering gas; this can help maintain the stoichiometry of the resultant film or incorporate important components (e.g., nitrogen). Other normal atmospheric species such as moisture and carbon dioxide can be present in the sputtering chamber. Such species can influence the deposition quality and properties [70]. This interaction is not well understood. Later, the sputtered films may also have similar interactions with atmospheres employed during post-deposition treatments and final use.

The basis for these atmospheric interactions is the large specific surface areas which occur in the films. It should be noted that one of the most reactive gas species is water vapor. The majority of oxide phases are unstable at room temperature with respect to their hydroxide phases [71]. Hence there is a thermodynamic driving force causing chemical interaction with moisture. Hydroxylated surfaces of oxide ceramics are the rule rather than the exception. When the grain and feature size is like that found in thin films, then the kinetics of this interaction may be quite rapid. Further analysis of the interaction between thin film materials with gas phase species will be crucial to our full understanding of microstructure-property-processing relationships; in this area, water vapor will be of special interest.

Phase Assembly: The phase present depends directly on the overall composition of the film and to some extent on the amount of post deposition processing that has occurred. Many films will be amorphous as-deposited. These can be crystallized by subsequent heat treatments. The tendancy to form amorphous films is probably inherent to many ceramic materials; they are often being deposited at temperatures far below their melting points, which suggests that atom mobilities will be low. Since the crystallization processes can be taking place at temperatures significantly below normal bulk reaction conditions, it is also possible to obtain crystalline phases that are metastable. The sol-gel and vapor phase processing routes can give very high levels of chemical uniformity. This may dictate the formation of crystalline phases (including solid solutions) that normally would not form with conventional processing.

Orientation: As discussed above, anisotropy can result from several aspects of the film deposition processes. These orientational features can be artifacts of the deposition process, as was found for columnar growth during physical deposition. They may also result from uni-directional shrinkage and densification during processing. There are many ways in which the substrate can influence this predisposition to orientation effects.

The sputtering variables play a direct role in the formation of anisotropic films. The deposition rate, angle, and gas phase atmosphere all can influence the structure. The deposition angle can cause the "shadowing" effect, yielding columnar structures, depending on the temperature of the substrate during deposition. If the deposition rate is higher, the columnarity will be reinforced: there will be less time for atom rearrangement before the next atomic layer has arrived at the surface.

If the columnar structure is comprised of crystalline material, then orientational anisotropy can arise through preferred growth directions. As an analogy, growth anisotropy has been used to advantage for lateral crystallization. A heater is passed over a thin film, melting a narrow zone. Existing crystalline material on the trailing edge then defines the final grain size. Faster crystallizing directions end up dominating the final structure ([40,41]). A similar phenomenon may be occuring in the coarsening of columnar structure during film deposition: many tiny nuclei exist in the first few layers of material that get deposited. Subsequent layers can then become attached preferentially on the sites with faster growth orientations. This should occur more when some reasonable atom mobility is possible. It is also possible even on amorphous substrates: the first layer of nuclei can be randomly oriented; later additions can provide some crystallographic texture to the film.

One interesting reinforcement of the anisotropy in physically deposited films has been found in multilayer films containing WO_3 ([72]). The layer sequence was: amorphous substrate, WO_3 deposited at non-normal incidence, SiO_2 deposited vertically, and Al on top. This structure was found to age through interaction with airborne water vapor. The water penetrated the top two layers through pinholes. Then the water molecules reacted with the WO_3 to form blue regions (WO_3 is known to color by proton insertion - in bulk samples, a H_xWO_3 phase arises). The blue regions were oval in shape with their major axes oriented perpendicular to the direction of tilting of columns; the water found better interconnected pathways through the structure in this direction. Similar anisotropic behavior has been seen in other optical interference filter film stacks ([73-75]). This illustrates both the anisotropy of the obliquely deposited layers and the fact that even thick vertically deposited layers can be relatively permeable to water vapor.

For sol-gel derived films, the unique sol-gel process variables also exert their influence. The bonding in the fluid during film deposition can impact how the molecules are incorporated into the structure. This is also correlated with the interaction of the solution with the vapor phase during preparation; evaporation of solvent and interaction of the film with atmospheric moisture can influence the level and homogeneity of drying and gel formation. These features are integrated with the dipping withdrawl rate as discussed above.

The substrate that is used for deposition can have some interesting influences which yield orientational anisotropy. These influences can be both structural and chemical. Structural orientation can easily occur on crystalline substrates. The substrate structure can provide a template for growth of film material. This epitaxial effect has frequently been used to advantage by depositing the new high T_c ceramic superconductors onto $SrTiO_3$ single crystal substrates ([76]). Some apparent epitaxial effects can also be found when the substrate is amorphous. One case was described above, where differential growth rates can give some orientations dominance as the thickness increases. It is also possible that even the first layer of material has some orientational preference: the crystal nuclei may form shapes which bias their orientation. This would be dictated by the surface tension of the deposited phase. Further substrate effects are described in the next section.

Substrate-Film Interactions: Interactions between the substrate and the film fall into three general areas: structural, chemical, and mechanical. Some of the structural effects were discussed above in the section on orientational effects. These structural effects resulted primarily from epitaxy of the deposited material. Apparent epitaxy was also described where higher energy nucleus facets could be eliminated by sticking directly to the substrate (either amorphous or crystalline).

Epitaxial effects are most common when there is some strong crystallographic relationship between the substrate and the film phase. Usually this means that two different crystallographic directions in each phase have similar interplanar spacings. These two directions should also have similar alignment with each other in the two phases. Sometimes this epitaxy can yield phases which would not normally form (metastable phases). This has been observed in alumina deposits on polycrystalline ferrite ceramic compositions: gamma alumina (defect spinel structure) phase has been observed epitaxially on some orientations of the ferrite ([70]). Other work has shown the occurance of various of the other metastable crystalline alumina phases ([77,78]).

Further structural effects may be found at more gross structures of the substrate. These could be scratches which exist from polishing the substrate before use, or dislocations that meet the surface which could provide special nucleation sites. It is frequently observed that metal films deposited on various cleaved salt crystals initially nucleate at cleavage steps or other structural flaws ([79]).

Chemical interactions between the substrate and the film abound. One major area of chemical interaction is capillary effects such as wetting and adhesion. Wetting is important because it can influence the shape and texture of deposited films. It is commonly observed in metals that the first few "monolayers" of material that arrive at the surface are not uniformly arranged on the surface. The metal atoms can have relatively good surface mobility; they are found clustered and in island-like shapes until enough material has been deposited to fill in the intervening spaces. The shape of the islands is determined by the action of a capillary force balance at the substrate-solid-vapor junction. Materials which are poorly wetting give hemispherical or spherical sections (like mercury). Materials which wet well form a continuous film immediately and never develop the islands. These initially-deposited island features may provide the starting surface perturbations that build into the final columnar microstructure.

For oxide materials on oxides, the capillary forces are more likely to favor complete wetting by the deposited material. Again, oxides will have lower atom mobility, which will inhibit the rearrangement of the deposited material. For oxide materials being deposited on metals, we would expect some of the same features found before for metals deposited on ionic materials.

A factor which is closely related to wetting is adhesion. Good adhesion is characterized by strong bonds between the substrate and the deposited material. These can be qualitatively correlated with a low interfacial tension between the material and the substrate. Delamination of the film requires the propagation of a crack along the substrate interface. If this interface has a high interfacial tension, then crack propagation will be easy since little energy will need to be added to advance the crack. However, if the interfacial tension is very low, then the crack advance must involve significant energy addition to create the two new free surfaces.

Sol-gel films are often well suited to forming good adherent films. Oxide substrates characteristically interact with atmospheric moisture, leaving dangling hydroxyl groups. These groups can participate in the condensation reactions described earlier, yielding primary bonding between coating and substrate. This helps promote strongly adherent coatings by sol-gel methods.

These capillary and adhesion characteristics can be strongly affected by the chemistry of the system. Surface segregation of impurities can change surface structure as well as interfacial tensions. Vapor species can also adsorb, changing the surface tensions, and, pretreatments of the substrate material may have a lasting effect on the properties of the deposited film. Chemical surface cleaning can leave molecular residues, change surface bonding, or remove chemical groups from the surface.

The substrate-film interface can also have more extensive chemical interaction in the form of interdiffusion and reaction to form interphases. This most commonly occurs during heat treatments of the deposited films. These reaction layers can have a beneficial effect on adhesion and electrical contact, as is found in metal silicide formation on silicon. However, the interdiffusion is often not uniform or does not give a useful reaction product. For polycrystalline films, the diffusion is expected to be enhanced along the grain boundaries. Detrimental grain boundary diffusion has been seen most commonly in the new ceramic superconductors, where the impurities present at the grain boundaries prevent superconduction ([80]).

The substrate and film also interact on a mechanical level. This is especially true when the deposited films are heat treated. Heating usually causes densification, which puts the film under tension as it shrinks. The crystallization process usually causes a net shrinkage, and would also cause tensile stresses. If shrinkage were to occur is some unconstrained film, then no stresses would result. However, the films are constrained by the substrate material. This requires that the densification occur unidirectionally; the material must become thinner and denser. The tensile stresses which arise from densification must never become too large or cracking at the film surface will occur. The film densifies through a shear deformation that relieves the stresses and causes shrinkage in only one direction ([30-35]). Some level of the densification-induced tensile stresses may be retained after heat treatment.

Not only is the film changing dimension because of structural relaxation, it is also changing dimension from simple thermal expansion effects. Since the substrate material is commonly of a different structure and composition, it is possible to develop stresses associated with the mismatch in the expansion coefficients between the film and substrate. These thermal stresses build up both on heating and on cooling. However, since the heat treatment is used to effect densification or crystallization, some atom rearrangement is possible at the high temperature. Hence, the stresses will, at least to some extent, be relieved by creep or other relaxation processes. At high temperature, after densification, the expansion mismatch effects are assumed to be nulled. However, on cooling back to room temperature, the mismatch will force the film-substrate pair to be left in a stressed state. If the film has a larger expansion coefficient than the substrate, then after cooling it will be in a state of tension. If the film has a smaller expansion coefficient, then it will be in a state of compression after cooling.

Sol-gel films may have an advantage in ability to withstand thermal expansion mismatches. The reactivity of the as-deposited structure and its potential ability to densify while still amorphous presents the attractive possibility of lower temperature processing conditions. The resultant stresses are lowered in proportion to the lowering of the processing temperature. They may also have an advantage whenever they generate fine grain polycrystalline structures. The columnar structures that result from physical deposition processes provide natural fracture initiation sites and fracture paths at the surface of the film. The homogeneity of the sol-gel film could make it more robust to fracture initiation.

Films can also have stresses which are directly associated with the deposition process. This is most common in physically deposited films. The source of these stresses is not well understood. The stresses can be either tensile or compressive and occur in films which have been deposited on room temperature substrates (the latter observation removes thermal expansion mismatch effects from consideration). The stresses have been attributed to condensation of point defects generated during deposition, where the defects may be included in supersaturation quantities due to the non-equilibrium nature of thin film deposition[81-83]. There is, however, no quantitative understanding at present. Other processes must surely also be active.

Multilayer Films: Many applications require multilayer stacks to be formed. The multilayers are used, e.g., for optical interference effects [68], optical data storage [84], and chemical protection and passivation [85], among other things. With multilayer materials the number and type of different interfaces increases rapidly. Each interface between two layers can have features like those described above. Because of the relatively poor understanding of simple single-layer films, it becomes increasingly difficult to understand multilayers.

Several difficulties are present in attempting to understand multilayers. One can approach the multilayer structure by simultaneously depositing single layers and characterizing them separately. It is recognized, however, that the structure and properties of the film can depend strongly on the nature of the substrate. Hence, dissection of the multilayer into its components may not give an accurate reflection of the structure of individual components when in the multilayer.

Much more careful characterization of these aspects of thin film microstructure is crucial to efforts at optimizing their behavior.

PROPERTIES OF FILMS

The previous section has focused on various detailed features of the microstructures of films and their dependence on processing history. Such microstructural features can have a profound effect on the properties of thin film materials.

As noted above, there will be some level of anisotropy in thin films deposited by different techniques. Sol-gel films acquire this anisotropy primarily from unidirectional shrinkage during drying and densification. Other orienting type physical processes can also leave artifacts in the structure. For sol-gel derived films, however, these cases should be more the exception than the rule. Physically deposited films will also have some depth-related non-uniformity and anisotropy in cases where there is a columnar microstructure. For vertically deposited material, the columns increase in diameter as the thickness of the film increases. The effect of these broadening columns on film properties is not known, although their occurrence implies that primary anisotropy will exist in many physically deposited films. The columnar features may be altered by depositing the films at an oblique angle, causing the columns to tilt away from vertical. This adds further anisotropy to the film properties.

All physical properties will be influenced by the structure of the film. Properties which are important for thin film applications include strength, electrical conductivity, dielectric constant, refractive index, magnetism, etc. The influence of the microstructure on these properties can be modeled (approximately) using effective medium theory ([86-89]). This can take into account mixtures of phases with various directional anisotropy. To date, however, detailed exploration of the relation between microstructure and properties of thin ceramic films remains in its infancy.

MICROSTRUCTURAL CHARACTERIZATION OF FILMS

From the earlier discussion of microstructural features of thin films, it is clear that many aspects of thin film structure must be carefully characterized to obtain a full understanding of the microstructure-property-processing relationships in any system. One needs quantitative information about micro-porosity, grain size, composition, bonding state, phase assembly, preferred orientations, and interface structure. None of these aspects of thin films is particularly easy to quantify: the thin film has only very small amounts of material, making all observations difficult. Various techniques for accessing thin film microstructures and properties include: transmission electron microscopy (TEM), scanning electron microscopy (SEM), scanning transmission electron microscopy (STEM), Auger electron spectroscopy (AES), X-ray photo-electron spectroscopy (XPS), secondary-ion mass spectroscopy (SIMS), X-ray diffraction (XRD), electron diffraction (ED), small angle X-ray scattering (SAXS), electron energy loss spectroscopy (EELS), Rutherford backscattering spectroscopy (RBS), and physical property measurements. These techniques are all relatively familiar and will not be covered directly here.

In addition, there exists a set of novel characterization tools which have been used much less extensively. Some of these techniques are newly developed; others simply represent applications of known technology in a new way. Some of these novel techniques are: spectroscopic ellipsometry ([90-95]), quartz crystal microbalances ([96]), surface acoustic wave (SAW) devices ([97,98]), nano-indentation ([99-101]), pump-probe laser spectroscopy ([102]), and surface second harmonic generation (SHG) ([103]). These techniques should be developed further and more frequently utilized to develop improved understanding of microstructure-property-processing relationships in thin films. Three of the above techniques (spectroscopic ellipsometry, nano-indentation, and SAW devices) are discussed briefly below, including particular areas of microstructure where their application is useful.

Spectroscopic ellipsometry is a more versatile variation of ellipsometry. Like normal ellipsometry, it reflects light off the thin film (perhaps multilayer) structure and senses variations in the state of polarization of the reflected beam. Its use of a broad spectrum of light is beneficial for several reasons. It can be used to monitor spectrum-specific processes occurring in the film, such as impurity absorption and band-gap features. Further, by looking at a wide portion of the optical spectrum, the thin film makeup can be more precisely determined (especially for multilayer stacks). This is partially due to the film being a specific physical thickness for all wavelengths.

Nano-indentation represents a more sophistocated version of simple indentation measurements. A diamond indenter is pushed into the surface of a film using an electromagnetic coil; the current defines the force being applied by the diamond on the film. Simultaneously the displacement is measured with a capacitance device. It is similar to the new scanning tip microscope systems. A displacement resolution of 2 to 3 Angstroms can be obtained. From the force-displacement curve, the loading and unloading behaviors of the film are determined; these are used to determine hardness ([101,102]). The strength of the film will be a function of the fraction of porosity and the strength of the solid microstructure present. From the loading-displacement curves, the property variation with depth can be determined. The technique has been used for understanding the densification behavior of sol-gel films where initially the gel structure is very soft and heat treatments cause densification, shrinkage, and crystallization ([12]).

Finally, surface acoustic wave (SAW) devices can be used as micro-porosimeters ([97,98]). This technique takes advantage of the sensitivity of the SAW device to differences in surface elastic properties. A thin film is deposited on the device by any given method. Then a controlled partial pressure of nitrogen gas is admitted to the sample. The surface wave velocity is sensitive to the mass of material on the surface and its elastic constants. As the gas adsorbs on the surface of pores in the material, the mass increases and the SAW resonance changes. Using the standard BET analysis ([104]), the pore size and size distribution can be determined. Using the SAW device as a micro-porosimeter is 3-4 orders of magnitude more sensitive than the normal BET technique. Since the porosity of thin films and its availability to reactive atmospheric species is so important to thin film structures, this technique will undoubtedly see wider application for both sol-gel and physically deposited films.

In addition to the specific novel techniques which can provide new insight into thin film microstructure and properties, the standard techniques mentioned earlier can also be used more effectively. This is especially true when the characterization techniques can be performed in-situ during the deposition process, so that the gradual addition of material to the substrate can be monitored and variations with thickness can be determined. Ellipsometry, even at one wavelength, is especially adaptable to deposition systems because it is a non-contact measurement([105,106]). Film deposition can also be interrupted at various stages and the samples can be examined ex-situ.

Caution must be used, however, when interrupting deposition or dissecting multilayers. Because of the sensitivity of thin film structure to the substrate, data taken from separated multilayers must be interpreted carefully to avoid artifacts derived from the substrate. This is also important when films are deposited in parallel onto specific substrates for TEM, RBS, or other observation techniques.

As indicated above, there exists a myriad of experimental techniques for characterizing thin film properties and microstructure, each with its particular strengths and weaknesses. It is suggested that a full understanding of film microstructure will only come through characterization of films using a wide variety of techniques on the same film stack. This broader characterization of carefully prepared materials might be achieved through wider inter-laboratory collaboration on research, as many of these techniques are expensive to implement.

CONCLUSIONS

The above discussion of the microstructure-property-processing relationships in thin films for microelectronics has emphasized two sayings which define the classic materials science paradigm. First, processing variables have a strong influence on microstructure, and second, microstructure influences properties. Microstructure is the common denominator. It is suggested that for thin films, understanding of microstructural features lags significantly behind understanding of either thin film processing or properties. Better microstructural characterization will lead to a better understanding of processing techniques and ways of achieving improved properties in thin films for microelectronics.

Typical thin film microstructures are known to have significant volumes of microporosity. They often exhibit columnarity or anisotropy. Any in-plane anisotropy of the films can also be related to processing details. Films are also prone to interactions with vapor species in surrounding environments, especially moisture.

It is suggested strongly that much more effort be devoted to the careful microstructural characterization of thin films. This will come from development of more sensitive structural probes with good spatial resolution. The potential also seems great for expanded collaboration between groups who specialize in various sensitive structural probes.

A stronger understanding of the microstuctures in thin films will lead to greatly improved understanding of the microstructure-property-processing triangle which represents the fundamental paradigm of Materials Science and Engineering. The suggested enhancement in our understanding will enable us to capitalize more effectively on the broad range of opportunities in the area of microelectronic thin films.

ACKNOWLEDGMENTS

The authors wish to express their appreciation to the Air Force Office of Scientific Research for their financial support.

REFERENCES

1) D. R. Uhlmann, in Ceramics: Today and Tomorrow, 1986, The Ceramic Society of Japan.
2) E. Ritter, Appl. Optics 15, (1976) 2318-2327.
3) D. R. Uhlmann and G. P. Rajendran, in Ultrastructure Processing of

Advanced Ceramics p241-253, J. D. Mackenzie and D. R. Ulrich Eds., John Wiley, 1988.

4) J. D. Mackenzie, J. Non-Cryst. Solids 100, (1988) 162-168.
5) K. H. Guenther, Appl. Opt. 23 (1984) 3612-3632, and references therein.
6) D. R. Uhlmann, B. J. J. Zelinski, and G. E. Wnek, in Better Ceramics Through Chemistry, MRS Proc. 32, Ed. C. J. Brinker, et.al., Elsevier, 1984.
7) B. J. J. Zelinski and D. R. Uhlmann, J. Phys. Chem. Solids 45, (1984) 1069-1090.
8) C. J. Brinker, A. J. Hurd, and K. J. Ward, in Ultrastructure Processing of Advanced Ceramics p223-240, J. D. Mackenzie and D. R. Ulrich Eds., John Wiley, 1988.
9) Plentiful citations can be found in volume 100 of J. Non-Crystalline Solids, (1988).
10) M. Yamane, S. Inoue, and K. Nakazawa, J. Non-Cryst. Sol. 48 (1982) 153-159.
11) G. Kordas, R. A. Weeks, and N. Arfsten, J. Appl. Phys. 57, (1985) 3812-3813.
12) B. D. Fabes and W. C. Oliver, to appear in J. Non-Cryst. Solids.
13) H. Hirashima and S. Kamimura, Mat. Res. Soc. Symp. Proc. 121 (1988) 779-784.
14) M. H. Francombe, in Basic Problems in Thin Film Physics p52-76, R. Niedermayer and H. Mayer, Eds. 1966.
15) L. I. Maissel and R. Glang, Ed. Handbook of Thin Film Technology, McGraw Hill, New York (1970).
16) H. A. Macleod, J. Vac. Sci. Techn. A4, 418-422.
17) R. Messier, A. P. Giri, and R. A. Roy, J. Vac. Sci. Technol. A2 (1984) 500-503.
18) M. Harris, H. A. Macleod, S. Ogura, E. Pelletier, and B. Vidal, Thin Solid Films 57, 173-178 (1978).
19) I. J. Hodgkinson, P. W. Wilson, CRC Crit. Revs. in Sol. St. and Mat. Sci. 15, 27-61 (1988).
20) A. G. Dirks and H. J. Leamy, Thin Solid Films, 47, 219-233 (1977).
21) O. Geszti, L. Gosztola, and E. Seyfried, Thin Solid Films, 136, L35-L38 (1986).
22) P. Meakin, CRC Crit. Revs. in Sol. St. and Mat. Sci. 13, 143-189 (1986).
23) H. G. Hakhodkin and A. I. Shaldervan, Thin Solid Films, 10, 109-122 (1972).
24) H. Fujiwara, K. Hara, M. Kamiya, T. Hashimoto, and K. Okamoto, Thin Solid Films, 163, 387-391 (1988).
25) R. Fiedler and G. Schirmer, Thin Solid Films 167, 281-289 (1988).
26) K. H. Guenther, Appl. Optics, 23, (1984) 3806-3816.
27) R. Messier, J. Vac. Sci. Techn. A4, 490-495 (1986).
28) R. Messier and J. E. Yehoda, J. Appl. Phys. 58, 3739-3746 (1985).
29) A. J. Bernoff and S. Lichter, Phys. Rev. B 39, 10560-10569 (1989)
30) G. W. Scherer, in Ultrastructure Processing of Advanced Ceramics p295-302, J. D. Mackenzie and D. R. Ulrich Eds., John Wiley,

1988.
31) G. W. Scherer, J. Non-Cryst. Solids 89, (1987) 217-238.
32) G. W. Scherer and T. Garino, J. Am. Ceram. Soc. 68, (1985) 216-220.
33) R. K. Bordia and G. W. Scherer, Acta Met. 36, (1988) 2393-2397.
34) R. K. Bordia and G. W. Scherer, Acta Met. 36, (1988) 2399-2409.
35) R. K. Bordia and R. Raj, J. Amer. Ceram. Soc. 68, (1985) 287-292.
36) R. L. Coble, J. Appl. Phys. 32(1961)787-99
37) R. L. Coble; J. Appl. Phys. 41 (1970) 4798-4807.
38) D. L. Johnson and I. B. Cutler, J. Am. Ceram. Soc. 46 (1963) 541-545.
39) W. D. Kingery and M. Berg, J. Appl. Phys. 26 (1955) 1205-1212.
40) H. I. Smith, M. W. Geis, C. V. Thompson and H. A. Atwater, J. Cryst. Gro. 63, (1983) 527-546.
41) H. A. Atwater, C. V. Thompson, H. I. Smith, and M. W. Geis, Appl. Phys. Lett. 43, (1983) 1126-1128.
42) J. M. Poate, J. de Physique, C4 (1980) C4-1.
43) W. Lukosz, and K. Tiefenthaler, Opt. Lett. 8 (1983) 537-539.
44) G. W. Dale, H. H. Fox, B. J. J. Zelinski, and L. A. Weller-Brophy, submitted to J. Non-Cryst. Solids.
45) R. L. Roncone, L. A. Weller-Brophy, B. J. J. Zelinski, submitted to J. Non-Cryst. Solids.
46) B. D. Fabes, G. W. Dale, and D. R. Uhlmann, in Ultrastructure Processing of Advanced Ceramics p883-890, J. D. Mackenzie and D. R. Ulrich Eds., John Wiley, 1988.
47) Y. Hidaka, M. Suzuki, T. Murakami and T. Inamura, Thin Solid Films, 106, (1983) 311-319.
48) I. J. Hodgkinson, Appl. Opt. 9, (1970) 1577-1586.
49) B. D. Cullity, Elements of X-Ray Diffraction, Second Edition, Addison-Wesley, 1978.
50) D. R. Uhlmann, B. J. Zelinski, L. Silverman, S. B. Warner, B. D. Fabes, and W. F. Doyle, in Science of Ceramic Chemical Processing, Eds. L. L. Hench and D. R. Ulrich, Wiley, New York, 1986.
51) C. S. Smith, Trans. AIME 175, 15-51 (1948)
52) W. D. Kingery, H. K. Bowen, and D. R. Uhlmann, Introduction to Ceramics, Second edition, Wiley, New York, 1976, p214.
53) G. W. Greenwood, J. Mat. Sci. 17, 2127-32 (1982).
54) I. V. Nicolaescu and F. Glodeanu, J. Nuc. Mat. 113, 253-255 (1983).
55) P. J. Wray, Acta Met. 24, 125-135 (1976)
56) W. C. Carter and A. M. Glaeser, J. Am. Ceram. Soc. Comm. 67(1984)C124-7
57) M. D. Drory and A. M. Glaeser, J. Am. Ceram. Soc. Comm. 68(1985)C14-5
58) D. J. Srolovitz, and S. A. Safran, J. Appl. Phys. 60 (1986) 247-254.
59) D. J. Srolovitz, and S. A. Safran, J. Appl. Phys. 60 (1986) 255-260.
60) K. Sekimoto, R. Oguma, and K. Kawasaki, Ann. Physics 176 (1987) 359-392.
61) K. A. Jackson in Progress in Solid State Chemistry Vol. 3, Pergamon Press, New York 1967.
62) H. K. Pulker and E. Jung, Thin Solid Films, 9 (1971) 57-66.

63) I. N. Shklyarevskii, R. I. Umerov, E. A. Lupashko, and V. I. Khramtsova, Opt. Spectrosc. 35, (1973) 206-208.
64) H. A. Macleod and D. Richmond, Thin Solid Films, 37 (1976) 163-169.
65) N. E. Holm and O. Christensen, Thin Solid Films, 67 (1980) 239-244.
66) J. R. Gee, I. J. Hodgkknson, and H. A. Macleod, Appl. Opt. 24 (1985), 3188-3192.
67) K. Kinosita, and M. Nishibori, J. Vac. Sci. and Techn. 6, (1969) 730-733.
68) D. R. Gibson and P. H. Lissberger, Appl. Opt. 22 (1983) 269-281.
69) K. Tiefenthaler and W. Lukosz, Thin Solid Films, 126 (1985) 205-211.
70) L. M. Gignac, Ph. D. Diss., U. of Arizona, 1988.
71) JANAF Thermochemical Tables, Third Edition, American Chemical Society, 1986.
72) T. Motohiro and Y. Taga, Thin Solid Films, 172, L71-L74 (1989).
73) C. C. Lee, M. Sikkens, I. J. Hodgkinson, H. A. Macleod, R. H. Potoff, R. Sprague, and M. R. Jacobson, J. Opt. Soc. Am. 73, 1871 (1983)
74) I. J. Hodgkinson, M. R. Jacobson, H. A. Macleod, R. H. Potoff, M. Sikkens, R. Sprague, and C. C. Lee, Thin Solid Films 138, 289-296 (1986).
75) R. F. Gledhill, I. J. Hodgkinson, and P. W. Wilson, J. Appl. Phys. 59 1453-1455 (1986)
76) X. D. Wu, d. Kijkkamp, S. B. Ogale, A. Inam, E. W. Chase, P. F. Miceli, C. C. Chang, J. M. Tarascon, and T. Venkatesan, Appl. Phys. Lett. 51, (1987) 861-863.
77) M. G. Mier and E. A. Buvinger, J. Vac. Sci. and Techn. 6, (1969) 727-730.
78) B. A. Movchan and A. V. Demchishin, Phys. Met. Metallogr. 28, (1969) 83-90.
79) M. Krohn, Thin Solid Films, 163 291-296 (1988).
80) B. Dam, T. S. Baller, G. N. A. Van Veen, H. A. M. Van Hal, J. W. C. DeVries and G. M . Stollman, in High-Temperature Superconductors, MRS Proc. 99, Ed. M. B. Brodsky et.al., 1988, p303-306..
81) J. A. Thornton and D. W. Hoffman, Thin Solid Films 171 (1989) 5-31.
82) J. Priest, H. L. Caswell, and Y. Budo, J. Appl. Phys. 34 (1963) 347-351.
83) F. M. d'Heurle, and J. M. E. Harper, Thin Solid Films, 171 (1989) 81-92.
84) D. P. Birnie, III, to appear in Proc 4th Intnl. Conf. on Ultrastructure Procg.
85) T. K. Hatwar, S. C. Shin, and D. G. Stinson, IEEE trans on Mag. MAG-22 (1986) 946-948.
86) D. Stroud, Phys. Review B, 12, 3368-3373 (1975).
87) R. W. Cohen, G. D. Cody, M. D. Coutts, and B. Abeles, Phys. Rev. B 8, 3689-3701 (1973)
88) M. H. Cohen, J. Jortner and I. Webman, Phys. Rev. B, 17, 4555-4557 (1978).
89) H. Seki, Appl. Phys. Lett., 43, 1000-1002 (1983).

90) D. E. Aspnes, J. Vac. Sci. Technol. 18, 289-295 (1981).
91) P. J. McMarr, K. Vedam, and J. Narayan, J. Appl. Phys. 59, (1986) 694-701.
92) B. Drevillon, Thin Solid Films 130, (1985) 165-170.
93) J. A. Woollam, P. G. Snyder, and M. C. Rost, Thin Solid Films, 166 (1988) 317-323.
94) B. Drevillon, Thin Solid Films, 163 (1988) 157-166.
95) A. M. Antoine and B. Drevillon, J. Appl. Phys. 63 (1988) 360-367.
96) V. M. Vecea, R. V. Bucur, and E. Indrea, Thin Solid Films, 171 (1989) 367-375.
97) G. C. Frye, A. J. Ricco, S. J. Martin, and C. J. Brinker, Mat. Res. Soc. Symp. Proc. Vol. 121, (1988) 349-354.
98) C. J. Brinker, G. C. Grye, A. J. Hurd, K. J. Ward, and C. S. Ashley, to appear in Proc 4th Intnl. Conf. on Ultrastructure Procg.
99) M. F. Doerner and W. D. Nix, J. Mater. Res. 1, 601-609 (1986).
100) W. C. Oliver, R. Hutchings, and J. B. Pethica, ASTM ATP899, Blau and Lawn ed. 1986 ASTM.
101) W. C. Oliver and C. J. McHargue, Thin Solid Films 161, 117-122 (1988).
102) C. Thomsen, H. J. Maris, and J. Tauc, Thin Solid Films 154 (1987) 217-223.
103) V. Mizrahi, F. Suits, J. E. Sipe, U. J. Gibson, and G. I. Stegeman, Appl. Phys. Lett. 51, (1987) 427-429.
104) S. Brunauer, P. H. Emmett, and E. Teller, J. Am. Chem. Soc. 60, (1938) 309.
105) A. M. Antoine, B. Drevillon and P. Roca i Cabarrocas, J. Appl. Phys. 61 (1987) 2501-2508.
106) C. Licoppe, Y. I. Nissim, and C. Meriadec, J. Appl. Phys. 58 (1985) 3094-3096.

INTERNALLY COOLED SUBSTRATES FOR HIGH POWER DENSITY PACKAGING

M. Kahn, Code 6374 Naval Research Laboratory, Washington, DC 20375-5000
R. Nelson, Microelectronics and Computer Technology Corporation,
12100 Technology Blvd. Austin, TX 76721

ABSTRACT

Localized power densities in high speed circuits can attain as much as 50 watts per square centimeter. A maximum thermal conductance of the mounting substrate is critical to obtain adequate cooling. This work describes numerical calculations leading to the design, as well as preparation techniques, of ceramic substrates that are liquid cooled through internal channels. The results of numerical calculations are also used to evaluate the differences in cooling capability between A1N and ore conventional substrate materials.

I. Introduction

TRENDS IN ELECTRONICS

In recent years, thermal control has become increasingly important in the packaging of electronic devices. This is due to two trends:

1. The power levels and power densities of modern chips keep increasing.
2. There is a trend to multichip modules with closer chip spacing, enabled by the switch from printed circuit boards to advanced interconnect technology such as copper/polyimide thin film interconnections.

Typical cooling methods have ranged from heat exchangers located above the substrate, as in the IBM Thermal Conduction Module [1] to cooling through the substrate, as in the Hitachi Silicon Carbide RAM module [2], the University of North Carolina internally cooled silicon carbide substrate [4], and the NTT internally cooled alumina substrate [3]. Cooling above the substrate potentially handles the highest power because of the lower thermal resistance to the coolant. Cooling with a fluid flowing through or under the substrate potentially provides a thinner module with easier electrical repair and a simpler mechanical structure, that is also easier to seal hermetically. However, electrical and fluid connections between substrates then compete for space, whereas they are separated in above-substrate-cooling.

The final choice in multichip module cooling design is, thus, a trade off between thermal, electrical, and structural factors. In this paper we will discuss the thermal and constructional aspects of one choice - an aluminum nitride substrate with cooling channels built into the substrate bulk. Also, we will restrict our analysis to that of an interconnection system made of multilayer structures of copper and polyimide deposited on the ceramic substrate surface. At some points we will compare performance to that of the IBM Thermal Conduction Module (TCM), primarily to put the results in a system perspective, although the TCM is based on a different cooling scheme and uses a cofired multi-layer ceramic substrate and interconnect.

Reasons for Choosing In-Substrate Cooling Our choice of placing cooling channels in the substrate is motivated by the desire to cool high power density chips in a thin, structurally simple module. A significant limitation in all cooling options is the thermal impedance of the layers and interfaces between the chip and the heat exchanger. In the limit, there is at least one such interface for above-substrate cooling, i.e. that between the chip top and the heat exchanger. In practice, several layers may exist there, such as module lids or heat spreaders. For thru-substrate cooling to a cold plate, the thermal path contains the attachment to the substrate, an interconnect layer on the substrate surface, the substrate bulk, and an attachment to the cold plate. Lack of substrate or cold plate flatness will reduce the area of contact and the heat transfer rate and will increase local temperatures. As a rule of thumb, it is difficult to achieve attachment specific thermal impedances less than a 0.1 to 0.2 W-cm^2/K [e.g. that of 50 micron thick silver epoxy] while those of interconnect layers can exceed 2 W-cm^2/K if no thermal vias are present. In critical cooling applications, it pays to reduce the number of interfaces and the amount of material between the chip and coolant; hence a system of in-substrate cooling will be second only to optimal above-substrate cooling, particularly if an interconnect system with good thermal properties is available. Furthermore, in-substrate cooling is potentially more compact then the other alternatives, because no additional structure is required. For instance, the IBM water cooled TCM uses a 5 mm thick substrate and a 5 cm thick cooling structure while the substrate we describe here, even if provided with a copper/polyimide interconnect, would be only 1.5 mm thick. Finally, cooling in the substrate is potentially the most simple of structural arrangements. Above-substrate cooling entails a host of mounting problems as good thermal contact must be made to every chip of a MCM. This in turn requires compliance, either in the chip mounting or at the heat exchanger interfaces or both, as the chip now lies between two relatively rigid elements - the substrate and the heat exchanger support structure. For in-substrate cooling there is only one such structure [the substrate itself] and the chips need only be attached to it. Since this attachment is face up (devices exposed) it is optimum for electrical testing and repair. Once the system is inspected, it may be encapsulated or sealed with a hermetic cover.

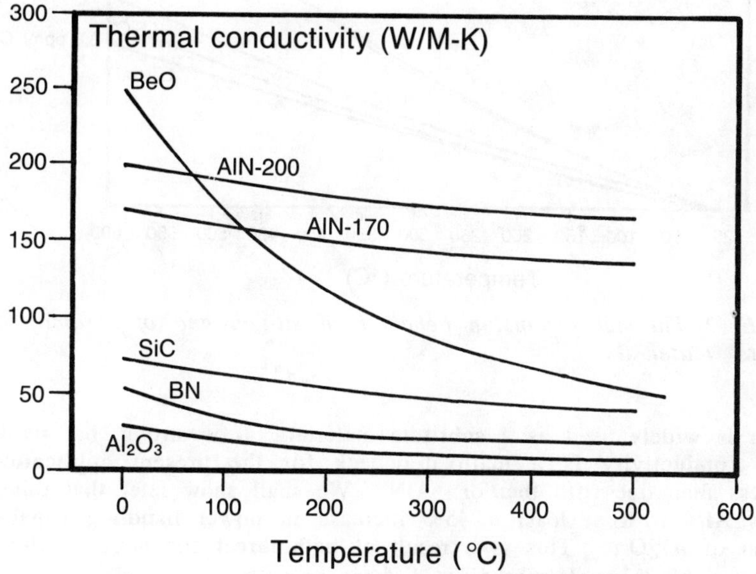

FIGURE 1 Thermal conductivity of ceramic substrate materials

Aluminium Nitride As A Substrate. Aluminum nitride (Al N) is a relatively new substrate material and it has several desirable properties which may lead to its choice over others. We shall discuss, its dielectric, mechanical, and thermal properties, comparing them to aluminum, aluminum oxide (Al_2O_3), silicon, and berylia: Figure 1 shows the thermal conductivities of a range of candidate substrate materials. It appears that above 70°C, aluminum nitride "200" has the highest thermal conductivity of the potential ceramic substrate materials shown. Figure 2 indicates that the coefficient of thermal expansion of aluminum nitride is quite linear and that at 4.4 ppm/°C it is much closer to that of silicon (3.8 ppm/C) than those of Al_2O_3 or BeO (~7 ppm/°C).

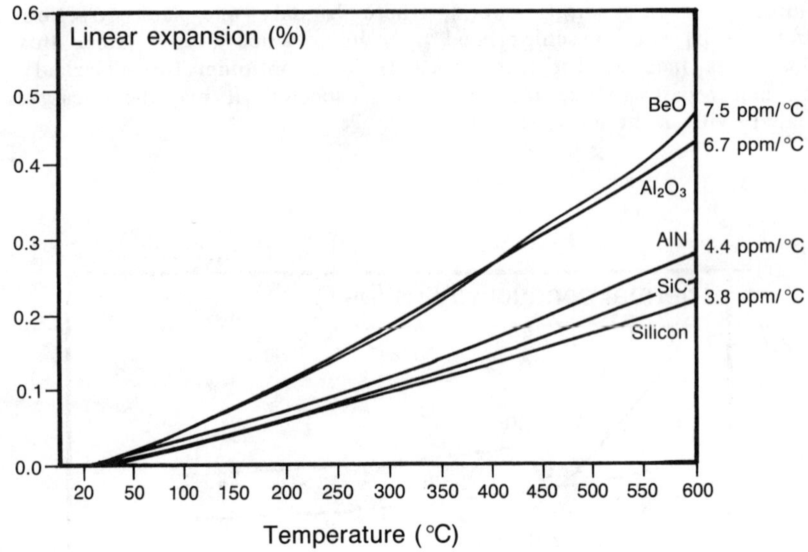

FIGURE *2 Thermal expansion behavior of silicon and of ceramic substrate materials*

Alumina is widely used as a substrate material. It is strong, but its lower thermal conductivity is a main drawback for the present application, being less than one fifth that of AlN. We shall show later that this provides AlN with at least a 45% increase in power handling capability over that of Al_2O_3. This is a result of both direct conduction differences as well as of thermal spreading differences.

Aluminum metal can be used as a substrate, but only where an electrically conductive material is acceptable. Thermally, it is good, but has a poor coefficient of thermal expansion (CTE) match to the silicon chips which must be attached to the substrate.

Beryllia is excellent thermally, but does not have as good a CTE match to silicon as Al N. There are also well known manufacturing hazards with this material.

Silicon as a substrate is about as good as AlN thermally, provides an identical CTE match to chips, and has been used with copper/polyimide interconnect in several studies. It has been used for the most advanced liquid cooling to date [5]. However, it is very brittle and requires most careful handling and mounting in larger sizes.

In summary, AlN presents new possibilities because of its attractive thermal and thermo-mechanical properties. The evolution of a thermal design and the development of a prototype substrate are the subjects of the remainder of this paper. We start with a thermal analysis leading to a design with 1mm x .191 mm [.040" x .0075"] channels on 3 mm [.120"] centers. Next we extend the analysis to discuss thermal spreading. Then we describe the preparation of prototype substrates using this design. Finally, we discuss our predictions of the thermal performance and the thermo-mechanics of substrates made of aluminum nitride and of alumina.

II. Substrate Design

Thermal System. In Figure 3a we show a layout of the proposed thermal system. The chip is attached directly to an interconnect layer on the substrate or to the substrate itself, if interconnection demands are not high. The die attach layer is an thermally important limit, and must be of low thermal resistance. Equally important is the minimization of the thermal resistance of the underlying interconnect layer. In figure 3b we show two layers of stacked cooling channels, the channels themselves running the width of the substrate that is 5.72 cm [2 1/4"] square. The channels are shown extending from a center distribution manifold to the edge, a length of 2.86 cm [1.125"]. This serves to reduce the total pressure drop.

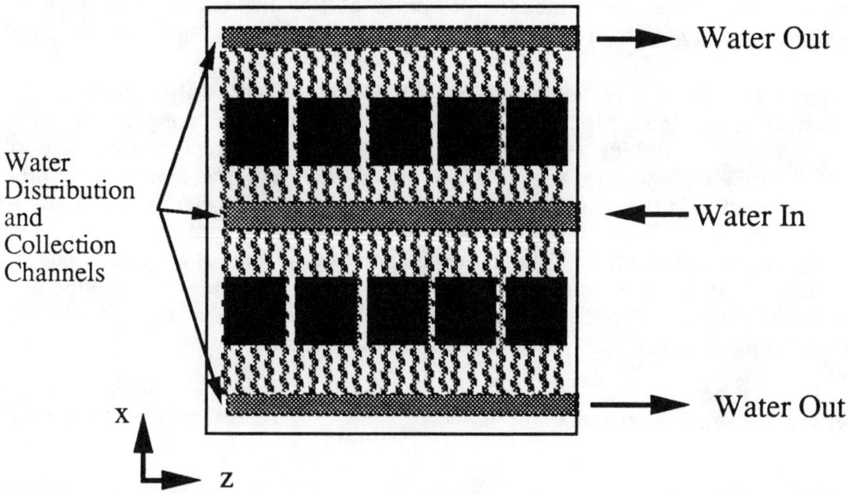

FIGURE 3A Typical chip layout on substrate. The underlying water channels and the manifolds are also shown.

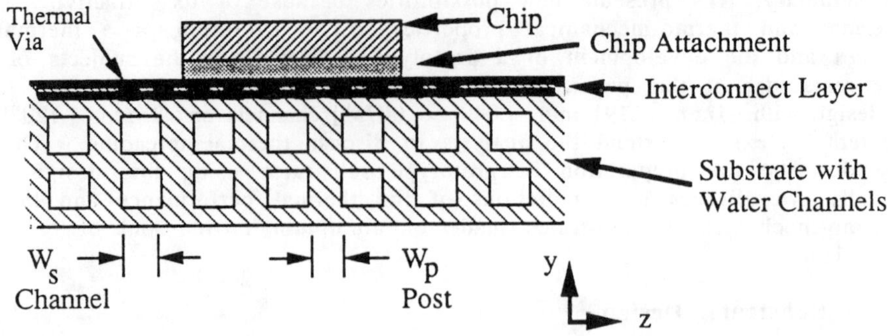

FIGURE 3B Enlarged cross section at one chip showing elements in the thermal path. Heavy vertical bars in the interconnect layer represent solid copper thermal vias.

Chip power densities of 20 to 50 watts/cm^2 are now appearing in chip sizes of 1 to 1.5 cm square and these are likely candidates for use in liquid cooled multichip modules. Chip density on the MCM substrate is limited by both interconnect and bonding technologies. Silicon coverage exceeding 50% (i.e. 50% of the substrate area occupied by the chip footprints) is aggressive, with 25% more common.

In this study we examine the temperature rises associated with ten, one centimeter square chips, each dissipating 50 watts. Thus the chip power density is 50 w/cm^2, the module power is 500 watts, the module power density is 15 w/cm^2 and the silicon coverage is 30%.

Temperature limits are usually derived from the allowable junction temperatures on the chip. With a maximum junction temperature of 75 °C and a 20 °C coolant supply, there is an allowable temperature rise of 55 °C. This temperature rise is made up of the rise associated with the heat conduction through the solid elements, plus the rise associated with the heat transfer from the substrate to the coolant, plus the temperature rise of the liquid itself. The latter is greatest at the downstream end of the chip. It is convenient to talk about three associated thermal resistances, each defined as the ratio of the temperature rise to the power being dissipated. These resistances are:

Internal Thermal Resistance, $R_{int} = \frac{DT_1}{Q_{total}}$, represents the total of all the thermal resistances between the heat source and the substrate - coolant interface.

Convective Interfacial Thermal Resistance, $R_{conv} = \frac{DT_2}{Q_{total}}$ represents the thermal resistance of the substrate-coolant interface and the liquid boundary layer.

Effective Caloric or Bulk Fluid Thermal Resistance,
$R_{cal} = \frac{DT_3}{Q_{total}}$ is equivalent to the reciprocal of thermal capacitance of the flowing coolant.

Notice that the total chip power is used in this definition and no reference to chip area is made. Thus the units of resistance are °K/Watt, and this resistance depends on the chip size and the fluid flow rate.

The usual techniques for treating heat conduction in solids can be used for finding the internal thermal resistance. In this study we use two, or three dimensional thermal resistance grids to find the temperature at nodal points in the solids.

The temperature rise at the interfacial convective boundary (the cooling tube wall) can be computed using the well known heat transfer theory for channel flow [6] [7]. In our case, the flow velocities are low and the channels small, resulting in laminar flow. The heat transfer coefficient is then found by determining the Nusselt number for the tube flow. To do this, two geometric effects must be taken into account: the tube cross sectional aspect ratio and the length of the heated section. These effects can be explained as follows.

There is a thermal boundary layer at the tube wall that becomes thicker as one moves downstream along the heated length of a tube (i.e. downstream in the substrate below the chips). This effect is particularly pronounced near the beginning of the heated portion of the tube and is known as the "entrance effect". As a result, the local interfacial convective thermal resistance increases, becoming highest at the chip's downstream edge. Ignoring this entrance effect can result in several hundred percent error in evaluating the upstream chip temperature. At the downstream chip edge the errors are smaller, but can be 15% to 30%, depending on the flow rate. These errors can be even higher when significant axial condition occurs in the chip or substrate. Because fluid velocities are low in the corner regions of rectangular tubes, the heat transfer is also affected by the geometric aspect ratio. For example, the Nusselt number varies by factor of over 2 for rectangular cross sections ranging from square to very short and wide.

Fortunately, the Nusselt numbers for rectangular tubes are available in the literature and are tabulated in several sources [6] [7]. We have used the tables in Phillips [7] to account for the combined effects of cross section and entrance effect in our spread sheet model of two and three dimensional heat flow.

The caloric component of temperature rise can be found at any point along a fluid channel by computing the amount of heat absorbed upstream and dividing by the product of fluid mass flow rate and specific heat.

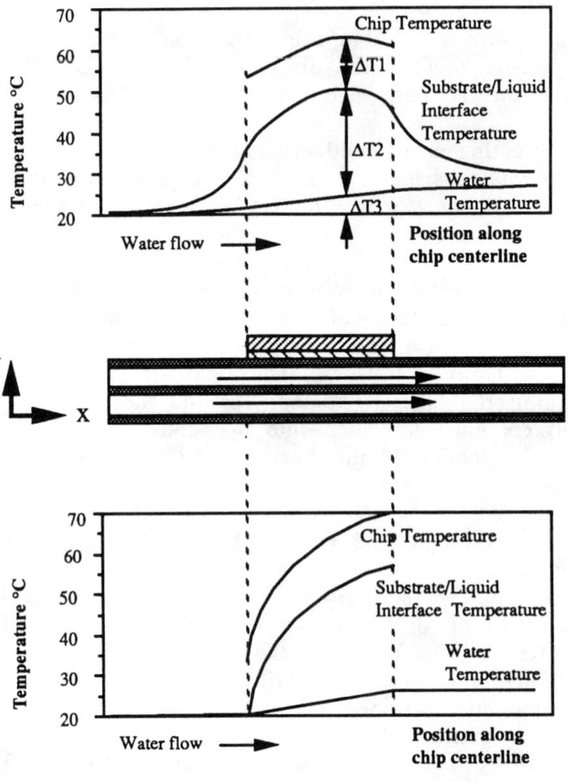

4a. Temperatures with axial heat spreading.

4b. Cross section of chip, substrate, and water channels.

4c. Temperatures with no axial heat spreading.

FIGURE 4 *Typical temperature profiles in a liquid cooled substrate, considering both water temperature rise and local Nusselt number variations.*

Figure 4a, shows schematically how the temperature of the chip varies locally. The hottest spot is closer to the downstream end. This is due to several phenomena:

1.) Heat is conducted by the substrate to heat transfer areas adjacent to the chip. This makes R_{conv} appear to be lower at the chip edges.

2.) Heat transfer is better at the upstream edge of the chip as the boundary layer is thin, i.e. R_{conv} is lower.

3.) The fluid heats up as it moves downstream; i.e. R_{cal} is higher.

Pressure Drop. In the design process to be described, we start with a specified coolant flow velocity (V). Thus the solution for the pressure drop in the channel is uncoupled from the thermal solution, except for the dependence of the viscosity on fluid temperature. Since the power input is fixed, the mean mixed fluid temperature at the exit is known, even without a thermal solution. Therefore, the average of the mixed inlet and outlet temperature can be used to determine the average viscosity (m) and density values. The pressure drop is then computed from the following equation for laminar flow:

$$DP = \frac{4\ \overline{C_f}\ m\ L\ V}{2\ D_h^2\ g}$$

where D_h is the hydraulic coolant channel diameter (4 x the channel area/channel perimeter) and g = the gravitational constant to be used in Newton's law, relating force units to mass units. $\overline{C_f}$ = friction coefficient, averaged for the length L of the channel.

The average friction coefficient is actually a function of the channel length, flow rate, and channel aspect ratio.

Phillips [7] has tabulated values of the friction coefficient for various dimensionless channel lengths and channel aspect ratios. We use this data, with the exception that Phillips considers an aspect ratio of 0.1 to be equivalent to "infinite parallel plates". We use 0.01 for the parallel plate equivalent.

The pressure loss is to be found considering the entire channel length to estimate the core loss. In addition an estimate of the fluid turning, expansion, and contraction losses should be added to obtain the total pressure loss. In our experience, designs using low velocities, laminar flow and small channels, are rarely dominated by additional losses. The total pressure loss generally becomes unacceptable at about the same flow rates at which the core loss equals the other losses. Thus, we calculated only the core loss and looked for geometries in which this remains below 10 psi.

Initial Design Process. A convenient way to start the design is to assume that heat flows mainly vertically in Figure 3b so that the temperature profiles of Figure 4c apply. Of course, heat does flow down the solid "post" sections and laterally into the sheets or "fins" bounding the channels. For now, however, we ignore axial conduction, in order to obtain a simple two dimensional network and an analytical solution for design. The symmetry in the dense packing of chips in the Z direction further reduces the problem to that of treating only half of a single set of channels and posts.

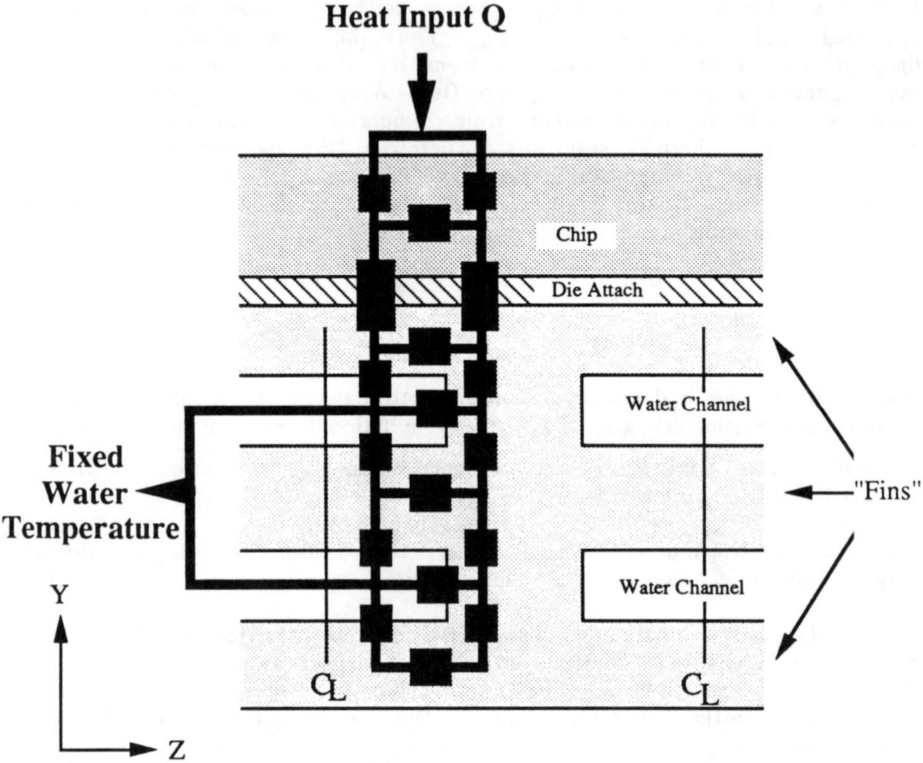

FIGURE 5 Schematic of equivalent resistance network used for two dimensional initial design calculations.

Solid boxes represent equivalent lumped thermal resistances between nodes. Resistances connected to water nodes include the interfacial convective heat transfer resistance. The water temperature is held fixed at its exit value as determined by the total heat input on the chip, the flow past the chip, and the fluid properties.

Figure 5 shows the network of thermal resistances which was used to analyze the problem. Resistance values were derived from calculations utilizing conventional equations for convective and conductive heat transfer [6] [7]. The nodes were located midway in or above each half-channel or midway in each half-post. Convective interfacial resistances were calculated from these and lumped with adjacent conductive resistances. A fin efficiency parameter [6] was not used. Instead, the resistances parallel to the Z axis in Figure 5 include the resistive contribution of one half the length of the ceramic "fin", i.e. of one lamina whose length is one quarter of the channel width.

This two dimensional model is amendable to simple resistance network calculations and was programmed into a spreadsheet. In the calculation we fixed the flow rate and chip power and determined the resulting temperatures. The maximum temperature occurs at the downstream edge; the temperature increase of the cooling fluid was based on the total chip power.

An upper bound on chip temperature comes from applying this two dimensional analysis to a known chip layout, looking at either the highest power density chip or the last chip downstream, and calculating the temperature at the chip's downstream edge. Here one ignores the possibility of heat spreading, and the resulting chip temperature can be significantly higher than if spreading had been accounted for. This approach is used for the initial stage of the design described below; the results are initial estimates of chip temperatures. These are about 1.5 times too high, as seen from the discussion of heat spreading given later. Still, it discloses the important relationships of the geometric variables and is sufficient to make a design choice. Subsequently we analyze the performance when heat spreading occurs.

In conformance with this discussion, we analyzed a variety of channel geometries for their thermal and pressure performance. In order to narrow the choices for experimental investigations, we first applied several criteria based on other considerations: We restricted the total thickness to 1.5 mm to conform with available interconnect processing equipment. We therefore chose 7 layers of 7 mil thick tape (after firing) with 2 layers of channels in the in-line geometry of Figure 2. We calculated the performance at a total coolant flow rate of 10 cc/sec, to cool 10 chips dissipating 50 watts each. For comparison, the IBM water cooled TCM uses 40 cc/s and cools 300 watts [1]. Our analysis includes the thermal resistance of the silicon die, a die attachment layer plus an interconnect. For convenience this is taken to be equivalent to a 0.051 mm [0.002"] thick layer of silver filled epoxy (K = 0.02 W/K-cm). This is expected to be characteristic of many applications where a copper/polyimide interconnect is deposited on the substrate, a provision for efficient thermal vias is made, and the die attach is a high performance epoxy.

With the above conditions, the only free geometric parameters left are the post and channel widths, Wp and Ws, shown in Figure 3b. We restricted our analysis to systems where the post was twice as wide as the channel, anticipating that this would give a design which could be made by punching ceramic green sheets. Thus the area occupied by Al N is in the form of a thermally conductive post, or rib with attached fins while the remainder is an open space for the water flow. The channel height is defined by the layer thickness. In the micro-machined silicon heat exchangers reported by Tuckerman [5], Wp represents a fin width, and Ws the channel width, each being on the order of 0.1mm, and these small sizes result in much lower thermal resistances than we have achieved, but with a significantly higher pressure drop.

To examine the effect of the channel size, we kept the ratio of post width to channel width at 2.0 as noted above, and changed the channel width. The calculation results in Figure 6 show that the pressure drop and chip temperature are both strong functions of the channel size in the range 0.25 mm to 0.75 mm. This is because the channel hydraulic diameter is affected by both the channel width and height in this region. For larger channel widths, the hydraulic diameter approaches twice the channel height, regardless of the width, and the pressure loss levels out. The chip temperature continues to show a dependence on channel width even beyond 1.0 mm. This is to be expected because of the increased thermal resistance the "fin layers, numbers 1, 4 and 7 in Figure 11.

The dotted line in Figure 6 indicates the design chosen for the preparation of the present samples, 1mm channel and 2 mm post. The thermal performance is adequate for our target specifications, but not as aggressive as those associated with smaller channel and post sizes. Moreover, there is little to be gained in lowering the pressure further by going to larger sizes than used here. Although slightly smaller channels might represent a better thermal compromise, the present sizes were chosen to minimize potential tooling and firing problems.

Figure 7 gives the predicted performance of the chosen system as a function of flow rate. The chip temperatures are calculated from the chip-based, conservative, one-dimensional estimate, using 50 watts for the power input. In addition, we show a plot of the predicted performance of an identical substrate made of (99.5%) alumina (K = 0.378 w/cmK). While the pressure drops are identical, the Al N substrate maintains lower chip temperatures as expected. In fact, the performance of each substrate will be 1.5 to 2 times better due to lateral heat spreading. This is the subject of the following section.

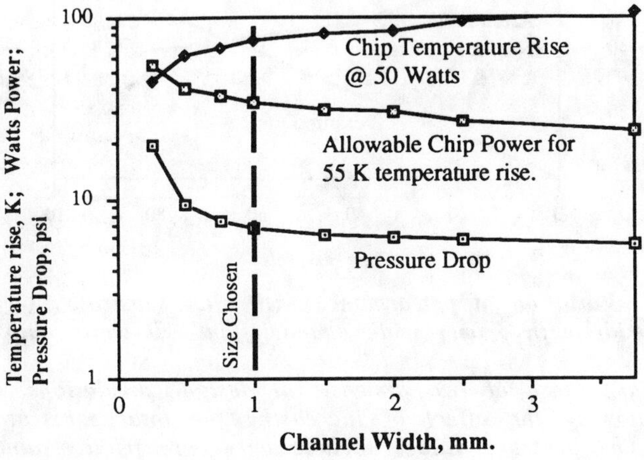

FIGURE 6 Variation of performance of an aluminum nitride substrate with different channel widths and the ratio of post width to channel width held at 2:1. The total substrate flow rate is 10 cc/sec in all cases. No heat spreading is included.

Corrections for Local Thermal Details. In order to evaluate the effects of thermal spreading, a three dimensional model was applied, that uses finite difference techniques. Figure 8 shows the equivalent network of thermal resistances. The model is equivalent to an axial section from the center of a 1 cm long chip on a 1.125 inch long substrate, with chips densely packed in the z direction of the substrate. This dense packing results in negligible heat spreading beyond the chip edges in the z directions and makes the centerline of each channel and post a thermal symmetry plane. Thus, the axial planes (xy planes) that delineate the extent of the model are at the post and channel centerlines which have zero flux boundary conditions.

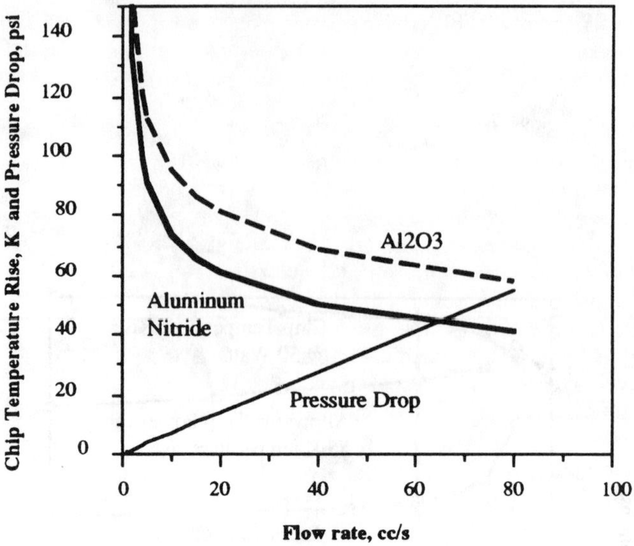

FIGURE 7 Variation of performance with total substrate flow rate for aluminum nitride with 1 mm wide channels and 50 watts chip power.

The results are those of two dimensional design calculations without heat spreading, showing the effect of increasing the total substrate water flow rate. The pressure values include only core friction and therefore somewhat underestimate the total pressure loss.

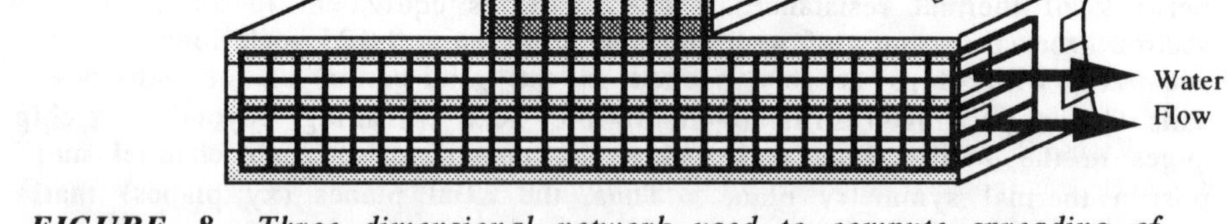

FIGURE 8 Three dimensional network used to compute spreading of heat along the substrate x axis by the finite difference method.

To account for the spreading of heat in the substrate along the water flow direction, parts of the thermal resistance grid of Fig. 5 was used at 48 points along the x axis. Resistances as shown as solid lines. Only one of the two sets of resistances i.e., those at the channels (see Fig. 5) are shown in detail. Every other set of nodes and resistances in the x direction is omitted for clarity. This calculation simulates the heat transfer when chips are densely packed in the z direction, as shown.

Note that the convective interfacial thermal resistances at the chip's upstream edge are negligible. Those even farther upstream are not defined by the conventional equations, since these only apply downstream from the start of the heated area. The first node below the chip is downstream from the chip edge, and its convective resistance is finite. For convenience, the other upstream convective resistances are taken to be equal to this value. These approximations are expected to have little effect on the conclusions since the hottest point is downstream. The convective interfacial resistances at and beyond the downstream chip edge are assumed invariant because by this point the boundary layer is well established. If anything, this is conservative as the heat transfer coefficient beyond the hot chip edge will in fact begin to increase as the thermal boundary layer becomes thinner.

The temperatures in the solid are calculated iteratively using the conventional over-relaxation scheme. Equating the heat flow in and out of each node discloses how much its temperature must be adjusted to obtain a better estimate on the next iteration. To accelerate convergence, the temperature is over-adjusted by a predetermined amount.

The temperature at each coolant node is determined by finding the total energy passing into all upstream coolant nodes, as determined by the current iteration. The channel flow rate and fluid properties then determine the node's new temperature. The calculation progresses from the upstream end to the downstream end.

For simplicity, all fluid properties are taken to be those of 20 °C water.

Convergence is assumed to be acceptable when the average of all the chip node temperatures changes by less than 0.001 K. The calculation is done in a FRAMEWORK spreadsheet on a IBM AT compatible personal computer and converges in about 200 iterations which takes about 1.5 hours.

In Figure 9 we plot the resulting temperature profiles for the chip surface and for the Al N substrate surface with silver-epoxy die attach [0.051 mm thick, K=0.02 W/K-cm] between the two. There is a 50 watt power input and 5 cc/s water flow. Several of the thermal phenomena which we have discussed are apparent. The Al N temperature clearly shows that the heat which enters through the die attach spreads both upstream and downstream from the chip edge. The steeper slopes at the upstream end are due to the improved heat transfer coefficient to the

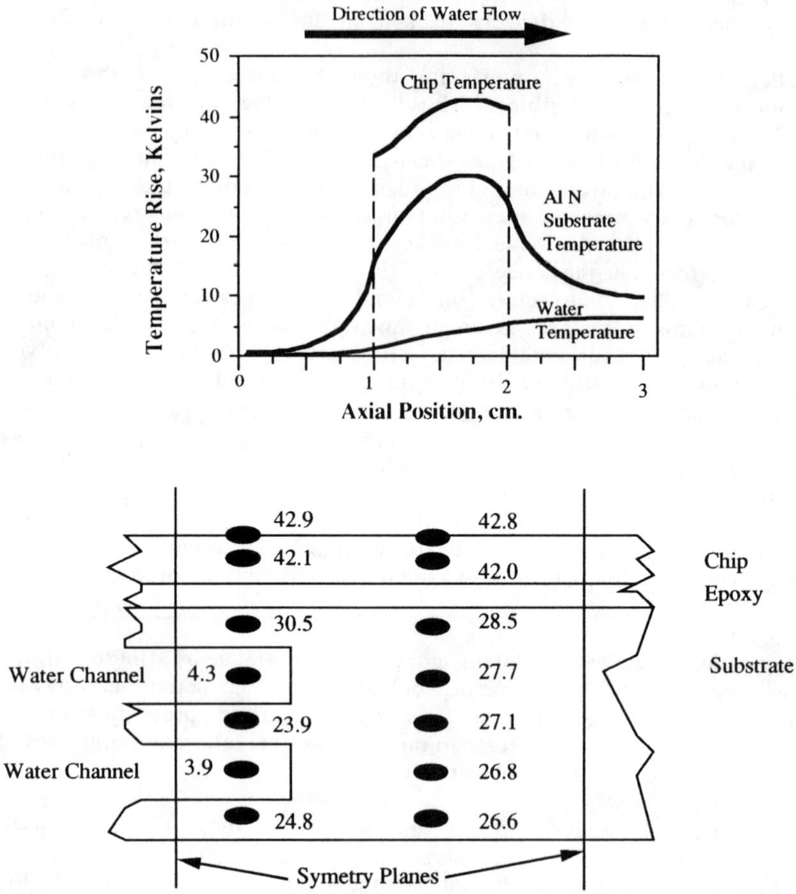

***FIGURE 9** Temperature increases computed for aluminum nitride substrate and chip using epoxy die attachment and 50 Watts power.*

The upper figure gives the temperature profiles of the chip, the aluminum nitride substrate, and the water along the water flow direction. The chip temperature increase is that directly over the water channel, and the substrate temperature increase is at the node just below this chip node. This is the maximum substrate temperature increase at each axial location. The lower figure gives the temperature increases of the nodes at the point of maximum chip temperature increase. The temperature increases of the water at this location are also shown.

liquid coolant near the entrance where the boundary layer is thin. The relative slopes of these curves in the first half of the heated area is a good indication of this: if the heat transfer coefficient were constant, the substrate and water temperature would be parallel, except for the effect of spreading upstream. There is significant heat conduction along the length of the chip also. This is seen from its temperature profile where it does not completely mimic the substrate temperature.

The temperatures at each of the calculation nodes are shown in Figure 9 for the cross section where the chip temperature is maximum. The important effect of the die attach layer is evident, as a major portion of the total temperature drop occurs across it. Similarly, the temperature drop between the channel walls and the coolant is large due to the convective interfacial thermal resistance. The water temperatures in the two channels differ by only 11% which justifies the approximation of identical temperatures used in the initial calculations. This is due to the high thermal conductance in the large post between water channels. The temperatures in the "fins" are somewhat lower than those in the post because of the adjacent coolant and because of the small width of the ceramic crossection there.

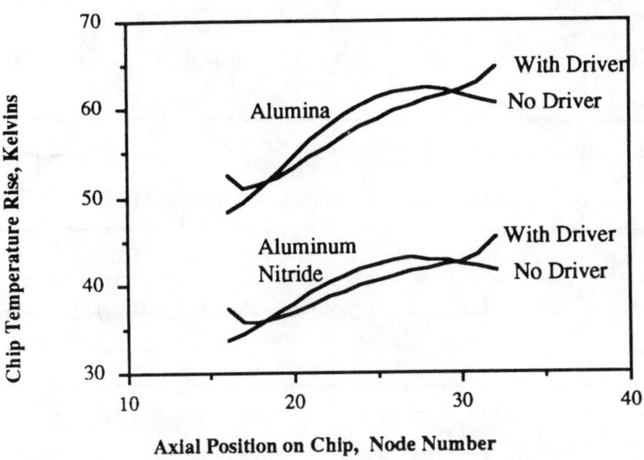

FIGURE 10 *Temperature profiles on the silicon chip with and without a driver at the chip edges.*

The curves labeled "with driver" have been calculated with the power density at the first and last nodes of the chip set at twice that at the other nodes. This simulates the presence of driver circuits in these regions. The total chip power is again 50 Watts, and the computed temperatures at 50 Watts without a driver are included. The central node temperatures with drivers present are lower because the local power density in the interior is then reduced.

Another case that has been modeled using the two-dimensional calculation is the case of a chip where the power density is higher at its edges than at its interior. Such a situation arises where the off-chip signal driving circuitry is clustered on the edges and is run at high clock rates. We have studied this by simply doubling the power dissipation at the nodes at the chip edge. The resulting chip temperature profiles are shown in Figure 10 for the case of epoxy chip attachment. Here the increased spreading in the substrate helps to limit the local temperature rises in the drivers located near the downstream chip edge. These temperature rises are bout the same for both alumina and aluminum nitride. Calculations like this are usually not made when thermal management is addressed, so that the heat spreading in the substrate provides an extra reliability safety factor.

III Layout Design and Sample Preparation

The proposed design calls for an aluminum nitride plate having a crossection similar to that shown in Fig. 11. Such a crossection is generated in a ceramic plate by using tape technology in combination with the application of fugitive inserts.

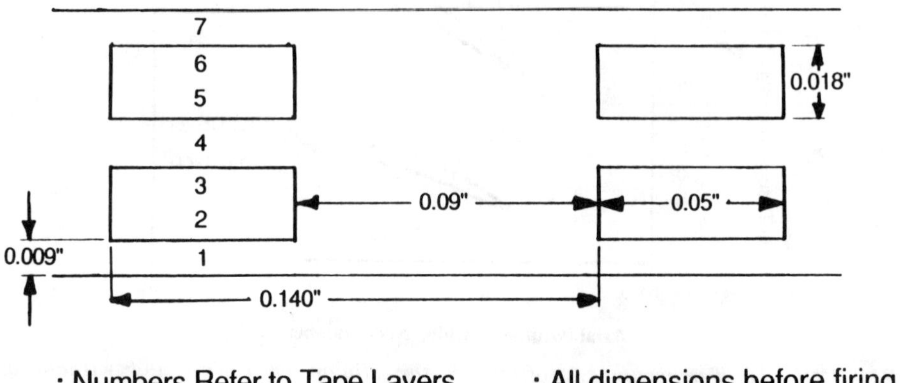

FIGURE 11 Partial cross section of laminate with cooling channels

Implementation of this design calls for the initial preparation of ~9 mil thick sheets of thermoplastic tape in which a high concentration (~60%) of aluminum nitride powder is suspended. Figure 12 shows the outline of such a sheet of tape into which 18 slots are punched. The dimensions in Figs. 11 and 12 are those of the green material. Shrinkage during sintering gives about 20% smaller dimensions than shown there. The two (only partially punched) sections on each side of the required 1.369" green width serve primarily to hold the punched sheet together. These punched sheets are then interleaved with solid tape in the sequence shown in Fig. 11, where two punched sheets are stacked one on top of another in order and then interleaved with an unpunched sheet. This gives the 0.05" wide slots an .018" green thickness.

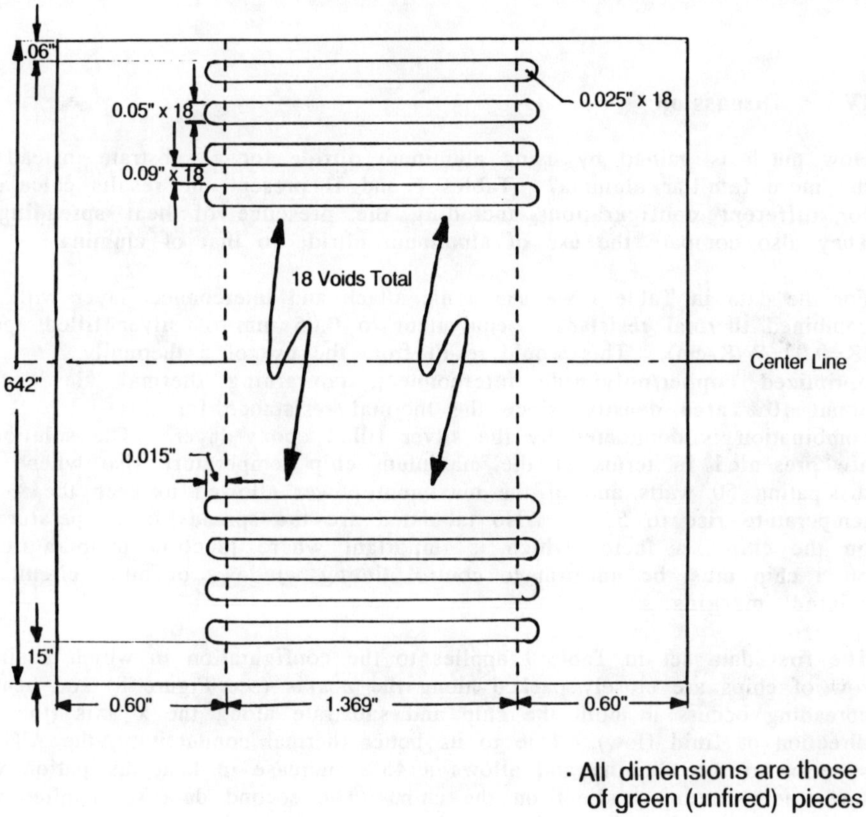

FIGURE 12 Configuration of punched tape

During stacking of the tape sheets, strips of cured, ceramic free, tape binder are inserted into the slots, so as to keep them from collapsing.

The stack is then laminated under heat and pressure into a solid compact, that essentially consists of loose ceramic powder in a matrix of an organic binder. The laminate is then cut ("diced") to the required 1.369 width along the two dotted lines shown in Fig. 12. This gives a finished laminate, equivalent to half of the substrate shown in Figure 3a, in which the cooling channels penetrate the two long edges, suitable for channelling cooling fluid into, through and out of the baseplate. The binder in the laminate is gradually driven off before the ceramic powder is sintered to itself. This "burnout" procedure has to be well controlled so that excess gas evolution rates and binder carbonization are prevented.

After the material is sintered its surfaces are prepared for the application and patterning of the required electrode paths.

IV Discussion

How much is gained by using aluminum nitride for a substrate instead of the more familiar alumina? Tables I and II present the results calculated for different configurations including the presence of heat spreading. They also compare the use of aluminum nitride to that of alumina.

For the data in Table I we use a die attach and interconnect layer with a combined thermal resistance equivalent to 0.05 mm of silver filled epoxy (K=0.02 W/K-cm). This would result from the use of a thermally optimized copper/polyimide interconnect, containing thermal vias, of about 10% area density, since the thermal resistance for this combination is dominated by the silver filled epoxy layer. The solutions are presented in terms of the maximum chip temperature rise when dissipating 50 watts and of the maximum power allowed to keep the temperature rise to 55K. Also tabulated are the spreads in temperatures on the chips, a factor which is important where junction temperatures on a chip must be uniform to control timing windows or other circuit related margins.

The first data set in Table I applies to the configuration in which a single row of chips are closely packed along the z axis (see Figure 8) and heat spreading occurs in both the chip and substrate along the x axis (the direction of fluid flow). Due to its better thermal conductivity, the AlN is superior to the alumina and allows a 45% increase in heat dissipation with a lower temperature gradient on the chip. The second data set applies when the chips are densely packed in both directions, thereby restricting spreading of the chip footprint. The aluminum nitride retains even then a 30% power advantage, caused by local heat spreading from the hotter area of the chip and of the substrate to the cooler areas upstream.

Table I. Summary of Computations, Including Thermal Spreading in Substrates Made of Aluminum Nitride and Alumina, Using Epoxy Die Attachment

Thermal System	Substrate Material	Chip Temperature Rise @ 50 Watts	Power Allowed for $\Delta T = 55K$	Chip Temperature Gradient
Full Axial Heat Spreading	Aluminum-Nitride	43	64	9.4
	Alumina	62	44	14.0
Axial Heat Spreading, Confined to Chip Region	Aluminum-Nitride	54	51	10.3
	Alumina	71	39	15.8
Solid Substrate Attached to Cold Plate, Full Axial Spreading	Aluminum-Nitride	78	35	6.7
	Alumina	115	24	12.5
Chips with drivers and Full Axial Heat Spreading	Aluminum-Nitride	45	61	9.7
	Alumina	65	42	13.5

The first column of results gives the chip temperature rise in Kelvins when 50 watts of power is dissipated in the chip. The second column gives the maximum power allowable if the chip temperature rise is limited to 55 Kelvins. The third column gives the difference between the maximum and minimum temperature on the chip. All computations were made using the model shown in Figure 8. Full axial heat spreading means the axial heat spreading was allowed in the entire substrate. Where noted, axial heat spreading was confined to the chip and the substrate region directly below it, simulating densely packed chips. The last data set applies to a substrate without internal channels, attached to a cold plate with thermal properties the same as those of the aluminum nitride substrate with internal channels.

The third data set in Table I gives the calculated performance of a 1.4 mm (55 mil) thick solid substrate attached to a cold plate, with the latter having the same performance as the internally cooled aluminum nitride substrate. To obtain this, the model of Figure 8 was used, but the thermal resistances at the third node level from the top of the chip were changed to account for the additional substrate material and for a 0.25 mm (10 mil) thick silver/epoxy glue layer. A thicker attachment is required here because of the larger module size and because the cold plate would probably be made of copper or aluminum with a higher CTE than either alumina or aluminum nitride.

The maximum allowed power figures shown in the third data set for solid substrates on a cold plate are only about 60% of the substrates with internal channels. A significant part of this degradation is due to the thick attachment layer needed. Thus, the additional effort required to make internal channels has significant rewards, not only in total module thickness, but also in thermal performance.

Table II shows results tabulated very similarly to those in Table I, except that a relatively poor die attachment layer is used. The model assumes here 0.05 mm (2 mils) of polyimide (K=0.002). This would result from a copper/polyimide thin film interconnect without the provision of low impedance thermal paths to the substrate. Again, the aluminum nitride is superior to the alumina, but the advantage permits only 12% of additional power. This because so much of the total thermal resistance is associated with the poor die attachment layer.

Table II. Summary of Computations, Including Thermal Spreading in Substrates Made of Aluminum Nitride and Alumina, Using Epoxy Die Attachment

Thermal System	Substrate Material	Chip Temperature Rise @ 50 Watts	Power Allowed for $\Delta T = 55K$	Chip Temperature Gradient
Full Axial Heat Spreading	Aluminum-Nitride	150	18	3.8
	Alumina	169	16	6.5
Axial Heat Spreading in Chip Region Only	Aluminum-Nitride	159	17	5.0
	Alumina	176	15	8.2

The first column of results gives the chip temperature rise when 50 watts of power is dissipated in the chip. The second column gives the maximum power allowable if the chip temperature rise is limited to 55 Kelvins. The third column gives the difference between the maximum and minimum temperatures on the chip. All computations were made using the model shown in Figure 8. Full axial heat spreading means that axial heat spreading was allowed in the entire substrate. When noted, axial heat spreading was confined to the chip and the substrate region directly below it, simulating densely packed chips. The last data set applies to a substrate without internal channels, attached to a cold plate with thermal capability equivalent to the aluminum nitride substrate with internal channels.

The thermal advantages of aluminum nitride for high power substrates are evident from these calculation. We should note that these are conservative estimates, since they are based a model which accounts for heat spreading only along the direction of fluid flow. In many applications, there will be considerable lateral heat spreading as well, and this will increase the power capability of each substrate. As a rough estimate for the case of well spaced chips, the substrate temperature is at or above half its maximum value for about 1.25 centimeters, as shown in Figure 9. If this amount of thermal spreading also occurs laterally, then the area of the substrate participating significantly in heat transfer to the fluid is about 1.5 times larger than the chip footprint area, and the acceptable power levels can be expected to be about 1.25 times larger than those reported here for axial spreading only.

The models discussed so far have been based on a ceramic substrate with a copper/polyimide interconnection system deposited on one surface. Typically this interconnect system would include two signal layers, a power plane, and a ground plane; for a thermally efficient system, thermally conducting vias would also be present. Another popular interconnection system is made of multiple layers of ceramic with cofired metallic signal, power, and ground layers. In general, the interconnection density of multi-layer ceramic is much lower per layer than for copper/polyimide. Thus multi-layer ceramic systems with comparable total interconnect densities will have 10 to 30 ceramic layers and be considerably thicker than the copper/polyimide substrate. For example, the substrate used with the early IBM TCMs was 5 mm thick [1] . Thus a multi-layer ceramic substrate with internal cooling that has interconnect layers under the chip will have more material between the chip and the first cooling channels and therefore a higher thermal resistance than the proposed copper/polyimide/ceramic substrate. This is somewhat ameliorated by the increased spreading offered by the thicker ceramic and by the fact that it contains thermally conductive metallic layers and vias. The present model, using a 5 mm thick interconnect layer, predicts a 42°C chip temperature rise for aluminum nitride, a very slight improvement over the thinner copper/polyimide system, and 93°C for alumina, a significant degradation.

Future ceramic substrates will utilize thinner greensheets, perhaps approaching 1 or 2 mils, as found in the ceramic capacitor industry. They will then approach the copper/polyimide/ceramic substrates in heat load capacity. Indeed, there are many alterations to the present design which would increase the thermal or interconnect capacity. For instance, both the water flow and the number of cooling layers can be increased to gain better thermal performance. The circuit density can be increased by providing interconnections for mounting chips on both sides of the substrate. The internal cooling channels provide a compact and efficient way of handling these double sided boards. Moreover, electrical vias can be placed between the cooling channels to provide short signal paths between the sides. In any case, the conclusions reached in the present work will hold as general rules for anticipating trends in thermal performance.

The substrate configuration used in this model appears on the surface amenable to carry two layers of interconnects only, one on each external surface. This does not preclude however the application and utilization of internal, patterned and cofired metallic electrode layers. Such layers would actually improve heat spreading within a plate. The utilization of thinner tapes in the lamination of such plates, as practiced in multilayer capacitor technology for instance would allow the inclusion of multiple conductor layers without increasing the overall plate thickness.

Although we have not treated them extensively, the thermomechanical aspects of aluminum nitride substrates should not be ignored, as they affect the loads on the chip attachment and on the chip-to-substrate interconnections. There are two conditions which should be considered: these are dimensional mismatches caused by internal temperature differences during operation and those occurring due to external temperature changes. The former can be found from the calculations described earlier. The latter are often specified for a system or component as "storage" temperature limits with -55°C to 150 being typical. In fact, even higher temperatures may encountered in processing when the chips are bonded to the substrate, say by soldering.

A rough estimate of the degree of stress due to environmental changes can be made by finding the products of the thermal expansion coefficient and the temperature change for the silicon chip and for the substrate under it. The differences between these are the displacement mismatches tabulated in Table III for a controlled and uniform -55°C to 150°C temperature change and for operational temperature changes. The operational temperatures are also given in Table III for internal power dissipations which raise the chip temperature by 55° Kelvin, using epoxy and polymide attachments. These substrate temperatures were calculated using the spreading model and are taken to be the maximum in the chip and substrate bulk; see for example Figure 9.

There is no straight forward answer as to which material combination is best thermomechanically. The displacement mismatch due to internal power dissipation depends on the chip as well as on the substrate temperatures which differ according to the die attachment material. Thus there is a trade-off depending on the thermal resistance of the interconnect/die attach layer and whether environmental or operational temperature transients are more frequent.

Substrate cost is a factor which will have a deciding influence on applications. At the present time, aluminum nitride is in the

Table III. Two Modes of Steady State Temperature Changes That Cause Potential Thermomechanical Displacement Differences

Temperature Change Chip	Substrate	Substrate Material	Attachment Material	Displacement Mismatch, microns
205°	205°	AlN	Epoxy Polymide	2.7 2.7
205°	205°	Al_2O_3	Epoxy Polymide	8.0 8.0
55°	31° 10°	AlN	Epoxy Polymide	-0.37 -1.3
55°	37° 13°	Al_2O_3	Epoxy Polymide	+0.87 -0.81

The first is a uniform temperature change from -55°C to 150°C. The second is operation of the chip at power levels sufficient to cause a 55K temperature rise of the chip. In this case the die attachment material affects the substrate temperature and therefore the resulting thermomechanical mismatch. As an indication of the displacement mismatch which will generate stress in the chip and its attachment, we show the difference in the product (CTE × Temperature Change) for a 1-cm long chip and for a 1-cm length of the substrate, using the maximum chip temperature.

developmental stage, while alumina is widely used. This is bound to make alumina the choice for cost sensitive applications in the near future. As development of aluminum nitride continues, its first uses will probably be for custom, high performance applications. If this development shows that it can be made with equal or better strength, porosity, surface finish, and processing simplicity as alumina, then aluminum nitride will increasingly find its way into lower cost, higher volume products.

VI Conclusions

1. The superior thermal conductivity of aluminum nitride always gives an internally cooled substrate better power handling capability than does alumina, with a similar geometry.

2. At chip coverages of less than 20% of the substrate area, Al N will handle at least 13% more power if the chip attachment is poor (polyimide) or at least 44% more power if the chip attachment is good (silver filled epoxy).

3. For tightly spaced chips, the advantages of thermal spreading will be mainly restricted to equilibrating temperatures on the chip and more effective thermal use of the area under the chip, reducing the advantages of AlN over Al_2O_3.

4. The low CTE of Al N makes it superior to alumina, regardless of attachment material, for matching thermomechanical deformations of the chip under conditions of ambient temperature changes.

5. Because the substrate temperature is a function of chip attachment material, there are conditions where the stresses generated by operational temperature changes will favor either alumina or Al N. In all cases, the operational mismatches are less than those associated with non-operational (storage) temperature changes.

VII Acknowledgements

The authors gratefully acknowledge the support of MCC (Microelectronics and Computer Technology Corporation) of the Naval Research Laboratory, ONR, and SDI and their permission to publish these results.

VIII References

1. A. J. Blodgett, "Microelectronic Packaging", Scientific American, July 1983.

2. A. Bar-Cohen, "Thermal Management of Air and Liquid Cooled Multi-chip Modules", IEEE Trans. Components, Hybrids, and Manuf. Tech., V9, 1986.

3. T. Kishimoto, and T. Okasaki, "VLSI Packaging Technique Using Liquid-Cooled Channels", IEEE Publication 0569-5503/86/0000-0595. 1986.

4. L-T. Hwang, I. Turlik, and A. Reisman, "A Thermal Module Design for Advanced Packaging", Journal of Electronic Materials, V16, N5, 1987.

5. D. B. Tuckerman and R. F. W. Pease, "High-Performance Heat Sinking for VLSI", IEEE Electron Device Letters, EDL-2, N5, May 1981.

6. W. Kays and M. Crawford, Convective Heat and Mass Transfer, McGraw-Hill, N.Y., 1980.

7. R. J. Phillips, "Forced-Convection Liquid-Cooled Microchannel Heat Sinks", Tech Rept 787, Lincoln Laboratory, MIT, Lexington Mass.

NEW MATERIALS FOR ADVANCED CONSUMER MICROELECTRONIC APPLICATIONS

W. Borland, E. I. du Pont de Nemours & Co. (Inc.)
Wilmington, DE 19880-0030

Abstract

The trend to higher circuit density at lower cost but with enhanced reliability is driven by the demand for less expensive, smaller, lighter, and more powerful products. This has placed additional demands on the circuit designer and manufacturer as appropriate materials systems or cost effective manufacturing techniques have not always been available.Consequently, the technology has moved backward and forward between printed circuit board and ceramic.

This paper discusses the future needs of the microelectronics circuit designer and provides a cost-effective thick film ceramic approach to more comprehensively address this trend.

1.0 Introduction

The demands of smaller, lighter, better, cheaper and more reliable circuitry have driven the consumer microelectronics industry for some considerable time and will continue to be dominant for the foreseeable future. Developments in both thick film ceramic and organic printed wiring technologies are geared towards meeting these demands and the ability of either to achieve their objectives will have a considerable impact on their future health and prosperity. In order for thick film ceramic technology to grow, improved process and materials technology have to be developed.

2.0 Density

The desire for faster digital processing has resulted in the packing of ever-greater numbers of functions onto each chip. The speed of circuit performance, however, is then limited by the time taken to transmit signals between chips. Closely spaced chips with short interconnections provide a solution to this problem.

The more complex the chip, the greater the I/O density resulting in the need for finer conductor lines and spaces. Cost effective thick film techniques are needed to satisfy the projected circuit density demands of the leading edge manufacturer.

Projected Circuit Density

Year	Line/Spaces (microns)	No. Layers	Via Size (microns)
1988	200	2	225
1989	150	2	200
1990	100	4	150
1991	80	4	125
1992	60	6	100
1993	40	8	50

As features become finer, high conductivity materials become critical to circuit performance. Traditional conductors become too resistive and signal speed is reduced. Cost effective, high conductivity material compositions based on silver and copper become the only viable solution.

The use of multilayer constructions will grow as interconnection density requirements increase. Increasing use of faster digital circuitry will mean more chips per circuit will be used leading to a greater need for interconnection and more layers.

As technological advances allow more IC's and their associated components to be packed into smaller spaces, the task of uniting the individual elements into a thick film circuit will become increasingly difficult. A single error can lead to reduced performance and reliability.

Complex high density multilayer hybrids quickly become a maze of interconnections. Designers approach the task by identifying individual circuit functions, designing them piece by piece, eventually bringing them together to create the completed design. Hybrid thick film CAD systems are now available that automate many of the steps that once depended upon an individual's know-how. Interactive placement of devices and conductor track routing for minimum interconnection lengths, optimized resistor pattern and value generation together with the capability to incorporate various via designs and chip and wire bonding to save surface real estate now allow a more cost effective thick film solution with maximized performance and reliability.

The direct attachment of chips will allow maximized density and performance by allowing the chips and their associated components to be physically close. Chip and wire technology will be the dominant attach technology. As land pads become smaller and smaller, pad adhesion requirements will favor ceramic technology.

3.0 Cost

It is recognized that the higher the interconnect density, the lower the connection cost. Hence, the trend to increased density on the chip pays off not just in speed but also in reduced cost per function. The same is true for thick film multilayer hybrids. It has been proposed that connection cost is more dependent on the interconnect distance than on the technology employed. Thus, as shown in comparative studies, silver multilayer thick film on alumina becomes more cost effective compared to not-in-kind technology as density and metal layers increase. In addition, thick film multilayers on alumina can be manufactured with no additional investment. This is not the case for alternative technologies.

Apart from design, the cost of a thick film circuit depends on materials cost, material used, processing costs and yield. The use of less expensive precious metals combined with reduced materials usage allows a lower overall cost position to be attained. High conductivity silver and copper systems can reduce systems costs by simplifying designs from 4-6 layer multilayers to double-sided circuitry. An overall cost reduction of as much as 30% can be attained from this approach.

Improvements in processing and yield can also significantly improve costs. Cofiring rather than the sequential firing process and faster processing is needed. Substantial cost reductions can be achieved, however, by only minor increases in manufacturing yields.

4.0 Reliability

Managing heat is a commonly shared problem across many electronic technologies. The currently used 96% alumina substrate is an inherently good heat conductor making it cost effective when compared to organic boards as external heat sinks can often be avoided. However, future needs may require improved systems such as aluminum nitride or silicon carbide systems.

Advanced ceramic hybrids demonstrate proven long life reliability in adverse environments under extremes of temperature and humidity. They have long been used in high reliability military and aerospace and medical applications. Comparable levels of performance are now economically available for silver and copper-based systems. The use of operating life reliability tests and statistical methods of treating data are being increasingly applied to consumer systems so that reliability and warranty periods can be extended.

5.0 **Materials**

Two materials systems have been developed to meet the circuit density, cost and reliability requirements of the immediate future. High conductivity, high resolution copper and silver bearing conductors, dielectrics proven to new and more stringent reliability levels and resistors demonstrating excellent post trim stability are elements of the new systems. Table 2 illustrates the two systems.

Table 2.

Series Q

Q-SIL

Conductors	
Silver/Platinum	QS 170
Silver	QS 175/QS 180
Copper	QS 190
Resistors	
Air Fired Conductors	QS 87 Series
Low temperature Copper Conductors	QS 90 Series
Crossover Dielectrics	QS 481/QS 482
Encapsulants	
Glass	QS 580
Organics	QQ 545/6

Q-PLUS

Copper Conductors	
General Purpose	QP 153
Through Hole	QP 161
Fine Line	QP 163
Resistors	QP 60 Series
Crossover Dielectric	QP 445
Encapsulants	
Organic	QP 545/6
Glasses	QP 501

5.1 The QSIL System

The QSIL system combines industry proven airfireable ruthenium based-resistor chemistry (1,2,3) with both air-fired silver-based conductors and postfired low temperature, nitrogen-fireable copper conductors (4). Integral to this system is a new highly reliable glass-ceramic crossover dielectric. This dual-atmosphere approach allows the circuit manufacturer to reliably use high-performance,air-fired resistors combined with a high conductivity copper conductor. QSIL materials are suitable for single and double-sided circuits with crossovers and are designed for fast, economical processing.

5.2 The QPLUS System

The QPLUS system is an all base-metal system (5,6,7) fireable in nitrogen at 900°C. Resistors based on lanthanum boride form a blendable sub-series (10-10K ohm/sq) while resistors based on doped tin oxide comprise a second blendable sub-series (6K-1M ohm/sq). Conductors are based on copper and cover general purpose, through-hole and fine-line requirements. QPLUS materials are suitable for single and doublesided circuits with crossovers.

6.0 Conductors

6.1 Metallurgy

Traditional Ag-Pd conductors employing high Pd content (>13%) have been widely used in the manufacture of thick film hybrids. Palladium is added to minimize silver migration and to improve solder leach resistance. However, resistivity of such conductors increases substantially with the addition of palladium, so that this approach is unsuitable when low resistivity is required.

To meet a conductivity criterion of <5 mohm/sq., both the conductor-alloy metallurgy and the dilution effect produced by the presence of the binder phase must be considered. A criterion of <5 mohm/sq. limits the alloy addition to a maximum of either 1% Pt or 5% Pd. Which of the two options is best depends on another functional property. Figure 1 compares the effectiveness of Pd and Pt in improving the solder leach resistance of a silver-bearing conductor. Clearly, the Ag-Pt metallurgy is preferable.

Figure 1.

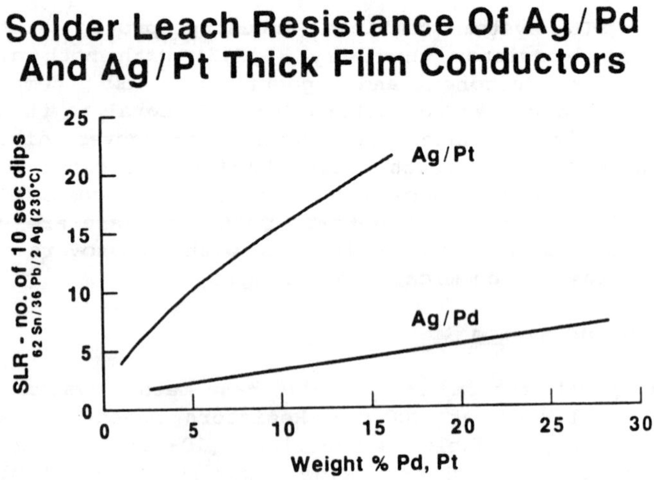

Alloying is not necessary for copper conductors because the copper itself possesses the desired combination of properties, i.e., conductivity, solder leach resistance, migration resistance and cost effectiveness.

6.2 Binder Content and Chemistry

Near the peak firing temperature, significant flow and substrate wetting can occur. The latter is a prerequisite for effective reactive bonding and can be enhanced by the addition of fluxing agents such as Bi_2O_3 and B_2O_3. Since the glass content needs to be as low as possible to retain good solder acceptance, reactive bonding is essential to achieve good adhesion. Reactive bonding occurs through the formation of compounds such as $CuAlO_2$ spinel phases (9-13) at the glass/alumina interface. These chemically bond the film to the substrate enhancing the overall adhesion.

Binder content and chemistry determine the adhesion of a thick film conductor to a substrate as well as conductor solderability. Substrate adhesion is achieved through a combination of mechanical and chemical means. Glasses, such as lead-bismuth-silicate or lead-silicate, wet the substrate, create a mechanical bond, and accelerate sintering of the metal. Oxides on the other hand (e.g., Cu_2O or PbO) are often added to form chemical or reactive bonds with the substrate, thereby enhancing adhesion.

During the firing process, conductor particles sinter by a combination of solid-state diffusion and reactive liquid phase sintering. Solid-state diffusion begins at relatively low temperatures by "neck" formation between individual metal particles. For significant densification, however, a densification temperature melting point temperature ratio, T/T_{mp} of 0.8 is required (i.e., T = 800°C for Cu and 700°C for Ag). Since the thick film process does not allow enough time at peak temperature for complete densification through solid state diffusion, low softening point glasses are used as sintering promoters. At higher temperatures, copper oxide or silver oxide can dissolve in the fluid glass phase which then acts as a transport agent for the metallic ions. This accelerates sintering of the metal, thereby allowing full densification to take place. Many of the enhanced properties of the new conductors result from their dense microstructures.

Figure 2.

Comparison of Fired Film Density
(Fired 850°C/30 min. cycle)

Conventional Ag/Pd X 2000

QS 170 X 2000

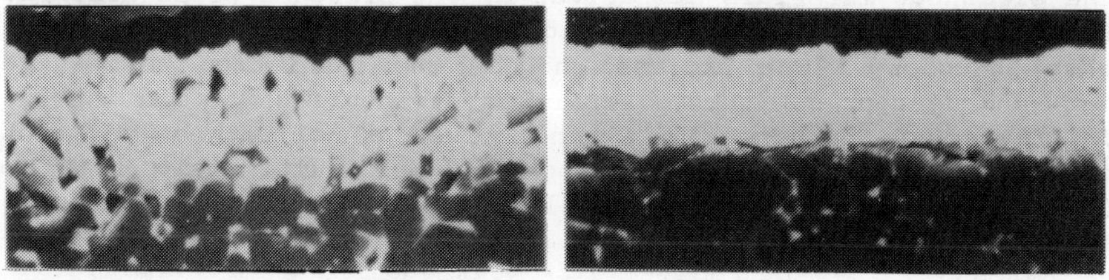

7.0 Dielectrics

Dielectrics are now expected to pass 1000 hour accelerated testing for both electrochemical and solid state migration. To achieve such a high level of reliability, a good understanding of the potential failure mechanisms is necessary.

7.1 Electrochemical Migration

A mechanism sufficiently general to apply to any conductor/dielectric crossover system is shown in Figure 3 (14). The initial step (I) is the formation of conductive pathways in the dielectric structure. This results from the condensation of atmospheric moisture in the structure and its adsorption of the walls of the dielectric pores. This moisture will hydrolyze and/or solvate some of the materials that form the pore walls. Atmospheric Co_2 and ionic contaminants such as chloride ions will subsequently dissolve the moisture film. The net effect is to increase the ionic strength and, therefore, the conductivity of the electrolyte formed.

In the second step (II), the ionic strength of the electrolyte increases to a point where the insulation resistance of the crossover begins to decrease. Once the electrolyte film completely penetrates the dielectric and contact is made between oppositely biased conductor tracks, a small current will flow. This current is supported by the migration of dissolved polar species and results in the formation of small quantities of hydrogen (cathode) and oxygen (anode) at the electrodes.

In the final step (III) of the process, metal ions are released at the anode and migrate to the cathode where they are electrodeposited. Ultimately a filament of electrodeposited material grows back through the pore towards the anode and a short circuit is formed.

The short circuit failure of horizontally adjacent conductors on a dielectric surface due to lateral migration occurs in a similar way. In this case, moisture adsorbed on the surface of the dielectric and small amounts of ionic solution of the electrode materials are driven by the presence of an electric field between conductor lines to form dendritic tracks between the conductors. The process reduces the insulation resistance between adjacent conductors separated by a surface of dielectric. This process, the growth of a conductive channel between adjacent conductors, is called "migration" and will ultimately lead to failure in conductors not suitably protected or chosen to resist this type of failure.

Figure 3.

General Electrochemical Migration Mechanism

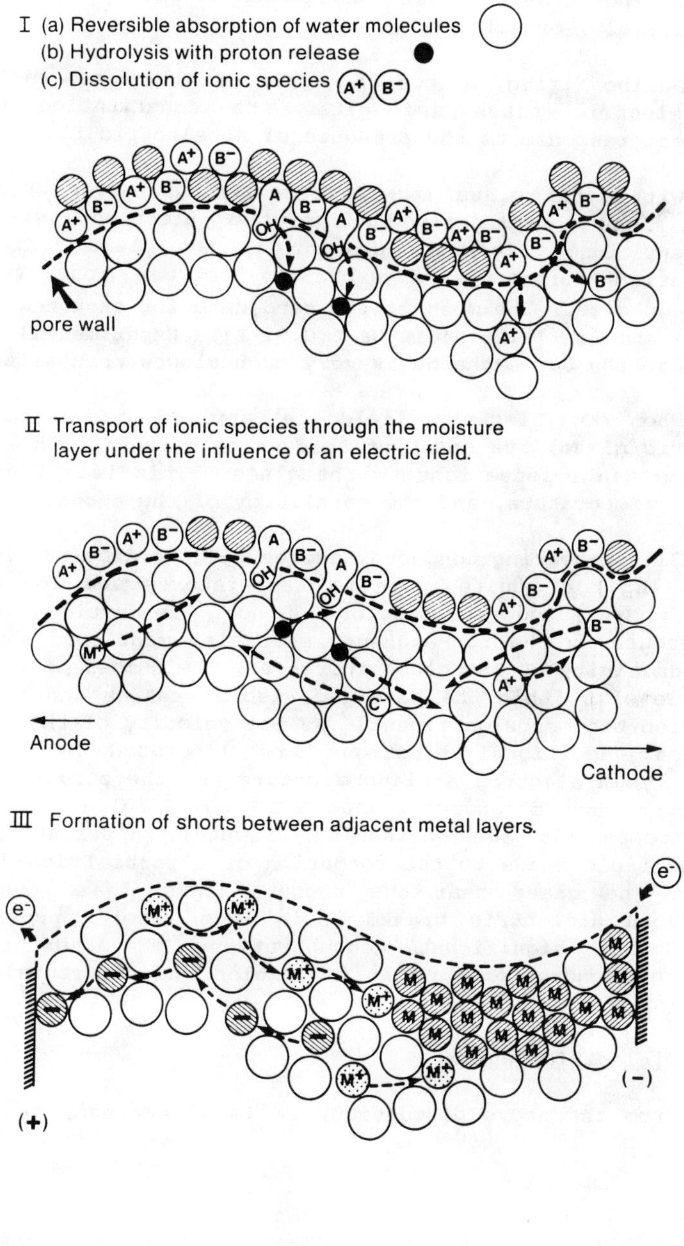

7.2 **Solid State Conduction**

Dielectric failure, due to the growth of a conductive channel between adjacent conductors, may result either as an indirect consequence of the changes that occur in the glass during firing or as a direct consequence of the application of an electric field, or both.

During firing, oxides of copper or silver can dissolve in the dielectric glass and alter its constitution and its subsequent response to the presence of an electric field.

Within the glass itself migration can occur by exchange between ions (15) derived from the electrode with other monovalent ions originally present in the glass. Strain and polarization forces favor exchanges between ions with near equivalent size. The most likely candidate for exchange with Cu^+ (0.96 A) and Ag^+ (1.26 A) is Na^+ (0.97 A). Experimental evidence shows that the Cu^+ exchange is very much slower than silver.

When an electric field is applied to a conductive dielectric glass, the onset of conduction depends on how readily the anode can release ions to the glass. This is a function of voltage, temperature, and the metallurgy of the anode.

Silver-bearing conductor anodes will readily supply mobile cations (Ag^+) at quite moderate temperatures and low electric fields, whereas copper conductors do not. Conduction within the glass occurs by the ion-exchange mechanism mentioned above. As the conductivity of the glass increases, the number of degenerate energy levels within the band-gap also increases and electronic conduction becomes significant. As the velocity of the electrons increases, additional electrons are liberated by collision; ultimately an electron avalanche occurs and the structure breaks down (16). When a conductor does not readily release ions to the glass (copper is an interesting example), application of an electric field leads to the formation of a partial ion-depletion zone in the glass near the anode. Under these conditions, ion-induced dielectric breakdown is less likely, provided, of course, that significant concentrations of other potential sources of conduction such as impurities and structural defects are absent.

Dielectric Design

From the above discussion, it is clear that, in order to

minimize the tendency to dielectric failure at elevated temperature and voltage, certain rules had to be observed. Using dielectric QS481 as an example, the glass was designed for low silver oxide solubility levels. This was achieved through use of a controlled alumino-borosilicate glass-ceramic composition. Critical elements, such as boron, that form solid solutions with silver oxide, were minimized. In this case, the level used was only that needed to form the required glass network. The glass-ceramic was also designed to crystallize rapidly on firing to give a high level of crystallized filler. This further reduced the dissolution tendency of the oxide because of the low residual "free" glass concentration.

The exchange mechanism of ionic migration was addressed through the use of high purity intermediates. Monovalent ions, such as Na^+, K^+ and Li^+, were eliminated from the composition.

Finally, the glass constituents were carefully balanced so that the material fully densified prior to crystallization during the firing process. The dense structure of the dielectric ensures that the electrochemical migration resistance of crossover structures is high.

Figure 4.

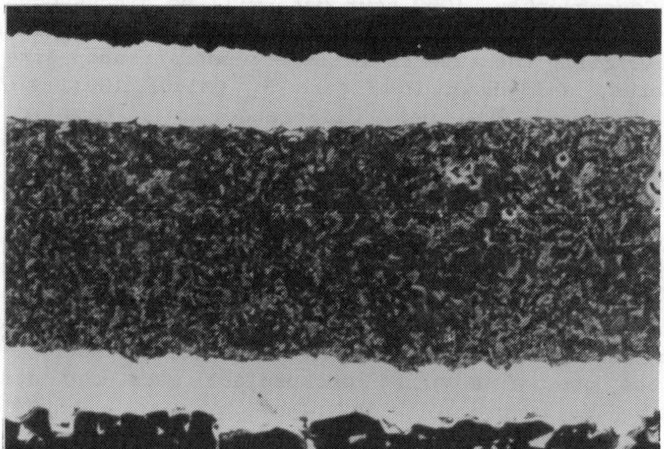

8.0 Series Q Reliability

The environmental reliability of Series Q materials has been studied using three different migration-resistance tests (17).

8.1 Lateral Migration Resistance Test (LMRT)

LMRT is used to assess the resistance to electrochemical migration of horizontally adjacent, closely-spaced conductor tracks in a high-temperature, high-humidity environment with a voltage bias present.

This test involved the exposure of interdigitated 125 or 250 um line and space patterns to $60^{o}C$, 90% RH for 1000 hours with a 48 Vdc bias between the lines. A 10 mohm resistor (current limited to 5 uA) was located in series with each circuit in order to prevent the destruction by any short circuits. Ten circuits were used per test. The total "adjacent" conductor track lengths represented by each of the test patterns were approximately 120 cm for the 250 um geometry and 150 cm for the 125 um geometry. Circuit failure was defined as an insulation resistance of $< 10^{9}$ ohms under the test conditions while the parts remained in the humidity chamber.

Relationships have not yet been developed that relate accelerated lateral migration test results with operating life yield. Consequently, 1000 hour data will be discussed. As shown in Figure 5, a traditional Pd-Ag conductor with a Ag/Pd ration of 2.5/1 will fail, the rate of failure dependent upon line spacing. As expected, and shown in Figure 5, QS170 (100/1 Ag/Pt) also fails this test. The only difference being rate of failure. However, as shown in Figure 6, the most effective way to eliminate potential lateral migration problems is through the use of suitable glass encapsulation.

Copper conductors are intrinsically more resistant to electrochemical migration than silver-bearing systems. One hundred percent yield was obtained for both the low and high temperature copper with the 250 um geometry. Some failures were obtained, however, with the finer 125 um geometry. As with Ag/Pt, the use of a glass encapsulant made the system 100% reliable.

From a practical viewpoint, it is not feasible to completely encapsulate a circuit with glass because the device attach pads must remain exposed. However, the use of a solder dam can provide the required protection.

Figure 5.

LMRT Test Results
Comparison of Ag/Pd (2.5:1) and
QS170 (Ag/Pt, 100:1) Conductors

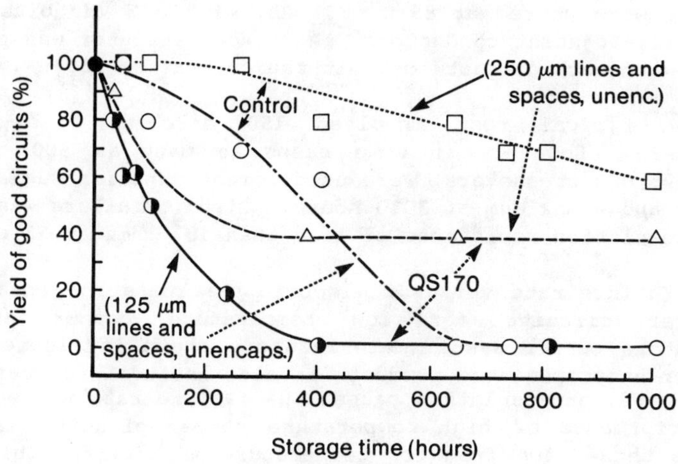

Figure 6.

LMRT Test Results for QS170 (Ag/Pt, 100:1)

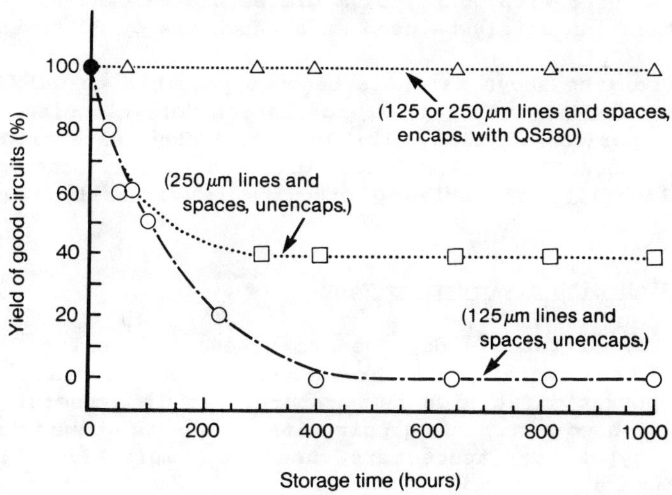

8.2 High Humidity Bias Temperature Test (HHBT)

This test monitors the ability of a dielectric to resist electrochemical migration when vertically-adjacent crossover conductor tracks are oppositely biased.

Crossover circuits, each containing 250 crossovers per circuit, were stored at 85°C, 85% RH, with a 5 Vdc bias between vertically-adjacent conductors. A 20 kohm resistor was placed in series with each circuit to limit the current to 250 uA.

A typical test involved 150 circuits, i.e., 37,500 crossovers. However, in some cases as many as 500 circuits, ie.., 125,000 crossovers, were used. Test duration was a minimum of 1000 and a maximum of 3000 hours. Circuit failure was defined as an insulation resistance of less than 10^{9} ohms at 25°C.

Failure rate versus dielectric thickness for encapsulated crossover circuits for low temperature copper and high temperature copper system materials are shown in Figure 7. The data for unencapsulated products is also included for reference. As expected, encapsulation causes the failure rate to be reduced. The performance of high temperature copper circuits is not as good as those made from low temperature materials. This is due to the lower hermeticity and inferior chemistry of the nitrogen dielectric (QP445). Reasons for this are beyond the scope of the present paper, but are related to the technical difficulties of nitrogen firing. It is recommended that, for optimum HHBT performance, encapsulated high temperature copper circuits should be manufactured with 50 to 60 um dielectric thickness. Air fired dielectrics can safely be used at a thickness of 30 to 45 um.

From the above data, it becomes possible to calculate the predicted reliability of a population of circuits over an extended period of time (18-20). With the exception of the non-recommended, unencapsulated QPLUS circuits, the predicted HHBT reliability of Series Q circuits is virtually 100% for 20 years.

8.3 High Bias Temperature Test

The HBT test is designed to assess how well a dielectric material can sustain its breakdown voltage during prolonged circuit operation at high temperature. It is a useful test for materials used in analog circuits such as power supplies. Similar types of tests are used by multilayer capacitor manufacturers.

Figure 7.

HHBT Failure Rates for QP153-QP445 Crossover Circuits Versus Thickness and Encapsulation

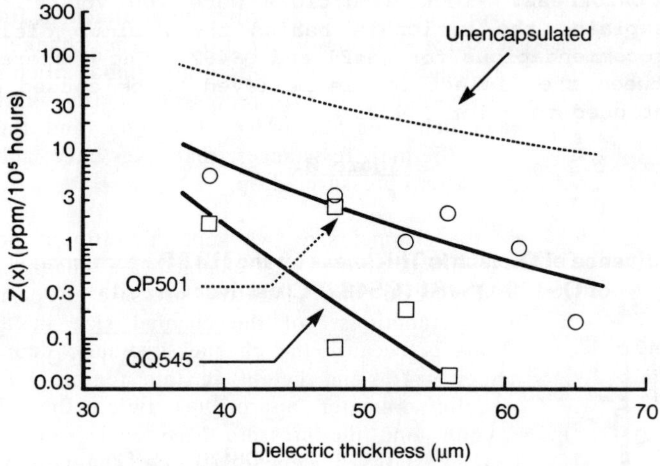

HHBT Failure Rates of Encapsulated QSII Crossover Circuits as a Function of Dielectric Thickness

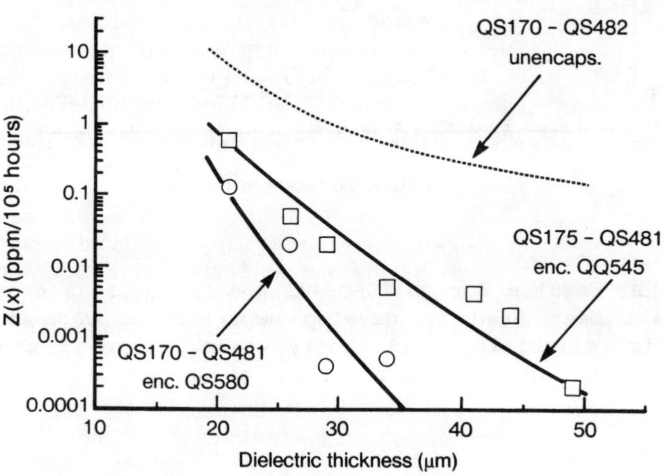

One of the strongest influences on the HBT reliability of a crossover circuit is the conductor material. For example, copper is considerably more inert than silver with respect to high-temperature solid-state migration (21).

The failure rate of QS481 and QS482 dielectrics with QS170 100/1 Ag/Pt conductor is shown in Figure 8 as a function of dielectric thickness. Test conditions were 200 Vdc at 150°C. The data explains the rationale behind the minimum dielectric thickness recommendations for QS481 and QS482. The difference in results between the dielectrics is believed to be caused by the blue pigment used to color QS482.

Figure 8.

Influence of Dielectric Thickness on the HBT Performance of QS170-QS481 (QS482) Crossover Circuits

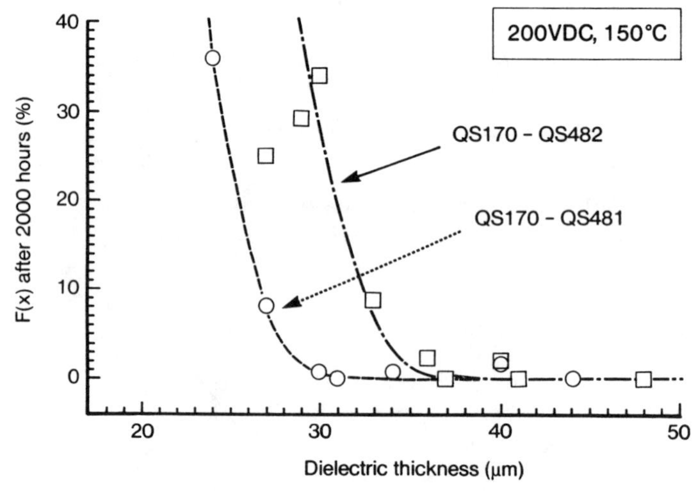

The HBT results for QS170-QS481 and QS170-QS482 crossover circuits have been used to develop empirical expression that predict their reliability over twenty years. This is shown in Table 3 for 30 and 40 micrometers at a variety of test conditions.

Table 3.

Twenty Year Environmental Yields of Unencapsulated QS170 -QS481 (482) - QS170 Circuits Following Exposure to Various HBT Conditions

300 crossovers per circuit

	VDC \ (°C)	50		100	
		QS481	QS482	QS481	QS482
30 μm	50	100	99.9996	99.9950	99.8876
	100	100	99.9981	99.9755	99.4429
	150	99.9998	99.9952	99.8784	98.5827
40 μm	50	100	100	100	100
	100	100	100	100	99.9997
	150	100	100	100	99.9991

At both 30 and 40 micrometers thickness, the reliability of QS481 is excellent. QS482, however, does show less than 99% yield at 30 micrometers for the more severe conditions. For this reason, it is recommended that QS481 is the preferred dielectric for circuits required to operate at very high temperatures and voltages.

The HBT performance of QPLUS circuits was found to be excellent. No failures were obtained during the standard HBT test and, therefore, as expected from previous consideration, copper circuitry has superior HBT performance.

9.0 Conclusions

Series Q materials have been designed to meet the immediate future needs of hybrid microelectronics. High conductivity silver bearing and copper conductors allow a choice of approaches in meeting circuit requirements. Dielectrics have been designed from first principles and offer outstanding reliability. Predictions based on test results indicate that circuits made from these materials will sustain high levels of reliability for as long as twenty years. In order for thick film hybrid ceramic technology to grow however, both process development and materials technology will have to be improved further.

REFERENCES

1 J. S. Shah and W. C. Hahn, "Material Characterization of Thick Film Resistor Pastes," Proc. of the 1978 Electronic Components Conference (IEEE), pp 234-243, 1978.

2 J. R. Larry, R. M. Rosenberg, and R. H. Uhler, "Thick-Film Technology: An Introduction to the Materials," IEEE Trans. Components, Hybrids, and Manufacturing Technology, Vol. CHMT-3, pp 211-225, 1980.

3 J. W. Pierce, D. W. Kuty and J. R. Larry. "The Chemistry and Stability of Ruthenium-Based Resistors," Proc. 1981 European Hybrid Microelectronics Conference, pp 283-302, 1981.

4 C. R. S. Needes, "Base Metal Thick Film Materials: A Review of Their Technology and Applications," Proc. 1982 International Microelectronics Conference, pp 94-101, Tokyo, 1982.

5 P. C. Donohue, J. Hormadaly, C. R. S. Needes, S. J. Horowitz, and J. F. Knaak, "Nitrogen-Fireable Resistors; Emerging Technology for Thick Film Hybrids." Proc. 1987 Electronic Components Conference (EEE), pp 299-307, 1987.

6 C. R. S. Needes, P. C. Donohue and J. Hormadaly, "Thick Film Materials for Copper Hybrid Circuits," Proc. 1984 International Microelectronics Conference, pp 239-248, Kobe, 1984.

8. R. C. Headley, "Laser Trimming is an Art that Must Be Learned," Electronics, June 1973, pp 121-125, 1973.

9. R. J. Bacher and V. P. Siuta, in Proc. 36th Electronic Components Conf. (IEEE), New York, 1986.

10. J. R. Larry, R. M. Rosenberg, and R. O. Uhler, IEEE Trans. CHMT-3, (2), 211-225, (1980).

11. R. W. Vest, Ceramic Bulletin 65(4), 631-636, (1986).

12. P. J. Holmes and R. G. Loasby, Handbook of Thick Film Technology, (Electrochemical Publications, Ltd., Glasgow, (1976).

13. R. G. Loasby, N. Davey and H. Borlow, Solid State Technology, May 1972.

14. C. R. S. Needes, "The Accelerated Life Testing of Copper Thick Film Materials." Proc. Int. Symp. on Microelectronics. pp 840-847, Atlanta, Georgia 1986.

15. D. E. Carlson, K. W. Hang, and G. F. Stockdale, "Electrode 'Polarization' in Alkali-Containing Glasses," Am. Ceram. Soc., Vol. 55, No. 7, July 1972.

16. W. D. Kingery, H. W. Bowen and D. R. Uhlmann, "Introduction to Ceramics," published by John Wiley, 2nd Ed. 1976.

17. C. R. S. Needes, K. M. Nair and M. V. Coleman. "The Environmental Reliability of Series Q: A Thick Film Materials System for the Manufacture of Advanced Hybrid Microcircuits." Proc. 1988 Int. Microelectronics Conf., Tokyo, May 25-27, 1988.

18. C. R. s. Needes, D. P. Button, "Reliability Testing of Thick Film Multilayer Materials," ECC Proceedings, 1985, pp 505-511.

19. R. R. Sutherland, I. D. E. Videlo, "A Comparison of the Reliability of Copper and Palladium-Silver Thick-Film Crossovers," ECC Proceedings, 1985, pp 498-504.

20. F. N. Sinnadurai, "Handbook of Microelectronics Packaging and Interconnection Technology," Electrochemical Publications, 1985, pp 135-174.

21. C. R. S. Needes, G. K. Boyce, "The Long-term Reliability of Thick Film Copper Multilayer Materials," HYBRID MICROTECH Proceedings, 1988, pp 178-205.

EVALUATION OF COMMERCIAL 96%ALUMINA SUBSTRATES AND MECHANISM OF PASTE-SUBSTRATE INTERACTION

Takashi Yamaguchi and Shuji Kato
Faculty of Science and Technology, Keio University, Yokohama, Japan 223

ABSTRACT

Seven commercial 96% alumina substrates and sapphire were evaluated. RuO_2-based paste and pyrochlore-based paste showed an opposite resistance dependending upon the substrate material. Enrichment of RuO_2 particles on the top surface and partial decomposition of pyrochlore particles have been suggested as factors responsible for the resistivity change.

1. INTRODUCTION

This work was undertaken to understand the effect of substrates on the resistivity of ruthenium oxide-based thick films in hybrid thick film technology. The resistivity dependence on the substrate materials is a well-known fact; identical paste and identical processing lead to a wide margin of resistivity depending on the manufacturer. This makes the exchange or application of substrates from different manufacturers difficult, and may decrease the reliability of thick film systems.

The mechanism by which the resistivity depends on the substrate has not been fully explored because of the large number of factors involved in the resistivity development.[1,2]

This work is divided into two parts; 1) measurement and analysis of resistivity of commercial pastes for correlation with substrate properties, and 2) efforts for understanding the mechanism by which resistivity depends on the substrate characteristics.

2. EXPERIMENTAL

2.1 Materials

Nominal composition, firing temperatures and density of seven commercial 96% alumina substrates are listed in Table 1. A sapphire plate was used as a reference material. Three series of resistor pastes were used. They are 1600, 1700 and 9000 series. They were designated as pastes X, Y and Z respectively in this paper. Each series consisted of 3 sub-series, 1kohm, 100kohm and 1Mohm. House-made pastes of a simple composition were prepared with just glass and conductive particles, RuO_2 or pyrochlore ($Pb_2Ru_2O_7$). The glass contained 62.5wt% PbO in addition to B_2O_3, SiO_2 and Al_2O_3.

2.2 Procedures

Pastes were printed on the substrates in 1x1mm pad shape with Au electrodes and fired at 850°C for 10 min. The sheet resistivity was normalized with respect to thickness. The film thickness was in the range between 10 and 15 microns after firing.

3. RESULTS AND DISCUSSION

3.1 Resistivity measurement with commercial 96% alumina substrates

Figure 1 shows the resistivities of thick films printed and fired on seven substrates and sapphire. The substrates are arranged in the order of increasing resistivity with paste Z-1Mohm and named A through G. The resistivity of paste Y-1Mohm, however, decreased from substrates A through G. Measurements with other paste materials showed that resistivity increased from A through G for some paste materials and decreased from A through G for others. Inspection of the paste materials such as by XRD and chemical analysis showed that the former paste group is based on RuO_2 and the latter, on pyrochlore. Hence, different mechanisms of paste-substrate interaction is assumed.

Table 1. Properties of alumina substrates

name	A	B	C	D	E	F	G
Firing temp.(°C)	1520	1500	1510	1560	1550	1590	1600
Density (kg/m^3)	3780	3770	3720	3720	3730	3670	3600
Al_2O_3 (wt%)	96.0	96.5	96.4	95.9	96.2	96.7	96.6
SiO_2 (wt%)	2.8	2.6	2.3	2.90	2.65	2.1	2.0
MgO (wt%)	0.6	0.8	0.9	0.87	0.52	0.9	1.0
CaO (wt%)	0.6	0.1	0.4	0.28	0.34	0.2	0.4

3.2 Effect of substrate properties on the resistivity
a) Figure 2 shows the relationship between the resistivity and density of the substrates. Increase in density increases the resistivity for paste A-1Mohm but decreases the resistivity for paste G-1Mohm. (With increasing density the resistivity increases for paste Y-1Mohm but decreases a for paste Z-1Mohm. No apparent correlation was observed with thermal expansion coefficient, surface roughness, and grain size of alumina[2].
b) Figure 3 illustrates the effect of SiO_2 content on the resistivity. Correlation is obvious. The resistivity was also dependent on the MgO content, but not on the CaO content.
c) Effects of surface properties were studied using polished substrates. Basically, polishing increased the resistivity, but the general tendency remained unchanged. Polishing of a substrate with a large amount of glass on the surface, as determined by the fluorescent chemical analysis, did not change the resistivity to a marked degree. Thus, thc surface glass has little effect on the resistivity.
d) In order to estimate the effect of pores on the resistivity, a series of substrates were obtained by firing green sheets with the same composition at different temperatures. Figure 4 shows the resistance of thick films as a function of firing temperature. The resistance increases with increasing firing temperature

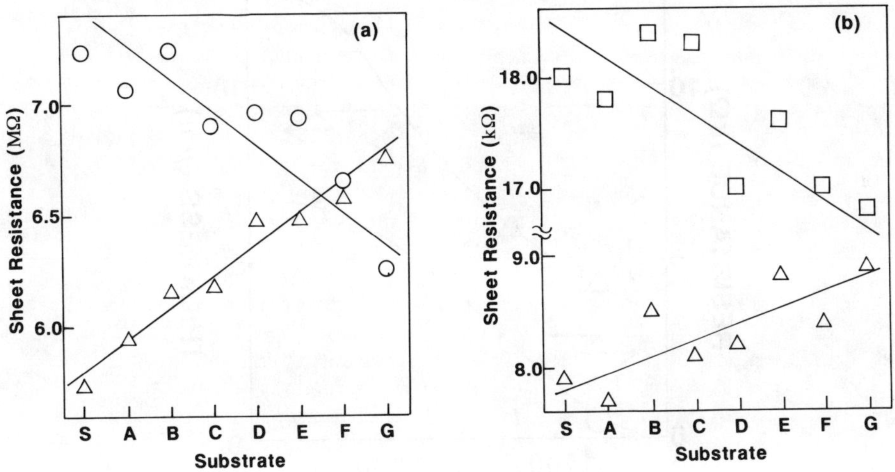

Fig.1 Sheet resistance on various substrates, (a) 1Mohm paste, (b) 1kohm paste; (□):X-paste,(○):Y-paste, and (△):Z-paste.

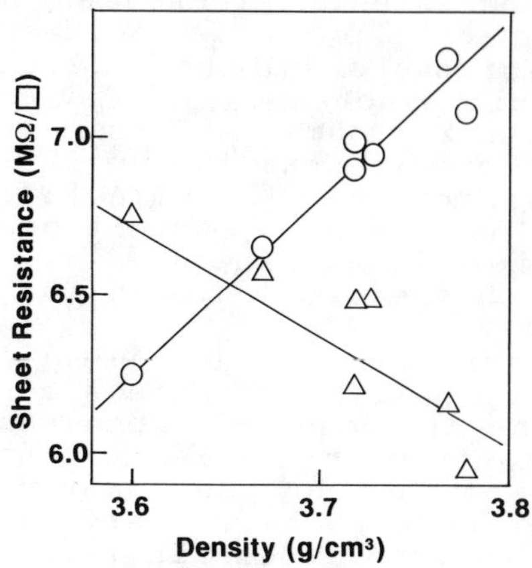

Fig.2 Relation between sheet resistance and density of substrate; (○):Y-1Mohm paste, (△):Z-1Mohm paste.

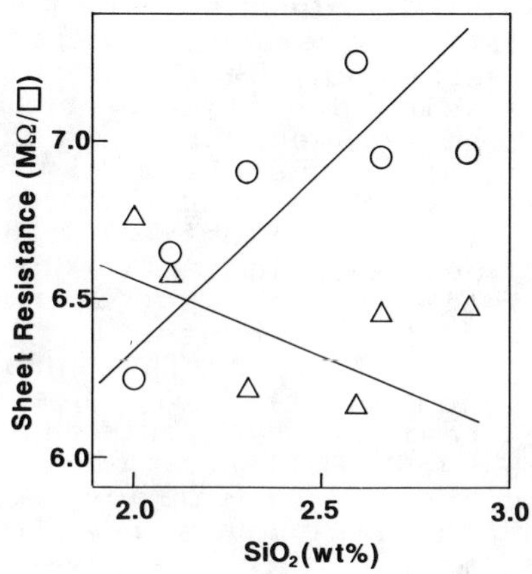

Fig.3 Relation between sheet resistance and SiO_2 content; (○):Y-1Mohm paste, (△):Z-1Mohm paste.

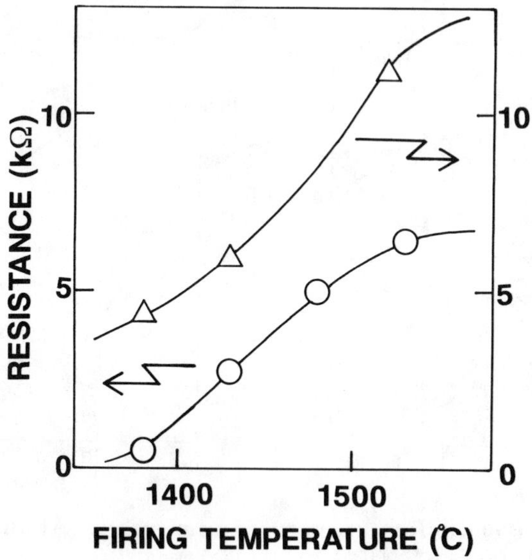

Fig. 4 Effect of substrate firing temperature on resistance

regardless of the paste type. This feature is understood because of the thickness of the film as shown on the right axis in Fig.4. Porous substrates absorbs glass resulting in the reduction of film thickness. Thus, the enrichment of conductive particles in the film leads to the decrease in the resistivity. Effect of pores is no longer obvious, once they are closed at high temperatures.

3.3 Origin of paste-substrate interaction

Several attempts were made to understand the observed dependence of resistivity on the substrate; effects of film thickness and glass penetration into the substrate, distribution of conductive particles in the film, and change in crystal structure in conductive particles on the resistivity were studied.

a) Effect of film thickness on the resistivity was studied using house-made pastes of a simple composition using RuO_2 and pyrochlore conductive particles. Substrates A and G were chosen referring to Figs. 1 , because they showed markedly different resistivities. Figures 5 show the thickness dependence of resistivity. Here, the conductance, not resistivity, is plotted on the ordinate for convenience of discussion. When pyrochlore particles are used, the same conductance was observed for substrates A and G, and the conductance is proportional to the film thickness. With RuO_2, on the other hand, the conductance-thickness relation is not a simple function of thickness and the conductance

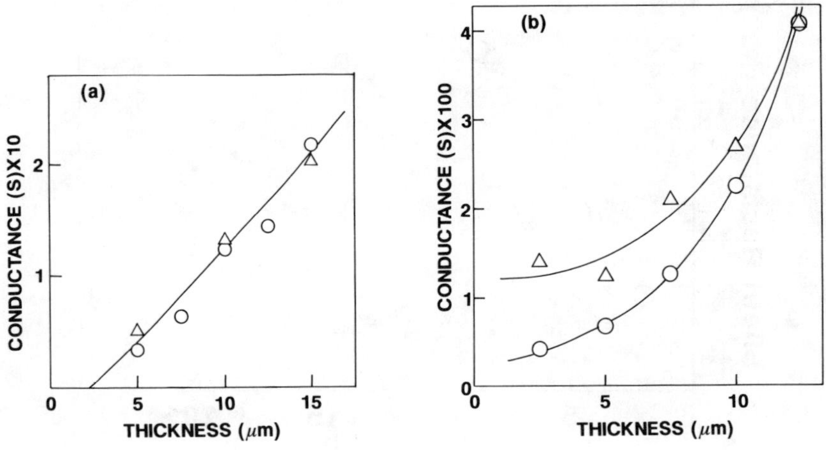

Fig. 5 Film thickness dependence of conductance, (a) pyrochlore-based paste, (b) RuO_2-based paste; (△):substrate A, (○):substrate G.

depends on the substrate. Glass-substrate interaction should be extensive for thinner films, and RuO_2-based thick films seems to suffer from such interaction.
b) Penetration of glass into the seven commercial substrates was estimated from quantitative determination of lead oxide using EPMA. Results are shown in Fig. 6. No apparent correlation can be claimed between penetration of lead and resistivity. Hence, the penetration of glass into substrates with closed pores does not influence the resistivity, unlike the behavior for substrates with open pores.
c) Distribution of conductive particles along the thickness direction was obtained using EPMA. As is shown in Fig. 7, RuO_2 conductive particles are enriched near the top surface. Such enrichment of RuO2 particles was remarkable when substrates which gave lower resistivities were used. For pyrochlore-based pastes, however, no enrichment of pyrochlore particles was noted regardless of the substrate material. A possible explanation for such difference in the distribution of conductive particles is that the particle size of pyrochlore is larger than that of RuO_2.
d) Fig.8 shows XRD patterns of pyrochlore-based thick films fired on substrates A and G, which gave rather large differences in resistivity(Fig. 1)[3]. Only the pyrochlore phase was observed at 700°C for both substrates. Firing at 850°C, however, gave a new peak, which was identified as RuO_2, with the substrate G.

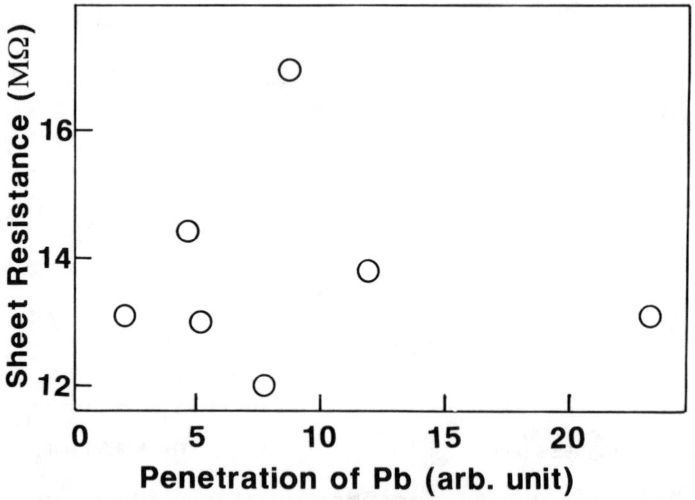

Fig. 6 Effect of Pb penetration into substrate on sheet resistance (Y-1Mohm paste).

Thus, the resistivity with substrate G higher than that with substrate A has been explained by the formation of more conductive RuO_2.

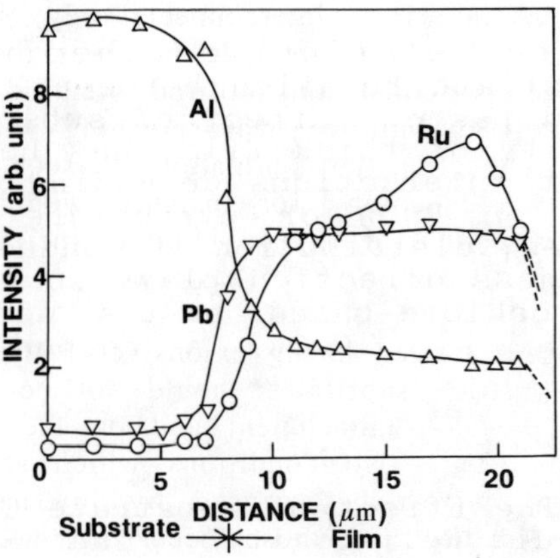

Fig. 7 Distribution of RuO_2 in resistor.

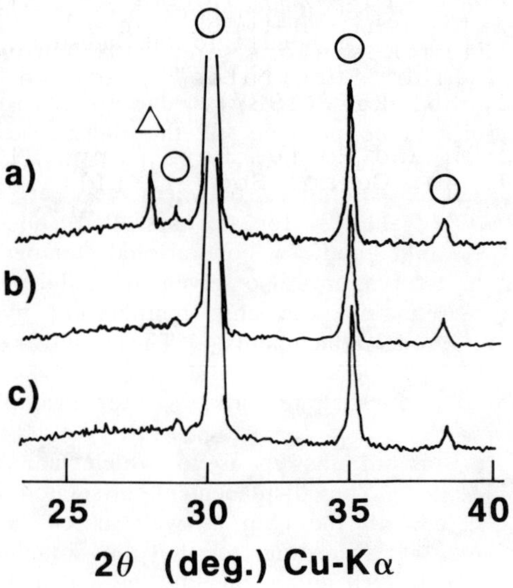

Fig. 8 Phase change in a pyrochlore-based resistor, (X-1kohm paste) heated at 850° C for 10 min., a) substrate G, b) substrate A, and c) unfired paste.

Here, it must be mentioned that almost the same resistivity was obtained for substrates A and G as shown in Fig. 5. The glass in this case contained a larger amount of lead than commercial pastes.

4. SUMMARY

Opposite correlation was observed in the resistivity-substrate relation between RuO_2-based and pyrochlore-based paste. Effect of substrate material on the resistivity of thick films has been understood by the different interactions depending on whether the paste is based on RuO_2 or pyrochlore particles. In RuO_2-based pastes, distribution of conductive particles along the thickness direction plays an important role, whereas in pyrochlore-based pastes decomposition of pyrochlore phase into RuO_2 is responsible for the resistivity.

5. REFERENCES

1. R.W. Vest. "The Effect of Substrate Composition on Thick Film Circuit Reliability", Final Report Contr. N00019-79-C-0240, Naval Air Systems Command, Feb. 1980.

2. Toshio Inokuma, Yoshiaki Tanaka and Miyoshi Haradome. "The Relationship Between The Electrical Properties of Thick Film Resistors and The Thermal Expansion Coefficient of The Substrates", Active and Passive Elec. Comp. 12, 251-257 (1987).

3. Eugene Aleshin and Rustum Roy. "Crystal Chemistry of Pyrochlore", J. Am. Ceram. Soc. 45 [1], 18-25 (1962).

SURFACE FINISH EFFECTS ON METALLIZATION OF ALUMINUM NITRIDE SUBSTRATES

E. Savrun and W. Lougher
Keramont Corporation, Tucson Arizona 85714 USA

ABSTRACT

As-fired, lapped, and polished aluminum nitride (AlN) substrates were metallized with molybdenum-manganese (Mo-Mn) paste. Metallized film properties were measured. Pull test results produced no metallization failures when the surface finish R_a ranged from 0.0254 to 0.813 microns indicating that metallization adhesion was not affected by surface finish. However, lapping decreased the pull strengths possibly due to surface damage introduced during machining.

Electron probe microanalysis (EPMA) studies were undertaken to clarify the nature of the interfaces. It appears that during the metallization process a chemical reaction forms a thin Al-O rich layer at the interface. Yttrium migrated from the grain boundaries of the AlN substrate through the interface layer and into the metallization probably as both an Y-Al-O phase and an oxygen poor yttria phase. It is shown that the bond between Mo-Mn metallization and AlN substrate is chemical and physical in nature.

INTRODUCTION

Aluminum nitride (AlN) is an attractive ceramic substrate for high power electronic circuit applications. However, in a majority of these applications the substrate has to be joined to a metal member. This can be accomplished by metallizing the substrate followed by plating and brazing to the metal member.

Metallizing ceramics with screen printable thick film refractory metal molybdenum-manganese (Mo-Mn) or tungsten (W) pastes is a popular means of joining metals to ceramics. Even though there exists a wealth of information on refractory metallization of oxide ceramics, alumina[1] and beryllia[2]; there is hardly any information on non-oxide ceramics, particularly aluminum nitride (AlN).

Fritless ceramic-metal joining via active metal brazing using Ti-Cu-Ag braze alloy was studied[3]. The process produced chemical bonding with the formation of TiN but at present is not useable for fine line circuit applications. In W paste cofired with an AlN green tape system, mechanical interlocking was found to be the primary bonding mechanism[4]. A major problem in metallizing AlN appears to be the absence of a glassy phase in the ceramic that can help interfacial bonding. This problem could be approached by adding a glass frit to the metallizing ink which becomes active at sintering temperatures. The glass frit must be both chemically and thermomechanically compatible with the ceramic. In case of BeO, it is common practice to add manganese oxide and titanium oxide to the frit to decrease the viscosity of the glass, and to promote chemical bond formation. However, in the case of AlN there is little information available.

It was claimed that in AlN substrates metallized with W-Mo-Mn paste, the mechanical interlocking was the primary bonding mechanism between the Mn-rich glass phase and the AlN ceramic via grain boundary penetration[5]. The glass phase produced amorphous and crystalline precipitates creating chemical bonding. It is also suggested that the possibility of AlN dissolution at the glass-ceramic interface could also contribute to the bonding mechanism.

It was shown that yttrium could diffuse into tungsten metallization and produce strong bonding between metallization and ceramic[6]. Pure tungsten coatings applied to polycrystalline yttrogarnet ($3Y_2O_3.5Al_2O_3$) were fired in wet H_2 atmosphere. Aluminum was not found across the interface indicating that yttrogarnet had broken into yttria and alumina and the yttria had diffused into the tungsten. A later study[7] by the same researchers suggested that similar reactions would be expected between molybdenum and yttria due to the similarity between Mo and W. These observations indicate that yttria, a sintering aid, can also play an important role in AlN-refractory metallization bonding.

The present study has been undertaken to investigate the nature of the ceramic-metallization interface and the effects of surface finishing on the AlN and Mo-Mn bond.

EXPERIMENTAL PROCEDURE

AlN substrates were produced in-house in size of 25.4 x 25.4 x 1.168 mm. High purity AlN powder (>98.5%) with fine particle size (d_{50} = 1.3 μm max) was spray dried, dry pressed, and sintered using yttria (Y_2O_3) to promote densification by liquid phase formation at a temperature of about 1800°C. Typical properties are given in Table I.

Table I. Properties of aluminum nitride.

PROPERTIES	KERAMONT 7R
Purity	>98.5 %
Density	3.28 g/cm^3
Flexural strength	25-35 kg/mm^2
Knoop hardness (500g)	1200 kg/mm^2
Thermal conductivity (RT)	160-190 W/mK
Thermal coefficient of expansion (RT~400°C)	4.3 X 10^{-6}/°C
Dielectric constant	8.6 at 1 MHz
Dielectric strength	15 kV/mm
Volume resistivity (RT)	>10^{13} Ohms·cm

The specimens were classified according to their surface preparations as follows:

(a) As-fired: These were as-sintered specimens without any surface preparation.

(b) Polished: These specimens were polished after sintering using a proprietary process.

(c) Lapped: The specimens were lapped through 400 grit SiC lapping slurry.

(d) Lapped and polished: These specimens underwent further polishing using a proprietary process after lapping.

Substrate surfaces were characterized by surface profilometry and scanning electron microscopy before they were metallized. Surface finish (R_a) values were obtained using a Mitsutoyo Surftest 401 analyzer with a 10 to 12 micron diamond stylus. Scanning electron microscopy was used to characterize the surface structures.

Specimens were metallized using molybdenum-manganese paste. The test pattern screen printed consisted of ten 2.032 x 2.032 mm squares for epoxy stud pull tests and a 360 square sheet resistivity pattern. A 325 stainless steel mesh screen with a 12.7 micron emulsion thickness was used. A heat lamp was used to dry the paste after printing and leveling. Specimens were fired together in a wet hydrogen/nitrogen atmosphere at 1250°C with a 30 minute soak time.

Metallized substrates were characterized for print thickness, sheet resistivity, and adhesion after firing. Print thickness was measured using a Dektak surface profilometer. Resistivity was calculated by dividing the resistance obtained with an ohm meter by the number of squares in the test pattern. Metallization adhesion was measured using a Sebastian IV coating adherence tester. The strength data was analyzed using a two-parameter Weibull expression. The Weibull model and its applications have been described in detail elsewhere[8]. Student t-tests were applied to the means of distributions to determine if they were statistically identical. Comparisons were done at the

98% confidence level. Failure sites were characterized using scanning electron microscopy.

Selected specimens were cross-sectioned, polished, and the interface characterized using electron probe microanalysis (EPMA).

RESULTS AND DISCUSSION

Surface Characterization

Surface finish values are shown in Table II.

Table II. Surface finishes obtained after different processes.

Group Tested	Surface Finish (R_a) (microns)	Thickness (millimeters)
As-Fired	0.406 - 0.813	1.168 ± 0.076
Lapped	0.635 - 0.787	0.762 ± 0.076
Polished	0.051 - 0.076	1.168 ± 0.076
Lapped and Polished	0.051 - 0.076	0.762 ± 0.076

As-fired specimens generally had the same surface finish R_a values of 0.406 - 0.813 microns when compared to lapped samples with R_a values of 0.635 - 0.787 microns. The surface finishes were consistently in the 0.051 - 0.076 micron range for "polished" and "lapped and polished" specimens. Scanning electron micrographs of as-fired, lapped, polished, and lapped and polished specimen surfaces are given in Figures 1 through 4. It is apparent that the lapping process has introduced some surface damage. This observation is also consistent with surface finish, R_a values. Even though average surface finish values were the same for both "polished" and "lapped and polished" specimens, Figures 3 and 4 show a distinct difference. Lapping caused grain pull out which could not be compensated for by polishing as seen in Figure 4.

Metallization

Metallized specimens were pull tested using the Sebastian IV pull tester and strength data were analyzed using two-parameter Weibull model. Weibull parameters and average pull strengths are given in Table III. As indicated by correlation coefficients, strength data showed excellent fit into the Weibull model. Weibull cumulative distribution function plots for all distributions are shown in Figure 5. We see that strength distributions shifted to the left after lapping and lapping and polishing. Student t-tests performed at the 98% confidence level on means of distributions (Table IV) show that: 1) The pull strengths of the as-fired and polished substrates were identical; 2) the pull strengths of the as-fired and polished substrates were higher than the lapped substrates; and 3) the pull strengths of the "lapped and polished" substrates were not significantly different than the lapped substrates. Therefore, polishing in itself did not reduce pull strengths of specimens.

Table III. Weibull parameters and average pull strengths.

GROUP	N	AVG. PULL STRENGTH (MPa)	$SIGMA_0$	m	R
As-Fired	20	50.3 ± 11.7	54.5	1.7	0.98
Lapped	17	37.9 ± 11.7	42.8	1.7	0.98
Polished	20	49.0 ± 13.8	53.8	2.0	0.96
Lapped & Polished	15	34.5 ± 7.6	37.9	1.1	0.97

Table IV. Results of t-tests performed on distribution means.

COMPARED GROUPS	t-VALUE	AVERAGE STRENGTH COMPARISON	PROBABILITY
As-Fired (1) & Lapped (2)	2.979	(1) > (2)	0.98
As-Fired (1) & Polished (3)	0.181	(1) = (3)	0.98
As-Fired (1) & Lapped & Polished (4)	4.267	(1) > (4)	0.98
Lapped (2) Polished (3)	2.523	(2) < (3)	0.98
Lapped (2) & Lapped & Polished (4)	0.920	(2) = (4)	0.98
Lapped & Polished (4) & Polished (3)	3.539	(4) < (3)	0.98

Failure analysis showed that all failures during the tests occurred in the substrates as shown in Figures 6(a) and 6(b). Since pull strengths of the as-fired and polished substrates were higher than lapped substrates, it was not clear whether the decrease in strength could be attributed to removal of YAG compounds from the surface of the ceramic, to machining damage induced from lapping, or to a combination of both. A study is currently under way to correlate the strength of AlN to the concentration of secondary (YAG) phase present and to the machining damage introduced during lapping.

It appears that substrate metallization bonding may also have a chemical nature since the failure modes did not change with surface finish, in particular with polishing. Scanning electron micrographs of specimen surfaces after metallization firings given in Figures

7(a) through 7(f) clearly show that the ceramic surface had changed in appearance. All fired ceramic surfaces displayed this phenomena but it was most apparent on the polished specimens. This "rumpling" of the ceramic surface was attributed to the oxidation, but x-ray diffraction studies could not identify any alumina. Several major peaks for alumina were missing. This indicates a poorly crystallized alumina or substitution of a trace element distorting the alumina lattice structure.

Print thickness was studied on fired specimens using a Dektak profilometer. The average print thicknesses are given in Table V. The as-fired group displayed the most substrate to substrate variance. The variance across the as-fired samples could be attributed to the camber in the substrates. Cambering prevented the squeegee from accurately traveling across the substrate. The random nature of the thickness distributions indicate that print thickness was independent of substrate surface finish. Hence, wetting of thick film paste was observed to be unaffected by surface finish.

Resistivities were measured on the metallized patterns to see the consistency in sintering between groups. Average resistivity values were generally the same indicating even sintering as shown in Table V. Resistivities could not be measured on all of the polished specimens since the metallized pattern was purposely offset in order to lay pull test patterns on the polished areas.

Ceramic-Metallization Interface Characterization

As-fired and "lapped and polished" substrates were cross-sectioned, polished, and examined using electron probe microanalysis (EPMA) in order to gain insight into the bonding mechanism between the AlN substrate and Mo-Mn metallization layer.

The specimens were analyzed at the outer edge, center, and inner edge of the metallized layer; the center of the interface, and 1 to 4 microns from the edge of the AlN ceramic for Al, Si, Ti, Mo, Mn, O_2, W, and Y.

Since nitrogen was not detectable on the machine, atomic percentages were calculated based on the total elements analyzed. Results are summarized in Tables VI(a) and VI(b).

Table V. Sheet resistivity (milliohms/square) and average print thickness (microns) data for specimens.

SUBSTRATE	N	THICKNESS	RESISTIVITY
As-fired	7	13.36 ± 1.14	36.67 ± 1.09
Lapped	6	13.33 ± 0.41	37.82 ± 1.10
Lapped & Polished	5	13.10 ± 0.42	39.72 ± 0.76
Polished	3	11.33 ± 1.15	41.39 ± 0.84

The interface phase always contained high oxygen and aluminum values indicating most likely the formation of alumina, Al_2O_3. The oxygen values in metallization were not surprising since the metallization paste contained oxides.

Yttrium was present in the metallized and interfacial regions in both specimens. This indicates a migration of yttrium from the AlN grain boundaries, through the interface layer and into the metallization. Low yttrium values in the polished substrate indicate that the analysis was performed in the center of an AlN grain. The yttrium values in the metallized layer of the as-fired substrate were higher than the lapped and polished substrate. This was expected since a high concentration of Y-Al-O phase initially existed on the surface of the as-fired substrate. A scanning electron micrograph of yttrium distribution of the cross-sectioned as-fired substrate are shown in Figures 8 and 9 respectively.

Table VI(a). As-Fired substrate EPMA results. All numbers are expressed in atomic percent.

	METALLIZATION				SUBSTRATE	
	OUTSIDE	CENTER	INSIDE	INTERFACE	GRAIN	GRAIN BOUNDARY
Al	0.080	0.635	1.721	44.770	55.900	50.545
Si	1.490	1.456	2.093	1.479	0.028	0.012
Ti	0.593	1.415	0.687	0.941	0.017	0.018
Mn	1.666	1.370	2.132	0.745	0.012	0.006
Mo	89.409	87.284	69.703	0.322	0.000	0.072
Y	0.834	1.370	1.601	0.680	0.035	6.859
W	0.531	0.082	0.050	0.345	0.009	0.254
O[1]	4.342	3.808	3.524	26.813	0.085	0.238
N[2]	1.055	2.580	18.489	23.905	43.914	41.996

Table VI(b). Lapped and polished substrate EPMA results.

	METALLIZATION			INTERFACE	SUBSTRATE
	OUTSIDE	CENTER	INSIDE		
Al	1.295	1.387	1.017	43.141	55.624
Si	1.550	1.612	2.142	3.960	0.102
Ti	1.034	1.064	0.543	1.520	0.011
Mn	3.113	4.213	1.703	1.394	0.019
Mo	73.626	72.111	70.001	0.053	0.053
Y	0.873	0.870	0.575	0.492	0.000
W	0.411	0.000	0.011	0.100	0.000
O[1]	12.126	12.979	12.809	47.678	2.015
N[2]	5.972	5.764	11.199	1.662	42.176

[1]Oxygen analyses were uncertain due to instrumental difficulties.

[2]Nitrogen atomic percentages calculated based on total elements analyzed.

Except for silicon, no significant migration of metallizing elements appeared to be present in the AlN grains in the substrates. In addition, this was probably true of the grain boundary phase, but only one data point from the as-fired substrate specimen was available at this time.

CONCLUSIONS

1. Since no failures were observed at the metallization-ceramic interface, metallization adhesion appears to be unaffected by surface finish.

2. It was not clear whether the observed decrease in pull strengths for lapped substrates could be attributed to the removal of YAG compounds from the surface of the substrates, to machining damage induced from lapping, or to a combination of both.

3. Substrate surfaces should be free of camber for accurate fine line printing. Also, no problems were encountered screen printing on substrates when the surface finish ranged from Ra = 0.025 to 0.813 microns.

4. Wetting of thick film paste was observed to be unaffected by surface finish.

5. During the metallization process, it appears that a chemical reaction forms a thin Al-O rich layer at the interface. Yttrium migrated from the grain boundaries of the AlN substrate through the interface layer and into the metallization probably as both an Y-Al-O phase and an oxygen poor yttria phase. These observations suggest that chemical reactions could contribute to metallization/ceramic bonding.

ACKNOWLEDGMENTS

The authors would like to thank Mr. Clarence Mercer for metallization and Mr. Robert Hoffman for EPMA work.

REFERENCES

1. J.T. Klomp, "Interfacial Reactions Between Metals and Oxides During Sealing", The American Ceramic Society Bulletin, 59 [8], 794-802 (1980).

2. D.E. Clark, L.L. Hench, and S.R. Bates, "Molybdenum Metallizing on Beryllia", The American Ceramic Society Bulletin, 53 [6] 473-477 (1974).

3. R.E. Loehman, "Infacial Reactions in Ceramic-Metal Systems", The American Ceramic Society Bulletin, 68 [4] 891-96 (1989).

4. Y. Kurokawa, C. Toy, and W.D. Scott, "Characterization of the AlN-W Interface in a Cofired Multilayer AlN Substrate", J. Am. Ceram. Soc., 72 [4] 612-16 (1989).

5. A.D. Westwood, M.R. Notis, "AEM Study of Thin and Thick Film Metallization on AlN Substrates", Mat. Res. Soc. Proc. vol. 108, 331-336 (1988).

6. R.E. Cowan, Stephen D. Stoddard, "Tungsten Metallizing Alumina-Yttria Ceramics", Informal Report #LA-6705-MS, Los Alamos Scientific Laboratory Report (March 1977).

7. R.E. Cowan, C.C. Herrick, and S.D. Stoddard, "Mechanism of Tungsten Adherence to Bodies containing Yttria", Informal Report #LA-6704-MS, Los Alamos Scientific Laboratory (March 1977).

8. K.V. Bury, <u>Statistical Models in Applied Science</u>, Wiley, New York, 1975, p.273.

Fig. 1. Typical as-fired surface. White grain boundary phase is YAG.

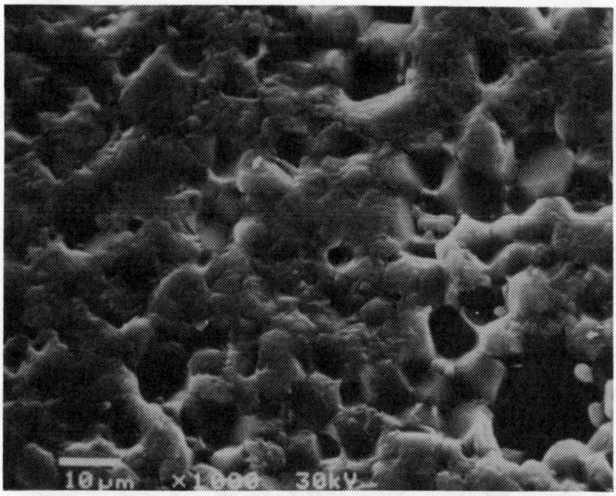

Fig. 2. Typical lapped surface. Note the extent of surface damage.

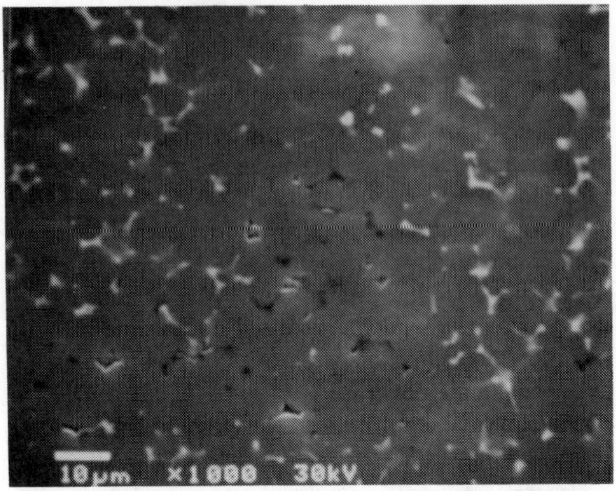

Fig. 3. Scanning electron micrograph of a polished surface. Note Y_2O_3 distribution and lack of grain pullout compared to those lapped and polished in Fig. 4.

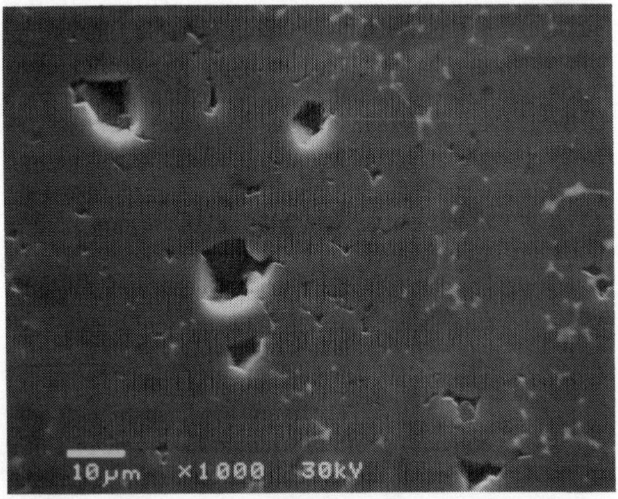

Fig. 4. Lapped and polished surface structure. Note the grain pullout and presence of Y_2O_3 (white phase).

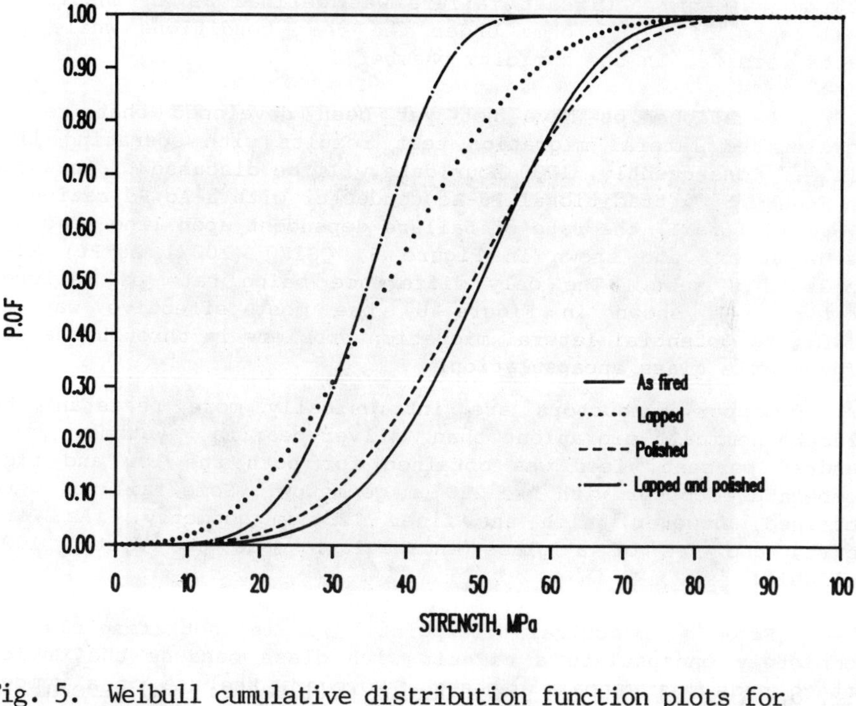

Fig. 5. Weibull cumulative distribution function plots for all specimens

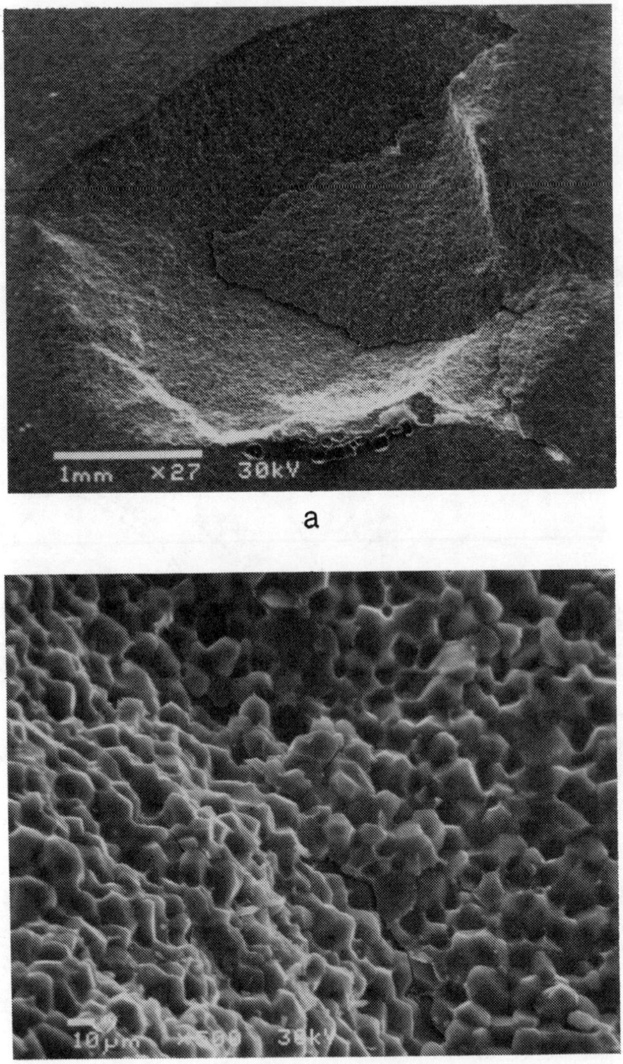

a

b

Fig. 6. Typical failure sites during pull tests: (a) general view, (b) mostly intergranular fracture

a

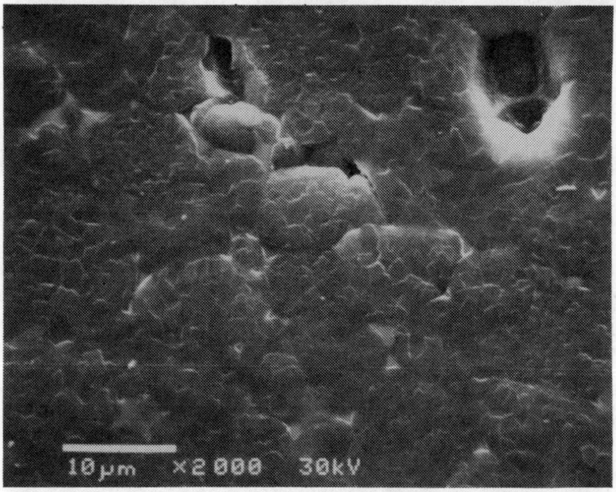

b

Fig. 7. Ceramic surfaces after metallization firings. Note "rumpling" in polished specimens: (a) lapped and polished; (b) high-magnification view of (a); (c) lapped surface after firing; (d) high-magnification view of (c); polished surface after firing; (f) high-magnification view of (e).

c

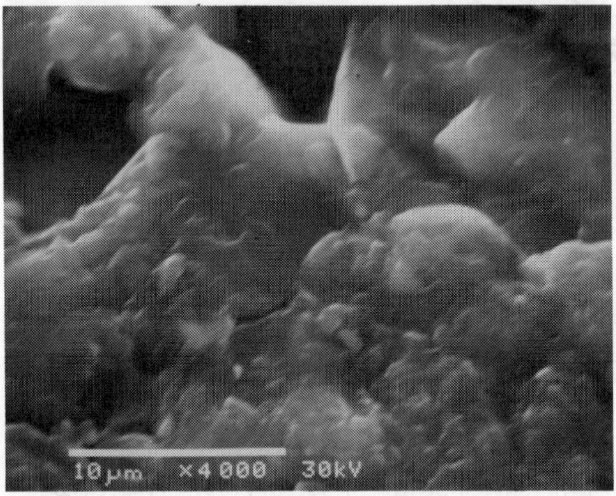

d

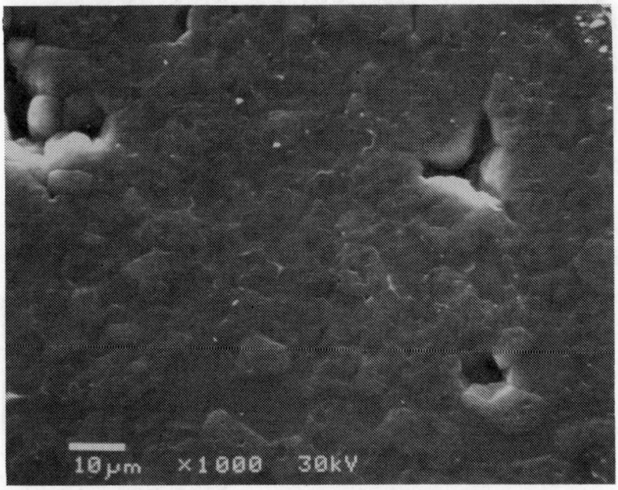

e

f

Fig. 8. Scanning electron micrograph of cross-sectioned as-fired and metallized substrate, X1000

Fig. 9. Yttrium X-ray dot map of the area shown in Fig. 8

EVALUATION OF SOLDERABILITY OF SILVER-GLASS ELECTRODE

Yasunori Wada and Takashi Yamaguchi
Faculty of Science and Technology, Keio University, Yokohama, Japan 223

ABSTRACT

Wetting behavior of silver-glass thick film conductors prepared under various conditions has been studied by the Meniscograph and solder ball methods, and correlated with the thick film properties such as surface concentration of silver and glass and microstructure. Different wetting properties obtained by the two methods were explained by the electrode properties.

1. INTRODUCTION

The purpose of this study was to analyze the solderability of silver-glass thick film electrodes. Two methods of evaluation were selected, Meniscograph [1,4,5] and solder-ball test[2,3], and the obtained results were compared and correlated with thick film properties such as surface concentration of silver and glass and microstructure. It is shown that the wetting behavior is not a simple function of the surface silver concentration but depends strongly on the pore structure.

2. EXPERIMENTAL

2.1 Materials

Silver-palladium electrodes were prepared by printing a commercial ink (TR4846 by Tanaka Matthey Corp.) on 96% alumina substrates and fired under the desired time-temperature conditions. Thickness was varied from 7.5 to 20.5 microns to modify the surface properties. Solder with the eutectic composition was used.

2.2 Properties of thick films

Surfaces and polished cross sections of thick films

were observed by SEM and optical microscopy. Surface concentrations of silver, palladium, boron, bismuth and oxygen were determined by Auger Electron Spectroscopy. Silver and oxygen were chosen as a measure of relative concentrations of metal and glass phases respectively.

2.3 Evaluation of wetting behavior

2.3.1 Meniscograph method

The Meniscograph method evaluates wetting properties from the wetting curve which is the time change of the force imposed on a specimen when it is dipped into a liquid, held at a certain depth, and then withdrawn. Figure 1 illustrates a typical example. Thus, the kinetics of wetting depend on many factors other than surface properties of materials. Results of our previous experiments revealed that the equilibrium attractive force(C in Fig. 1) does best describe the wetting properties of a material and that other two force parameters (A and B in Fig. 1) depend strongly on the shape of specimens and conditions of measurement. The wetting curve was obtained at 250°C. Flux was applied just before the measurement.

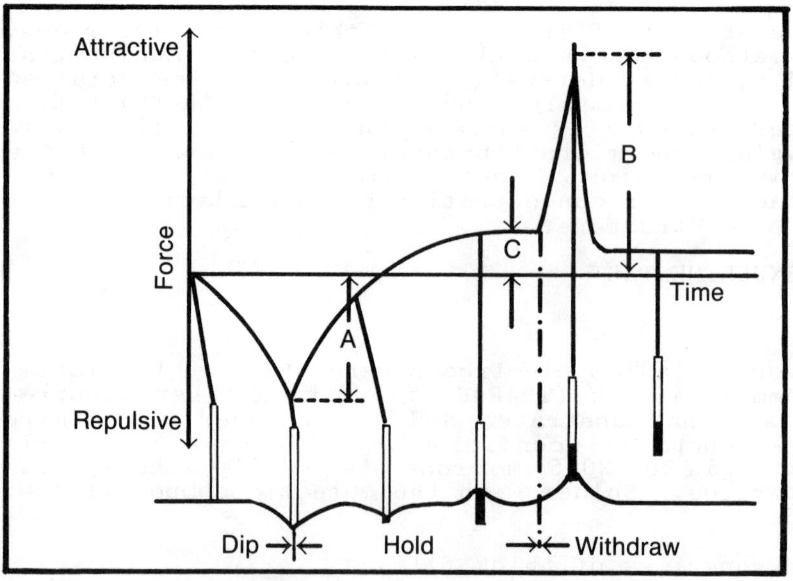

Fig. 1 Typical Wetting Curve

2.3.2 Solder-ball test

A solder ball with a diameter of 1mm was placed on a thick film at 250°C and kept for 15sec. The height was recorded after cooling. The recorded data are the average of eight determinations. Flux was also applied just before placing a solder ball on a thick film.

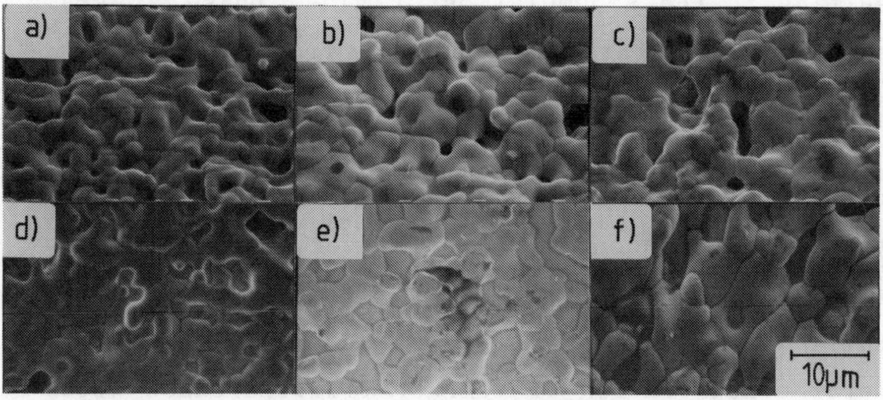

Fig. 2 Surface views of thick films.
Thickness:(a),(b) and (c)=11.5μm(d),(e) and (f)=20.5μm
Firing time:(a) and (d)= 40min, (b) and (e)=120min
(c) and (f)=360min

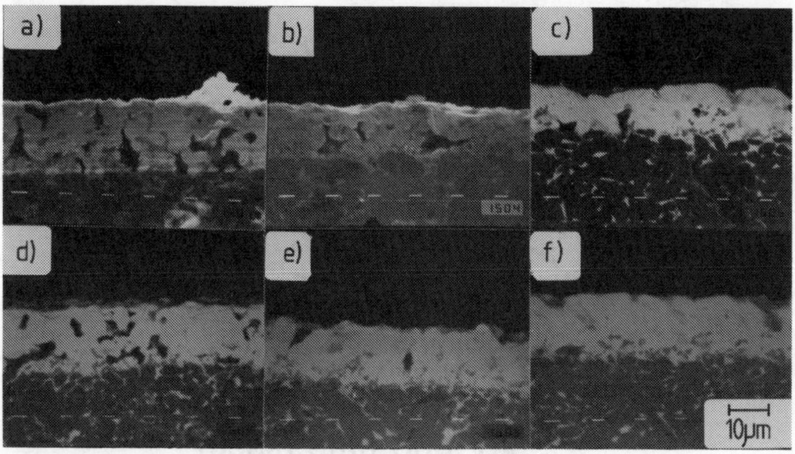

Fig. 3 Cross sectional views of thick films.
Thickness:(a),(b) and (c)=11.5μm(d),(e) and (f)=20.5μm
Firing temperature: (a) and (d)=760°C,
(b) and (e)=850°C,(c) and (f)=920°C

3. RESULTS AND DISCUSSION

3.1 Characterization of thick films

Figure 2 shows surface views of thick films fired for different time periods and Figure 3 shows cross sectional views of thick films fired at different temperatures. Grain growth proceeds with time resulting in the accumulation of glass phase at grain junctions. Extended heating causes roughening of the surface due to extensive material transport. Observations of cross sections showed that the glass is squeezed into grain boundaries and absorbed by the substrate material. Thinner films are more porous than thicker ones, probably due to the smaller amount of glass incorporated in the former. Thicker films retain the glass phase.

Figure 4 illustrates the effect of firing time on the surface concentration of silver and oxygen for estimating the relative distribution of metal and glass phases respectively. The glass concentration decreases with firing time regardless of the thickness. This is interpreted by the accumulation of the glass phase at grain junctions. The surface metal concentration of thinner film increases with firing time but that of thicker one remains unchanged. The difference is explained by the porous nature of thinner films. Thus, glass is depleted from the films resulting in an increase of metal concentration.

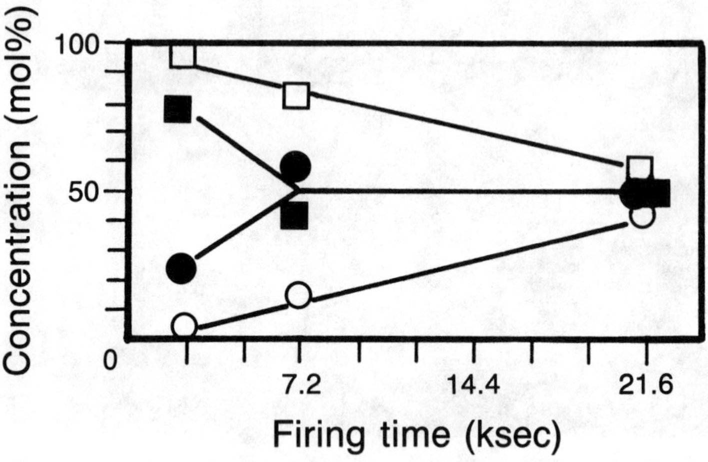

Fig.4 Effect of firing time on surface concentration.
Metal :(○ and ●), not-Metal:(□ and ■)
thickness:(○ and □)=11.5μm, (● and ■)=20.5μm

3.2 Mechanism of microstructure development

The sintering of silver-glass thick films depends on the silver-glass interaction. Probably, glass will act as a liquid phase in so-called liquid phase sintering. When glass softens, silver particles are wetted and activated for particle rearrangement leading to densification. Bonding and neck growth of silver particles squeezes the viscous glass to grain junctions through grain boundaries. Furthermore, glass-substrate interaction causes the depletion of glass from the film, leading to porosity. The extent of depletion depends on film thickness, because the glass volume per unit film area is proportional to the film thickness. This is the main reason for the increased surface silver concentration on extended heating(Fig.3).

3.3 Wetting behavior of thick films

Figure 5 illustrates the results of wetting measurements. As mentioned already, the equilibrium attractive force was plotted as a function of firing time in the Meniscograph method. In the solder-ball

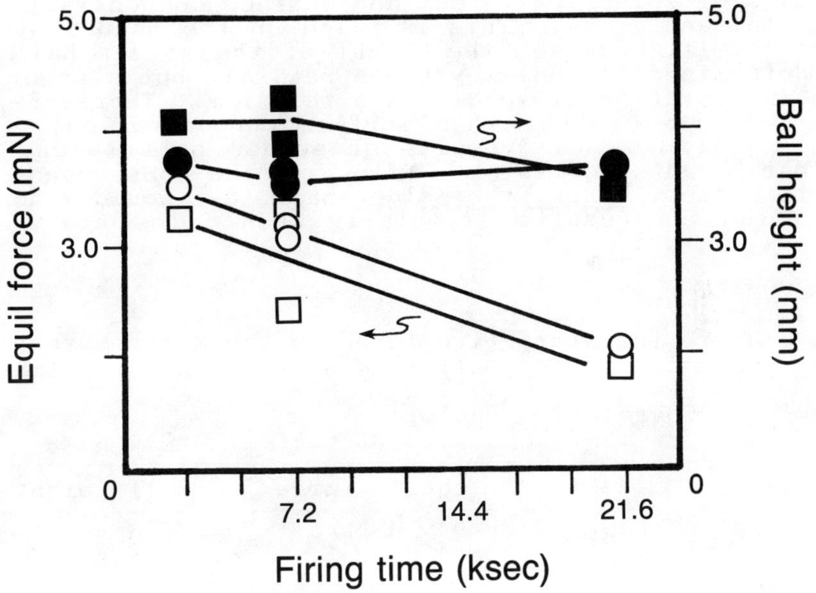

Fig. 5 Effect of firing time on wettability test. Thickness: (○ and ●) = 11.5μm, (□ and ■) = 13.5μm

test, the height is recorded as a measure of wetting. In the Meniscograph test, wetting of thick films degraded with increasing firing time with a more pronounced effect in thinner films. In the solder-ball test, on the other hand, no correlation was observed between wetting and firing time irrespective of the film thickness. Basically, thinner films gave better wetting in both tests.

The results of the Meniscograph test showed that the equilibrium attractive force is related to the surface concentration of silver, especially for thinner films. However, wetting decreases with increasing surface concentration of silver. This is rather puzzling, but surface distribution of silver and glass is probably responsible for this behavior. Thus, accumulation of glass at grain junctions would prevent uniform and fast wetting of solder on the film surface. This is supported by the degraded wetting for reduced surface oxygen concentration. Table 1 shows the correlation factors between wetting by Meniscograph and surface concentrations of silver and glass. Better correlation was obtained for thicker films than for thinner films. A possible explanation is the porous nature of thinner films.

No systematic correlation was obtained between the results of solder-ball test and surface concentration of metal and glass. This is based on the nature of test itself, because the height of the solder ball is determined not only by the spreading but also by the infiltration of solder into the film. Therefore, the porosity, rather than surface concentration, would determine the results of solder ball test. Larger heights for thicker films are the consequence of this fact. Thus, the solder-ball test would lead to erroneous results especially when films are porous.

4. SUMMARY

The equilibrium attractive force in the Meniscograph test was found to better describe the wetting

Table 1. Correlation factors

thickness	atom	equil. force	ball height
11.5 μm	Ag	0.665	0.238
	O	0.585	0.574
20.5 μm	Ag	0.924	0.200
	O	0.969	0.508

behavior of silver-glass thick films than solder-ball test. The solder-ball test is influenced by porosity of films more heavily than by surface concentration of metal and glass.

5. Reference

1. J. ten Duis and E. van der Meulen, "Measurement of Solderability of Components", Philips Technical Review, 28(12),362-365(1967).

2. B. Allen, "The Kinetics of Soldering", Welding Metal Fabrication, 50(1),47-50(1982).

3. J. Barranger, "Setting Standards for Solderability", Welding Metal Fabrication, 48(5),267- 271(1980).

4. I. Okamoto, T. Takemoto, M. Mizutani and I. Mori, "Physical Meaning of Wetting Curve Traced by Meniscograph Tester", Transactions JWRI, 14(1),21- 27(1985).

5. I. Okamoto, T. Takemoto, M. Mizutani and I. Mori, "Physical Meaning of Wetting Curve Traced by Meniscograph Tester(Report II)", Transactions JWRI, 14(2),201-207(1985).

6. J. Jellison, D. Johnson and F. Hosking, "Statistical Interpretation of Meniscograph Solderability Tests", IEEE Transactiuons on Parts, Hybrids, and Packaging, PHP-12(2),126-133(1976).

DIMENSIONAL CONTROL IN LOW-TEMPERATURE CO-FIRED CERAMIC MULTILAYERS

H. T. Sawhill, R. H. Jensen, K. R. Mikeska,
The Du Pont Company Experimental Station Laboratory,
Wilmington, DE 19880

ABSTRACT

Tight dimensional tolerances are required in manufacturing of low temperature co-fired ceramic packaging systems. Material and processing factors important to dimensional control are reviewed. The relative influence of ceramic powder surface area and binder/plasticizer/ceramic ratios on dielectric tape shrinkage and mechanical properties was determined. Distortion analysis was used to determine the contributions to shrinkage variability associated with the blanking, handling and lamination processing steps. The dimensional issues surrounding signal continuity, surface land pattern geometries and controlled electrical impedance are discussed.

INTRODUCTION

The worldwide electronics industry continues to demand packages with higher electrical density, improved heat dissipation and lower cost. Ceramic and organic approaches are used routinely; reliability, density and integrated passives are cited advantages for ceramics and cost is the chief factor cited by those favoring organic approaches. [1,2] Multilayer ceramics packages include leaded and leadless chip carriers, pin grid arrays, side-brazed dual-in line (DIP's) packages and flat paks. Ceramic fabrication technology combines cast dielectric tapes or prefired substrates with thick film metal conductor and passive (e.g. resistors and capacitors) pastes. Co-fired ceramic tape based systems are frequently categorized as either high temperature (e.g. alumina based with refractory W or Mo-Mn metallization fired at temperatures of 1375-1600°C) or low temperature (precious or base metal (Ag, Au, Cu) conductors fired at temperatures of 850-950°C).

Co-fired ceramic tape systems exhibit lateral shrinkage values ranging from 10 to 17 percent between the green and fired state. A high degree of dimensional control in co-fired packages is required for via registry, surface mounted component placement, post-fired printing and testing alignment, and cutting of multi-up parts. Controlling the reproducibility of dimensions to small variance is critical for alignment during post printing or surface mounting com-ponents and for accurate positioning of circuit features such as vias, landing pads, wire bonding pads and cavities. Reproducibility of fired thickness dimensions is important in achieving consistent circuit impedance in final packages. As co-fired packages grow to larger sizes for multi-up configurations or larger number of components, the issue of dimensional control of features increases in importance. Typical shrinkage tolerance values for high temperature tapes are ± 0.5% while those for low temperature systems are ± 0.2 - 0.3%.[3] Recent improvements in processing control have narrowed this gap.[4] Factors influencing dimensional control can be classified as either materials or processing related. Materials factors include the particle size and surface chemistry of ceramic powders, ratios between pigment-binder-plasticizer, particle morphology and degree of dispersion in a paste or casting slip. Materials interactions during co-firing can also play a large role in resulting lateral shrinkage values. Processing factors include tape casting, tape handling, via formation and filling, printing, laminating and firing conditions.

This paper will review the material and processing factors important to control of dimensions in low-temperature co-fired ceramic packaging systems. Examples from commercial systems and research materials will be chosen to illustrate the capacity for particular variables to influence dimensions in fired parts.

DIMENSIONAL TOLERANCE REQUIREMENTS

Circuit features on the scale of 250μm (10 mils) for lines/spaces and via diameters are common in todays multilayer ceramic packages. Current thick film printing technology allows 125μm (5 mil) lines/spaces to be routinely achieved in high volume manufacturing. Reduction to half the current resolution poses difficult technical challenges; reducing resolution to a 25μm (1 mil) scale will very likely require alternate technologies such as photoimaging in place of screen printing. Ceramic packages range in size from tens to hundreds of mm on a side. Point to point shrinkage variations scale as the product of the diagonal package dimension and the shrinkage tolerance of the materials system. Thus, variations on a 25 mm x 25 mm (1 in. x 1 in.) scale would tend to increase by a factor of 7 ($\sqrt{2}$ x 5) on a 125 mm x 125 mm (5 in. x 5 in.) scale. Alignment of 250μ (10 mil) features at opposite ends of the diagonal of a

76 mm x 76 mm (3 in. x 3 in.) multilayer requires lateral shrinkage control on the order of 0.25%. For operations such as post printing or gang bond probe testing, it is the overall point to point dimensional variation across the top surface that is important. For automated production operations, image recognition systems are often used, and the dimensional variations only within a given feature [e.g. chip land pattern] is the important factor. In this case, the common reference for dimensional variation is the diagonal feature length rather than the overall package diagonal. Dimensional variations observed in practice are not always uniform across the top surface of a part. Shrinkage values quoted by manufactures denote edge to edge measurements but surface feature locations may vary in a different fashion than edge to edge variation. A square array of points may distort slightly during firing and end up with either random or systematic shifts in position resulting in rectangular or hourglass shapes instead of squares. Therefore, differences in measurements for shrinkage (dimensions between exterior surfaces of a multilayers) and feature dimensions (point to point locations across the top surface of a multilayer) will be distinguished.

Impedance requirements vary by circuit design but typical design and tolerance values run 50 Ω ± 5 Ω. The impedance is a function of metal geometry and conductivity, dielectric constant, and thickness. Shrinkage in the out of plane (or "z") direction for commercial systems often differ from the lateral shrinkage values, and tolerance values are in the ± 0.5 - 1.0% range. These shrinkage tolerance values are sufficient to meet existing impedance requirement of circuit designers. Since the impedance varies with overall geometry of the circuit, the out of plane shrinkage is only one of several factors important in providing consistent circuit impedance values.

DIMENSIONAL MEASUREMENT TECHNIQUES

In this study, edge to edge shrinkage measurements were made using electronic calipers with a measurement accuracy of 5µm (0.2 mil). Point to point surface feature dimensions were measured using an automated traveling microscope (accuracy of ±5µm (0.2 mil)) focused on fine metal crosses (20µm line width) on the top layer of a ceramic multilayers. The metal crosses were patterned on the dielectric tape using a proprietary photo forming technology. A schematic of the template pattern consisting of a 5 x 5 array of metal crosses (20µm lines, 2.5 mm long) is shown in figure 1. An example of a distortion plot for a fired sample made without rotating layers of tape is shown in figure 2. There is a slight difference in shrinkage between machine cast and transverse directions in the tape. This results in a distortion of the overall square template pattern to one of rectangular shape.

LOW TEMPERATURE CO-FIRED CERAMIC SYSTEMS

Low temperature co-fired ceramic systems are designed for use with precious or base metal conductors (viz. Au, Ag, Cu). The melting points of these metals limit the processing temperatures to below 1050°C. Most systems are designed to fire between 850°C and 925°C to allow compatibility with existing thick film technology. Dielectric tapes designed for these purposes include ceramic filled glass systems, crystallizable glasses and crystalline ceramics. Table 1 and Table 2 show a list of firing conditions and properties used in a number of different commercial dielectric tape systems. The differences in sintering behavior and properties of different systems are discussed elsewhere. [5]

MATERIALS VARIABLES

A typical breakdown by volume of materials components of a dielectric tape and thick film conductor are shown in figures 3 and 4 respectively. The figures show the volume percentages of each component in a wet, dried and laminated configuration. The dried to fired shrinkage value is associated with the removal of porosity and organic components during firing. For both the tape and thick film, this corresponds to the elimination of approximately 35 v/o of the green compact. Factors influencing the volume fraction or morphology of the porosity or organic components therefore have the capacity to influence shrinkage values.

Particle size distributions of the inorganic components and degree of dispersion in the paste or tape influence the products green density. Figure 5 shows an example of the influence of surface area of a glass powder on the lateral shrinkage of a dielectric tape. In this case, there is a inverse linear relation between the green density of the tape and the BET surface area of the glass powder.[3]

Particle size ratios between inorganic components can influence the overall packing factors. There are many theoretical and experimental studies published on the packing factor effects for multicomponent powder mixtures. To the knowledge of the authors, there have been no demonstrations of the required quantitive variations in overall particle size distributions of a multicomponent system needed to maintain specified green densities within the tolerances required (<1%) to achieve shrinkage control in the 0.2% range. In practice, shrinkage reproducibility is commonly achieved by qualifying the inorganic powders by surface area specification rather than by specification based on particle size averages or other indicators of the particle size distribution. Surface area is specified rather than particle size for several reasons. The amount of polymer adsorption onto powders surfaces is more sensitive to surface area than to particle size per se. Surface area and particle size change differently during milling operations. Also, consistency in polymer adsorption levels is important in maintaining consistency in tape mechanical properties.[6]

The ratio of inorganic to binder and plasticizer in the tape dielectric can influence the lateral shrinkage of a tape. Figure 6 shows a contour map of shrinkage as a function of the ceramic to polymer and polymer to plasticizer ratio of a dielectric tape. The contours represents measurements of 15 separate compositions within this field. For a given ceramic to polymer ratio, increased amounts of polymer increase the average distance between ceramic particles, causing lower green densities and higher shrinkage values. The mechanical strength of these tapes is illustrated in the contour map of figure 7. Strength is represented by the stress required to achieve 3% elongation. For this materials system, a strong statistical correlation exists between the stress values at 3% elongation and the stress values at failure (typically >25% elongation). Strength increases significantly with higher polymer to plasticizer ratios as expected [6]. A contour plot of the tape modulus is shown in figure 8. The modulus is the measured slope of the initial linear portion of the stress strain curve representing strain values between 0 and about 1% elongation. The modulus increases both with increasing polymer to plasticizer ratios and with increasing ceramic to polymer ratios. The best combination of mechanical properties is found in the region centered around ceramic/polymer ratios of 11 and polymer plasticizer ratios of 7. Higher values for both strength and modulus can be achieved but the tapes start to become brittle and lose the flexibility required during handling.

The high shear stresses experienced by the ceramic slip as it passes below the doctor blade during casting can lead to particle alignment in the cast tape. It is this phenomena that is thought to be responsible for the difference in tape shrinkage values observed between parallel and transverse orientations to the casting direction. Tapes made by spraying slips onto the carrier do not experience these shear stresses and show no difference in shrinkage values with tape orientation. Tapes made by doctor blade casting of slips comprising spherical powders as shown in figure 9, show no difference in shrinkage value with tape orientation.

CO-SINTERING EFFECTS:

Interactions between the ceramic tape dielectric and thick film metal pastes during firing can influence shrinkage of multilayer packages. The activation energy for diffusion and stress relaxation behavior is different for metals and ceramics. Designing the sintering characteristics of metal pastes to closely match the path of the tape dielectric is a common approach to controlling co-sintering shrinkage. Suppressing the sintering onset of these metals by addition of sintering inhibitors can compromise the final electrical properties . Recently published research [7] has shown that sintering path matching for metals and dielectrics is not necessarily a prerequisite for compatibility. In cases where the metal densifies much faster than the dielectric, the metal has reached near full density before the onset of dielectric shrinkage. In this case, the creep behavior of the metal is a determining factor on the overall shrinkage of the multilayer. The behavior depicted in figure 10 where the metal-ceramic multilayer shrinkage lies outside the range defined by the individual tape dielectric or metal components is explained by the influence of metal creep. However, the differences in the sintering paths between the metal and ceramic can strongly influence the resulting camber of co-fired multilayers. Camber variation have been observed with heating rate differences in co-fired multilayers.[3] These effects have been largely attributed to the suppression of metal densification onset with increased heating rates relative to the glass softening temperature.

PROCESSING VARIABLES:

A flow chart of the different processing steps used in the manufacture of ceramic multilayers is shown in figure 11. Dimensional variations introduced by the screening and tooling steps indicated in figure 11 are associated with specific equipment. Replacement schedules for equipment such as screens helps control and minimize the dimensional variations introduced during these processing steps. In this paper, we will report measurements of dimensional variations associated with selected processing steps of figure 11.

The blanking step consists of removing the cast tape from the film carrier and cutting it into blanks using a precision blanking die. The die simultaneously cuts the tape and punches the registration holes used for alignment during collation and lamination. For some processes, the tape is blanked while still attached to the carrier film. Results of distortion measurements of feature dimensions associated with the removal of the tape from the carrier film are shown in figure 12. Dimensional variations associated with this processing step appear to be random (a slight bias toward the stretching side) with average distortions in the range of 12 μm (0.5 mils) over 50 x 50 mms (2 in. x 2 in.).

Vias in the tape dielectric can be formed by a number of different methods, such as die set rigid punch arrays, numerically controlled punches, drills and lasers. Dimensional tolerances for each of these methods depends on the accuracy of the equipment used. In general, rigid punch arrays are suitable for high volume production of a proven design and tolerances for the process are a function of the individual punch. For drills, numerically controlled punches, and lasers, the table holding the part or the mechanical tooling is moved into position and the accuracy of the equipment in this regard determines the variations associated with this step. High via densities can increase the potential for further deformation of the tape during subsequent handling. Tolerance values for high quality moving "x-y" tables are of the order of 2μm or less.

After via filling, conductor printing and inspection, the individual tape layers are collated and aligned so that corner registration holes are concentric for placement on the registration pins in the lamination die as shown in figure 13. Experiments were run to determine the degree of tape distortion incurred through routine handling of the tape during this step. Patterned layers of tape were repetedly removed and replaced onto the registration pins. Measure-ments of dimensional positions of the patterened crosses (see section on dimensional measurement technique) were made with no external reference point in order to determine the distortions associated with tape handling and placement on the pins. A control was run in which the patterned layer was measured 4 times without removing or touching the tape to determine the variations associated with the xy table mechanism and, more importantly, the operator placement of cross hairs. The results are shown in figure 14. Average variations associated with this step are on the order of 12 μm but the measurement variations in the control were a significant part of this total. Similar measurements were made using external (to the tape) feature locations to determine the variations associated with movement of the tape while on the registration pins. In this particular case, the average errors were on the order of 17μm indicating that 5μm movements of the tape sitting on the alignment pins was possible for this particular lamination die.

The lamination step has the largest potential of any of the processing steps to influence the shrinkage in multilayer co-fired packages. Application of heat (~ 70°C) and pressure (~21 MPa) are used to laminate individual tape layers together. A graph showing the relationship between the lamination pressure and lateral shrinkage values is shown in figure 15. Previous reports [3] have shown this relation is associate with the removal of additional porosity in the green tape with additional applied pressure. During the lamination step, the v/o porosity in the tape is reduced from 33% to approximately 14% (depending on the applied pressure). Similar relationships between shrinkage and lamination [8] or calendering pressures [9] have been reported.

CONCLUSIONS

Dimensional control is becoming more important in co-fired ceramic multilayers as metallization feature sizes decrease and package sizes grow to incorporate multi-up configurations. Current commercial materials systems that deliver 0.2-0.3% shrinkage tolerances work well for 25μm (10 mil) features sizes on 76 x 76 mm (3x3 inch) boards or larger boards where dimensional variation within a feature (eg. chipland pattern is the limiting factor. Impedance requirements 50Ω ± 5Ω can be met with "z" direction shrinkage tolerances on the order of 1%, but "z" shrinkage is only one of several factors important in providing consistent circuit impedance values.

Powder surface area has a strong influence on green packing densities and consequently on the lateral shrinkage in dielectric tapes. Binder/Plasticizer/ Ceramic ratios in the slip influence the shrinkage with increased polymer yielding higher shrinkage values. The mechanical modulus of the tapes increase both with increasing polymer to plasticizer ratio and with increasing ceramic to polymer ratios. Anisotropy in shrinkage between the lateral and transverse directions in tape cast dielectrics originates from the high shear stresses experienced as slip passes below the doctor blade. Such shrinkage aninotropy is not observed in tapes made by slip spraying or casting with slips comprised of spherical powders.

Differences in sintering kinetics of the metal and dielectric components during the firing of the multilayer can significantly influence shrinkage and camber. In cases where metal densification precedes that of the dielectric, the creep behavior of the metal becomes the determining factor in the overall shrinkage of the multilayer.

Dimensional errors associated with removal of cast tapes from carrier films appear to be random with a slight bias toward the stretching side. Errors associated with movement of tape sitting on alignment pins prior to lamination were on the order of 5μm in a particular experimental design. A strong correlation exists between the tape shrinkage and the pressure used during lamination. Exaggerated distortion plots were found to be helpful in visualizing point to point distortions associated with given processing steps.

ACKNOWLEDGEMENTS

The authors gratefully acknowledge the assistance of Milli Kahn, Dan Schaeffer, Mary Ann Barsoum, and Mike McDevitt.

REFERENCES

(1) W. A. Schroen and J. M. Pankratz, VLSIC Assembly and Packaging, VLSI Electronics Microstructures, Academic Press, p 185, 1985

(2) D. W. Bergman and G. L. Ginsberg, Design and Layout, in Printed Circuits Handbook, C. F. Coombs, Jr., Ed., McGraw-Hill, p 1, 1979

(3) H. T. Sawhill, "Materials Compatibility and Co-Sintering Aspects in Low Temperature Co-Fired Ceramic Packages", Ceramic Engineering Science Proc., 9 [11-12] pp. 1603-1617, 1988

(4) Kenneth J. Hamilton, U. S. Patent 4,774,221, September 27, 1988, "Methods for Predicting and Controlling the Shrinkage of Ceramic Oxides During Sintering"

(5) H. Bhedwar, H. T. Sawhill, Volume 1 Packaging, Electronics Materials Handbook, C. A. Dostal, ASM International, Metals Park OH, November 1989, (To be published)

(6) W. H. Morrison, Jr., R. E. Johnson Jr., Adv. in Ceramics, "The Effect of Particle Surface Chemistry on the Mechanical Properties of Unfired Ceramic Tape", Vol. 26, pp. 1-14, 1989

(7) Tsungnan Cheng, PhD Thesis, Cornell University, August 1989, "Co-Sintering Behavior of Ceramic-Metal and Glass-Metal Multilayer Films: Modelling and Experiments"

(8) J. C. Biggers, T. R. Shrout, and W. A. Schulze, "Densification of PZT Cast Tape by Pressing", Am. Ceram. Soc. Bull., 58 [5] 516-21, 1979

(9) B. R. Schat and H. J. Engel, Calendering as a Method of Forming Thin Ceramic Sheets, Proc. Brit. Ceram. Soc., No. 18, pp. 281-92 (1970)

(10) M. Takabatake, J. Chiba, and Y. Yokubu, Composition for Multilayer Printed Wiring board Comprises Borosilicate Glass, Alumina, and Forsterite Powders, U.S. Patent 4,593,006, 1986

(11) H. C. Bhedwar, A. L. Eustice, S. J. Horowitz, F. K. Patterson, and H. T. Sawhill, "Low Temperature Co-Fired Ceramic Tape System '_An Overview", Proceedings of the SAMPE Electronic Materials Conference, Society for the Advancement of Material and Process Engineering, p 720, June 1987

(12) K. Niwa, N. Kamehara, H. Yokoyama, K. Yokouchi, and K. Kurihara, "Multilayer Ceramic Circuit Board With Copper Conductors", Multilayer Ceramic Devices, Advances in Ceramics, Vol 19, Eds., J. B. Blum and W. R. Cannon, American Ceramic Society, p 41, 1986

(13) Y. Baba, K. Higashiyama, and S. Segawa, "Co-Fireable Copper "Multilayer Ceramic Substrates", Proceedings of the International Society for Hybrid Microelectronics, p 405, Oct 1988

(14) H. Mandai, K. Sugoh, K. Tsukamoto, H. Tani, and M. Murata, "A Low Temperature Co-Fired Multilayer Ceramic Substrate Containing Copper Conductors", Proceedings of the IMC, p 61, May 1986
(15) S. Nishigaki, S. Yano, H. Kawabe, J. Fukuta, T. Nonomura, and S. Hebishima, "LFC III _ A New Low temperature Multilayered Ceramic Substrate With Au (top) _Ag (internal) _Tl/Mo/Cu (bottom) Conductors System", Proceedings of the International Society for Hybrid Microelectronics, p 400, Sept. 1987
(16) Y. Shimada, Y. Yamishita, Y. Shiozawa, M. Suzuki, and H. Takamizawa, "Low Dielectric Constant Multilayer glass-Ceramic Substrate and Ag-Pd Wiring for VLSI Packaging", Proceedings of the 37th Electronic Components Conference, Institute of Electrical and Electronics Engineers, p 398, May 1987
(17) N. Yoshida, T. Endo, and M. Sakamoto, "The Low Temperature Co-Fireable BaO-ZrO_2-B_2O_3 Ceramic green Sheet System", Proceedings of the IMC, p 98, May 1988
(18) S. Tosaka, K. Hoshi, T. Honda, K. Onigata, and N. Yamoaka, "Properties of Low Temperature Co-Fired Multilayer Substrates with Buried Passive Components", Proceedings of the IMG, p 83, May 1988

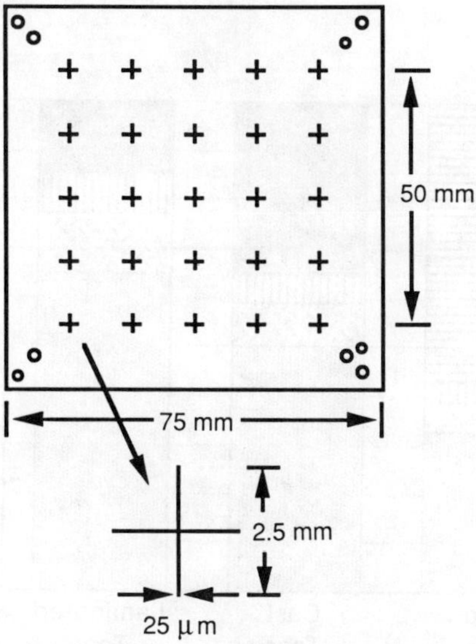

Fig. 1. Template of fine crosses used as markers for measurements of distortion during shrinkage

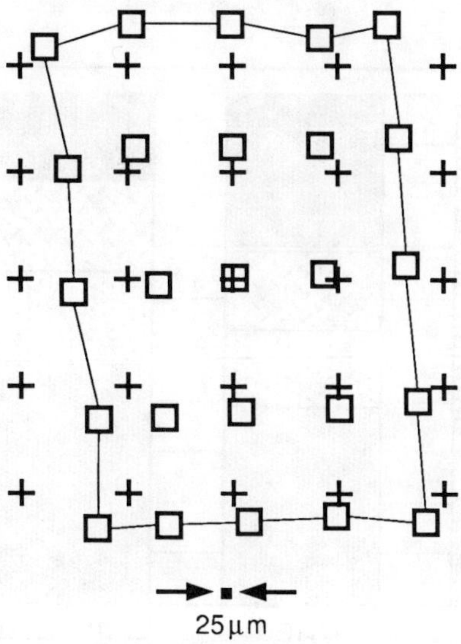

Fig. 2. Exaggerated (x50) distortion plot of cross positions before and after firing a multilayer with no rotation between adjacent tape layers. Data values were corrected for 12% shrinkage before exaggeration of distortions.

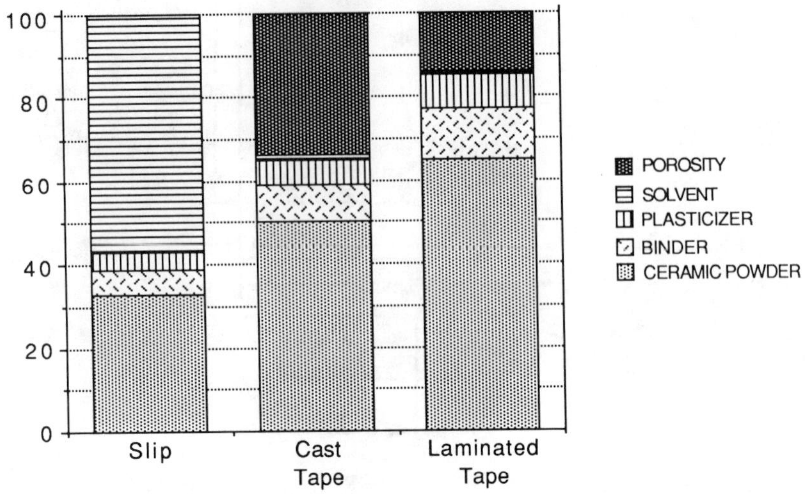

Fig. 3. Volume fraction of components in a representative tape dielectric in the slip, cast and laminated form

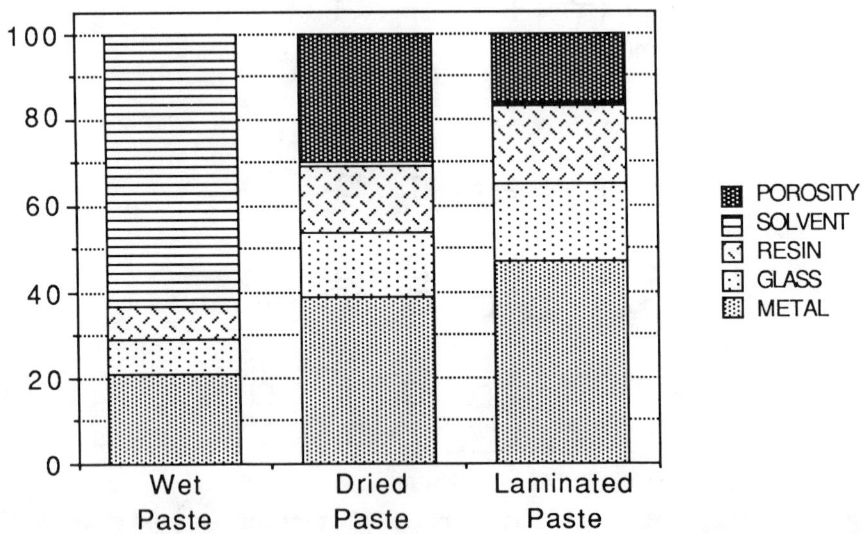

Fig. 4. Volume fraction of components in a representative thick film conductor in the wet, dried, and laminated form

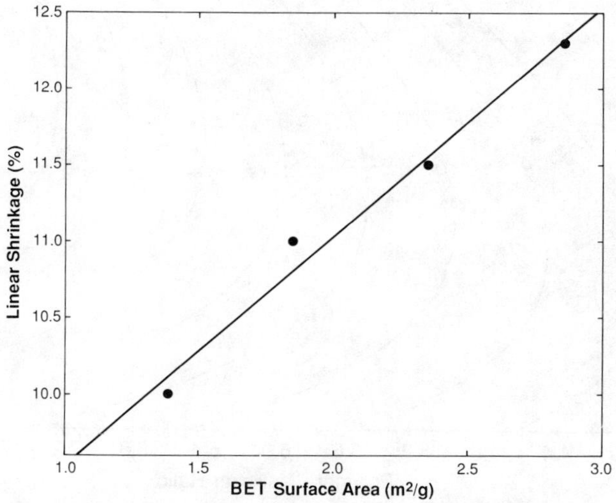

Fig. 5. Linear shrinkage of a tape dielectric as a function of BET surface area of the glass component

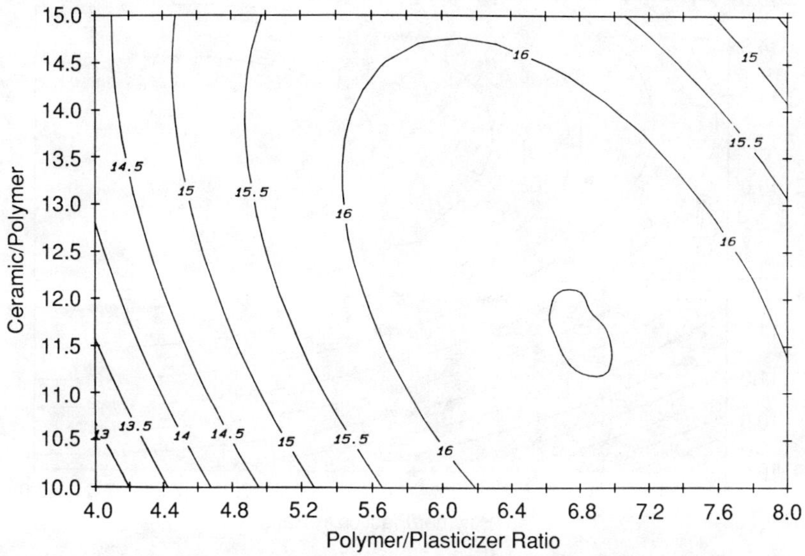

Fig. 6. Contour plot of linear shrinkage of a tape dielectric as a function of ceramic/polymer and polymer/plasticizer ratios

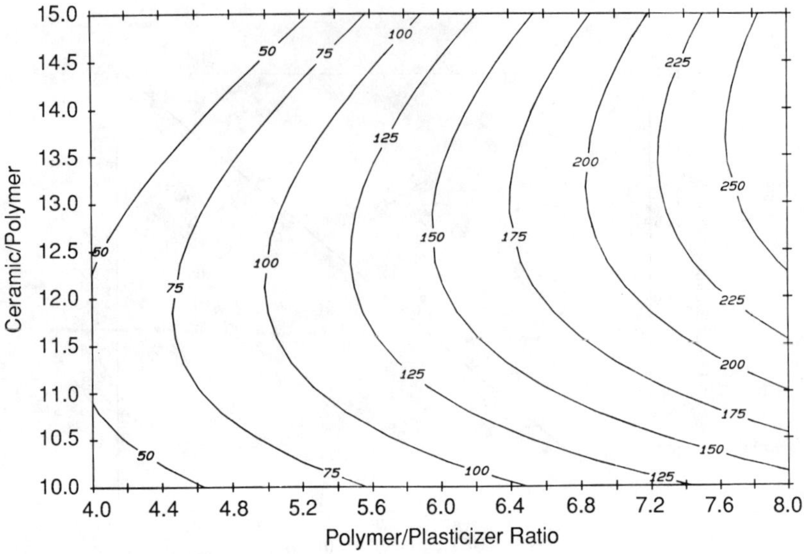

Fig. 7. Contour plot of mechanical strength of tape dielectric (stress at 3% elongation) as a function of ceramic/polymer and polymer/plasticizer ratios

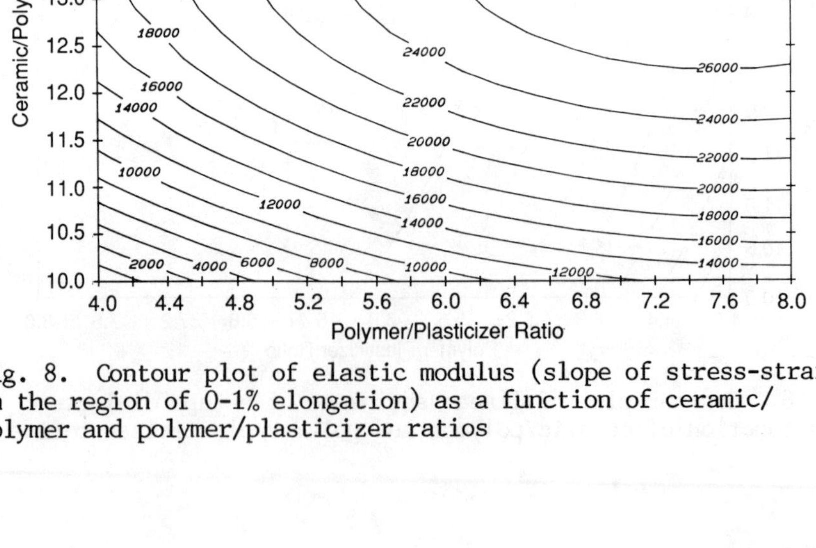

Fig. 8. Contour plot of elastic modulus (slope of stress-strain in the region of 0-1% elongation) as a function of ceramic/ polymer and polymer/plasticizer ratios

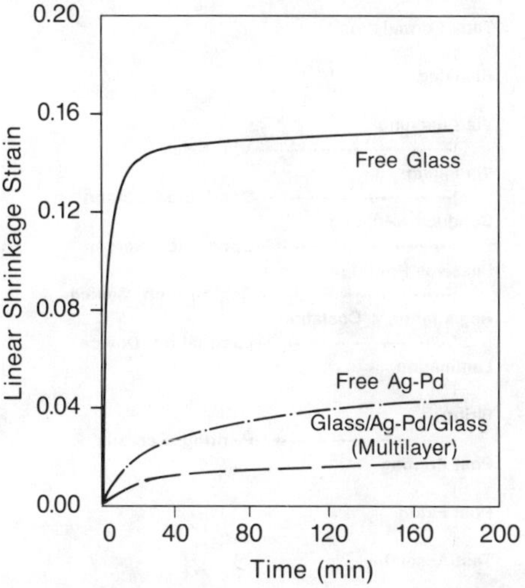

Fig. 9. SEM micrograph of fractured tape dielectric containing spherical powder. No variation in shrinkage with tape orientation was observed in this case.

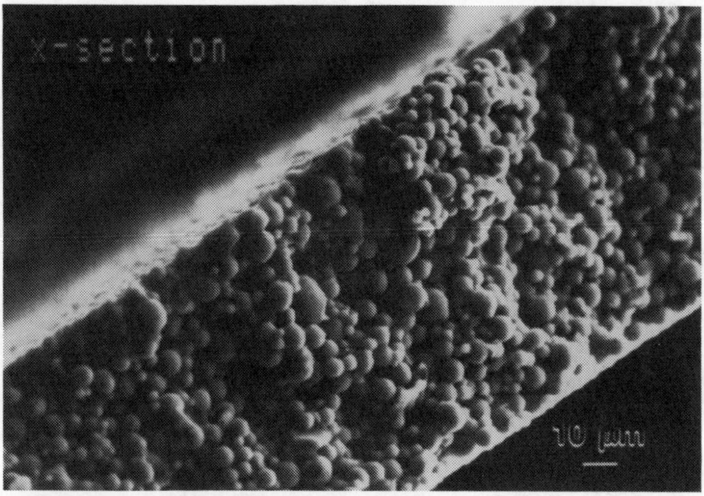

Fig. 10. Linear shrinkage strain vs time of composite and individual components of a ceramic multilayer. Reprinted with permission of the authors in reference 7.

Tape Formulation
|
Blanking
|
Via Punching
|----------------------------Via Mask
Via Filling
|----------------------------Conductor Screen
Conductor Printing
|----------------------------Conductor Screen
Passives Printing
|----------------------------Registration Device
Registration & Coalation
|----------------------------Lamination Device
Lamination
|
Firing
|----------------------------Printing Screens
Post Printing
|
Post Firing
|
Test/Assembly

Fig. 11. Fabrication process flow for co-fired ceramic multi-layers

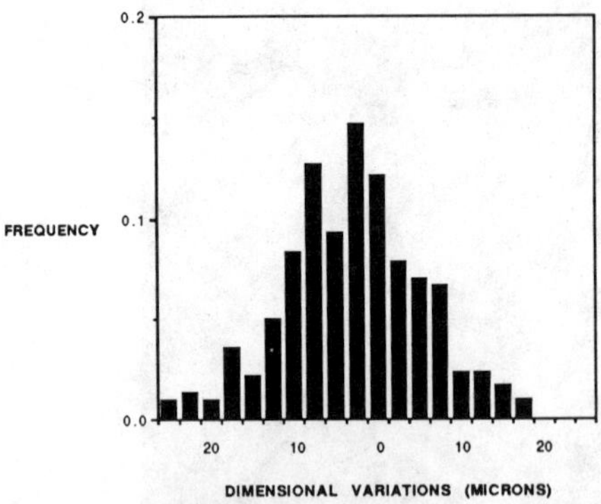

Fig. 12. Dimensional variations associated with process of removing cast tape from carrier layer

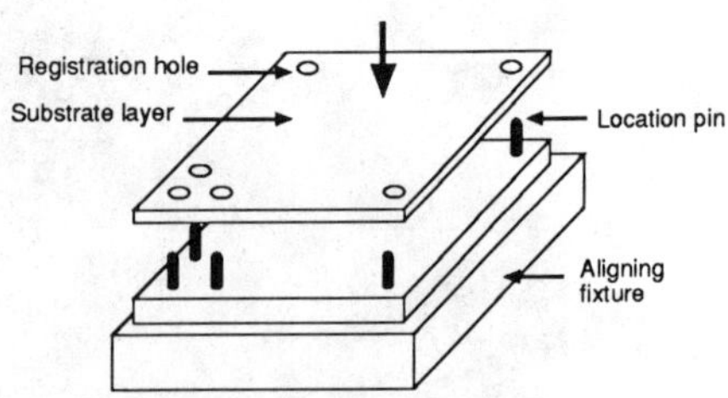

Fig. 13. Schematic of alignment fixture for tape layer registration and collation

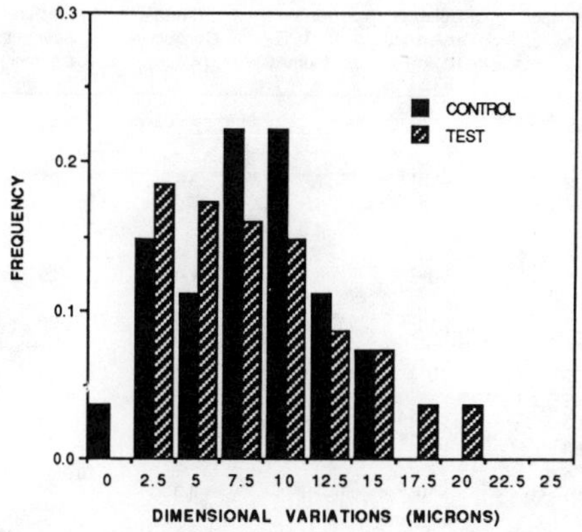

Fig. 14. Dimensional variations associated with process of tape layer registration and collation

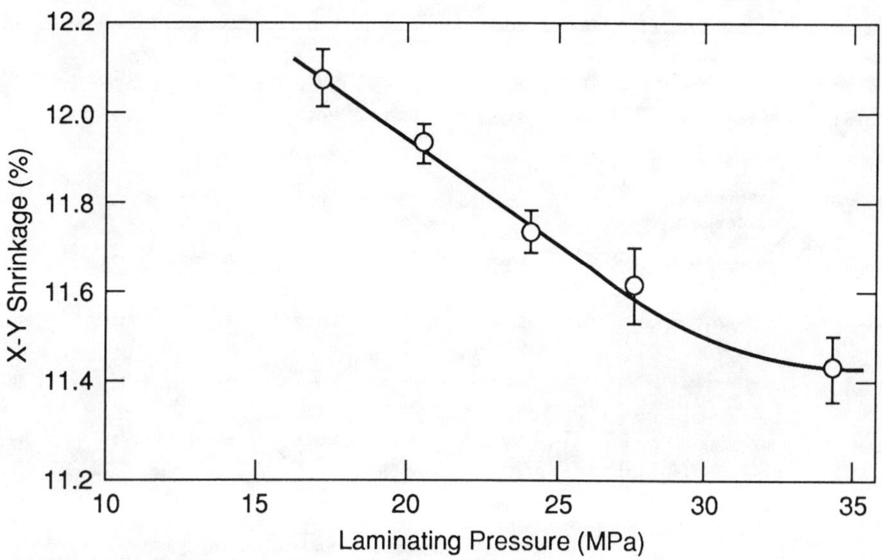

Fig. 15. Lateral shrinkage of tape dielectric as a function of lamination pressure

System	Dielectric Constant	Dissipation Factor	Insulation Resistance (Ω.cm)	TCE (ppm/°C)	Thermal Conductivity (W/mK)	Fracture Strength (kg/mm^2)	Breakdown Voltage (kV/mm)	Reference
Asahi Glass	6.3	0.001	-	3.8-6.8	-	25	-	10
Du Pont	7.8	0.002	$>10^{13}$	7.9	2.9	21	40	11
Fujitsu	4.9	-	-	4.0	-	-	50	12
Matsushita	7.4	0.002	10^{14}	6.1	2.9	25	-	13
Murata	6.1	0.0007	$>10^{14}$	8.0	4.2	20	20	14
Narumi	7.7	0.0003	10^{14}	5.5	2.5	20	88	15
NEC	7.8 3.9	0.003 0.003	$>10^{14}$ $>10^{13}$	7.9 1.9	- -	35 14	- -	16
Shoei	7.0	0.001	$>10^{14}$	7.7	3.3	20	-	17
Taiyo Yuden	6.7	-	-	5.8	6	25	-	18

Table 1. Physical Properties of Some Commercial Low-Temperature Co-Fired Ceramic Material Systems

System	Matrix	Filler	Metallization	Burnout (Temperature, °C/Time, h)	Firing (Temperature, °C/Time, h)	Atmosphere	Shrinkage (%)	Ref.
Asahi Glass	Ba-Al_2O_3-SiO_2-B_2O_3	Al_2O_3	-		900/1	Air	12	10
Du Pont	Alumino Silicate	Al_2O_3	Silver, gold	350/1	850/0.3	Air	12	11
Fujitsu	Borate glass	Al_2O_3	Copper	-/-	950-1000/-	Reducing	-	12
Matsushita	0.35 $NaAlSi_3O_8$- 0.65 $CaAl_2Si_2O_8$	-	Copper	550/4* (in air)	900/0.3	Nitrogen	-	13
Murata	BaO-B_2O_3-Al_2O_3- CaO-SiO_2	-	Copper	-/-	950/-	Reducing	13.5	14
Narumi	CaO-Al_2O_3-B_2O_3-SiO_2	Al_2O_3	Silver Gold (top)	-/-	880/0.3	Air	16	15
NEC	Lead Borosilicate	Al_2O_3 SiO_2	Silver/Pallad.	-/-	900/-	Air	13	16
Shoei	$BaZr(BO_3)_2$	-	Copper	600/- (in air)	980/2.5	Nitrogen	12	17
Taiyo Yuden	CaO-MgO-Al_2O_3-SiO_2- B_2O_3	-	Copper	-/-	950/-	Reducing	-	18

* followed by a reduction step in N_2/H_2, 350/0.3

Table 2. Components and Processing Conditions for Some Commercial Low-Temperature Co-Fired Ceramic Material Systems

Pressure-Assisted Sintering of Multilayer Packages

K.R. Mikeska and R.H. Jensen
E.I. du Pont de Nemours & Co. (Inc.)
Experimental Station
Wilmington, DE 19880-0304

ABSTRACT

A low-temperature low-pressure-assisted constrained sintering (PAS) process was developed that eliminates in-plane shrinkage during firing, allows for the cofiring of ceramic multilayers on rigid prefired substrates to tight dimensional tolerances, and eliminates the need for complex hot-pressing equipment. Examples of complex multilayer parts are shown along with linear dimension diagrams. Numerical calculations based on an analysis of viscous sintering under a uniaxial load show that the uniaxial loads required to eliminate radial contraction are greater than those encountered in practice. Friction between the sintering substrate and load-bearing constraining dies apparently influences densification.

I. Introduction

The formation of complex monolithic multilayered ceramic packages (such as the one shown in Fig. 1) by traditional multilayer fabrication techniques is difficult. The figure shows a multiplicity of dielectric layers attached to a prefired rigid substrate with a cavity for integrated circuit chip mounting. Individual layers are connected by conducting via pathways. The rigid substrate adds mechanical strength and, if made of a high thermal conductivity material such as AlN, it can be used to remove heat from the package. An IC chip would be mounted in the cavity on the AlN and sealed with a cover to provide hermeticity. Such a package cannot be formed conveniently by methods such as thick film and cofiring. Thick film methods are sequential and require multiple firings. Cofired technology relies on the past processing history of the starting materials to control densification.[1-3] In addition, sintering on a rigid substrate requires that all shrinkage occur normal to the substrate,[4,5] which limits the number of dielectric layers or thickness. If unlike materials are combined, such as high dielectric constant and low dielectric constant materials to create buried capacitors or perhaps layers of resistor materials to create buried resistors, homogeneous densification is even more difficult to achieve. Areas of high conductor metal density often lead to warping and flaw generation during firing since the metal densifies at different temperatures and rates than the surrounding ceramic.[6] The free sintering (i.e., sintering in the absence of any external constraint) of complex dielectric systems and complex geometries to tight dimensional tolerances is extremely difficult.

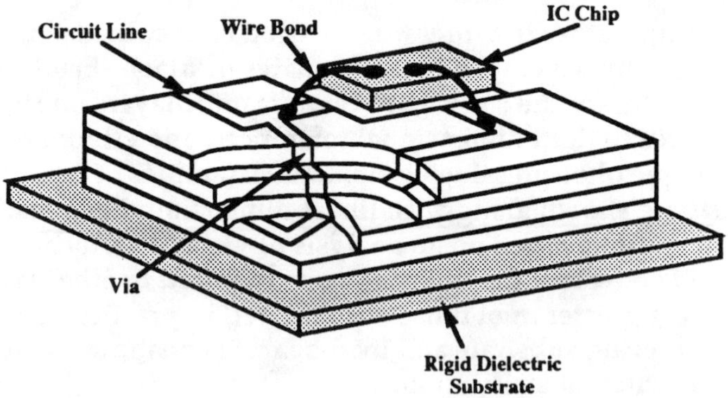

Fig. 1. Schematic of ceramic multilayer electronic package showing multiple layers of dielectric tape on a rigid substrate with a cavity for integrated circuit mounting.

This paper discusses the use of unidirectional pressure-assisted constrained sintering as a method of uniformly densifying low-temperature glass-filled dielectric systems to form complex multilayers such as the one shown in Fig. 1. Pressure-assisted sintering (PAS) offers a number of practical engineering advantages for fabricating advanced multilayer ceramic packages. The process forces all densification in the direction normal to the plane of the sample, which eliminates in-plane shrinkage and provides a means of fabricating multicomponent multilayers to tight dimensional tolerances. Eliminating in-plane shrinkage eliminates planar location problems during automated component placement. Combining unlike materials to form multicomponent packages increases the level of package integration. Viscous sintering theory is reviewed in an effort to determine the mechanisms of constrained densification during the sintering of relatively large thin multilayer substrates under a uniaxial load. A surprising feature of the sintering analysis is that the uniaxial load required to eliminate in-plane shrinkage is greater than that required in practice. Apparently, friction between the sintering sample and the load-bearing constraining dies plays an important role in the densification process.

II. Experimental Procedure

(1) Materials Preparation

The dielectric composition used throughout this study was an alumina-filled silicate glass tape system composed of 70 v/o glass† and 30 v/o Al_2O_3‡. This material is a typical low-temperature cofiring dielectric composition. Tapes were made by dispersing the glass/Al_2O_3 mixture in an acrylic binder§/dioctyl phthalate plasticizer¶/tricloroethane solution and milling in Al_2O_3 mills containing Al_2O_3 milling balls for 24 hours. After milling, the slurry is cast under a doctor blade to form a flexible tape with a dried thickness of ~115 μm (0.0045 inch). Tapes were subsequently metallized, stacked, and laminated to form multilayer samples.

III. Results and Discussion

(1) Multilayer Substrates

Fabricating multilayer circuit parts by a flexible tape process consists of stacking a multiplicity of unfired metallized tape layers, laminating the unfired layers under pressure and temperature to form an unfired ceramic/metal monolith, and subsequently firing the package to sinter the metal and ceramic. In this study, the unfired multilayer samples

† Ferro Corp., Cleveland, OH

‡ Alcoa Industrial Chemical, Bauxite, AR,

§ E.I. du Pont de Nemours & Co. (Inc.),Wilmington, DE

¶ Aldirich Chemical Co., Inc., Milwaukee, WI

were fabricated by cutting individual layers of tape to a blanking size, punching vias, screen printing the individual tape layers with conductor metallization and via fill metallization, stacking the metallized tape layers, and uniaxially laminating the stacked layers at 20.86 MPa (3000 psi) and 70 °C for 10 minutes. Lamination fuses the individual tape layers. The final thickness of the unfired multilayer can be adjusted by adding or subtracting layers of tape. After lamination, the relative density of the unfired tape compact was ~0.67 v/o as determined by Archimedes methods. During heating, the binder and plasticizer components pyrolyze between 300 °C and 400 °C. Incomplete polymer removal during firing often results in blistering and delamination, especially over conductor patterns, and a degradation of electrical properties. The tape compacts can be free sintered (i.e., no external load) to full density at ~850 °C, as shown in the constant heating (5 °C/min) plot in Fig. 2. The isothermal heating plot in Fig. 3 shows that the material can be free sintered to full density in less than 10 minutes at 850 °C. Figure 4 is an electron micrograph of a polished cross-section of a sintered tape.

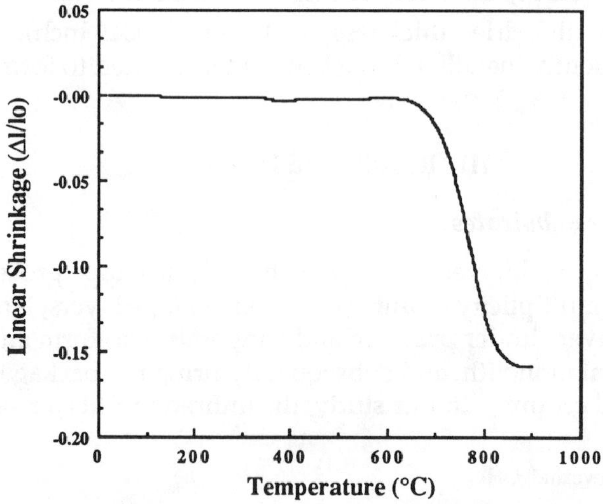

Fig. 2. Constant heating linear shrinkage (5°C/min) of free-sintered (i.e., no external load) multilayer tape.

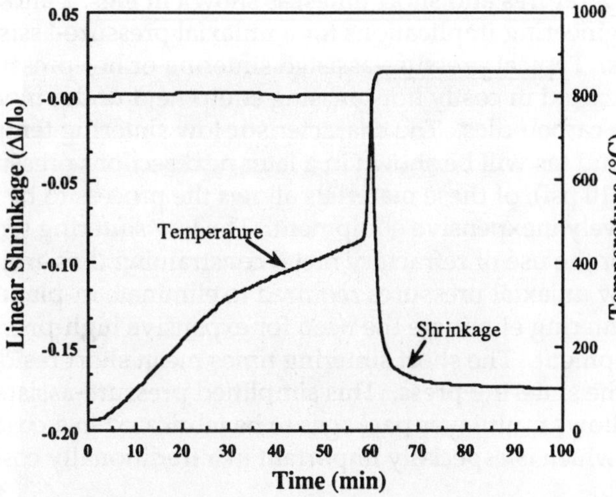

Fig. 3. Isothermal linear shrinkage (850 °C) of free-sintered (i.e., no external load) multilayer tape.

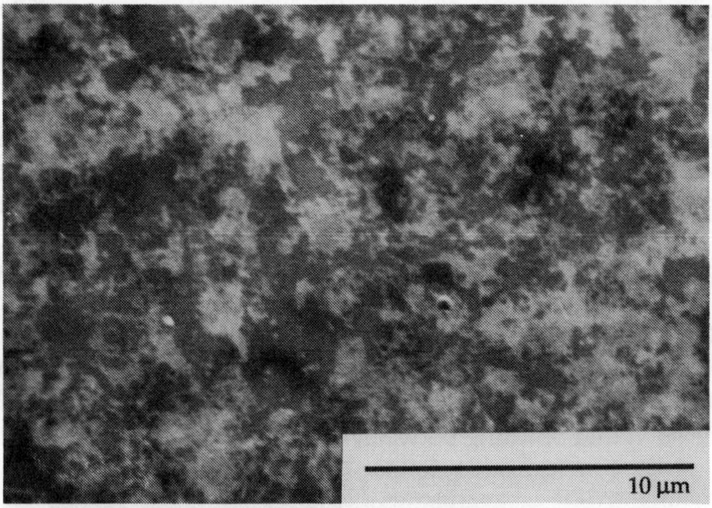

Fig. 4. Scanning electron micrograph of a polished cross-section of a sintered multilayer tape.

The ability to sinter these glass-filled dielectric systems to full density at low temperatures and short times, as shown in Figs. 2 and 3, has important engineering implications for a uniaxial pressure-assisted sintering process. Typical pressure-assisted sintering or hot-pressing processes are practiced in costly hot-pressing equipment under inert atmospheres using carbon dies. The characteristic low sintering temperatures, times, and (as will be shown in a later next section) pressures, < 0.068 MPa (< 10 psi), of these materials allows the process to be practiced in relatively inexpensive equipment. The low sintering temperatures allow for the use of refractory metal constraining dies and pressure rams. The low uniaxial pressures required to eliminate in-plane shrinkage during sintering eliminate the need for expensive high-pressure hot-pressing equipment. The short sintering times mean short residence times inside the sintering press. This simplified pressure-assisted sintering process allows multilayer packages to be fabricated in a cost-effective manner, which is especially important in a traditionally cost-driven industry.

Examples of multilayer ceramic parts made by pressure-assisted sintering are shown in Figs. 5-10. The specimens were prepared by sandwiching the unfired laminated multilayer compacts between constraining dies fabricated of Haynes Alloy*, a porous plate, and a proprietary release layer. Uniaxial pressure is applied to the setup by a ceramic pressure ram. The Hanes Alloy constraining dies distribute the applied uniaxial pressure evenly across the faces of the specimen. The porous plate permits for the escape of polymer decomposition products during heatup. The release layer isolates the sintering sample from the load bearing constraining dies. All pressure assisted sintered samples were fired in a box furnace fitted with pressure rams capable of continuous pressure adjustments during heating. Samples were heated at 3 °C/min to 500 °C, 5 °C/ min to 850 °C, and held at 850 °C for 0.5 hour.

Figure 5 shows photographs of the top surface of a prototype test circuit package before and after pressure-assisted sintering. Figure 6 is an electron micrograph of a polished cross section of the final fired prototype test package shown in Fig. 5. This package consists of four layers of dielectric tape with buried Ag circuitry and post-fired Ag/Pt and Au top surface circuitry. The tape layers are co-sintered on the rigid Al_2O_3 substrate such that bonding between the dielectric tape and the Al_2O_3

* Haynes International, Kokomo, IN

substrate occurs during processing. The dielectric tape in this circuit package is offset from the rigid substrate to allow for direct connections to the substrate level conductor. The unique features of this multilayer package are that the entire package was co-sintered, except for the top surface metal; there was no in-plane shrinkage during firing; and the multiple layers of tape were sinter-bonded to a rigid Al_2O_3 substrate during the co-sintering pressure-assisted process. Notice from the figure that the in-plane dimensions have not changed in going from the green state to the fired state. Dimensional control is quantified and discussed in the next section. This specimen is a completely functional test circuit. It should be noted that the thermal expansion of the dielectric tape was matched to the rigid Al_2O_3 substrate (8.8×10^{-6} in/in/°C) to prevent warping and possible cracking of the dielectric/Al_2O_3 composite during cooling.

Figure 7 shows photographs of the top surface of a 30-layer dielectric sinter-bonded to a rigid Al_2O_3 substrate before and after pressure-assisted sintering. The rigid substrate is not visible in the photographs. Figure 8 is an electron micrograph of a polished cross section of the final fired sample shown in Fig. 7. Again, all in-plane shrinkage was eliminated. This example shows that pressure-assisted sintering can be used for many layers (>30 layers) of dielectric tape on rigid substrates. The dark area (blue coloring) of the layers in the cross-section micrograph is due to Co additions. This coloring allows individual layers to be distinguished.

Figure 9 shows photographs of the top surface of two samples with dielectric sinter-bonded to rigid AlN substrates. The first photograph is of a specimen with a cavity made from ten layers of dielectric tape. The AlN is visible inside the cavity. An IC chip would be mounted in the cavity directly on the AlN. The heat generated by the chip during operation would be removed through the AlN substrate. A lid could be placed over the cavity to provide hermeticity. The second photograph in the figure is of ten layers of dielectric tape with buried Au circuitry. The rigid AlN substrate is not visible in this photograph. The thermal expansion of the dielectric tape in these samples is matched to AlN (4.5×10^{-6} in/in/°C).

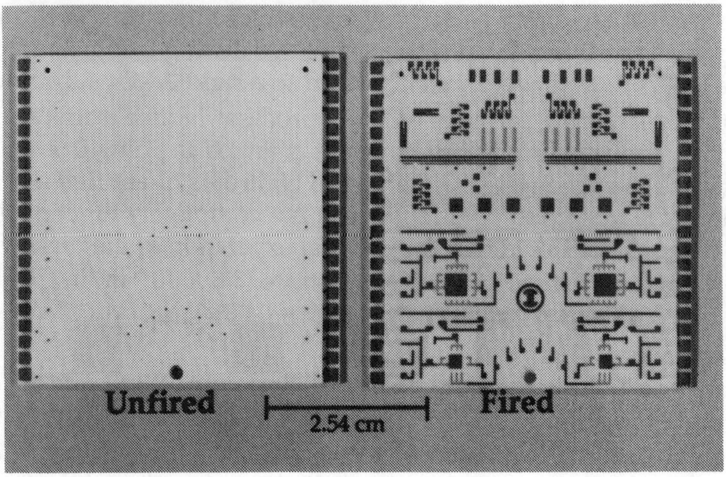

Fig. 5. Photographs of the top surface of a test prototype multilayer circuit before processing and after pressure assisted sintering. The sample consists of four layers of tape sinter-bonded on to a rigid alumina substrate. x-y shrinkage was eliminated during processing.

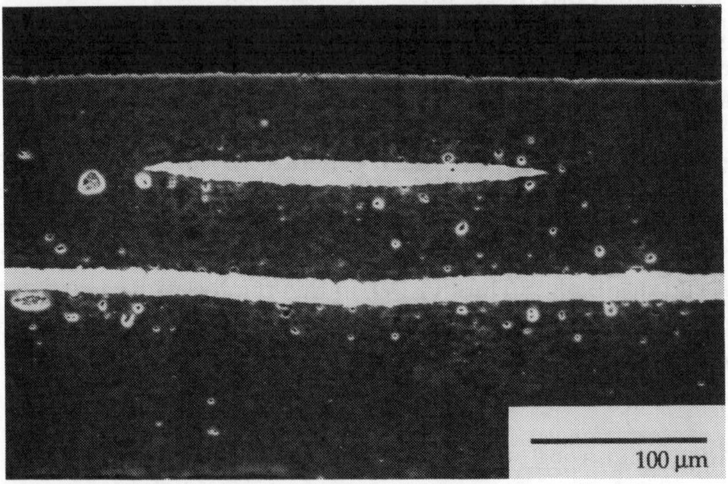

Fig. 6. Scanning electron micrograph of a polished cross-section of the prototype test package shown in Fig. 5.

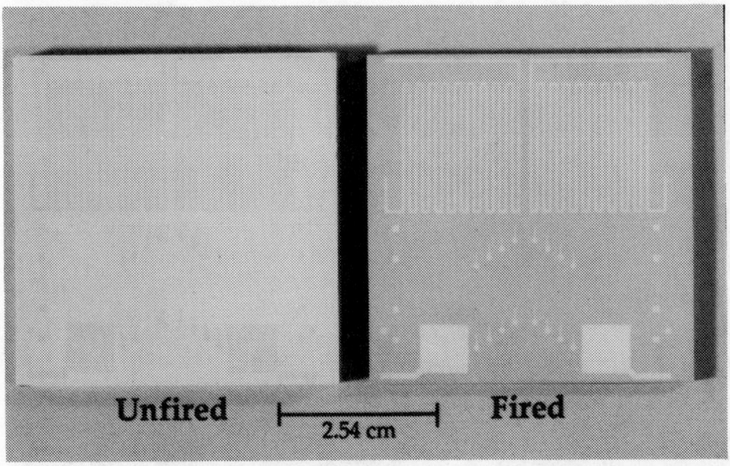

Fig. 7. Photographs of the top surface of a 30-layer dielectric tape sample sinter-bonded on to a rigid alumina substrate before processing and after pressure-assisted sintering.

Fig. 8. Scanning electron micrograph of a polished cross section of the 30-layer sample shown in Fig. 7.

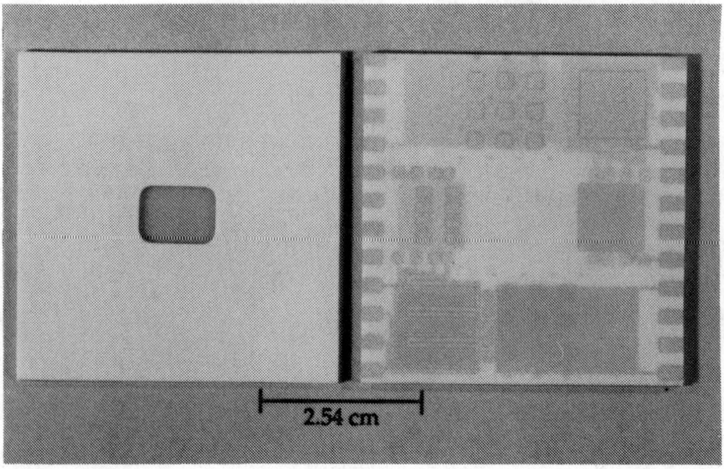

Fig. 9. Photographs of the top surface of two test samples with dielectric sinter-bonded to rigid AlN substrates. The first photograph shows ten layers of dielectric tape on AlN with a cavity. The second photograph shows ten layers of dielectric tape on AlN with Au circuitry.

(2) Dimensional Control

Linear dimension errors large enough to cause problems with electrical continuity and automated component placement (i.e., die attached or solder attached) in multilayer packages are often too small for the human eye to detect. Moreover, the particular shape of a distorted ceramic part may offer clues to the origin of the distortion. Meaningful measurement of linear dimensions on large uniquely shaped multilayer circuit parts, however, is difficult. A technique was developed for enhancing or magnifying linear dimensional errors in a graphic presentation so that small magnitude errors can be seen directly and the general shape or character of the linear distortions becomes apparent. Linear dimension errors are presented in Figs. 11-14 as exaggerated distortion plots. The plots are created from data generated by measuring distances between relatively high-resolution surface features (crosses) positioned systematically on the top surface of a multilayer sample. In this study, a proprietary photolithographic printing technique was used to place 25 individual crosses with 25 μm line widths on the top surface of standard eight-layer 7.5 cm x 7.5 cm multilayer samples in a simple 5 cm x 5 cm array as shown in Fig. 10. An automated traveling optical microscope was used to measure the linear distances between individual crosses to an accuracy of ± 5.0 μm. By placing the points in a symmetrical array, distortions in localized regions could be measured.

The exaggerated distortion plots were created by comparing the actual surface feature locations in the x-y plane with a standard or desired set of feature locations. Deviations from the standard were then multiplied by a factor sufficient to make them readily apparent to the eye. A multiplication factor of 75X was used here. To get a feel for the magnitude of error associated with processing, the actual image (multiplied by 75X) was superimposed over the desired image. A scale reference marker was multiplied by the same multiplication factor so that the viewer can easily determine the magnitude of error. To determine the linear shrinkages and possible distortions associated with firing, the cross locations were measured before firing and then again after firing. The after-firing errors were multiplied by 75X and superimposed on the standard or desired firing data. The standard or desired sample dimensions are represented by crosses and the actual sample dimensions by squares. In addition to distortions associated solely with firing, this technique can also be used to measure distortions at any point along the fabrication process. By knowing where in the fabrication process distortions are occurring, steps can be taken to eliminate the cause.

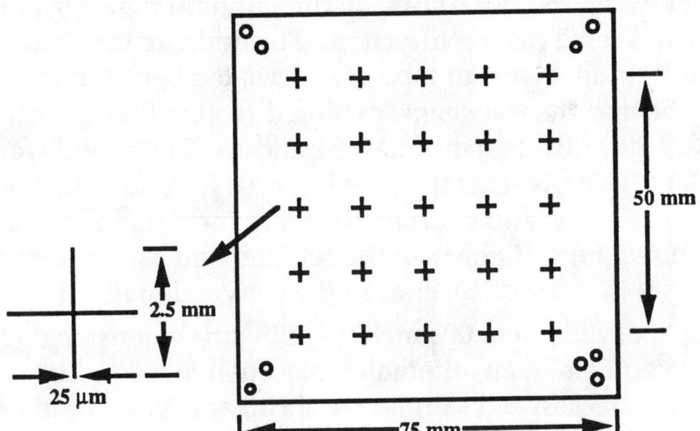

Fig. 10. Cross matrix used to measure linear dimensions of multilayer samples.

Figures 11-14 are exaggerated distortion plots for various sample configurations comparing conventionally sintered (free sintered) multilayer parts with pressure-assisted sintered multilayer parts. The conventional or free sintered samples were prepared in a manner identical to the pressure assisted samples, except, that the conventionally sintered samples were fired in a belt furnace to simulate production multilayer firing practices. As discussed, the squares represent the actual dimensions of the sintered sample (errors multiplied by 75X to allow the viewer to see distortions), and the crosses represent the desired or ideal dimensions (i.e., the dimensions of the sintered part if no distortions occurred during firing). For the conventionally sintered samples, the data in all cases were expanded by 12.0 % to take into account the expected linear shrinkage of the sample since the part shrinks in the x, y, and z dimensions during firing. The pressure-assisted data were not expanded since x-y shrinkage is eliminated during densification. Expanding the dimensions of the conventionally sintered sample by 12.0 % allows the viewer to compare the conventional and the pressure-assisted sample on a one-to-one basis.

Fig. 11 shows distortion plots for 5 x 5 cm square, alternated, eight-layer dielectric tape samples with no conductor metal. Alternation refers to the stacking orientation of the individual tape layers. During tape casting, high shear stresses align the individual particles in the casting suspension as they pass under the doctor blade. In an individual layer of tape during firing, this particle alignment causes an anisotropy in shrinkage between the lateral casting direction (in the direction of casting) and transverse casting direction. To eliminate the shrinkage anisotropy, individual layers of tape are alternated between the lateral casting direction and the transverse casting direction (i.e., alternating layers of 0° and 90°) during a multilayer buildup. The multilayer configuration in Fig. 11 represents the simplest case (i.e., no conductor metal, simple geometry, and alternated). Only the data for one set of samples are shown here. Other sample sets showed similar behavior. The figure shows that the distortions in the conventionally sintered sample are in excess of 125 μm (0.005 inch). A 125 μm dimensional error would be unacceptable in an automated placement or bonding operation. In the pressure-assisted sample x-y shrinkage was eliminated, and the sample shows limited distortion.

Figure 12 shows distortion plots for 5 x 5 cm square, nonalternated, eight-layer dielectric tape samples with no conductor metal. The figure shows that the conventionally sintered sample was badly distorted.

Comparing Figs. 11 and 12 shows that the effect of tape orientation on shrinkage is pronounced. In the case of the pressure-assisted sample, again, x-y shrinkage was eliminated, and the sample shows limited distortion.

Figure 13 shows distortion plots for 5 x 5 cm square alternated, eight-layer dielectric tape samples with two layers of Ag conductor metallization. The conventionally sintered sample was distorted, and the metal layers caused the sample to shrink more than the expected 12.0 %. The pressure-assisted sample shows limited distortion and x-y shrinkage was eliminated. The conventionally sintered sample shows how combining materials with different sintering characteristics, such as metal and ceramic, can affect shrinkage. The pressure-assisted sintering method allows materials with different sintering characteristics to be combined with no adverse densification effects as shown by the distortion plots.

Figure 14 shows distortion plots for 5 x 5 cm square, nonalternated, eight-layer dielectric tape samples with six layers of Ag conductor metallization. Again, the conventionally sintered sample was distorted and shrank more than the expected 12.0% due to the presence of metal while the pressure-assisted sample shows limited distortion and no x-y shrinkage. A comparison of the conventionally sintered two metal layer and six metal layer samples shows that increasing the amount of metal causes an increase in shrinkage. Increasing the amount of metal does not affect the pressure-assisted sintered samples.

The distortion plots show that pressure-assisted sintering allows multilayer ceramic parts to be fabricated to dimensional tolerances unattainable by conventional processing.

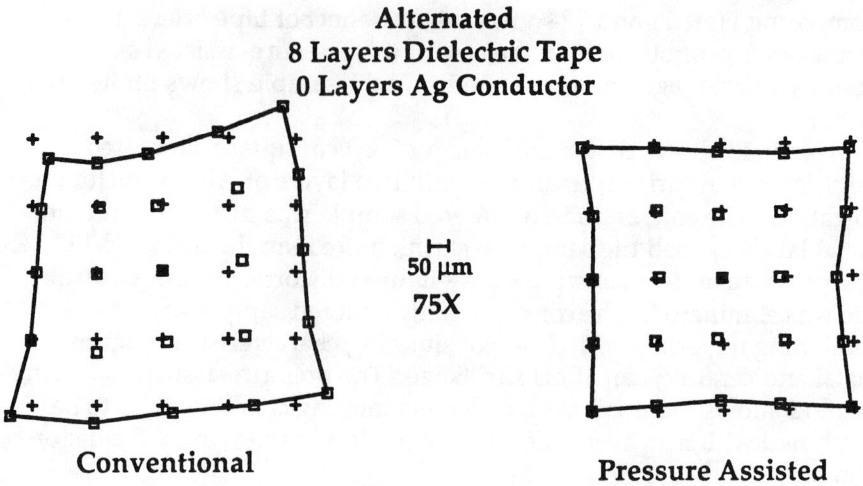

Fig. 11. Distortion plots of alternated, eight-layer dielectric tape, no Ag conductor layers. Conventionally sintered sample is distorted in excess of 125 µm (0.005 inch). The pressure-assisted sample shows limited distortion.

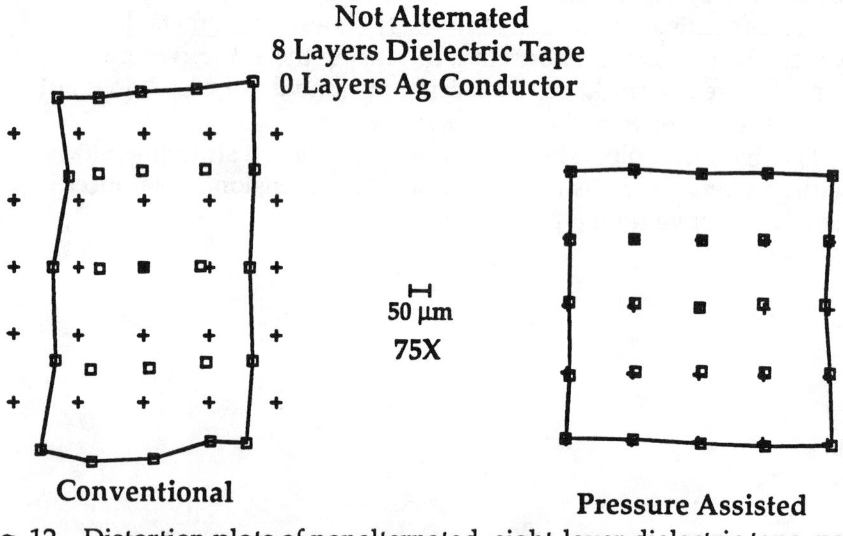

Fig. 12. Distortion plots of nonalternated, eight-layer dielectric tape, no Ag conductor layers. Conventionally sintered sample is very distorted. The pressure-assisted sample shows limited distortion.

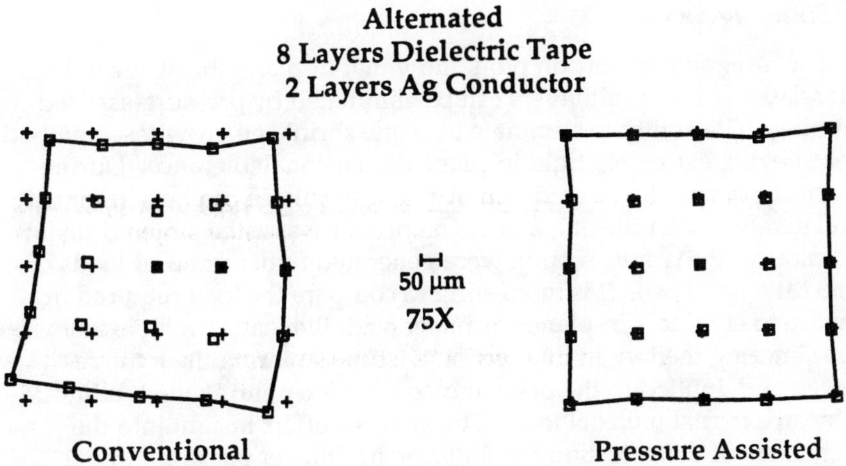

Fig. 13. Distortion plots of alternated, eight-layer dielectric tape, two Ag conductor layers. Conventionally sintered sample is distorted. The pressure-assisted sample shows limited distortion.

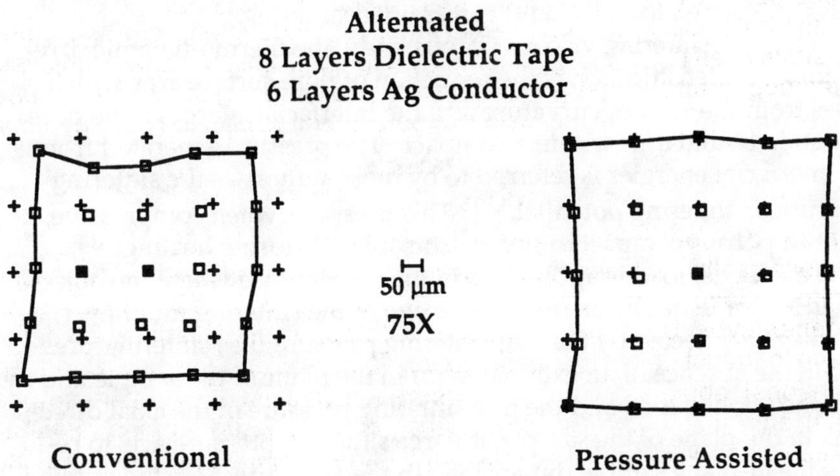

Fig. 14. Distortion plots of alternated, eight-layer dielectric tape, six Ag conductor layers. Conventionally sintered sample is distorted by the metal layers. The pressure-assisted sample shows limited distortion.

(3) *Sintering Analysis*

The foregoing distortion plots show that in-plane shrinkage in large, flat, relatively thin multilayers can be eliminated by pressure-assisted sintering. The ability to eliminate in-plane shrinkage provides a method for achieving extremely tight in-plane dimensional tolerance. During this study, it was discovered that in-plane shrinkage can be eliminated at relatively small uniaxial loads. The pressure-assisted sintered distortion plots in the results section were generated under uniaxial loads of < 0.068 MPa (< 10 psi). It is interesting to compare the load required in practice to eliminate in-plane shrinkage with the loads predicted by viscous sintering theory. In this section, viscous sintering theory is reviewed and applied to the present case where x-y shrinkage is eliminated by an external uniaxial load. The analysis offers insight into the mechanism of densification for sintering multilayer packages under uniaxial pressure.

Glass sintering theory originated with the viscous analysis by Frenkel[7]. It was extended by Mackenzie and Shuttleworth[8] for closed spherical pores and by Scherer[9] for open nonspherical pores. More recently the theory has been extended to include rigid inclusions,[3] sintering on rigid substrates,[4,5] bimodal pore size distributions,[10] sintering under an applied load,[11,12] and constrained sintering.[13]

For a free-sintering viscous porous body, the thermodynamic driving force for densification is the reduction of pore surface area which arises from the surface curvature and the interfacial energy of the pores and grain boundaries within a compact. The pressure generated from the interfacial energies is referred to by most authors as the sintering pressure or sintering potential.[13-15] This pressure, when compressive, causes a porous compact to shrink isotropically during heating.

If a uniaxial load is applied, as in the present case where the objective is to eliminate shrinkage in the plane of the sintering multilayer, shrinkage is affected by the free sintering pressure (i.e., sintering pressure in the absence of an external load) in the plane of the sample and by both the applied load and the free sintering pressure in the axial direction. In the plane of the sample, the free sintering pressure acts to reduce the in-plane sample dimension. In the axial direction, the applied load and the free sintering pressure act to reduce the sample height. For the process to be effective (i.e., no in-plane shrinkage) when making a multilayer, these effects must balance. If the applied axial load is too large, the load would "squash" the sample. If the load is too small, the sample would shrink in the plane of the part.

This problem has been reviewed by Scherer[4] and can be analyzed by considering a cylindrical sample, as shown in Fig. 15, which is under uniaxial stress σ_z, with the axis of rotation defined in the z direction and the height H, radius R, and circumference θ. The following constitutive equations, in cylindrical coordinates, have been proposed for a porous viscous body as:[9]

$$\dot{\varepsilon}_r = \dot{\varepsilon}_f + E_m^{-1}\left[\sigma_r - \nu_m(\sigma_\theta + \sigma_z)\right] \tag{1}$$

$$\dot{\varepsilon}_z = \dot{\varepsilon}_f + E_m^{-1}\left[\sigma_z - \nu_m(\sigma_r + \sigma_\theta)\right] \tag{2}$$

$$\dot{\varepsilon}_\theta = \dot{\varepsilon}_f + E_m^{-1}\left[\sigma_\theta - \nu_m(\sigma_z + \sigma_r)\right] \tag{3}$$

where $\dot{\varepsilon}_r$, $\dot{\varepsilon}_z$, and $\dot{\varepsilon}_\theta$ represent the strain rates (i.e., shrinkage rates) and stresses in the radial, axial, and circumferential axis, respectively. For a viscous material, E_m is the apparent Young's modulus, or axial viscosity, which can be approximated by

$$E_m \approx \frac{3\eta\rho}{3\rho_T - 2\rho} \tag{4}$$

where ρ is the sample density, ρ_T is the theoretical density, and η is the viscosity.

The term ν_m is the apparent Poisson's ratio and is approximated by

$$\nu_m \approx \left(\frac{1}{2}\right)\left[\frac{\rho}{(3\rho_T - 2\rho)}\right]^{1/2} \tag{5}$$

As porosity goes to zero ($\rho \rightarrow \rho_T$), $E_m \rightarrow 3\eta$ and $\nu_m \rightarrow 1/2$ and equations 1, 2, and 3 reduce to the standard constitutive equations for a noncompressible viscous fluid.

The free strain rate $\dot{\varepsilon}_f$ is the contraction rate of an unconstrained sintering body. For a body with open porosity ($\rho \leq 0.94\,\rho_T$), $\dot{\varepsilon}_f$ according to Scherer's cylinder model[9] is

$$\dot{\varepsilon}_f = -\frac{\gamma n^{1/3}}{\eta}\frac{(3\pi)^{1/3}}{6}\frac{2-3cx}{x^{1/3}(1-cx)^{2/3}} \tag{6}$$

where η is the bulk glass viscosity, c is a constant equal to $8\sqrt{2}/3\pi$, x is equal to a/l (according to Scherer's cylindrical model, a is the radius of the solid cylinder and l is the length), and $n^{1/3} = (l/l_o)(\rho_T/\rho)^{1/3}$ where l_o is the initial model cylinder length.

For a body with closed porosity ($\rho \geq 0.94\,\rho_T$), the free sintering rate according to Mackenzie and Shuttleworth's model [8] is

$$\dot{\varepsilon}_f = -\frac{1}{2}\left(\frac{4\pi}{3}\right)^{1/3}\left(\frac{\gamma n^{1/3}}{\eta}\right)\left(\frac{\rho_T}{\rho}-1\right)^{2/3} \tag{7}$$

During the densification of a multilayer where it is desirable to eliminate the in-plane shrinkage $\sigma_r = \sigma_\theta = 0$, equation 1 reduces to

$$\dot{\varepsilon}_r = \dot{\varepsilon}_f - \nu_m\sigma_z/E_m \tag{8}$$

Since, in the present case, no radial contraction is occurring, $\sigma_r = 0$, and when the load is compressive both $\dot{\varepsilon}_f$ and σ_θ are negative; therefore,

$$\sigma_z = \frac{E_m}{\nu_m}\dot{\varepsilon}_f \tag{9}$$

Thus, equation 9 can be used to calculate the uniaxial stress necessary to eliminate radial or in-plane contraction. When a constant compressive load, W, is applied, the axial stress as a function of time is

$$\sigma_z(t) = \left(\frac{W}{A_o}\right)e^{-2\dot{\varepsilon}_r} \tag{10}$$

where A_o is the initial cross-sectional area of the sample and since $\dot{\varepsilon}_r = 0$,

$$\sigma_z(t) = \frac{W}{A_o} \tag{11}$$

Combining equations 9 and 10 allows the load necessary to eliminate radial contraction to be calculated by

$$W = A_o \frac{E_m}{\nu_m} \dot{\varepsilon}_f \tag{12}$$

Applying the analysis to determine the load required to eliminate radial contraction in a viscously sintering sample requires the knowledge of material surface energy, sample pore radius, and sample density. Using a value of 300 ergs/cm^2 for the surface energy of the glass used in the present system, Fig. 16 shows the axial load necessary to eliminate radial shrinkage as a function of sample density, as predicted by the Scherer model for $\rho \leq 0.94\ \rho_T$ and by the Mackenzie and Shuttleworth model for $\rho \geq 0.94\ \rho_T$ for various sample pore sizes. For the present system the relative density of an unfired tape compact after lamination and polymer pyrolysis was ~0.67 as determined by Archimedes methods. The average void (pore) radius of the unfired tape after lamination and polymer pyrolysis is ~100 nm as shown in Fig. 17, as determined by Hg penetration porosimetry. It should be noted that porosimetry measures the size of the pore entrance channel and the actual pore radius may be larger than indicated by the data. Fitting the measured values to the theoretical curve predicts that the initial load required to eliminate radial shrinkage is > 2 MPa (> 300 psi).

In practice radial shrinkage is eliminated at a uniaxial load of < 0.068 MPa (< 10 psi), which is less than predicted by the analysis. This absence of radial shrinkage implies that friction between the sintering sample and the external constraining dies plays an important role in the densification process. This implication is not surprising since, during the pressure-assisted sintering process, the relatively thin (< 1000 μm) multilayers have large cross-sectional areas in contact with the load bearing constraining dies. The friction generated during the process can be calculated by using a viscous friction analysis proposed by Jagota.[16] The friction analysis is applied to the present case in a forthcoming paper.

Another feature of the sintering analysis is that the load required to eliminate radial shrinkage decreases with increasing sample density. Since the viscosity terms cancel in equation 12, the load required dependents on the capillary pressure as the sample densifies. This implies that the weight should be removed from the sample when full density is reached to avoid "squashing" the sample.

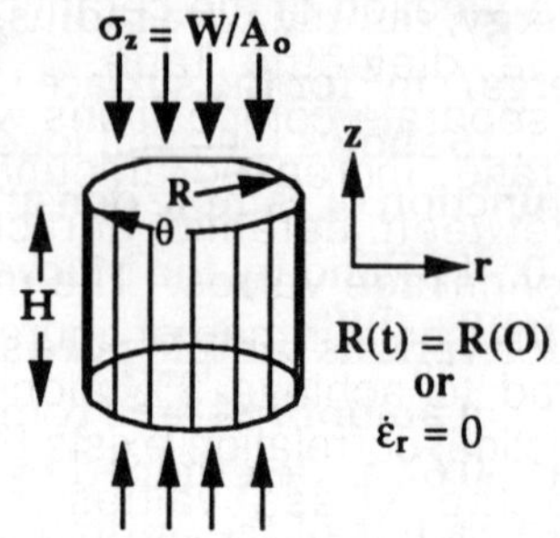

Fig. 15. Schematic of cylindrical sample with radius R, height H, circumference θ, and initial cross-sectional area A_o.

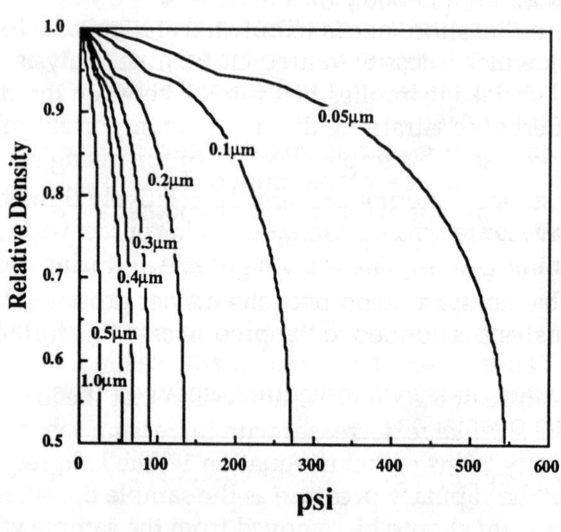

Fig. 16. Relative sample density as a function of uniaxial load for various pore radii.

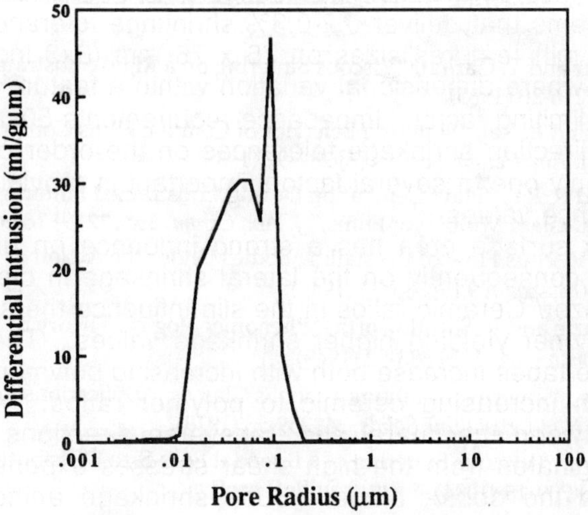

Fig. 17. Hg penetration porosimetry pore size distribution of a multilayer tape compact laminated at 20.86 MPa (3000 psi) after binder pyrolysis.

IV. Conclusions

Pressure-assisted sintering can be used as a cost-effect method of eliminating in-plane shrinkage during firing to produce complex low-temperature multilayers to tight dimensional tolerances. A review of viscous sintering under a uniaxial load shows that the uniaxial load required to eliminate in-plane shrinkage is higher than that required in practice. Interfacial friction between the sintering substrate and load-bearing constraining dies plays an important role in densification.

Acknowledgments: We acknowledge with thanks the contributions of D.T. Schaefer and M.A. Barsoum, DuPont, for their care and ingenuity in sample preparation.

References

[1]A. G. Evans, "Considerations of Inhomogeneous Effects in Sintering," *J. Am. Ceram. Soc.*, **65** [10] 497-501 (1982).

[2]C.H. Hsueh, A.G.Evans, and R. M. Cannon, "Viscoelastic Stresses and Sintering Damage In Heterogeneous Powder Compacts," *Acta metall.*, **34** [5] 927-36 (1987).

[3]G. W. Scherer, " Sintering with Rigid Inclusions," *J. Am. Ceram. Soc.*, **70** [10] 719-25 (1985).

[4]G. W. Scherer and T. Garino, "Viscous Sintering on a Rigid Substrate," *J. Am. Ceram. Soc.*, **68** [4] 216-20 (1985).

[5]R.K. Bordia and R. Raj, "Sintering Behavior of Ceramic Films Constrained by a Rigid Substrate," *J. Am. Ceram. Soc.*, **68** [6] 287-92 (1985).

[6]T. Cheng and R. Raj, "Flaw Generation During Constrained Sintering of Metal-Ceramic and Metal-Glass Multilayer Films," *J. Am. Ceram. Soc.*, **72** [9] 1649-55 (1989).

[7]J. Frenkel, "Viscous Flow of Crystalline Bodies Under the Action of Surface Tension," *J. Phys. (Moscow)*, **9** [5] 385-91 (1945).

[8]J. K. Mackenzie and R. Shuttleworth, "Phenomenological Theory of Sintering," *Proc. Phys. Soc., London*, **62** [12B] 833-52 (1949).

[9]G.W. Scherer, "Sintering of Inhomogeneous Glasses: Applications to Optical Wave Guides," *J. Non-Crystalline Solids*, **34**, 239-56 (1979).

[10]G. W. Scherer, "Viscous Sintering of a Bimodal Pore-Size Distribution," *J. Am. Ceram. Soc.*, **67** [11] 709-15 (1984).

[11]G. W. Scherer, "Viscous Sintering Under a Uniaxial Load," *J. Am. Ceram. Soc.*, **69** [9] C-206-C-207 (1986).

[12]M.N. Rahman, L.C. DeJonge, G.W. Scherer, and R.J. Brook, "Creep and Densification During Sintering of Glass Powder Compacts," *J. Am. Ceram. Soc.*, **70** [10] 766-74 (1987).

[13]R.K. Bordia and G.W. Scherer, "On Constrained Sintering -I, II, and III," *Acta metall.*, **36** [9] 2393-416 (1988).

[14]R. Raj, "Analysis of the Sintering Pressure," *J. Am. Ceram. Soc.*, **70** [9] C-210-C-211 (1987).

[15]R. Raj and R.K. Bordia, " Sintering Behavior of Bimodal Powder Compacts," *Acta metall.*, **32** [7] 1003-019 (1984).

[16]A. Jagota, "Mechanics of Constrained Thin Films," *Acta metall.*, To be published (1990).

MANUFACTURING PIEZOELECTRIC SUBSTRATES

J. Anthony Brewer
GE Lighting
Advanced Technology Department
Nela Park, Building 336
Cleveland, Ohio 44112

ABSTRACT

A process has been developed for manufacturing .195 +/- .017 mm thick piezoelectric substrates for use as electrical actuators. This paper discusses the complete manufacturing process from qualifying powder vendors through final quality checks. Process steps include powder specifications, ball milling, tape casting, green punching, stack firing, electrical testing and inspection. This work takes a laboratory process and upgrades to a piloting operation, manufacturing over 1000 substrates per week.

INTRODUCTION

Piezoelectric materials, such as lead zirconate titanate ceramics, have unique electromechanical properties such that an electrical potential is generated when a force is applied to the material and a force is generated when electric potential is applied across opposite faces of the material. The desirable electromechanical properties of these materials can be tailored for specific applications by adjusting the Zr/Ti ratio, use of one or more dopants, and processing techniques. Common dopants used are niobium, lanthanum, strontium, iron, antimony, nickel and others. Processing steps that influence properties include milling and fabrication methods, organic removal rates, furnace ramp rates, peak temperature and soak time, packing powder, and oxygen atmosphere control. Uses of piezoelectric ceramics range from ultrasonic cleaners to ink jet printers and to sonar devices.

This paper deals with a specific application where an electric potential is used to displace the piezoelectric material for use as a switching device. This displacement causes contact to an adjourning surface thereby switching the electrical path. The powder used was $Pb(Zr_{.52}Ti_{.48})O_3$ modified with oxides of strontium, lanthanum and niobium. The material properties needed here include extremely high permittivity, high coupling and high piezoelectric constants.

The complete PZT production process flow sheet is shown in Figure 1.

POWDER SPECIFICATIONS

Powders with the desired piezoelectric properties were ordered from various vendors. Once received, these powders were analytically tested for surface area, particle size distribution, X-ray fluorescence (XRF), X-ray diffraction (XRD) and thermal gravimetric analysis (TGA). XRF tested for powder composition while XRD tested for chemical composition, free lead and Zr/Ti ratio. TGA showed if any absorbed moisture or organic materials were present.

The powder would then go through standard processing of ball milling, tape casting, firing, and testing for electrical properties, which included dielectric constant (K), piezoelectric constant (D31) and coupling coefficient (kp). This procedure was repeated for three different lots of the same material from each vendor to test for material consistency. The approved powder vendors are shown in Table 1, along with the specific powder types tested.

Channel Products' 5700 powder was the main powder used in this work. This powder has an average particle size of 0.64 microns. The major problem using the Channel powder was the lot to lot variation in free PbO, which ranged from .60 to 2.40 weight percent. The desired range for free PbO would be between .50 to 1.00 weight percent.

Table 1. Qualified Powders

Company	Powder Type
Channel Products Inc.	5700
Vernitron	5H
Piezoelectric Products Inc.	G-1355

SLURRY PREPARATION

Channel Products' 5700 powder was used exclusively in this manufacturing process even though other powders were qualified. This process flow sheet for ball milling and tape casting is shown in Figure 2. Powder was ball milled in a 13.5 gallon mill using half inch zirconia media, for four hours, to mostly break up agglomerates with little particle size reduction.

The composition of the slurry formulation is shown in Table 2. A two component solvent system of 1,1,1 trichloroethane/ethanol was effective in producing a good tape with acceptable drying characteristics.

The milling operation contained both solvents and ten percent of the total binder, by weight. There was no so called dispersant used in this formulation but the binder material, polyvinyl butyral, acts as a dispersant when used in this low amount.

A six hour mixing step produces a homogeneous slurry ready for casting. The six hour mixing time is the time needed to dissolve all of the dry binder material. If all the binder does not go into solution the cast tape will have hundreds of tiny clumps of undissolved binder on the surface. One way around this is to add the binder already predissolved in solvent which would eliminate three to four hours of milling. Materials added in this mixing operation are the remaining binder, solvents, two plasticizers and a flow control agent. The plasticizers control the cast tape physical properties such as strength and handling ability. A flow control agent, liquid polyethylene, acts as a lubricant during casting and also keeps the surface of the cast tape from drying allowing uniform drying. A viscosity check was tested on each ball mill prior to emptying the mill, with an acceptable range of 22 to 27 poise.

It was determined that degassing the slurry was not necessary if the containers were slow rolled for a period of twelve hours prior to casting.

Scrap and rejected tape, too thick or thin, was reused in the slurry operation as either ten percent of the total batch or separately. The latter process was the better of the two methods.

Table 2. Slurry Formulation

Material	Function	Weight %
Piezoelectric Powder	Ceramic Matrix	67.8
1,1,1 Trichloroethane	Solvent	18.7
Ethanol	Solvent	7.8
Polyvinyl Butyral	Binder	2.9
Polyalkylene Glycol	Plasticizer	1.7
Polyethylene Glycol	Plasticizer	0.9
Liquid Polyethylene	Flow Control Agent	0.2

TAPE CASTING

A continuous casting process was used to manufacture green tape of .241 +/- .014 mm thickness. The casting unit was thirty four feet long with a heated drying area of twenty-four feet. The slurry was cast onto an organic carrier film, up to 38 cm wide. A wet thickness gage was used to set the initial wet thickness at .56 mm, with adjustments on both sides of the doctor blade to control thickness. The speed of the carrier film ran at 30 cm/hour.

A drying schedule was developed that allowed for uniform drying over the length of the unit. Heat was applied only over the final eight feet, by radiant heaters. The most critical area of drying was within the first ten feet where cracking would start due to rapid solvent removal. This was improved with the addition of the flow control agent which kept the surface of the cast tape from drying too rapidly.

Dried tape was cut to the size of a template while on the carrier film and stacked in plastic trays. The template size was determined from the size of the die used in punching out green substrates.

GREEN TAPE SHAPING

Due to differences in material shrinkage, three sets of tooling were made, each corresponding to a specific shrinkage range. Shrinkage samples were punched out, fired and measured for its specific shrinkage and the appropriate tooling was used. Shrinkage values ranged from 18.5 to 21.5 percent.

The tooling was designed to punch out two-by-two arrays, four parts, each part being 6.8 cm square. Scrap tape was kept in a separate container for re-use.

STACKING

Prior to stacking, one side of each part was lightly coated with PZT powder. The PZT powder used for stacking is fired scrap material, ground to <325 mesh, or <44 microns. This powder does not react with the green tape during firing and allows for separation of the layers after firing. Sheets were normally stacked 24 layers high on alumina setters.

The alumina setters were coated with zirconia powder to stop reactions between the samples and setters. A 140 mesh sieve was used to sprinkle the zirconia powder onto the setters. The bottom two layers of a stack were scrap material. There were two stacks of parts per setter with six setters per row, and three rows high, totalling 840 parts per firing. Parts for electrical testing were fired in identical position in every firing.

FIRING

A 10.2 centimeter diameter tube furnace was used, which had a 99 centimeter long hot zone, with processor controlled ramp rates and gas flows. The process tubes are made of high alumina material.

One of the most critical steps in the firing process is controlling the balance of lead vapors during peak temperatures between the excess PbO in the tapes and the atmosphere powders (1). Atmosphere powder originally used in firing was the same unfired PZT powder. An alumina boat was filled with 360 grams of powder and placed in front of the samples on the gas inlet end. This method worked fine in controlling the atmosphere during firing. A second atmosphere powder combination was also used, PZT with $PbZrO_3$, in a weight ratio of 25/75. This powder mixture was placed in three locations within the tube, before the first, third and fifth rows of parts with decreasing weight percents, 50, 30, and 20, respectively. This powder worked just as well if not better than the PZT alone.

The firing profile used is shown in Figure 3. This profile shows a long organic removal step needed for elimination of all carbon residue, especially during stack firing. The amounts and type of gases used were critical at all steps of the firing cycle. Low temperatures, 200 to 350°C, where organics were burning off needed high gas flows with a slight nitrogen rich atmosphere. The exit end of the furnace had five ports with tygon tubing attached to carry the organic material to a reservoir for collection. These tygon tubes were removed at 700°C.

From 350 to 700°C a high flow rate of oxygen was used to burn off all residue carbon. Low oxygen flow rates were used above 700°C up to peak temperature and down to 900°C, when nitrogen was used to help in cooling. Above 700°C just enough oxygen was used to produce densities of greater than 98 percent.

A peak temperature of 1290°C for two hours was used to produce densities of greater than 98 percent.

INSPECTION

After firing each stack had to be separated by hand or a razor blade if needed. Every piece in each stack was inspected and resulted recorded. The inspection criteria used is shown in Table 3.

Table 3. Inspection Criteria

1. Separation losses; broken pieces, edge cracks
2. Visual losses:
    - cracks
    - bumps; greater than 25 microns
    - holes
    - discoloration
        - black/brown spots
        - excessive yellow spots
        - orange spots
        - white spots
3. Edge curl/flatness
4. Shrinkage
5. Thickness; acceptable range 18.8-22.3 mm

If parts were not coated properly, successive layers would fire together and be rejected. Flatness was tested by placing parts between two flat plates and applying slight pressure. Parts were rejected with edge cracks greater than 5 mm from the edge. Shrinkage was measured by calibers for outer dimensions and by placing parts onto a fixture with proper hole dimensions for proper hole placement.

ELECTRICAL MEASUREMENTS

All samples for electrical testing are fired in identical positions in every firing. Sputtered nickel, two to three hundred angstroms thick, was used for electrodes in all electrical measurements. Round discs for measuring K and kp are prepared by sputtering a 1.3 cm diameter masked area of nickel onto 1.5 cm diameter sample. The D31 samples are cut from sputtered nickel sheets into 1.3 by 2.5 cm sizes. All samples are poled at 1.27 volts/mm (50 volts/mil). The minimum electrical values allowed are shown in Table 4.

Table 4. Acceptable Electrical Property Values

Property	Minimum Value	Standard Value
K	3000	3200
kp	.55	.60
D31	3.50	4.00

SUMMARY

The transition of a laboratory process to a piloting operation manufacturing 1000 substrates per week was attained by upgrading the entire process. Many process improvements were needed, especially in firing, to manufacture flat substrates passing all quality checks and electrical properties. All process steps have to be characterized and understood to learn processing limitations. This process developed has shed light on understanding some of the processing parameters needed in manufacturing piezoelectric materials with more fundamental experimentation needed.

LITERATURE

1. A. Kingon, B. Clark, J. AM. CERAM. SOC. 66 (1983) 253-260.

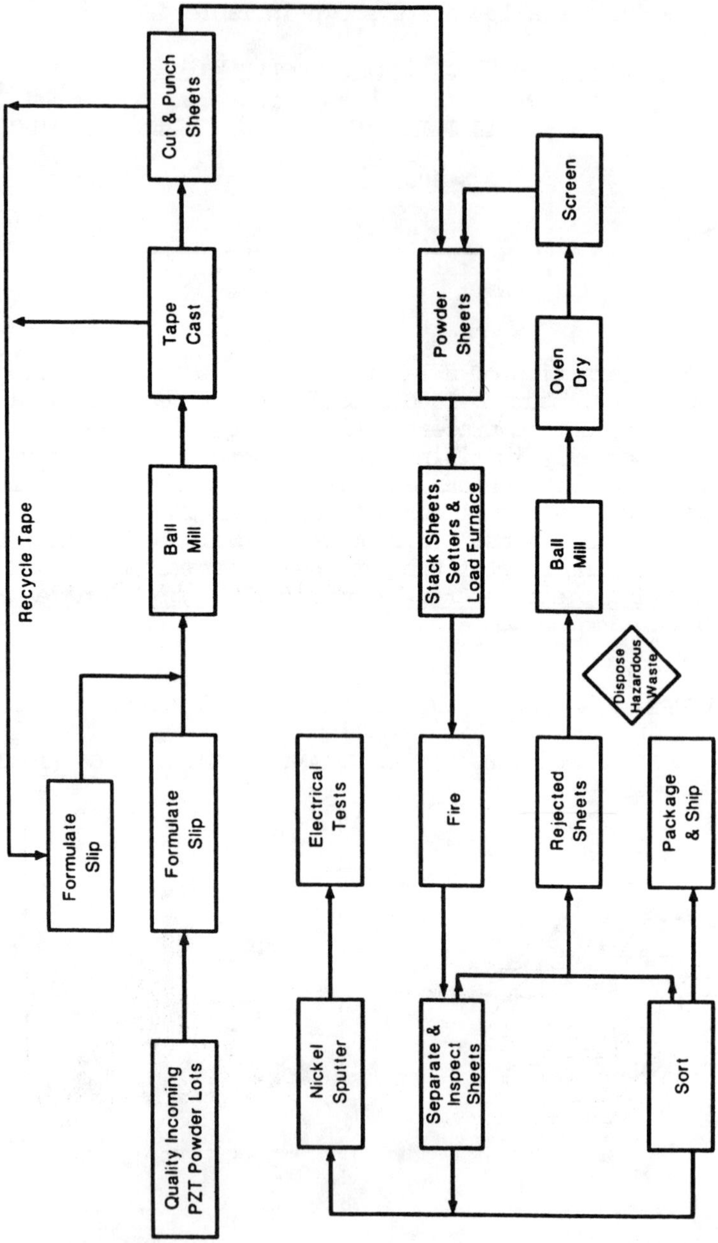

Fig. 1. PZT production process

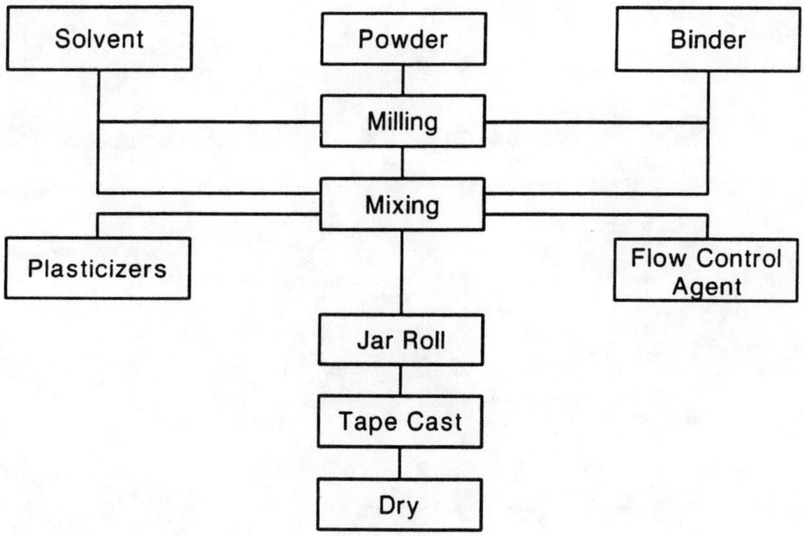

Fig. 2. Ball-milling and tape-casting flow sheet

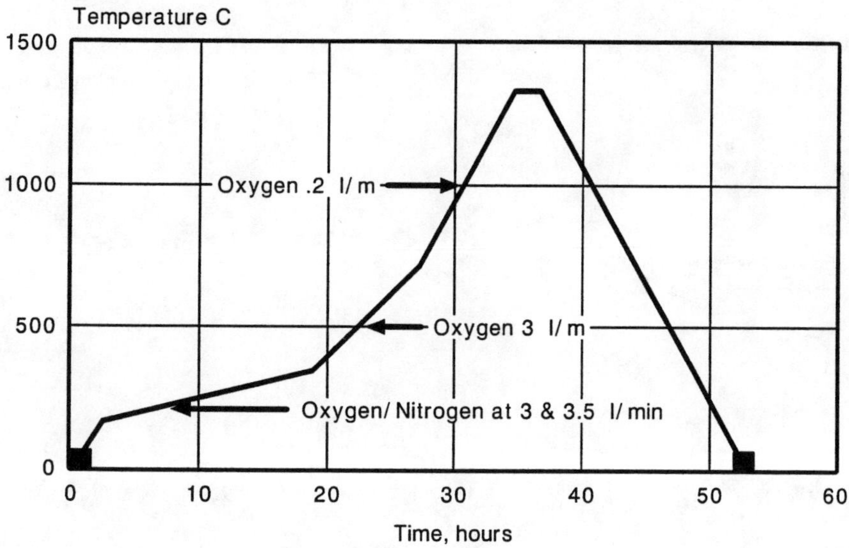

Fig. 3. PZT firing schedule

FAST RESPONSE TO INFRARED RADIATION IN SiC FIBERS

N. Muto, M. Miyayama, H. Yanagida, RCAST, Univ. Tokyo, 4-6-1,Komaba, Meguro-ku,Tokyo 153, JAPAN; T. Kajiwara, N. Mori, Sogo Keibi-Hosho Co. Ltd., 1-6-6, Motoakasaka, Minato-ku, Tokyo 107, JAPAN; H. Ichikawa, H. Harada, Lavoratory, Nippon Carbon Co. Ltd., 1-1, Shin-urashima-cho, Kanagawa-ku, Yokohama-shi 221, JAPAN

ABSTRACT

Thermal response to infrared radiation was measured for SiC fiber detectors. Commercial SiC fibers(Nicalon) with different resistivities (10-3000 Ωcm) and diameters(7-14 μm) were used as thermistor materials. The SiC fibers detected IR radiation at room temperature with a steady output voltage and short response time. In a detector made of separately arranged fibers with a low resistivity (10 Ω cm) and of one number with a short length(1 mm), a small thermal time constant of 1.8 msec was obtained. Decreases in fiber length and fiber number were found to be effective in improving the response time and the responsivity, respectively.

INTRODUCTION

The importance of IR(infrared) detectors has been increasing recently in many fields. The IR detectors can be classified into two categories photon detectors and thermal detectors. The thermal detector responds to changes in electrical properties caused by temperature changes due to absorbed IR radiation. The pyroelectric-type detectors, using electrical polarization, and the thermistor bolometer, using electrical resistance changes, are included in the category of thermal detector. The pyroelectric-type detectors are used widely today, but they can't detect stationary IR radiation from immobile targets. The conventional thermistor-type detectors have slow thermal responses. Fast response with thermal time constant of tens of msec is required for detection of moving IR radiation such as a walking human body. Therefore, a new type of thermistor is needed for the detection of stationary and moving IR radiation.

Since the thermal capacitance of the fiber-shaped material with a small volume/surface ratio is very low, its resistance value is changed by a small amount of IR radiation. From our preliminary experiments, SiC fiber behaves as an n-type semiconductor and shows NTC resistance-temperature characteristics with a good reproducibility.[1] We found that SiC fibers and Si-Ti-C-O fibers could be used as new type of thermistor materials with fast thermal time constants.[2],[3] The SiC fibers have been developed for structural materials, especially for reinforcing composite materials. The object of this paper is to measure the effect of structure of detector element containing SiC fibers on IR response properties.

EXPERIMENTAL PROCEDURE

Semiconducting SiC fiber, with four kinds of resistivities (10, 80, 480, 3000 Ω cm) and two kinds of diameters(7, 14 μm), were used as the IR detector material. The fibers were commercial SiC fibers (Nicalon) manufactured by Nippon Carbon CO. Ltd.. The SiC fibers are synthesized from a polycarbosilane, and has an excess carbon and SiO_2 even after heat treatment.[4] A typical composition used in the present study is Si:57.3, C:32.2, O:9.8, H:0.2 wt%, respectively.[5)]

Electrical resistivity-temperature characteristics, current-voltage characteristics and IR detecting properties of semiconducting SiC fibers were measured using a fiber holder made by a plastic substrate with copper electrode plates. The fibers were placed over the window of the holder with different arrangements(separately and randomly arranged) and both ends of the fibers were connected to the electrodes by graphite paste.[3)] The thermistor constant B of SiC fibers were calculated using the following formula,

$$B= (T_1T_2/T_2-T_1)\log(R_1/R_2) \quad [K] \qquad (1)$$

where R_1 and R_2 are electrical resistance of the fibers at temperature of T_1 and T_2, respectively.

The IR detecting property of SiC fibers was determined by measuring a change in output voltage of the SiC fiber by IR radiation. The apparatus used for measurement of the IR detecting properties is shown schematically in Fig.1. A blackbody furnace of 600K was used as an IR source. The radiation energy W of an IR ray was calculated to be 55 mW/cm^2 at the detector position(11cm from the backbody furnace).[2)] An opened and shutted light was produced by using a fast shutter with a speed of about 290μ sec.

In order to detect a small resistance change of the fiber and eliminate the effect of variation of ambient temperature, a bridge circuit was constructed in a sensor box. The electrical circuit is shown schematically in Fig.2. Fibers with the same resistance were used for the IR detection and for the reference element. The change of the output voltage E_{out} under biasing E_b is expressed in the following formula,[7)]

$$E_{out} = -\Delta R E_b / 4R \qquad [v] \qquad (2)$$

where R is the electrical resistance of the fiber without IR radiation, $\Delta R(<0)$ is the resistance change by IR radiation. The E_{out}-time curve was memorized by a storage scope and was recorded by a X-Y recorder. The thermal time constant of the IR detector was measured from E_{out}-time curve as the period in which E_{out} reaches 63% of the final value. The responsivity R_v was measured from the following formula,

$$R_v = E_{out} / WA \qquad [V/W] \qquad (3)$$

where W is the radiation energy, and A is the receiving radiation area.

RESULTS AND DISCUSSION

The surface in semiconducting SiC fiber has a smooth profile, and the fibers have a wide range of diameters(8-20 μm for the 14 μm fibers reported in this study). The electrical resistivity of SiC fibers can be freely controlled between $10^0 \Omega$cm and $10^6 \Omega$cm by selection of the final heat-treatment temperature.[5)] The SiC fibers consist of cubic β-SiC crystal, and the electrical resistivity of the SiC fiber decreases as the heat-treatment temperature increases due to the crystallization of β-SiC.[6)] With an increase in temperature of final heat treatment, concentrations of excess C and O decrease, and the crystalline size increases, and as a result, electrical resistivity decreases, but we don't know in detail. The electrical resistivity as a function of temperature in SiC fiber is shown in Fig.3. Calculated thermistor constant B is also indicated in Fig. 3. All fibers showed NTC resistivity-temperature characteristics, and the fiber with a larger resistivity showed a larger thermistor constant B.

In general, when large electric power is applied to a semiconducting material, the temperature increases from self-heating(Joule heat), and the current becomes larger than that in a linear I-V relation. For all SiC fibers, current was proportional to the applied voltage under an electric power below 4mW, and self heating of SiC fiber didn't occur during the measurement of IR response (applied voltage E_b:8V).

Figure 4 shows a typical output voltage curve (amplified 100 times) to IR radiation of the detector using one SiC fiber with fiber length of 1.5 mm and resistivity of 80 Ωcm. The curve showed a steady state with a S/N ratio at 9.0 with IR radiation turned on and off. The thermal time constant was found to be 8.0 msec.

The IR response properties are affected markedly by the structure of detector element(length and number of SiC fibers). Figure 5 shows the influence of fiber length(the distance between electrodes) on thermal time constant and responsivity of SiC fiber detector. As the fiber length decreased, the thermal time constant decreased, and the responsivity increased slightly. It indicated that a decrease in fiber length is effective for improving the IR response time. Fig.5 shows that the fiber diameter has negligible effect on thermal time constant.

The thermal time constant and the responsivity are expressed in the following equations,[3)]

$$\tau = H/G = H/(G_e + G_s) \quad [sec] \qquad (4)$$

$$R_V = -\alpha\eta E_b/4G = -\alpha\eta E_b/4(G_e + G_s) \quad [V/W] \qquad (5)$$

where H is the heat capacity and G is the thermal conductance, α is the temperature coefficient of resistance($-B/T^2$) and η is the absorption efficiency of the SiC fiber, respectively. In the present study, thermal conductance can be expressed by $G = G_e + G_s$, where G_e and G_s are the thermal conductances of the SiC fiber to the electrodes/substrate, and to the surrounding atmosphere, respectively. Since H and G_s are proportional to the volume and the surface area of the fiber, respectively, H and G_s must be proportional to the fiber length L. On the contrary, G_e would have no or slight dependence on L. When fiber length is short, thermal conductance to the electrodes and the substrate will be dominant, and as a result, a marked dependence of the thermal time constant on the fiber length would appear as shown in Fig.5.

Figure 6 shows influences of the fiber number on the thermal time constant and the responsivity of SiC fiber detector. As the fiber number decreased, the responsivity increased but the thermal time constant was constant. It shows that G and H are affected by the fiber number. As the fiber number decreases, G_e, G_s and H decrease proportional to the fiber number. As a result, the thermal time constant is independent on the number as expected from Eq.4, and the responsivity is inversely proportional to the number as expected from Eq.5.

From above results, it is found that a faster response and a higher responsivity is achieved by decreasing length and number of fibers. Figure 7 shows a typical output voltage time curve with a good IR response at τ:1.8 msec and R_V:206 V/W in a detector with resistivity:10 Ωcm, n:1, l:1 mm).

CONCLUSION

By measurement of IR detecting properties in SiC fibers, the following results were obtained.

1. The SiC fibers detected IR radiation at room temperature with a steady output voltage and short

response time reaching to 1.8 msec.

2. Decreases in fiber length were effective in improving the thermal time constant and decreases in fiber number were effective in improving the responsivity.

REFERENCES

1) N. Muto, M. Miyayama and H. Yanagida," Thermal Response of SiC fiber", J. Ceram. Soc. Jap., **97**(3)403-05(1989).

2) N. Muto, M. Miyayama, H. Yanagida, K. Kajiwara, N. Mori, H. Ichikawa and H. Harada," Infrared detection by SiC-Fiber Thermistor", J. Ceram. Soc. Jap., **97**(10)1304-07(1989).

3) N. Muto, M. Miyayama, H. Yanagida, K. Kajiwara, N. Mori, H. Ichikawa and H. Harada," Infrared Detection By Si-Ti-C-O Fibers", J. Amer. Ceram. Soc., in press.

4) S. Yajima, K. Okamura, T. Matsuzawa, Y. Hasegawa and T. Shishido," Anomalous characteristics of the microcrystalline state of SiC fibres", Nature, **279**(21), 706-07(1979).

5) H. Ichikawa," Effect of curing conditions on Mechanical Properties of SiC fiber(Nicalon)", J. Mater. Sci. Lett., **22**(6), 420-22(1987).

6) H. Teranishi, H. Ichikawa and T. Ishikawa, " Properties of Silicon Carbide Fiber "Nicalon" and Its Aluminum Composites", New materials and New Processes, **20**, 379-85(1983).

7) M. Inaba, M. Miyayama and H. Yanagida," Infrared sensing Properties of $BaTiO_3$ PTC Thermistor", J. Ceram. Soc. Jap., **97**(10)1250-55(1989).

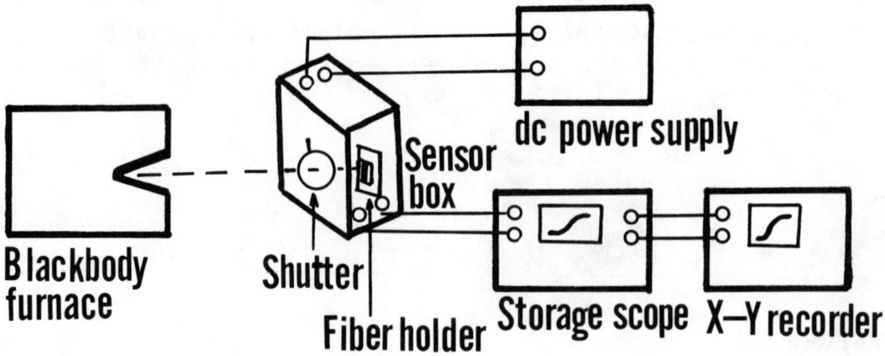

Fig.1. Schematic drawing of apparatus used for IR detection.

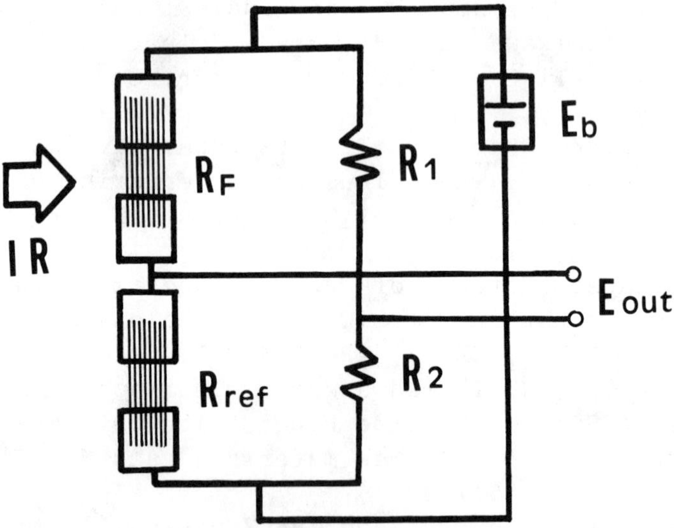

Fig.2. Schematic drawing of electric circuit used for IR dtection.

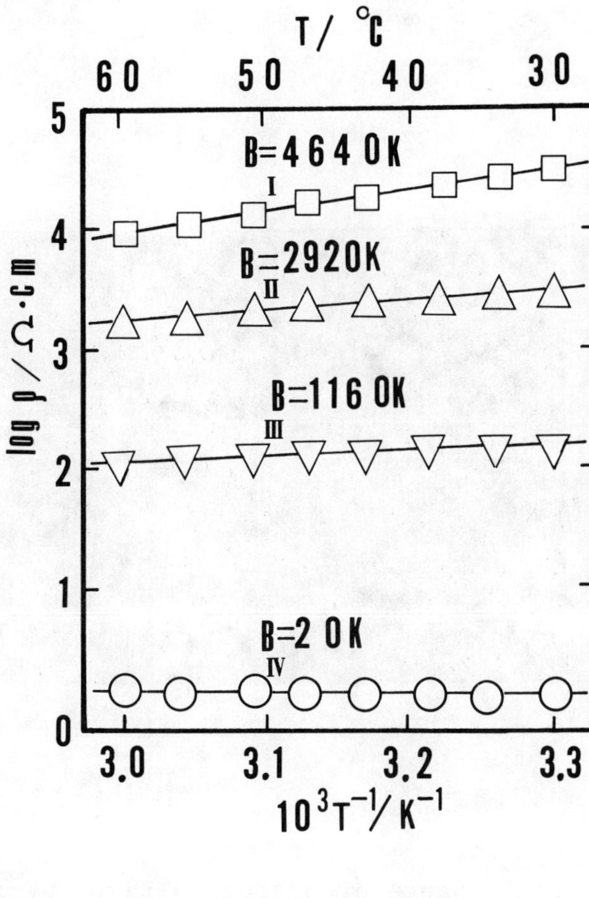

Fig.3. Electrical resistivity and thermistor constant B between 30°-60°C in SiC fiber.

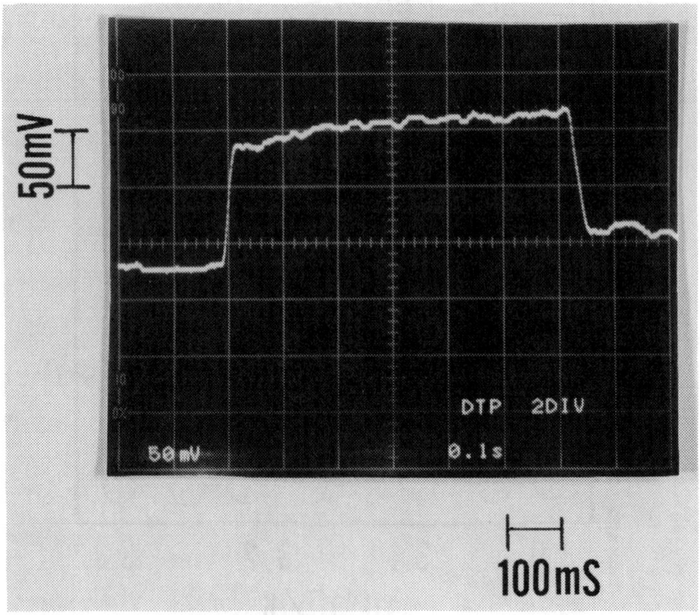

Fig.4. A typical change of output voltage by IR radiation in SiC fiber(thermal time constant:8.0 msec, S/N:9.0; resistivity:480 Ωcm, n=1, l=1.5mm).

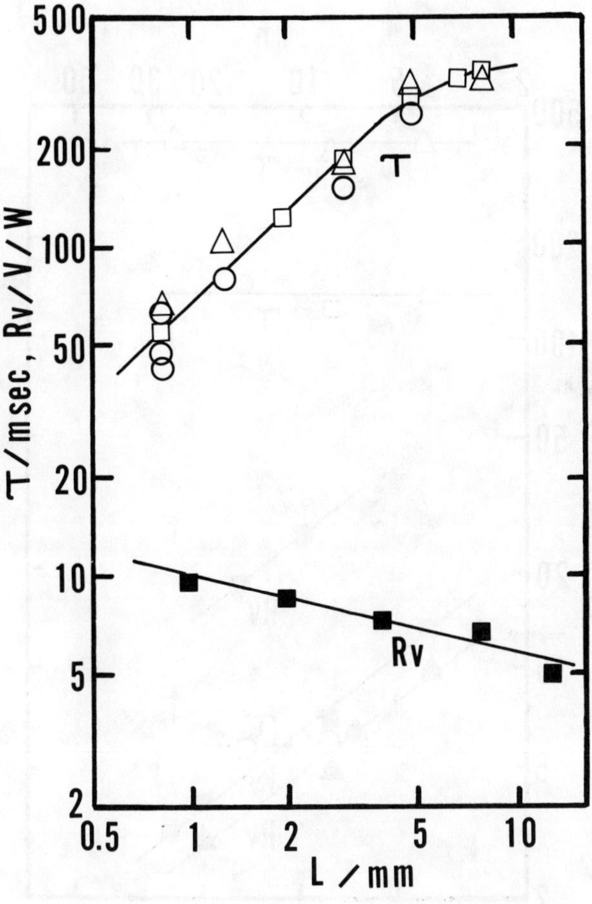

Fig.5. Influences of the fiber length on thermal time constant and responsivity of SiC fiber{resistivity(Ω cm), number, length(mm), diameter(μ m) are of the fiber, (○):3000 Ω cm, n=500, d=7 μ m, (△):3000 Ω cm, 500, 14 μ m, (□):480 Ω cm, 800, 7 μ m}.

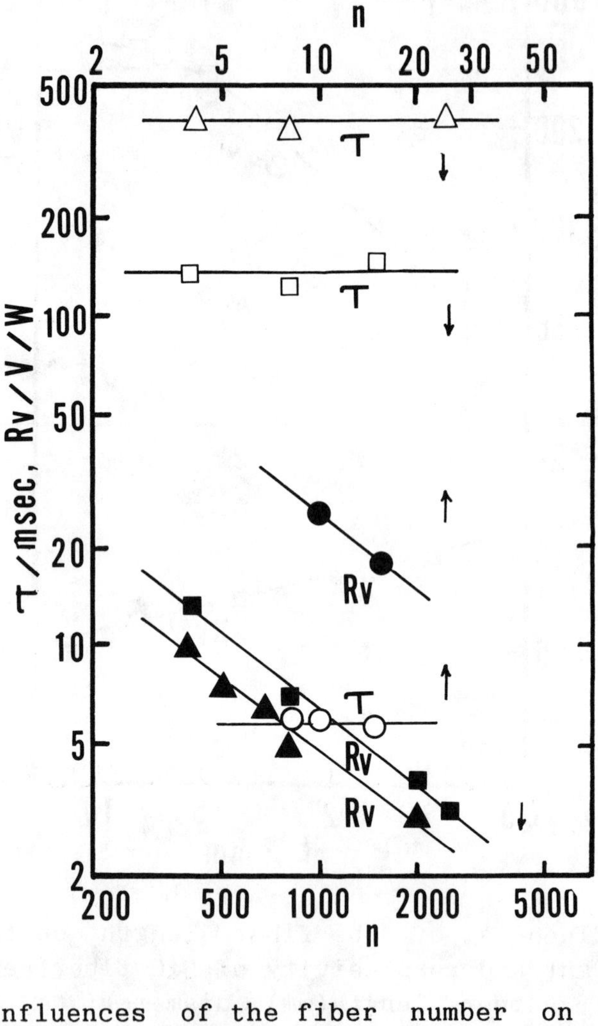

Fig.6. Influences of the fiber number on thermal time constant and responsivity of SiC fiber{ (○):80 Ωcm, l=1 mm, (□):480 Ωcm, 2 mm, (△):3000 Ωcm, 5 mm}.

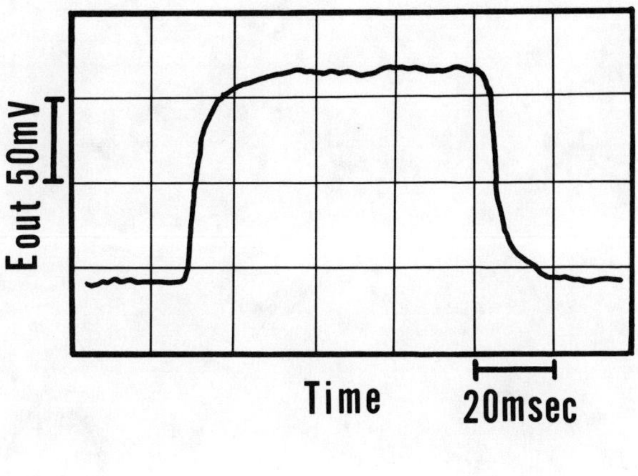

Fig.7. A typical change of output voltage by IR radiation in SiC fiber{ thermal time constant:1.8msec, Responsivity:206 V/W; resistivity:14.9 Ωcm, d=7.0 μm, L=0.5 mm, n=1}.

CHARGE TRANSPORT AND RELIABILITY OF DIELECTRIC COMPONENTS

Larry C. Burton
College of Engineering
Virginia Polytechnic Institute and State University
Blacksburg, VA 24061

ABSTRACT

This is a review type of paper dealing with materials and measurement parameters that influence charge transfer and degradation in dielectrics, with particular emphasis on MLC capacitors. The nature and causes of leakage current are reviewed, and a number of pertinent materials-related parameters are discussed. Reference is made to work done by us and by others.

LEAKAGE CURRENT

The main signature of dielectric degradation is increased leakage current. We have reviewed this elsewhere in some detail [1]. The nature of the physical mechanisms may be determined if the current type or types can be identified. There are several types of current commonly seen. These are: a) Ohmic—seen usually at low fields. Carriers can be ionic and electronic. b) Tunnelling current—can occur only in thin layers (≲100Å or so), or through Schottky barriers and grain boundaries. The charge carrier must be electronic. c) Schottky emission—dominated by thermionic emission over a potential barrier at an interface. d) Poole-Frenkel emission—similar to Schottky emission, only occurring in the bulk of the solid. e) Space charge limited current (SCLC)—usually occurs in low mobility materials, where current is limited more by transport than by availability of carriers. This is an electronic current. f) Hopping—charges move by hopping between centers in the band gap; prevalent in defect or amorphous materials. This is also an electronic current.

There are many materials—related parameters than can effect charge transport in, and reliability of, dielectrics. Those that we will discuss in varying degrees are:

- dielectric thickness and its variation
- dielectric constant and uniformity
- charge carrier species
- grain boundaries and grain size
- composition, coping, and homogeneity

- trapping at deep states

- dielectric/electrode interface

- local effects such as pores and electrode asperities

There are also several stress parameters that can influence degradation. Well known ones are electric field strength (voltage), temperature, time, and ambient (humidity, for example). Other lesser studied stress parameters include signal characteristics (frequency, amplitude, risetime, etc.), and mechanical stress. The latter one certainly deserves greater attention, with the growing use of surface mount technologies and embedded components.

DIELECTRIC THICKNESS (d)

Several key parameters related directly to dielectric thickness can influence both current and degradation. Some of these are

- Electric field and capacitance, both of which vary inversely with thickness at constant voltage;

- Resistance, proportional to thickness

- Electric field can subsequently determine carrier type and mode of transport

- Degree of severity of interface effects and electrode roughness will be dependent on thickness

DIELECTRIC CONSTANT

The dielectric constant ($K = K' + jK''$) and its uniformity can also bear directly on currents and reliability. High K automatically means high loss, as dictated by, for example, the Kramers-Kronig dispersion relations [2]. Hence, the reliability of high-K capacitor materials is determined in part by the physics of the coupling between K' and K''.

Some leakage currents depend directly on K. Examples are SCLC (linear dependence) and Schottky emission (exponential).

Non-uniformity of K is also of interest. In general, the electric field $\boldsymbol{\varepsilon}$ increases in low K regions. This effect can be striking in a high K material with pores, delaminations, etc. where K = 1. The electric field in these regions can be magnified by thousands [3]. If electrical conductivity is also discontinuous, which can occur for multi-phase materials, then a charge density Q_S must exist at the interface:

$$Q_S = \varepsilon_O \boldsymbol{\varepsilon}_2 (K_2 - K_1 \sigma_1 / \sigma_2) \tag{1}$$

where $\boldsymbol{\varepsilon}$ is electric field and σ is conductivity, the subscripts pertaining to the two regions. An example is the existence of a low K second phase at grain boundaries

(GB), with nearly all of the voltage being dropped there due to enhancement of the field. Schomann has based a degradation model on this phenomenon [4].

Other examples where enhanced currents and/or dielectric degradation have been attributed to various types of inhomogeneities are SiO_2 electrode asperities [5], point emission [6], MLC capacitor electrodes [7], and charge trapping effects [8]. Localized conduction in dielectric films due to the barrier height lowering by defects has been discussed by Albin [9].

CHARGE CARRIER SPECIES

Charge carrier species include electrons, holes, ions, and vacancies, all of which can move under the influence of electric fields or their own gradients. Modes of conduction can be intrinsic band, extrinsic band, polarons, etc. Control of transport can be by phonon or impurity scattering, hopping or ionization potentials, grain boundary or Schottky barriers, Coulomb screening, and exchange/correlation effects.

All of the above-named species are important in dielectrics—a single one usually, but not always, dominating. Examples are as follows:

1. Electrons dominate intrinsic conduction in SiO_2, and conduction electron energy disbutions have been reported by Brorson, et al. [10]. Conduction electrons in SiO_2 form the current, occupy traps which modifies the field, and become hot enough to actually generate new traps (with kinetic energies to about 40 eV).

2. Both p-type and n-type behavior are observed in ferroelectric ceramic, depending largely on oxygen partial pressure [11].

3. Due to oxygen vacancy diffusion in $SrTiO_3$, the formation of a p-n junction has been proposed [12]. This is caused by an oxygen rich region on the anode side, and a vacancy rich region on the cathode side.

4. Electrons have been proposed as the dominant carrier in commercial X7R ceramic, based on current versus time, Seebeck, and galvanic measurements [13].

5. There is much additional evidence for ionic movement in ferroelectrics, which has been seen in many ways (color gradients, resistivity gradients, gravimetric measurements, etc.) [14,15].

IONIC TRANSPORT AND DEGRADATION

Electronic degradation for many dielectric materials is almost certainly caused by ionic movement. Waser has recently treated polycrystalline $SrTiO_3$ degradation using an electrochemical model. In this model, GB play a dominant role in stability, by virtue of their impedance to ionic transport [12]. The situation for ferroelectric ceramic, in general, is not so clear, and more on the role of GB will be discussed later. Harmer et al. [16] find little difference between samples with substantially different grain sizes. Of course, even though leakage currents may be identical at some zero state, degradation rates could still vary considerably, since the former would depend on electronic current, the latter on ionic.

It is widely accepted that the degradation of $BaTiO_3$ based ceramic is due to the movement of oxygen and oxygen vacancies [17]. We have proposed two distinct models, both of which are in substantial agreement with the empirical degradation behavior originally reported by Prokopowicz and Vaskas [18], and subsequently by Minford [19]. In the first model [13], the degradation equation is derived from basic principles of ionic diffusion. The final result of the model is that oxygen vacancy concentration increases with time as

$$[V_{\ddot{O}}] = K \exp(\beta t) \tag{2}$$

where

$$\beta = \frac{\mu_i V}{W^2} \tag{3}$$

where μ_i = oxygen ion mobility, W = dielectric thickness, V = voltage. Since conduction electron concentration is proportional to $[V_{\ddot{O}}]$, leakage current is predicted to increase exponentially with time, the degradation rate constant equal to β. Temperature dependence is implicitly contained in μ_i. Note the strong inverse square dependence on thickness.

Our second model is based on passivation of GB charge by ionic motion, and gives a similar result [20]. Filled GB states are neutralized by oxygen vacancies, reducing the GB barrier height and increasing current. It thus seems possible that both bulk and GB dominated processes may be involved with ionic-motion-induced degradation.

Additional evidence for ionically induced degradation is the near-reversible behavior of DC degradation. This is seen in Figure 1 for a Z5U device. In this figure the initial exponential increase in leakage current is clearly seen, under accelerated conditions (100 V, 150°C). When the DC polarity is reversed, leakage current decreases, and then again degrades. Such an oscillatory behavior can be continued almost indefinitely, indicating the "sloshing" of ions back and forth across the layer. Such behavior has been seen for MLC capacitors [13], and thick film capacitors [20].

ROLE OF GRAIN BOUNDARIES (GB)

The role of GB in current conduction has been unambiguously established for some materials. Examples are polycrystalline silicon, thin semiconductor films, ceramic varistors and thermistors, and GBBL capacitors. Such is not the case for high resistance ceramic; the situation is still not clear as to the role of GB in current transport and degradation. As noted earlier, Harner et al. [16] found no significant dependence of leakage current of stoichiometric $BaTiO_3$ on grain size (1 to 10 μm). Waser, on the other hand, reports striking differences between coarse-grained and fine-grained $SrTiO_3$ ceramics. He reports that both leakage current and lifetime depend on the number of GB between sample electrodes [12].

We have discussed the role of ceramic GB in some detail [21–23], and have reported several experimental results that seem attributable to GB conduction. These are as follows:

1. Thermal activation energies decrease with bias voltage [23], in fact, in a manner very similar to that of the GB controlled ZnO varistor.

2. Thermal activation energies decrease as degree of degradation increases [1]. This is consistent with a model whereby GB states are passivated by trapped ions [20].

3. We have seen I-V behavior at large voltages (much higher than rated voltage) whereby the curves, following non-ohmic behavior, become ohmic. Examples are shown in Figures 2 and 3. Note the ohmic (power law voltage exponent of near unity) behavior at high voltages. This behavior is striking in that it is again reminiscent of ZnO varistor behavior, a GB-controlled device.

CONCLUSIONS

Much progress has been made in recent years toward improved understanding of current and degradation mechanisms. The variety of materials systems has made this task even more difficult. There are some key areas where charge transport and electrical reliability will be ever more important in coming years, and must be better understood. Some of these areas, related to dielectric components, are:

- All MLC structures
- IC packaging
- Transducers
- Sensors
- Thick film electronics
- MOS/MIS devices. New developments will include even thinner dielectric layers (< < 25 μm in MLC), thin films, and low K materials, in all of which improved reliability will be an issue.

ACKNOWLEDGEMENTS

The help of H. Y. Lee, S. Villamil, N. Schunke, T. Zhang, E. Ellis, K. C. Lee, I. Yoo, A. Vaseashta and K. Taszarek is gratefully acknowledged. Sponsors of the research includes ONR, DuPont, Ford, Engelhard, and Virginias' Center for Innovative Technology, and they are also gratefully acknowledged.

REFERENCES

1. H. Y. Lee, K. C. Lee, J. N. Schunke and L. C. Burton, "Leakage Currents in Multilayer Ceramic Capacitors," IEEE Transactions CHMT-7, 443 (1984).

2. A. K. Jonscher, Dielectric Relaxation in Solids, Chelsea Dielectrics Press (1983).

3. W. L. Robbins, "Ceramic Capacitor Failures - The Phenomena," Workshop on MLC Capacitor Reliability, National Academy of Sciences (1982).

4. K. D. Schomann, "Electric Breakdown in Barium Titanate: A Model," Appl. Phys. 6, 89 (1975).

5. R. M. Anderson and D. R. Kerr, "Evidence for Surface Asperity Mechanism of Conductivity in Oxide Grown on Polycrystalline Silicon," J. Appl. Phys. 48, 4834 (1975).

6. T. Hibma and H. R. Zeller, "Direct Measurement of Space Charge Injection from a Needle Electrode into Dielectrics," J. Appl. Phys. 59, 1614 (1986).

7. H. Y. Lee and L. C. Burton, "Influence of Electrode-Ceramic Interface on MLC Leakage Current," Advances in Ceramics 19, 219 (1986).

8. D. J. DiMaria, R. Ghez and D. W. Dong, "Charge Trapping Studies in SiO_2 Using High Current Injection from Si-rich SiO_2," J. Appl. Phys. 51, 4830 (1980).

9. S. Albin, "Localized Conduction in Dielectric Films," Thin Solid Films 86, 369 (1981).

10. S. D. Brorson, et al., "Direct Measurement of the Energy Distribution of Hot Electrons in Silicon Dioxide," J. Appl. Phys. 58, 1302 (1985).

11. W. Huebner, H. V. Anderson and D. E. Day, "Reliability Studies of Ceramic Capacitors," Progress Report, ONR Contract N00014-82-K-0294 (October 1984).

12. R. M. Waser, "Electrochemical Boundary Conditions for the Resistance Degradation of Coped Alkaline Earth Titanates," Multilayer Layer Capacitor Reliability Symposium Proceedings, p. 241, Penn State University (1989).

13. H. Y. Lee and L. C. Burton, "Charge Carriers and Time Dependent Currents in $BaTiO_3$-Based Ceramic," IEEE Transactions CHMT-9, 469 (1986).

14. D. D. Glower and R. C. Heckman, "Conduction-Ionic or Electronic - in $BaTiO_3$," J. Chem. Phys. 41, 877 (1964).

15. K. Lehovec and G. A. Shirn, "Conductivity Injection and Extraction in Polycrystalline Barium Titanate," J. Appl. Phys. 33, 2036 (1962).

16. M. P. Harmer, Y. H. Hy, M. Lai and D. M. Smyth, "The Effects of Composition and Microstructure on Electrical Degradation in $BaTiO_3$," Ferroelectrics 49, 71 (1983).

17. C. Schaffrin, "Oxygen Diffusion in $BaTiO_3$ Ceramic," Phy. Stat. Sol. (a) 35, 79 (1976).

18. T. I. Prokopowicz and A. R. Vaskas, "Research and Development, Intrinsic Reliability, Subminiature Ceramic Capacitors," Final Report, ECOM-907050-F, NTIS AD-864068 (1969).

19. W. J. Minford, "Accelerated Life Testing and Reliability of High K Multilayer Ceramic Capacitors," IEEE Transactions CHMT-5, 297 (1982).

20. I. K. Yoo, L. C. Burton and F. W. Stephenson, "Electrical Conduction Mechanisms of Barium Titanate Based Thick Film Capacitors," IEEE Transactions CHMT-10, 274 (1987).

21. L. C. Burton, "Models for Electronic Conduction Across Ceramic Grain Boundaries," MRS Proc. 60, 179 (1986).

22. S. S. Vallamil, H. Y. Lee and L. C. Burton, "The Resistance of Grain Boundaries in $BaTiO_3$ Based Ceramic," IEEE Transactions CHMT-12, 482 (1987).

23. L. C. Burton, "Voltage Dependence of Activation Energy for Multilayer Ceramic Capacitors," IEEE Transactions CHMT-8, 517 (1985).

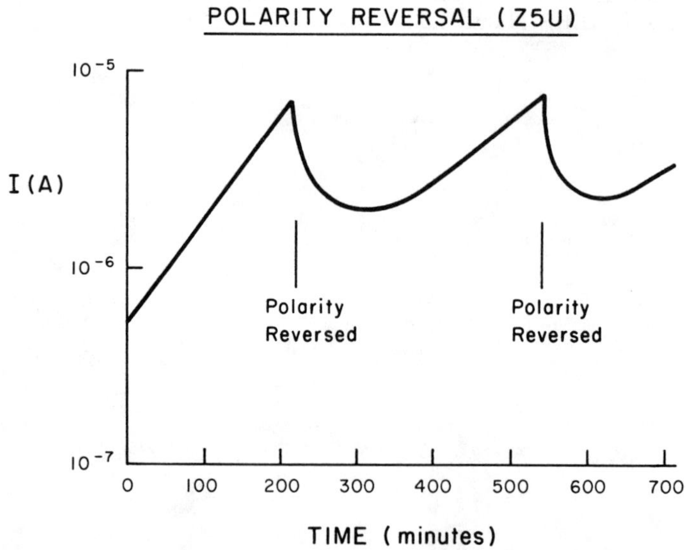

Figure 1. DC polarity reversal healing effect for Z5U MLC capacitor being tested at ±100V, 150°C.

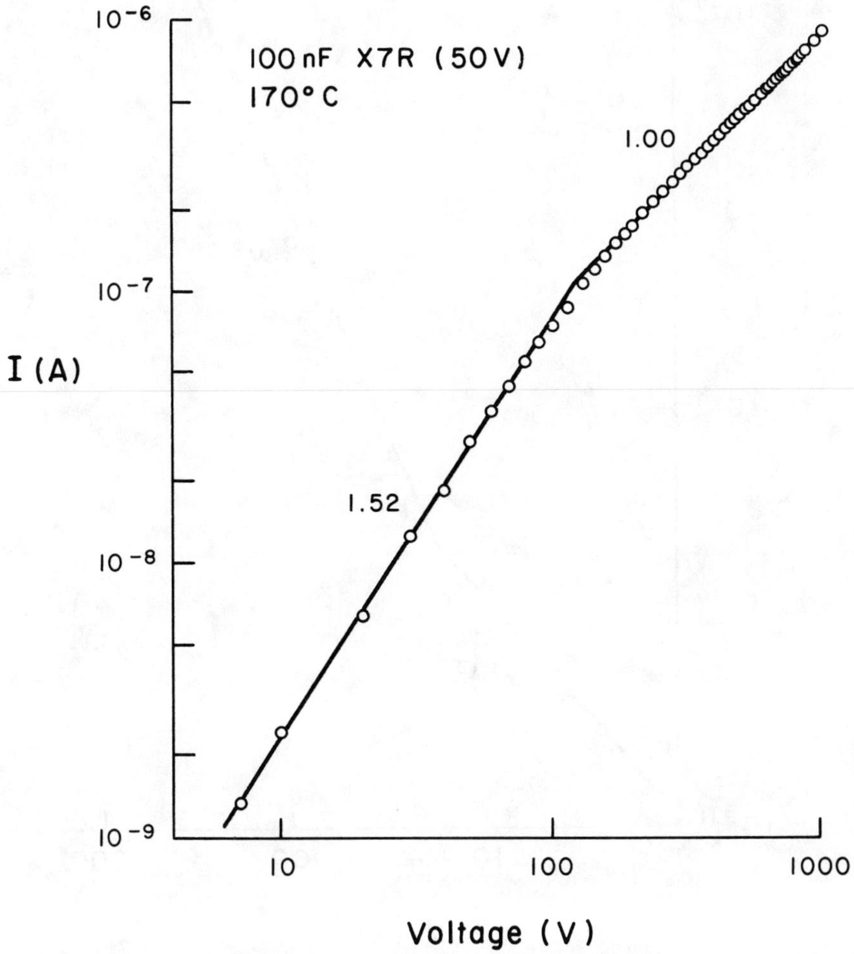

Figure 2. I-V behavior for X7R MLC capacitor at 170°C, showing high voltage ohmic region.

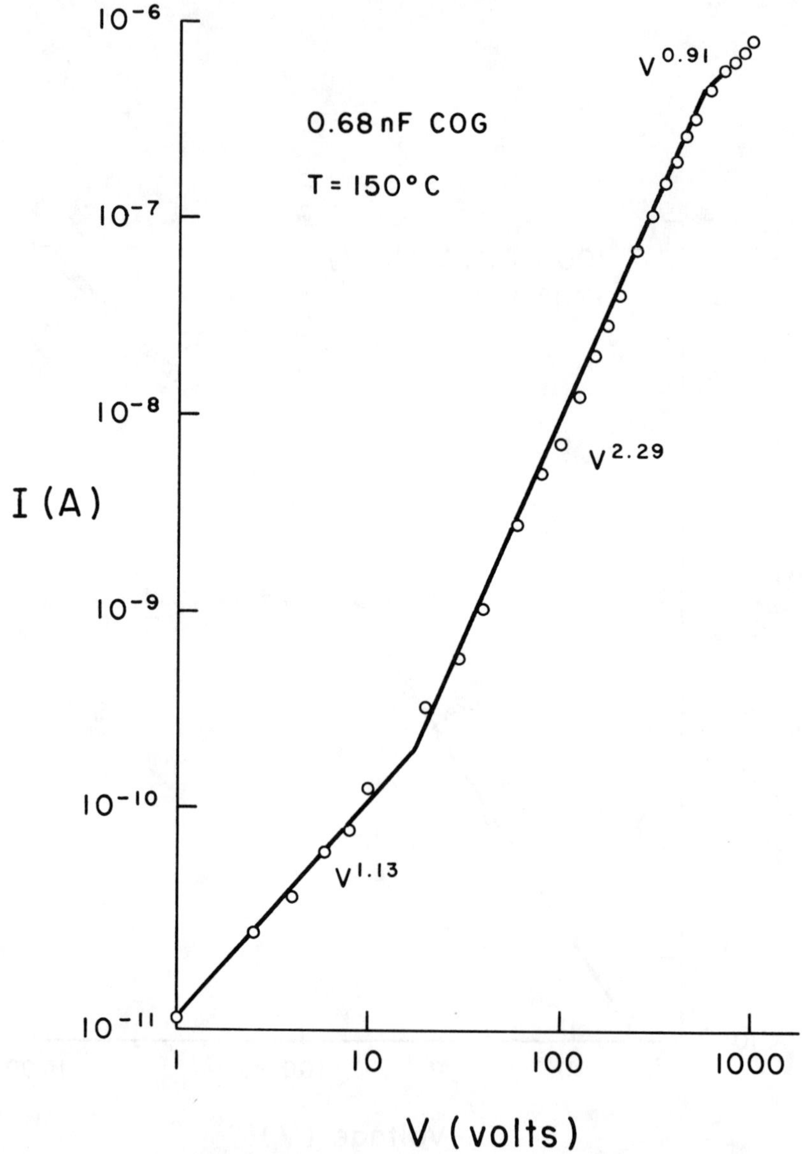

Figure 3. I-V behavior for COG MLC capacitor at 150°C, showing high voltage ohmic region.

THERMAL CONDUCTIVITY CONSIDERATIONS FOR AlN INTERFACES

Alistair D. Westwood & Michael R. Notis,
Department of Materials Science & Engineering,
Whitaker Lab #5, Lehigh University, Bethlehem, PA 18015

ABSTRACT

The microstructure and microchemistry of bulk AlN substrates containing different sintering aids (CaO, Y_2O_3, WO_3 and $CaWO_4$), as well as thin film ~1μm Ti, and thick film ~10 μm metal(MoMn or W)-AlN interfaces were studied, using a variety of electron microscopy techniques. A newly developed laser flash technique was used to measure the thermal conductivity of the substrates and metal-AlN interfaces. The thermal conductivity results are interpreted with respect to the different microstructures observed. The microchemistry and crystallography of boundaries observed in AlN are described and their possible effect on thermal conductivity is discussed.

INTRODUCTION

The drive towards higher circuit densities with their associated higher heat output hasled the industry to develop a number of high thermal conductivity substrate materials, and to introduce a new design criteria in packaging technology, that of thermal management[1,2]. However, other criteria must also be satisfied by the substrate, such as: thermal expansion matching close to that of silicon, good electrical insulation, adequate mechanical properties, non-toxicity, low dielectric loss (as the demand for higher switching speeds increases), and ease of metallization. New advanced ceramic substrate packaging materials which are now available offer at least an order of magnitude improvement in thermal conductivity compared to alumina (Al_2O_3), which up to the beginning of the 80's had been the standard substrate material for the ceramic packaging industry. The reason for the wide use of alumina is that its processing technology was well understood, packages with high reliability hermetic seals could easily be made, and the thermal conductivity offered by alumina was adequate to cope with the heat generated in the lower circuit density devices.

Of the three new candidates that have emerged as high thermal conductivity ceramic substrate materials, aluminum nitride (AlN) is the most favored over silicon carbide (SiC), and beryllia (BeO). This is because SiC exhibits poor dielectric behavior, and BeO is extremely toxic, causing severe processing restrictions. AlN substrates are now available with high thermal conductivity, and onto which high circuit density devices can be mounted. Recently, another driving force has emerged besides good thermal conductivity for the introduction of AlN to replace existing Al_2O_3 substrates. As logic and memory chips become progressively larger the need for close thermal expansion match between die and substrate becomes increasingly important. AlN has a close thermal expansion match to silicon (α_{AlN} = $4x10^{-6}$ $^{O}C^{-1}$, α_{Si} = $3.5x10^{-6}$ $^{O}C^{-1}$); therefore AlN is being used in surface mount applications to minimize failures associated with delamination at the device/substrate interface due to thermal cycling. Other areas of application are for multilayer ceramic packages using green tape technology with cofired tungsten metallization, and in the military sector where reliability and high performance are of paramount importance.

For the new generation of ceramic substrate materials to be fully utilized they must be easily metallized using a variety of different metals. Initial studies of metallization on advanced ceramic substrates were phenomenological in nature and evaluated the metallization process primarily by adhesion testing[3,4]. More recent work has concentrated on microscopic analysis of the bonding mechanism at the interface[5,6].

The thin and thick film metallization processes presently being applied to these newceramic materials have been directly adopted from technology developed for metallizing alumina. This approach may severely restrict the potential benefits offered by these materials, because the metallizing technology developed for alumina focussed only on achieving high bond strength and good hermeticity. Although these two considerations are essential for metallizing any substrate, two other equally important factors to be considered when metallizing AlN, SiC or BeO, are maintaining high thermal conductivity across the metal-ceramic interface, and controlling the dielectric constant of the ceramic at the metal-ceramic interface, which affects the switching speed of the circuit.

If thermal conductivity is poor across the interface, then a local temperatureincrease at the interface will result if the heat is not rapidly dissipated via the substrate. One consequence of this could be the increased probability of device failure from stresses produced by thermal expansion mismatch across the interface due to thermal cycling. Development of a high thermal conductivity interface requires study of the microstructural and microchemical features that will maximize heat flow across a metal-ceramic interface without compromising other necessary properties. Information is available to determine the factors that are important in producing good thermal conductivity in the bulk ceramic substrate[7,8], but the role of interfaces (ceramic-metal, ceramic-ceramic, and metal-metal) in thermal transfer has drawn little attention.

The dielectric constant should be as low as possible because this property has a directaffect on the switching speed of the circuit, being linked to the delay time of the circuit by the equation:

$$t_d = \{(\varepsilon)^{1/2}.l\}/c$$

where t_d is the time delay in the signal, l is the distance signal travels, c is the speed of light, and ε is the dielectric constant. Therefore, reducing ε decreases the time delay, allowing faster switching times. The dielectric constant of the material adjacent to the metallized circuitry will dictate the delay time; therefore, the metallizing process and resultant interface microstructure should not increase the dielectric constant of the adjacent substrate material. This will is important parameter, particularly in multilayer ceramic packages because of the long path lengths involved.

Thermal Conduction in Non-Metallic Systems

The thermal conductivity of a material is governed by the frequency with which the heat transport process encounters a scattering mechanism or site, causing a disruption in the heat flow. In non-metallic solids heat transport is by phonons (a phonon is a quanta of thermal vibrational energy generated by atoms oscillating around their atomic sites within the lattice) propagating through the lattice. If the system is a perfectly periodic structure at thermal equilibrium, then all the atomic displacements are the same. Thus, all phonons possessing the same wavelength, amplitude, and frequency, can be described by perfectly harmonic atomic vibrations and can be represented by a set of standing waves. This condition minimizes phonon interactions, but is an idealized situation, because real crystals are not perfect periodic structures. In real crystals, interactions between phonons and defects occur, which destroy the harmonic wave, replacing it with an anharmonic wave. The greater the degree of anharmonicity present, the more defect-phonon or phonon-phonon interactions that will have occurred, and the lower the thermal conductivity. High thermal conductivity is obtained when phonons can travel through the material without being scattered; the fewer scattering centers the longer the mean free phonon path length and the greater the thermal conductivity. Four parameters exist which increase the magnitude of the thermal conductivity for a non-metallic material: 1) low atomic mass, 2) strong interatomic bonding, 3) simple crystal structure and 4) low anharmonicity[9].

Thermal conductivity in non-metallic solids shows a strong temperature dependence[10] Figure.1. This behavior arises from the relationship between phonon wavelength (λ_{ph}), temperature and scattering sites. The dominant phonon wavelength is related to the temperature by θ/T, where θ is the Debye temperature, (950°K for AlN)[11]. The phonon wavelength at room temperature is therefore of the order of a few interatomic spacings. The relationship between phonon wavelength, defect size, and defect distribution determines which phonon scattering site will control thermal conductivity. Structural defects in ceramic materials are predominately point defects (vacancies, interstitial atoms, substitutional atoms, isotopes), planar boundaries (interphase boundaries, inversion domain boundaries (IDB), antiphase boundaries (APB), grain boundaries), and pores, all of which can act as phonon scattering sites. As the phonon wavelength approaches the size of the defect, then that defect will become the dominant scattering site.

The dominant scattering sites for advanced packages operating between room temperature and 200°C are point defects. At these operating temperatures the phonon wavelength is ~1nm, i.e., the wavelength is of similar magnitude to the size of the point defect, ~0.3nm, (the interatomic spacing), causing these to be strong scattering sites. The formation of point defects in AlN is related to the affinity which it shows towards oxygen. Oxygen is easily incorporated into the nitrogen sublattice, and aluminum vacancies are generated in the aluminum sublattice to maintain charge neutrality. Phonon scattering by these defects is due to the mass difference, causing a change in the amplitude and frequency of the phonon and thus introducing anharmonicity into the phonon wave. Other point defects, nitrogen or aluminum interstitials, would have a similar affect on the thermal conductivity but are energetically unfavorable[11]. Aluminum vacancies will be stronger scattering sites than oxygen vacancies because of their larger mass difference. Slack[12] showed that incorporation of oxygen into the AlN lattice produces a marked decrease in the thermal conductivity.

The other scattering sites commonly encountered in AlN are boundaries. Any boundary scatters phonons either diffusively or specularly, and can change the phonon frequency and propagation mode. The atomic disarray, and the presence of segregant species at the interfaces will produce a stronger scattering factor than the individual point defects. Therefore, the density and type of boundaries present, and also the type of second phases present at these boundaries are important. In the majority of cases the grain boundary density is far less than the point defect density, so their effect is expected to be small. For example, assuming a grain size of ~4μm and a dissolved oxygen content of 0.05wt%, the number of linear point defect scattering sites across the grain is ~1000, compared to the two boundary scattering sites(the sides of the grain). However, this example does not take into account the quite considerable effect of impurities at or near the boundary.

Either prevention or reduction of point defects appears to be the key in maximizing interface thermal conductivity. Removal of oxygen within the matrix substantially improves thermal conductivity because the associated aluminum vacancies are removed. Insoluble metallizing products should be used to prevent point defect formation. Solid state displacive reactions used in thin film metallization produce both intermediate phases and an interdiffusion zone in the adjacent AlN. Both of these contribute to degradation in the thermal conductivity, because the intermediate phases formed will almost certainly have a thermal conductivity lower than AlN, and the interdiffusion zone is a region with high point defect density.

A microcrystalline phase will possess a larger number of interfaces per unit volume than a crystalline film, and therefore will be more detrimental to the thermal conductivity. A glassy phase will be worse than a crystalline film owing to the loss of periodic structure, causing the mean free phonon path to decrease to the interatomic spacing, ~0.3nm. However, the difference in behavior between a microcrystalline and glassy phase will depend upon the number of interfaces, and their effect on thermal conductivity. The presence of a second phase will tend to decrease the thermal conductivity both because it possesses a lower thermal conductivity than the matrix phase and its presence creates an interphase interface. The volume fraction of this phase and also its distribution through the matrix will have an effect.

EXPERIMENTAL

Characterization of bulk AlN and metal - AlN interfaces was carried out using Conventional Transmission Electron Microscopy (CTEM), Analytical Electron Microscopy (AEM), Convergent Beam Electron Diffraction (CBED), Scanning Electron Microscopy (SEM) and Auger Electron Spectroscopy (AES). Depending on specific analysis requirements, the electron microscopy was carried out either on a Philips EM400T microscope operated at 120KeV, fitted with a x-ray energy dispersive spectrometer (EDS) and Tracor Northern TN2000 computer, or a Philips EM430 microscope operated at 250KeV, or a Vacuum Generators HB-501 FEG dedicated Scanning Transmission Electron Microscope (STEM) operated at 100KeV with a windowless EDS and Link 10000 computer system. AES was conducted on a Physical Electronics instrument. Thermal conductivity measurements were conducted using a newly developed laser flash technique which removes many of the ambiguities associated with previous methods[13]. Unmetallized AlN substrates, Ti-thin film and MoMn-thick film coated substrates were supplied by Hitachi. Shapal™ (SH-04) substrates, Ti-thin film on Super Shapal™ and cofired W-thick film on Shapal™ (SH-15) substrates were supplied by Tokuyama Soda. "Black" AlN substrates doped with Y_2O_3-WO_3 or Y_2O_3-$CaWO_4$ were obtained from Narumi China Co. Hot pressed undoped AlN pellets sintered using Tokuyama Soda Grade F (>350ppm cation impurities) powder were used for defect characterization.

RESULTS & DISCUSSION

Microstructures of Bulk AlN

The Hitachi AlN substrate was observed to possess an ~5μm equiaxed grain matrix,

Figure.2. The substrates were known to have been manufactured using CaO as a sintering aid. However, there was no evidence of second phases, inclusions, precipitates or grain boundary films in CTEM mode, and neither Ca nor O could be detected using the Philips EM400T AEM in STEM mode. A very small number of curved defects were observed. These results indicate that if Ca and/or O were present at the grain boundaries, then their concentration was very low and/or confined to within a few tenths of a nanometer from the boundary. The surface sensitive technique of AES allowed in-situ intergranular fracture of the samples, and did reveal Ca and O to be present at the boundaries. However, this technique did not allow any information to be obtained concerning the concentration and segregation profiles of these species near the boundary region.

Grain boundary composition profiles were obtained using the high spatial resolution HB-501 FEG dedicated STEM, which was fitted with a windowless detector capable of light element analysis. The trace across the grain boundary interface revealed the presence of Ca and a very high oxygen concentration associated with a decrease in Al and N. To determine how the Ca and O were incorporated at the boundary, two simple techniques were employed to image the boundary. A defocus imaging technique[14] showed a ~1nm thick film to be present at the grain boundaries. The second technique, using SAED, allowed imaging of the film at the boundary; analysis of the extra spots present in the SAED patterns identified the phase to be calcium aluminate, $3CaO.\ Al_2O_3$, Figure.3. Another diffraction pattern from a different grain boundary region was not indexable as $3CaO.\ Al_2O_3$, indicating that other calcium aluminate phases were present. Previous work[15,16] has reported $CaO.\ 6Al_2O_3$ and $CaO.\ 2Al_2O_3$ at grain boundaries. The unidentified phase did not correspond to these two calcium aluminate phases either. Calcium aluminate phases are known to be extremely volatile and evaporate from AlN. A possible mechanism for the weight loss of Al_2O_3 is the formation of a volatile Ca-Al-O species, in which far greater amounts of Al_2O_3 are removed than CaO[17].

The morphology of the calcium aluminate phase is particularly important when considering the thermal conductivity of bulk AlN sintered with CaO. The continuous grain boundary film would consist of an AlN-xCao.yAl_2O_3-AlN, sandwich. This grain boundary configuration will obviously produce a greater thermal resistance than the simple AlN-AlN boundary, because phonons will be scattered at two interfaces rather than one, and the calcium aluminate phase has a very much lower thermal conductivity ($\sim 1.2 Wm^{-1}K^{-1}$)[18] than the AlN.

Three different bulk AlN substrates, were supplied by Tokuyama Soda, Shapal™ SH-04 and SH-15 both of which contain Y_2O_3 as a sintering additive, and Super Shapal™ which is doped with CaO. Analysis of the Super Shapal substrate revealed an equiaxed microstructure, with a grain size of ~5µm, Figure.4a. No second phases have been detected using CTEM, AEM or SAED techniques. Therefore, the CaO has most likely been removed from the sample by heat treatment. The SH-04 substrate contained a large quantity of three different yttria-alumina second phases with two different morphologies, Figure.4b. Quantitative AEM[19] indicated second phases with angular morphology (labelled 1 in Fig.4b) to be $2Y_2O_3.\ Al_2O_3$ (2:1) or (not shown) $Y_2O_3.\ Al_2O_3$ (1:1). The second phase exhibiting a smooth rounded morphology (labelled 2 in Fig.4b) was identified as $3Y_2O_3.5Al_2O_3$ (YAG). The SH-15 substrate contained only one type of second phase, possessing an angular morphology and with composition $2Y_2O_3.\ Al_2O_3$ (2:1), Figure.4c. This substrate appeared to contain a lesser amount of second phase than the SH-04 substrate. The grain size of the AlN matrix in both the SH-04 and SH-15 substrates was ~5µm. The different morphologies exhibited by the yttria-alumina compounds suggests different grain boundary wetting characteristics, which have a direct effect on the thermal conductivity of the bulk substrate.

Two different AlN packages were obtained from Narumi China Co.: 1) X01B doped with Y_2O_3 and WO_3 and sintered at 1800°C; 2) X02B, doped with Y_2O_3 and $CaWO_4$ and sintered at 1750°C. Both packages were dark gray/black. The X01B microstructure consisted of an AlN matrix with an angular second phase dispersed evenly throughout the substrate. Quantitative analysis of this second phase using AEM showed it to be 1:1 Y_2O_3. Al_2O_3. Another phase was present in the form of small spherical or facetted particles, Figure.5a. These were found primarily at the grain boundaries; however, a small number were seen inside the AlN grains. AEM (EDS, EELS) and CBED were used to identify these particles as metallic tungsten. Clustering of tungsten particles, some of which were very large, appeared in a small number of areas in the sample. These areas were later identified as pores because of their smooth curved surfaces, Figure.5b.

The X02B sample exhibited a typical liquid phase sintered microstructure. Figure.6 shows rounded AlN grains encapsulated by a grain boundary region, which also contained small spherical particles. The encapsulating grain boundary region was found to be comprised of three different phase compositions, which were morphologically indistinguishable. Quantitative AEM revealed a ternary oxide phase, $3Al_2O_3$. 2CaO. Y_2O_3, with a tetragonal crystal structure, and two calcium aluminate phases, CaO. Al_2O_3 and CaO. $2Al_2O_3$. The spherical particles were analyzed to be metallic tungsten, and were slightly larger than the in the X01B sample. The presence of CaO can be seen to have a dramatic effect on the microstructure with respect to second phase formation. Analysis of these substrates also helped explain the observed interfacial microstructure in the tungsten metallized AlN (see later). In both Narumi China samples an extremely small number of curved defects were observed (see following).

Boundaries in AlN

Crystallographic defects appearing as curved or planar boundaries have both been observed in AlN, Figure.7. The crystallographic structure, formation process and the effect on thermal conductivity of these boundaries has been a debated topic.

AlN possesses a slightly distorted wurtzite crystal structure, consisting of two interpenetrating hexagonal close packed lattices of Al and N atoms. The Al and N atoms are tetrahedrally coordinated with respect to one another. AlN exhibits the $P6_3mc$ space group (#186), which is a noncentrosymmetric structure. As a result, two adjacent AlN regions separated by a boundary could exist, such that one region would have all upward pointing tetradehral sites filled, while the other region would have all downward pointing tetrahedral sites filled, and consequently inverted with respect to the other side. A polarity change therefore exists across the boundary. This structural inversion and polarity change across the boundary can be demonstrated using a 2 beam CBED technique[20]. This structure could be obtained in three different ways, i.e, by three different crystallographic operations. In one case, by an inversion operation across the boundary, with the resulting defect defined as Inversion Domain Boundary (IDB). In the second case, by an 180° rotation operation parallel to the boundary plane with the resulting defect defined as an Antiphase Domain Boundary (ADB). The final case is one in which a translation of the cation sublattice from one tetrahedral orientation to the other occurs with the anion lattice being undeviated across the boundary. This boundary configuration is known as an Inversion Twin[21].

A number of authors[22-25] have analyzed both curved and planar boundaries, and argument exists as to whether they are IDBs or ADBs. Analysis of these defects using CTEM has revealed the planar boundaries to lie on either the basal or {1011} planes. Our analysis so far of the curved defects indicates that it does not possess a habit plane; however, portions of the boundary do lie on the {1011} plane as suggested by other authors[24,25]. Fault vector analysis has been carried out using the standard g.R=0 criterion; the curved boundaries possess R=[0001], the magnitude of **R** has not been determined. No **R** value has been found for the planar boundaries. Fringe contrast experiments comparing BF and DF micrographs show the fringe contrast to be symmetrical. This type of contrast behavior is exhibited by ADBs. Dislocations are present at the intersection of planar and curved faults and also between two planar faults.

Microchemical analysis of the planar fault indicated oxygen always to be present. Analysis of the curved defects showed them to have much less oxygen associated with them and in some instances, none at all[22], Figure.8a & b. Recent annealing experiments on undoped AlN have shown the planar faults as embryonic polytypes, Figure.9. The manner in which they grow, with oxygen either being incorporated from the lattice, or by diffusing from the grain boundary and down the fault, is not clear. The early stages of polytype precipitation on the curved boundaries has been observed. A large amount of strain appears to be associated with this process, Figure.10. Dislocation loops have also been seen to nucleate on the curved portions of these defects. All of these observed processes would have an affect on the thermal conductivity.

The concentration of planar and curved defects is high in undoped AlN, which contains a high residual oxygen concentration within the grains and at grain boundaries. Only a very small number of curved defects were seen in the AlN substrates containing sintering aids. In these substrates the oxygen is gettered from the matrix and confined to second phases at the grain boundaries. Oxygen removal from the grains therefore coincides with the disappearance of the planar defects, which are known to have high oxygen concentration. It therefore appears that oxygen is required for planar defect formation. However, it is not clear if oxygen is required for the formation of the curved defects, as these are still present in reduced numbers even after oxygen removal from the grain.

Nucleation of the planar faults is believed to occur during sintering. During initial densification, a high oxygen concentration exists at the newly forming grain boundaries. The final grain size is larger than the initial powder size, indicating grain growth to have occurred. It is therefore reasonable to assume that during grain growth, grain boundary mobility is more rapid than the rate at which oxygen dissolves and diffuses away from the original powder surface region. This causes the grain boundary and the oxygen rich region to become separated. At separation, the oxygen rich region of the crystal has a high concentration of aluminum vacancies such that charge neutrality is maintained. It is proposed that the high oxygen and aluminum vacancy concentration causes nucleation of the planar fault. Energetically, it may be more favorable for a defect to form and incorporate the imbalance in the cation-anion ratio, rather than maintaining a solid solution region containing a high oxygen and aluminum vacancy concentration within the AlN crystal structure. It is presumed that the basal plane is a low-energy plane which allows this defect formation to occur. The mechanism by which N and Al atoms change tetrahedral sites on crossing the boundary is not yet clear.

Microstructure of Metal-Ceramic Interfaces/Adhesion Mechanisms

Thin Film Metallization

The thin film metallization system on which most research has been conducted is the Ti-AlN system. The Ti layer is usually ~0.1 µm thick; layers of Au, Pt, Pd, or Ni can then be deposited over the Ti. The thin film systems rely on solid state displacive reactions to provide adhesion between the metal and ceramic substrate. This is most readily accomplished using reactive metals e.g. Ti, Sc. A number of researchers[5,6,26,27] have studied the reaction between Ti and AlN and have characterized the interface chemistry and structure. The results indicate the formation of TiN and $TiAl_3$ layers. Because of the affinity shown by Ti and Al for oxygen, the presence of titanium oxide of varying stoichiometric composition (Ti_xO_y) and Al_2O_3 is highly likely. The reported experimental results are in agreement with the reaction products predicted from thermodynamic calculations. Figure.11a shows a 0.1μm thick reaction layer formed at the Ti-AlN interface. Figure.11b shows a depth profile through the reaction layer obtained using AES. The leveling out of the intensities of Al, N, and Ti at ~400min is associated with compound formation. Also, note the diffusion of Al and N through the reaction layer, and the association of oxygen and titanium profiles.

Thick Film Metallization

Early published studies of thick film metallization on AlN concentrated on the use of Molybdenum-Manganese (MoMn) fritted pastes; these provided good bond strength and hermeticity. Recently, the drive toward multilayer AlN packages requires the application of cofired metallization technology. The major candidate is tungsten because it can be cofired and it has a close thermal expansion match to AlN. A comparison between interface chemistry and structure produced in the case of MoMn-AlN and W-AlN is therefore instructive in order to highlight the effect of ceramic-metal interfaces on thermal conductivity.

Our previously reported[5,6] study on the MoMn-AlN system indicates that the adhesion mechanism consists of two mechanical interlocking mechanisms. The first occurs between the continuous Mo layer and the glass frit layer, and the second one occurs between the glass frit and the AlN substrate via grain boundary penetration, Figure.12. The presence of W in the Mo layer is believed to reduce the thermal expansion of the layer. The glass frit forms two phases after the metallization process is complete, an amorphous phase and a crystalline phase, having a composition close to that of mullite ($3Al_2O_3 . 2SiO_2$). Mullite has a thermal expansion coefficient between Mo and AlN, and it appears that the glass composition has been formulated to produce a graded thermal expansion coefficient across the interface. A further contribution to adhesion is through chemical bonding via dissolution of the glass-AlN interface, forming a microcrystalline region and grain boundary penetration of crystalline precipitates. These precipitates varied in composition from almost pure alumina to Al-N-O-Ca quaternary compounds. The precipitation produced a continuous grain boundary penetration up to 20-30μm into the substrate, causing encapsulation of the surface AlN grains.

In the W - AlN system that was studied the tungsten metallized layer consisted of an ~8-10mm thick continuous layer of tungsten, on top of which was a 1mm thick plated nickel overcoat. No glassy phase was observed at or near the W-AlN interface. The interface structure, Figure.13, was very different from that of the MoMn-AlN sample, and a smooth but wavy interface was observed at the W-AlN interface. Interlocking between the tungsten and AlN was observed in the form of "tunnelling" of W into the AlN grains. A thin crystalline grain boundary film ~1-2nm thick was found between AlN grains in the region just below the W-AlN interface. Within the AlN grains near the interface and along the AlN grain boundaries a large number of faceted and non-faceted precipitates were found. The precipitates ranged in size from 10nm to 200nm, the larger ones being present at the grain boundaries. This precipitate-rich region existed to a depth of 15-20μm into the substrate. Chemical analysis of the precipitates using EDS indicated only W; EELS indicated no oxygen or nitrogen; and CBED revealed a simple b.c.c structure. The presence of metallic W in the AlN indicates that a reaction does occur between the W-metallization and the AlN, contradicting previous work[28] in which it was noted that the particles were believed to be an artifact due to TEM sample preparation. In the present study stereomicroscopy was conducted which allowed a three dimensional image of the foil to be formed. The precipitates were clearly visible throughout the foil thickness, thereby confirming that they are a reaction product and not an artifact. These precipitates exhibit very similar morphology to those observed in the Narumi China samples.

The adhesion of the tungsten metallization on AlN is believed to be due to a combination of interlocking of the metal layer and the substrate, and by reaction with AlN which results in the formation of pure tungsten precipitates in the AlN substrate. The cofire metallization process consists in the use of AlN green-tape (AlN ceramic powder with its associated organic vehicle) onto which the tungsten paste (powder suspended in an organic medium) is screen printed. The AlN powder possesses an oxide surface coating and the metallization system which must be considered is therefore Al-N-O-W plus organics rather than just Al-N-W.

During the cofiring process, the organic materials will be broken down and driven off primarily as CO, CO_2, and H_2O. Thermodynamic calculations predict that tungsten can react with CO and H_2O to form tungsten trioxide:

$$W + 3CO \text{ -----> } WO_3 + 3C(s) \qquad DG_{700K} = -138.8 \text{ kJmol}^{-1}$$

$$W + 3H_2O \text{ ----> } WO_3 + 3H_2(g) \qquad DG_{400K} = -63.9 \text{ kJmol}^{-1}$$

Therefore, a thin oxide coating can form on the tungsten powder particles. The Al_2O_3 - WO_3 phase diagram indicates that a low melting temperature eutectic exists at 1190°C between WO_3 and $Al_2O_3.3WO_3$. The WO_3 and Al_2O_3 on the surface of the AlN can therefore react and form a liquid phase which can penetrate into the AlN green tape, enhancing sintering in the near surface region. Thermodynamics predicts that the reduction of WO_3 contained in the liquid phase by AlN is a favorable reaction:

$$2AlN + WO_3 \text{ ----> } Al_2O_3 + W(s) + N_2(g) \qquad DG_{1500K} = -425.9 \text{ kJmol}^{-1}$$

Thus, as the liquid penetrates into the AlN, the WO_3 is reduced as it reacts with AlN to W which then precipitates out. Those particles seen within the grains are likely to have been associated with a grain boundary which migrated, moving away from the particle, and leaving them behind in the matrix.

Thermal Conductivity Analysis of Bulk AlN and Metal Ceramic Interfaces

Thermal Conductivity of Bulk AlN

The measured thermal conductivity for each of the bulk AlN samples is shown in the table below:

Sample	Sintering Aid	Thermal Conductivity ($Wm^{-1}K^{-1}$)
Hitachi	CaO	159
Shapal SH-04	Y_2O_3	167.5
Shapal SH-15	Y_2O_3	182
Super Shapal	CaO	208
Narumi X01B	Y_2O_3, WO_3	162
Narumi X02B	Y_2O_3, $CaWO_4$	103.5

As mentioned previously, there are a number of microstructural and microchemical features which contribute to the degradation or enhancement of thermal conductivity in AlN. These include: the initial oxygen concentration and its removal; the presence of a second phase(s), the structural and chemical identity of the second phase and its thermal conductivity, the shape (morphology) and distribution of the second phase and its wetting characteristics with respect to AlN grain boundaries, and the total amount of each type of second phase; the grain size of the AlN; and the presence of both curved and planar defects within the AlN grains. The importance of a number of these contributions can be ascertained by examination of the above table.

In the case of the CaO doped materials (Hitachi and Tokuyama Soda's Super Shapal), the high vapor pressures exhibited by the variety of calcium aluminate second phases allows removal of the grain boundary phase by heat-treatment of the substrate. Time and temperature affect the grain boundary phase composition and reduce its thickness, and hence improve thermal conductivity. Holding the substrate at high temperature for prolonged periods also decreases the bulk AlN oxygen concentration by allowing diffusion of oxygen to AlN grain boundaries (and hence to form second phases) and to crystalline defects within the AlN grains; as oxygen is removed from the bulk AlN grains the concentration of associated point defects is also reduced. Prolonged heat-treatment also allows grain growth and thus enhances thermal conductivity by decreasing the number density of grain boundary scattering sites. However, grain growth is extremely slow in AlN, and the economics associated with a long heat-treatment at elevated temperatures would tend to negate the relatively small expected improvement in thermal conductivity. A larger grain size would also be detrimental to mechanical properties and would increase the problem related to grain pullout and surface finishing. Although the thermal conductivity value for Super Shapal is somewhat lower than expected from supplier information, it is high compared with the Hitachi material, as expected in relation to its clean microstructure, having no visual evidence for the presence of a second phase or measurable Ca at the grain boundaries.

The difference in thermal conductivity between the two Y_2O_3 doped substrates (Shapal SH-04 and SH-15) can be explained in terms of the different grain boundary wetting characteristics, volume fraction of second phase present, and the different thermal conductivities of each of the yttrium aluminate phases present, all of which will play a role in determining the overall bulk thermal conductivity[29]. The SH-15 substrate contained only one type of second phase having an angular morphology which did not wet the grain boundaries. Because this phase gettered the oxygen and other impurities into discrete pockets, it left only clean AlN-AlN grain boundaries, which enhances the thermal conductivity. The SH-04 substrate contained three types of second phase, one of which wetted the grain boundaries, Figure.4b. This produces a similar effect to that described for the CaO doped substrates, and produces a low thermally conductive phase sandwiched between two interfaces. This is believed to be the predominant reason for the difference in bulk thermal conductivity between SH-04 and SH-15 substrates.

The effect of curved or planar defects on thermal conductivity will depend upon their interface structure, chemistry and density. The planar boundaries are expected to have a larger effect on thermal conductivity due to the high oxygen concentration associated with these boundaries. The nucleation of polytypes from these planar boundaries will also decrease the thermal conductivity of the AlN. Thermal conductivity measurements have been reported for some polytypes[30]; the thermal diffusivity of a 27R polytype is $0.024 cm^2 s^{-1}$ compared to AlN (SH-04) with $0.70 cm^2 s^{-1}$. The lower the polytype number, the higher the oxygen concentration, and the lower the thermal conductivity. The curved defects possess little or no oxygen and their effect on thermal conductivity is expected to be less than the planar fault, unless an annealing treatment has been carried out, causing nucleation of polytype material on the curved boundary.

Except for the presence of tungsten particles in the Narumi material, the microstructure of the Narumi X01B sample is similar to that of the SH-15 sample. The difference in thermal conductivity between the samples (of about 12.5%) can be attributed either to the presence of these particles, which increases the number of scattering centers, or to an increase in Al_2O_3 at the grain boundaries, due to the reduction of WO_3 by AlN, or to a higher initial oxygen concentration in the starting powder. The extremely large volume fraction of various second phases present at the grain boundaries in the Narumi X02B sample is the most probable cause of its low thermal conductivity. It should be noted that the Narumi materials were primarily intended for use in black light-excluding packages and not necessarily for their thermal conductivity; the presence of W in both Narumi materials certainly acts to provide effective light absorption by optical scattering and also aids metallization.

Thermal Conductivity of Metal-Ceramic Interfaces

The presence of a metal-AlN interface was detrimental to heat flow in all cases, as indicated by the table below, which compares the thermal conductivity across a metal-AlN interface with that of the corresponding bulk substrate:

Substrate	Metallization	Thermal Conductivity ($Wm^{-1}K^{-1}$)
Hitachi	MoMn Thick Film	139
Hitachi	None	159
SH-15	W Thick Film	166
SH-15	None	182
Super Shapal	Ti Thin Film	185
Super Shapal	None	208
SH-04	0.1mm Ti as Deposited	167.5
SH-04	None	167.5

The reduction in thermal conductivity due to the presence of a MoMn thick-film, W thick-film, or a Ti-thin film interface is ~16%, ~9%, and ~12%, respectively. The degradation of thermal conductivity across each of the three interfaces can be attributed to the different microstructures and microchemistry present at each of the interface regions.

The MoMn-AlN interface microstructure was expected to degrade the heat flow more than the other two interface morphologies. This is because of the presence of a glass frit containing a 3:2 mullite phase, a microcrystalline phase of unknown structure and chemistry having a high internal boundary interface density, deep grain boundary penetration to ~30μm, and precipitates of complex structure and chemistry. These are all structures known to be detrimental to heat flow. In contrast, the W-AlN interface does not contain any glassy or microcrystalline phases. Only a thin crystalline grain boundary film extends from the interface to the maximum depth of W penetration of the tungsten particles themselves. This interface microstructure was predicted to be less detrimental to heat flow than that of the Mo-Mn interface, and this was found to be the case.

Two cases must be considered when interpreting the effect of a Ti-AlN interface upon thermal conductivity: no change of thermal conductivity was found for the case of as-deposited Ti on Shapal SH-04, but a significant decrease is observed across the Super Shapal Ti-AlN interface. In this latter case a post deposition thermal treatment results in the formation of a reaction layer which enhances adhesion. The magnitude of the decrease in thermal conductivity across the Ti thin film is quite significant, considering the thickness of the reaction layer in this sample (~0.1μm) compared to that of the reaction layers in the other samples, (10-30μm). A similar decrease in thermal conductivity occurs between the thin- and thick-film interfaces, even though a large difference in reaction layer thickness exists. This highlights the effect of microcrystalline phases and complex crystal structures which exhibit low thermal conductivities and the likely nonstoichiometry of these structures.

CONCLUSION

1. There are a variety of possible microstructural defects present in AlN which can directly influence the thermal conductivity. These include planar and curved crystalline domain boundaries, grain boundaries, and second phases present both as discrete precipitates and along grain boundaries. The presence of a second phase can give rise to two possible effects which degrade thermal conductivity: the second phase material itself, and the interphase interfaces which are created between the second phase and the AlN matrix.
2. The adhesion in W-AlN thick film metallization is due to both interlocking of the tungsten with the AlN, and the formation of a chemical bond via reduction by AlN of WO_3 which forms during sintering by reaction of W with the breakdown products of the various organics. Formation of metallic tungsten particles in the AlN and at grain boundaries is a result of this process.
3. The modified laser flash technique is appropriate for the measurement of thin AlN samples without correction, and can be applied to the measurement of thermal conductivity across metallization interfaces in samples with the appropriate geometry.

REFERENCES

1. W. Werdecker and F. Aldinger, "Aluminum Nitride an Alternative Ceramic Substrate for High Power Applications in Microcircuits", IEEE Trans. Components Hybrids, Manuf. Technol., CHMT-7, p.399-404, 1984.
2. A. J. Blogett, "Microelectronic Packaging", Scientific American, vol.249, p.86-97, 1983.
3. N. Kuramoto, H. Taniguchi and I. Aso, "Translucent AlN Ceramic Substrate", IEEE Trans. Components Hybrid, Manuf. Technol., CHMT-9, p.386-390, 1986.
4. W. Werdecker, "Metallizing of Aluminum Nitride Substrates", Proceedings ISHM 1985, Stressa Italy.
5. A. D. Westwood and M. R. Notis, "AEM Study of Thin and Thick Film Metallization on AlN Substrates", Mat. Res. Soc. Symp. Proc., vol.108, p.331-336, 1988.
6. A. D. Westwood and M. R. Notis, " AEM of AlN Substrates and Metallization Interfaces", to be published in Advances in Ceramics 26, Ceramic Substrates and Packages, edited by M. F. Yan, ACerSoc, Westerville OH, 1989.
7. C. L. Julian, "Theory of Heat Conduction in Rare Gas Crystals", Phys. Rev., vol.137, A128-137, 1965.
8. J. E. Parrott and A. D. Stuckes, "Thermal Conductivity of Solids", 1975, Pion Press.
9. G. A. Slack, "Nonmetallic Crystals with High Thermal Conductivity", J. Phys. Chem. Solids, vol.34, p.321-335, 1973.
10. R. Berman, "Thermal Conductivity of Solids", 1976, Oxford University Press.

11. A. N. Cormack, "Intrinsic Disorder in Aluminum Nitride", J. Am. Ceram. Soc., vol.72, p.1730-1732, 1989.
12. G. A. Slack, R. A. Tanzilli, R. O. Pohl and J. W. Vandersande, "The Intrinsic Thermal Conductivity of AlN", J. Phys. Chem. Solids, vol.48, p.641-647, 1987.
13. R. C. Enck, R. D. Harris and R. A. Youngman, "Measurement of Thermal Diffusivity of Translucent Aluminum Nitride", to be published in Ceramic Transactions, Advanced Characterization Techniques for Ceramics, ACerSoc, Westerville OH, 1989.
14. D. R. Clarke, "On the Detection of Thin Intergranular Films by Electron Microscopy", Ultramicroscopy, vol.4, p.33-44, 1979.
15. N. Komeya, A. Tsuge, H. Inque and H. Otha, "Effect of $CaCO_3$ Addition on the Sintering of AlN", J. Mat. Sci. Let., vol.1, p.325-326, 1982.
16. Y. Kurokawa, K. Utsumi and H. Takamizawa, "Development and Microstructural Characterization of High Thermal Conductivity Aluminum Nitride Ceramics", J. Am. Ceram. Soc., vol.71, p.588-594, 1988.
17. L. Weisenbach and Y. M. Chiang, "The Role of Ca in Oxygen Gettering during Sintering of High Thermal Conductivity Aluminum Nitride", MRS Fall Meeting Boston, 1987.
18. J. H. Enloe, R. W. Rice, J. W. Lau and R. Kumar, "Microstructural Effects on the Thermal Conductivity of Polycrystalline Aluminum Nitride", to be published in Advances in Ceramics 26, Ceramic Substrates and Packages, edited by M. F. Yan, ACerSoc, Westerville OH, 1989.
19. D. B. Williams, "Practical Analytical Electron Microscopy in Materials Science", 1984, Philips Electronic Instruments.
20. J. Tafto and J. C. H. Spence, "A Simple Method for Determination of Structure-Factor Phase Relationships and Crystal Polarity using Electron Diffraction", J. Appl. Cryst., vol.15, p.60-64, 1982.
21. S. B. Austerman and W. G. Gehman, "The Inversion Twin: Prototype in Beryllium Oxide", J. Mat. Sci., vol.1, p.249-260, 1966.
22. A. D. Westwood and M. R. Notis, submitted to ACerS for Publication.
23. S. Hagege, Y. Ishida and S. Tanaka, "HREM and HVEM of Interfaces in Aluminum Nitride Ceramics", J. de Phys., vol.45, C5-189, 1988.
24. S. McKernan and C. B. Carter, "Planar Defects in AlN", Proceedings of 47th EMSA meeting San Antonio TX, p.432, edited by G. W. Bailey, San Francisco Press, 1989.
25. M. F. Denanot and J. Rabier, "Extended Defects in AlN", J. Mat. Sci., vol.24, p.1594-1598, 1989.
26. R. K. Brow, R. E. Loehman, A. P. Tomsia and J. A. Pask, "Interface Interactions during Brazing of Non-oxide Ceramics for Electronic Packaging", to be published in Advances in Ceramics 26, Ceramic Substrates and Packages, edited by M. F. Yan, ACerSoc, Westerville OH, 1989.
27. R. Beyers, R. Sinclair and M. E. Thomas, "Phase Equilibria in Thin-Film Metallizations", J. Vac. Sci. Technol., B2, p.781-784, 1984.
28. Y. Kurokawa, C. Toy and W. D. Scott, "Characterization of the AlN-W Interface in a Cofire Multilayer AlN Substrate", J. Am. Ceram. Soc., vol.72[4], p.612-616, 1989.
29. T. Takahashi, N. Iwase, A. Tsuge and N. Nagata, "Properties and Reliability of AlN Ceramics for Power Devices", to be published in Advances in Ceramics 26, Ceramic Substrates and Packages, edited by M. F. Yan, ACerSoc, Westerville OH, 1989.
30. T. Yagi, K. Shinozaki, N. Ishizawa, N. Mizutani, M. Kato and A. Tsuge, "Effect of Silicon Dioxide on the Thermal Diffusivity of Aluminum Nitride Ceramics", J. Am. Ceram. Soc., vol.71[7], C334-338, 1988.

ACKNOWLEDGEMENTS

The authors wish to thank the Semiconductor Research Corporation (SRC) for funding this work; BP International, in particular Drs. R. A. Youngman, R. C. Enck and J. H. Harris, for allowing us the use of their thermal conductivity measurement facility; Bethlehem Steel Co., especially Dr. J. R. Michael, for the use of the HB-501 AEM; and Dr. S. Witek for supplying the undoped hot pressed AlN sample.

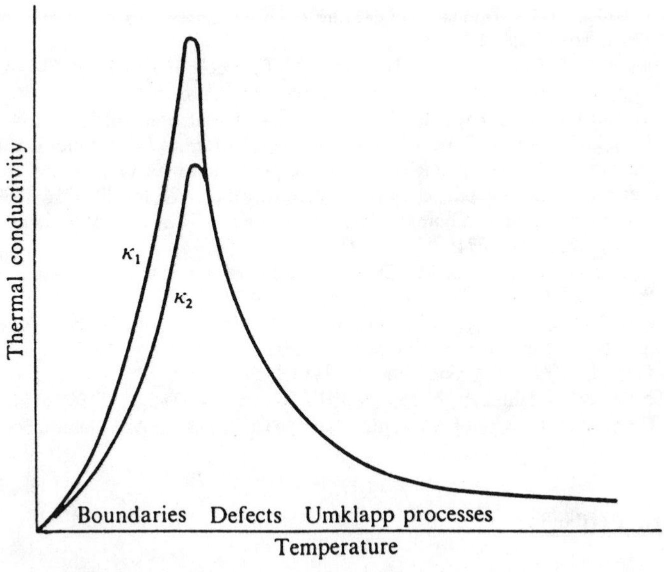

Figure.1 The temperature dependence of thermal conductivity for non-metallic crystals. The dominant phonon scattering mechanisms are indicated along the x-axis. Curve κ_1 has a larger grain size than curve κ_2.

Figure.2 BF TEM micrograph of the microstructures in the Hitachi CaO doped substrate

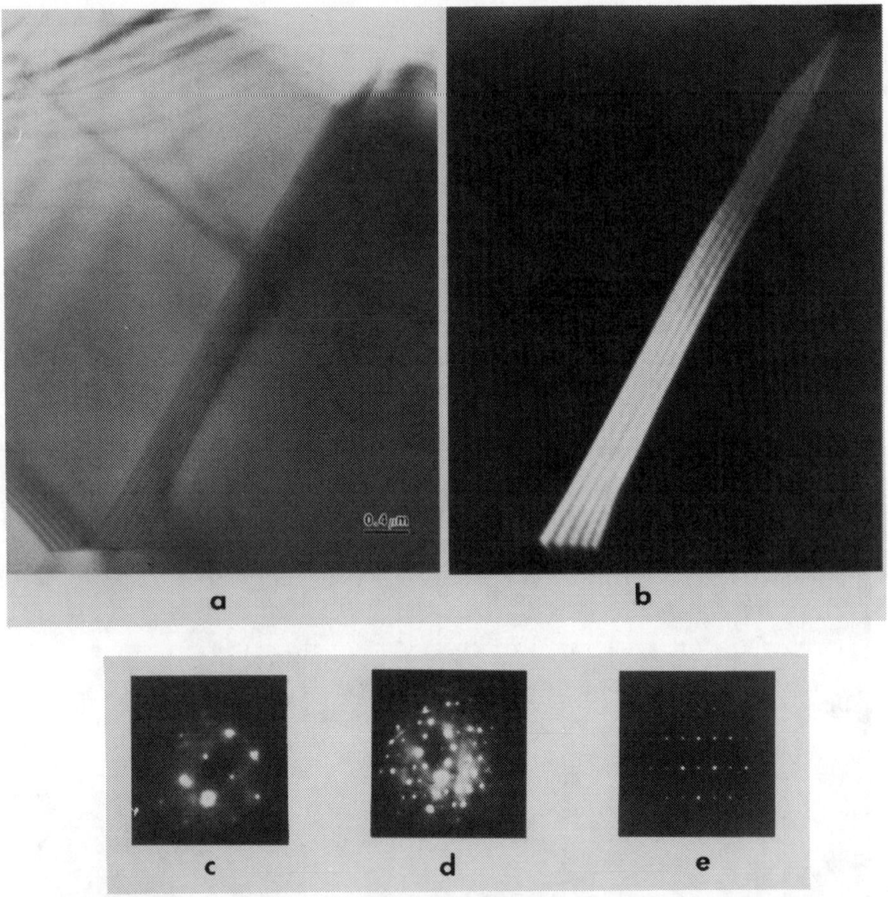

Figure.3 Identification of a second phase at a grain boundary in Hitachi CaO doped AlN.
a) BF TEM micrograph of grain boundary, b) CDF of grain boundary, second phase grain boundary film highlighted using diffracted spot in (d), c) SAED pattern from LHS grain, d) SAED pattern obtained across inclined boundary, array of extra spots due to grain boundary phase, pattern indexed as $3CaO.Al_2O_3$, e) SAED pattern from RHS grain.

a

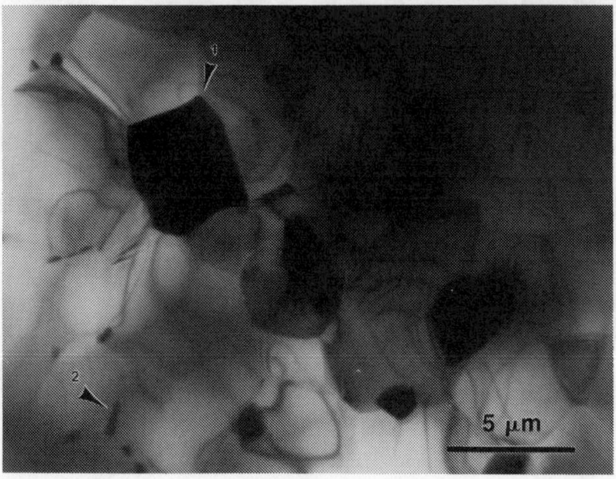

b

Figure.4 BF TEM micrographs of microstructures observed in samples studied,
a) BF Super Shapal CaO doped, b) BF SH-04 Y_2O_3 doped, note the two different second phase morphologies, 1- angular $2Y_2O_3.Al_2O_3$ or $Y_2O_3.Al_2O_3$ non-wetting and 2- smooth/rounded $3Y_2O_3.5Al_2O_3$ wetting grain boundary phase, and c) BF SH-15 Y_2O_3 doped with a second phase with an angular morphology and single composition $2Y_2O_3.Al_2O_3$

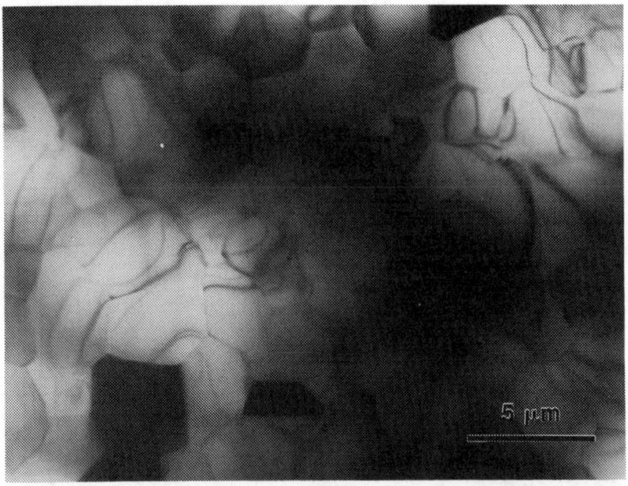

c

a

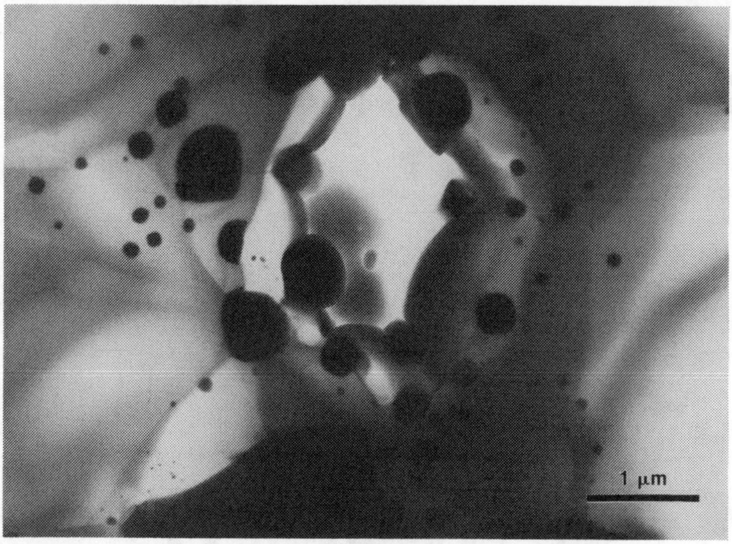

b

Figure.5 a) Microstructure of Narumi X01B, the dark grains are $Y_2O_3.Al_2O_3$ second phase (arrowed), the small dark particles at grain boundaries and within grains are metallic tungsten.
b) Clustering of tungsten particles on and around a pore.

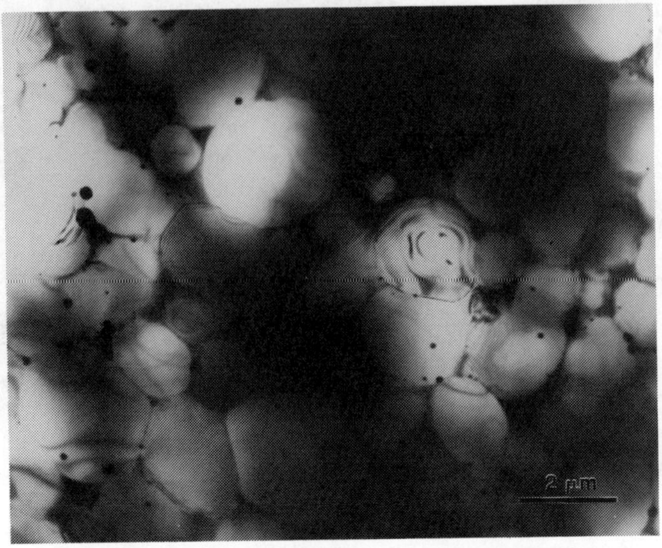

Figure.6 Microstructure of Narumi X02B exhibiting a typical liquid phase sintered morphology, with rounded AlN grains encapsulated by a second phase. The second phase comprised four distinct phases, $CaO.Al_2O_3$, $CaO.2Al_2O_3$, $3Al_2O_3.2CaO.Y_2O_3$ and metallic tungsten particles (dark spheres).

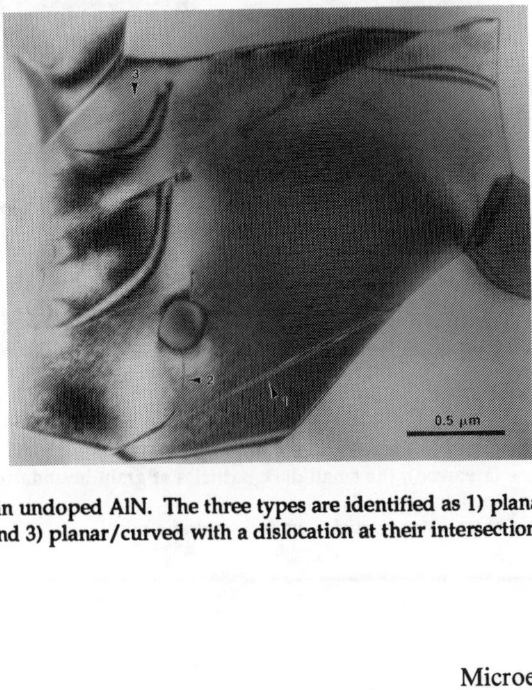

Figure.7 Defe cts in undoped AlN. The three types are identified as 1) planar, 2) curved and 3) planar/curved with a dislocation at their intersection point.

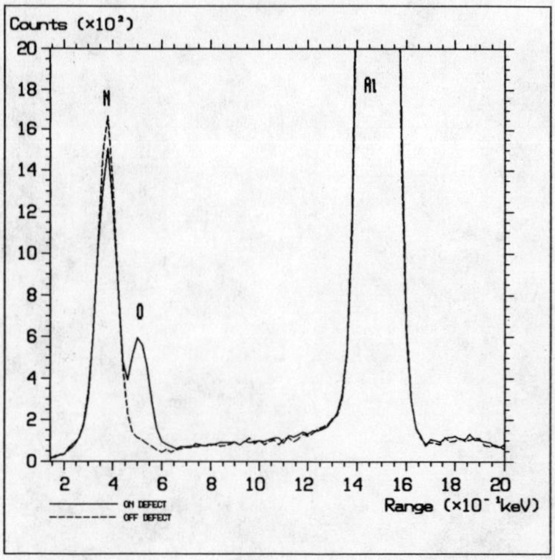

a

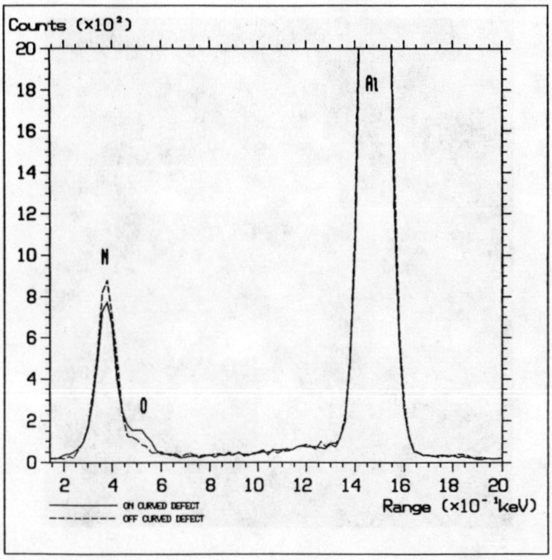

b

Figure.8 EDS spectrum from a) planar defect and b) curved defect.
Dashed line in spectrum was taken from a point in the AlN matrix next to the boundary. A high oxygen concentration was observed in all planar boundaries analyzed. The oxygen concentration on the curved boundaries varied from a small amount to none at all.

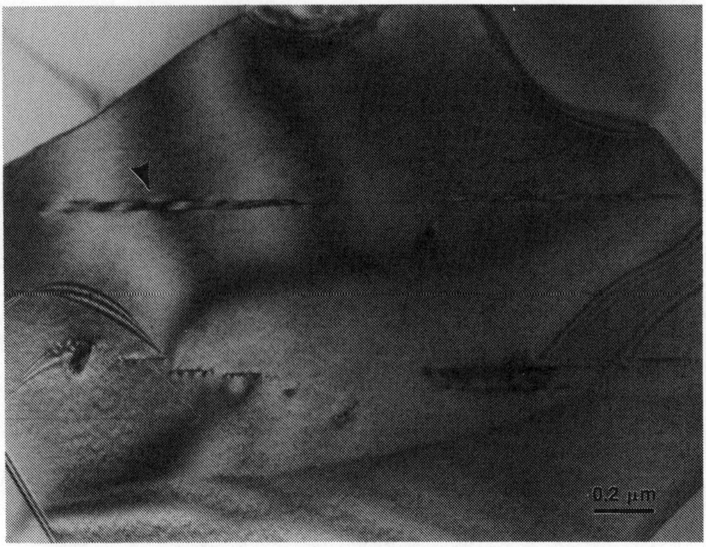

a

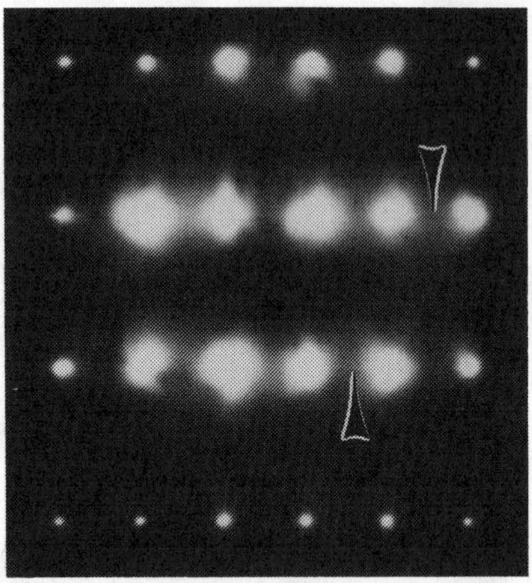

b

Figure.9 An annealed (45mins/1950°C/N_2) undoped AlN sample showing broadening of the planar boundaries. The SAED pattern was obtained from the arrowed region; small discrete spots (arrowed) can be seen in the pattern between the fundamental spots. This type of diffraction pattern is commonly seen in polytypoid structures. It appears that the planar boundaries can nucleate polytypes by diffusion of oxygen to these sites.

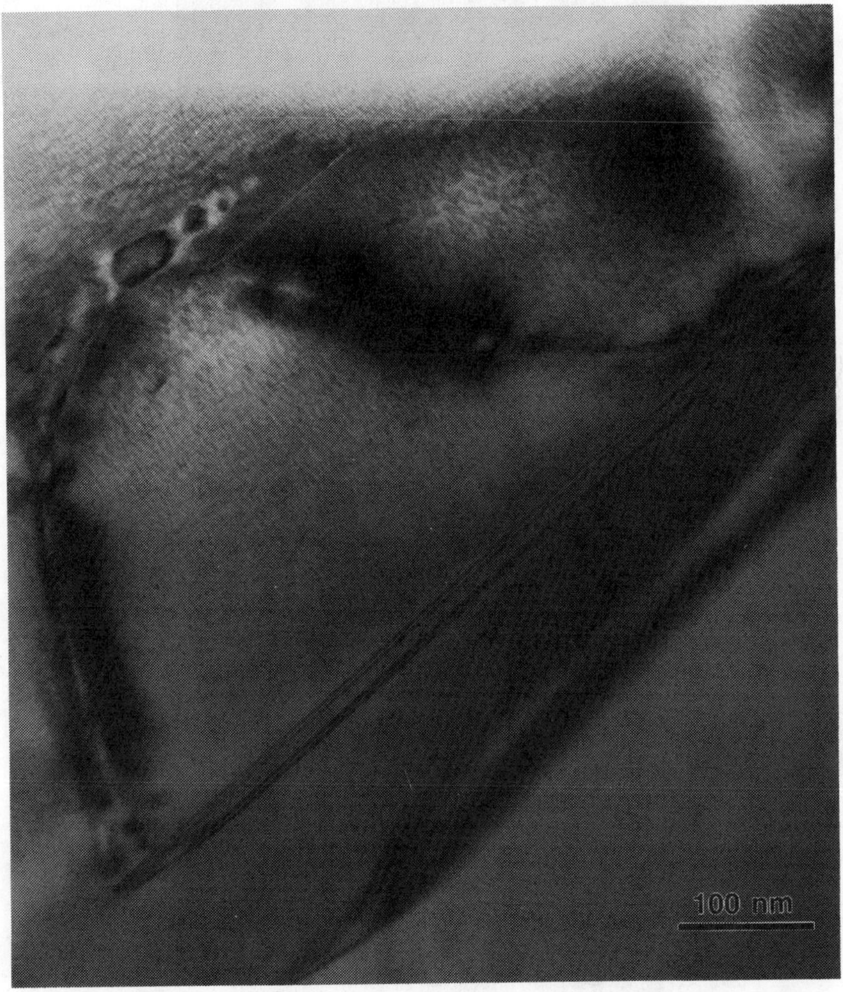

Figure.10 An annealed (30min/1950°C/N_2) undoped AlN planar/curved defect exhibiting precipitation of material along a section of the curved portion. Significant strain appears to be present along other portions of the curved boundary. These precipitates are believed to be AlN polytypes.

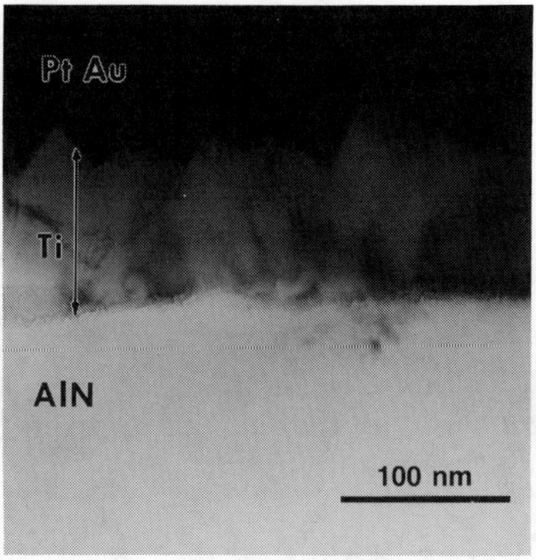

a

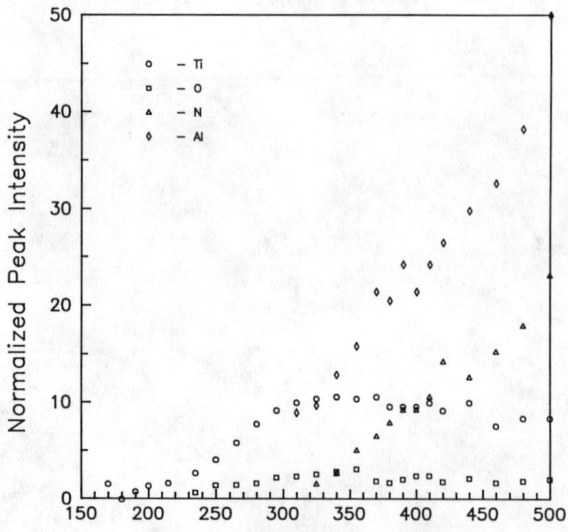

b

Figure.11 a)Titanium thin film AlN interface, the 100nm thick Ti rich reaction zone can be clearly seen at the interface, b) AES depth profile; Al and N have diffused through the Ti layer, oxygen is associated with the Ti film. At ~400mins Al,N and Ti profiles appear to level out briefly, we believe this is a result of compound formation.

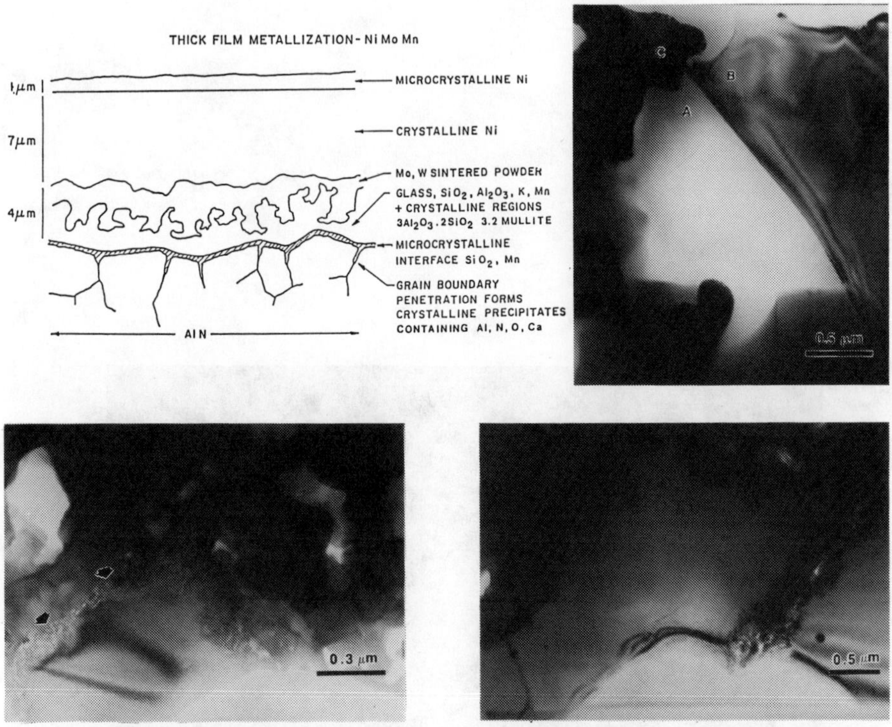

Figure.12 (Clockwise) Schematic of MoMn-glass frit -AlN interface on Hitachi substrate.
Mo sintered layer interpenetrates the glass frit, A - amorphous aluminosilicate glass containing Mn and K, B - 3:2 mullite which crystallized from the glass on cooling, C - Mo sintered layer.
Grain boundary penetration from the microcrystalline phase at the glass-AlN interface.
Microcrystalline phase (arrowed) at interface between glass and AlN.
Interpenetration between the glass and Mo is clearly seen in this micrograph.

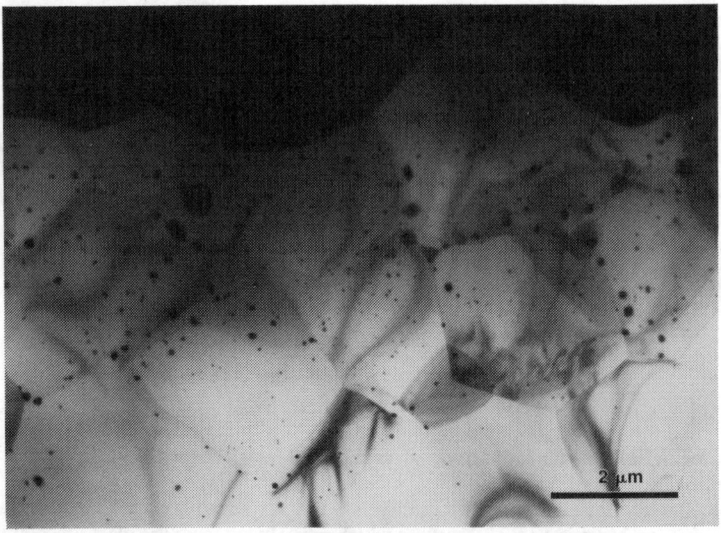

Figure.13 The cofired tungsten-AlN interface with a smooth but wavy morphology. A large number of metallic tungsten particles are seen in the AlN near the interface due to the proposed oxidation-liquid phase penetration-reduction reaction of tungsten.

EFFECT OF ANNEALING ON THE GRAIN BOUNDARY PROPERTIES OF ZnO VARISTORS

Tapan K. Gupta
Alcoa Technical Center
Alcoa Center, PA 15069

ABSTRACT

The processing of a ZnO varistor must insure stability of the device under an externally applied voltage stress over the life of the device. For a long time, in order to assure stability, the manufacturers have been subjecting the varistor to a post-sintering heat treatment in air, known as annealing. Furthermore, a temperature of 600-700°C is emerging as the most critical temperature of annealing. This paper compiles results of several experiments to demonstrate the criticality of this annealing temperature. It is shown that at this temperature, the varistors become highly stable as a result of the elimination of a so-called metastable barrier known to be responsible for the instability of the device according to the grain boundary defect model proposed earlier.[14] Simultaneously, many grain boundary properties are shown to alter drastically at this temperature, irrespective of compositions, sintering conditions and the type of manufacturers. The annealing can thus be regarded as a complex process where many grain boundary properties are concurrently changed at the critical annealing temperature to impart stability to ZnO varistors.

1. INTRODUCTION

When a varistor is formed by sintering, a barrier voltage, similar to that of a p-n junction, is formed at the grain boundary which gives rise to its nonlinear property.[1-5] However, a number of experiments indicate that an "as-sintered" varistor has a poor life expectancy due to the presence of a so-called metastable component in the barrier.[6] The as-sintered varistors are in general subjected to an annealing treatment to eliminate the metastable component of the barrier,[6-8] and thus, make the device stable to externally applied voltages as a function of time.[6] Recent evidence[9-11] indicates that there is a critical annealing temperature at which dramatic changes in several grain boundary properties occur simultaneously in conjunction with device stability. This critical temperature appears to lie between 600 and 700°C, irrespective of compositions, sintering conditions and the type of manufacturers. Varistors show varying degrees of instability when annealed above and below this temperature range. Several studies[12-13] have concluded that at this temperature, a phase transformation (α to γ) occurs in Bi_2O_3 phase that is present at the grain boundary, and that the varistor becomes stable as a result of this phase transformation. This explanation fails to explain how the metastable component is removed from the barrier. An alternative explanation is provided by invoking a grain boundary defect model[14] which predicts the removal of metastable barrier as a result of annealing. A still different chemical approach[15] to increase the varistor stability has further affirmed the concept of metastable barrier being composed of zinc interstitials in the depletion layer.

For the past several years, new characterization techniques were applied not only to understand the details of these grain boundary defects, but also to determine why the annealing temperature of 600-700°C is so critical to varistor stability. To study this effect, deep level transient spectroscopy (DLTS)(9-10) and positron annihilation spectroscopy (PAS)(11) has been used to probe the nature of defects at the grain boundary as a function of annealing temperature. A study of the activation energy, estimated from the Arrhenius plot of leakage current vs. temperature in the prebreakdown region, and as a function of annealing temperature has been completed. When these results are combined with the previous studies on the grain boundary defect model for the varistor, the importance of a critical temperature (e.g., 600-700°C) of annealing becomes increasingly valid. The purpose of this paper is to highlight the criticality of this annealing temperature as it emerges from the results of these studies.

2. GRAIN BOUNDARY DEFECT MODEL AND THE METASTABLE BARRIER

In order to understand the metastable component of the barrier which gives rise to instability in ZnO varistor, it is desirable to briefly review the grain boundary defect model(14) for the varistor. This model assumes the presence of donor-like positively charged defects in the depletion layer, compensated by the acceptor-like negatively charged defects at the interface (figure 1). The key feature of the depletion layer is that it consists of two components:

(1) A stable component consisting of positively charged ions, such as oxygen vacancies, $V_o^{\bullet}, V_o^{2\bullet}$, and the donor ions $D_{zn}^{\bullet}$, which are located on the lattice sites and hence are relatively immobile and are spatially fixed.

(2) A metastable component consisting of zinc interstitials, $Zn_i^{\bullet}, Zn_i^{2\bullet}$, which are located on the interstitial sites and are considerably mobile as a result of available empty interstitial sites. It is this metastable component that is responsible for instability of the ZnO varistor.

During annealing, the zinc interstitials are permanently out-diffused from the depletion layer via diffusion and subsequent chemical reaction with the oxygen at the grain boundary. The key factor in restoring the stability is the elimination of the metastable component by forming stable ZnO lattice at the grain boundary.

Figure 2, taken from Reference 6, shows how two devices originally made in an identical manner differ in their responses to an externally applied electrical stress. The unstable device which contains both the stable and the metastable barriers degrades with time; the stable device obtained by removing the metastable component of the above varistor by heating at 600°C, shows no degradation in barrier height with time. In this figure, the barrier height is represented by the nonlinear voltage(14) designated as $E_{0.5}$ and is the voltage at 0.5 mA/cm^2. Simultaneously, the leakage current is dramatically stabilized(10) for the annealed device (figure 3), but increases with time for the unannealed or poorly annealed device. As discussed earlier[6], the constancy of the leakage current for the annealed device has been attributed to the elimination of the metastable barrier as a result of annealing. Several recent studies[9-11,14,16] have now brought additional features for discussion regarding the criticality of annealing temperature. These are summarized in the following sections.

3. DLTS STUDY VS. ANNEALING

The deep level transient spectroscopic (DLTS) studies[9-10] were carried out on ZnO varistors as a function of annealing temperatures between 400 and 1000°C. It was found that there was a pronounced effect of annealing on a characteristic trap (L_2) which appears at the energy level of E_c-(0.20-0.26)eV below the conduction band. The concentration of this trap[10], as shown in figure 4, decreases as the annealing temperature is increased, reaching a minimum value at 600°C and then rises again at the annealing temperature of 800 and 1000°C. Although the DLTS data are not available at 700°C anneal, it is interesting to note that the lowest trap density occurs at a temperature (e.g., 600°C) where the ZnO varistors are normally annealed to achieve a high degree of stability. The trap level in figure 4 has been associated[9] with the doubly charged $Zn_i^{2\bullet}$ in the depletion layer described in figure 1. Thus, the stability of the varistor is enhanced with the minimization of zinc interstitials which is known to constitute the metastable component of the barrier.

4. PAS STUDY VS. ANNEALING

The positron annihilation spectroscopic (PAS) study[11] was undertaken to identify the negatively charged zinc vacancy, V'_{zn}, at the grain boundary interface (figure 1), also as a function of annealing temperature. An important feature of this study is presented in figure 5, which illustrates the relative change in negatively charged zinc vacancy (measured in terms of the so-called S-parameter) at the interface as a function of annealing temperature. The relative S-parameter (S/S_O) initially increases with annealing temperature, but shows a steep decline at temperatures between 600 and 800°C, then again increases beyond 800°C. Unfortunately, as in DLTS study, no measurement was taken at 700°C anneal, so the bottoming-out of S-parameter is not precisely known. In spite of this uncertainty, it is clear that a drastic reduction in S-parameter occurs at temperatures above 600°C, but below 800°C. According to the defect reactions presented in Reference 11, the reduction in S-parameter is associated with the reduction of negatively charged V'_{zn} described in figure 1. Since $[V'_{zn}]$ and $[Zn_i^{2\bullet}]$ are complementary balancing charges at the grain boundary, the simultaneous reduction of both charges as a result of annealing, as determined by DLTS and PAS experiments, are consistent with the hypothesis that the metastable component is removed by annealing.

It follows from the above discussion that if one could measure the charges that exist at the interface, there will also be a reduction of such interface charges as a result of annealing. This has already been demonstrated in an earlier study[14], where the effective surface state density, N_S, is plotted against the barrier height ϕ_B and is shown in figure 6.

It is seen that the effect of annealing at 600°C is to shift the surface state density to the left, resulting in a permanent decrease in the total number of charge sites (Q_S) at the grain boundary interface. The total number of charge sites per cm^2, Q_S, is a measure of the area under the N_S-ϕ_B curve, and is a function[14] of $N_S(cm^{-2}eV^{-1})$, ϕ_B (eV) and the Fermi energy E(f).

5. ACTIVATION ENERGY STUDY VS. ANNEALING

On reflecting the results of DLTS and PAS experiments, it becomes clear that the transport characteristics of the grain boundary leakage current will also be different between an annealed and an unannealed varistor. This assumption has led to a detailed study on the current transport mechanism of the varistors in the prebreakdown region, as a function of annealing temperature. In this study, described elsewhere,[16] the resistive component (I_R) of the leakage current was measured as a function of the ambient temperature (T). The activation energy was then determined from I_R-T Arrhenius plots in the prebreakdown region. The samples were annealed at temperatures between 500 and 800°C at 100°C interval, and I_R measurement was conducted from room temperature to 175°C.

According to this study,[16] the experimental data fall in two distinct temperature regimes in I_R-T curves: a low temperature regime between 50 and 100°C, and a high temperature regime between 125 and 175°C for both unannealed and annealed samples. The transition between two regimes occurs at temperatures between 100 and 125°C. This is illustrated in Figure 7 for an unannealed and a 600°C-annealed sample. Note the change in slope in the low temperature regime for the 600°C-annealed sample. The varistors annealed at 700°C also showed a similar change in slope in the low temperature regime. The detailed statistical analysis indicates that in the low temperature regime, the activation energy remains constant with a value of ~ 0.183 eV up to 500°C anneal. It then suddenly increases to a value of ~ 0.244 eV at both 600 and 700°C annealing temperatures; it then decreases at higher annealing temperatures. The results thus dramatically highlight the criticality of 600-700°C where the activation energy is drastically increased as a result of annealing. As shown before, this is the same temperature range where other grain boundary parameters are also changed simultaneously in conjunction with improved device stability.

6. SUMMARY

During the fabrication of a ZnO varistor, a barrier voltage consisting of a stable and metastable component is formed at the grain boundary of the ZnO ceramic. Under an externally applied voltage stress, the varistor life is reduced due to the degradation of the metastable component. To alleviate this problem, the varistors have always been subjected to a post-sintering heat treatment in air, known as annealing. The metastable component is thus removed and the varistor life is enhanced. This paper presents several results which show that

there is a critical annealing temperature at which the varistors become highly stable combined with concurrent changes in several grain boundary properties. This critical temperature lies between 600 and 700°C, and is independent of compositions, processing conditions, and the type of manufacturers. The evidences concerning critical temperature are as follows. First, there is the grain boundary defect model, developed earlier, that attributes this metastable component to the presence of excess zinc interstitials in the depletion layer. It has been shown that at the critical temperature of annealing, the excess interstitials are out-diffused to the grain boundary where they are annihilated by reacting with oxygen to form ZnO lattices. Then there is a trap density at the energy level of E_C-(0.20-0.26)eV (as determined by DLTS) which shows a minimum at the annealing temperature of 600°C. It is at the same annealing temperature that the negatively charged zinc vacancy at the grain boundary interface starts to decline drastically (PAS study), accompanied also by a reduction in the total charge density at the grain boundary interface (N_S-ϕ_B curve). Finally, the activation energy for the leakage current (resistive) transport increases at the annealing temperature of 600-700°C when the measurements are conducted in the so-called low temperature regime
(50-100°C). Convergence of all these occurrences at this critical temperature range leads one to conclude that annealing is a complex process that is accompanied by a simultaneous change in many grain boundary parameters that are conducive to varistor stability.

REFERENCES

1. M. Matsuoka, "Non-ohmic Properties of Zinc Oxide Ceramics," Jpn. J. Appl. Phys. 10 (6) 736-746 (1971).

2. L. M. Levinson and H. R. Philipp, "The Physics of Metal Oxide Varistors," J. Appl. Phys. 46 (3) 1332-1341 (1975).

3. R. Einzinger, "Grain Boundary Phenomena in ZnO Varistors," in Grain Boundaries in Semiconductors, Ed. by H. J. Leamy, G. E. Pike and C. H. Seager, Elsevier, New York, 1982, pp. 343-355.

4. G. E. Pike, "Electronic Properties of ZnO Varistors: A New Model," Ibid, pp. 369-79.

5. P. Emtage, "The Physics of Zinc Oxide Varistors," J. Appl. Phys. 48, 4372-4384 (1977).

6. T. K. Gupta, W. G. Carlson and B. O. Hall, "Metastable Barrier Voltage in ZnO Varistors," in Grain Boundaries in Semiconductors, edited by J. H. Leamy, G. E. Pike and C. H. Seager, Elsevier, New York, 1982, pp. 393-98.

7. K. Eda, A. Iga and M. Matsuoka, "Degradation Mechanism of Non-ohmic Zinc Oxide Ceramics," J. Appl. Phys. 51 (5), pp. 2678-2684 (1980).

8. A. Iga, M. Matsuoka and T. Masuyama, "Effect of Heat Treatment on Current Creep Phenomena in Non-ohmic ZnO Ceramics," Jap. J. Appl. Phys. 15 (9), pp. 1847-1848 (1976).

9. A. Rohatgi, S. K. Pang, T. K. Gupta and W. D. Straub, "The Deep Level Transient Spectroscopy Studies on a ZnO Varistor as a Function of Annealing," J. Appl. Phys. 63 (11), pp. 5375-5379 (1988).

10. T. K. Gupta, "Varistor Stability and Its Relation to Trap Density," Second Int. Conf. on Varistor," held at GE Research Corp., Schenectady, NY, 1988 December 05-06.

11. T. K. Gupta, W. D. Straub, M. S. Ramanachalam, J. P. Schaffer and A. Rohatgi, "Grain Boundary Characterization of ZnO Varistors by Positron Annihilation Spectroscopy," to be published in the J. Appl. Physics, 1989.

12. A. Iga, M. Matsuoka and T. Masuyama, "Effect of Phase Transformation of Intergranular Bi_2O_3 Layer in Non-ohmic ZnO Ceramics," Jap. J. Appl. Phys. 15 (6), pp. 1161-1162 (1976).

13. T. Takemura, M. Kobayashi, Y. Takada and K. Sato, "Effects of Bismuth Sesquioxide on the Characteristics of ZnO Varistors," Am. Ceram. Soc. 69, pp. 430-436 (1986).

14. T. K. Gupta and W. G. Carlson, "A Grain Boundary Defect Model for Instability/Stability in a ZnO Varistor," J. Mat. Sci. 20, pp. 3487-3500 (1985).

15. T. K. Gupta and A. C. Miller, "Improved Stability of ZnO Varistor via Donor and Acceptor Doping at the Grain Boundary," J. Mat. Res., 3 (4), pp. 745-754 (1988).

16. T. K. Gupta, to be published.

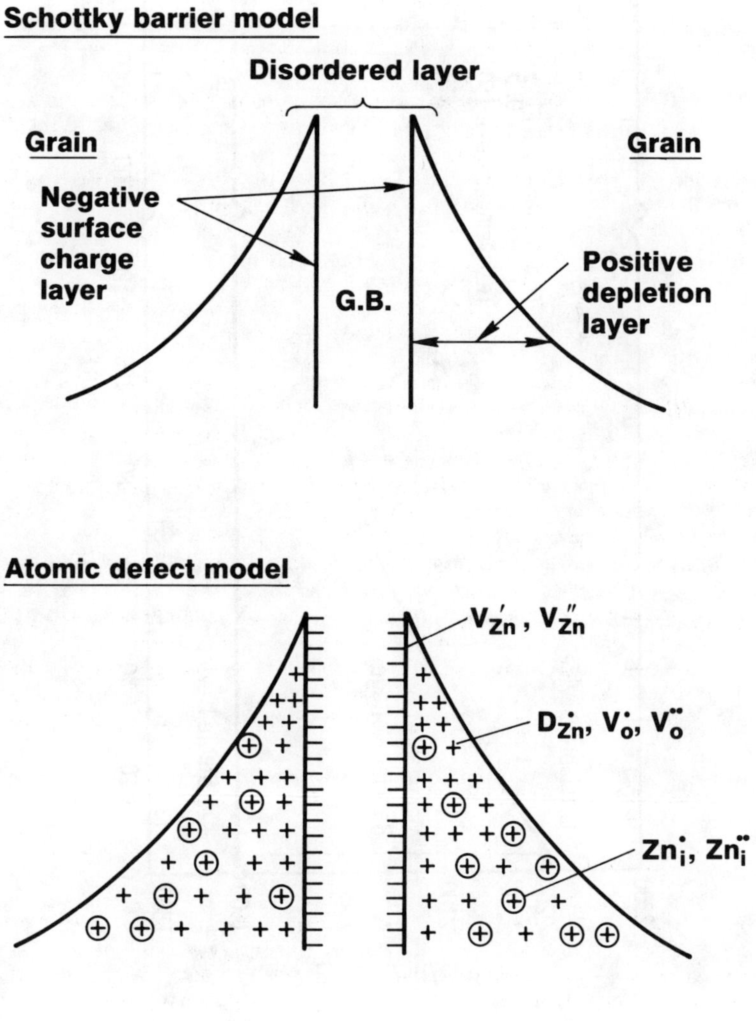

Fig. 1. The atomic defect model proposed for the Schottky barrier at the grain boundary; also shown in the figure is the analogy with band model (Reference 14).

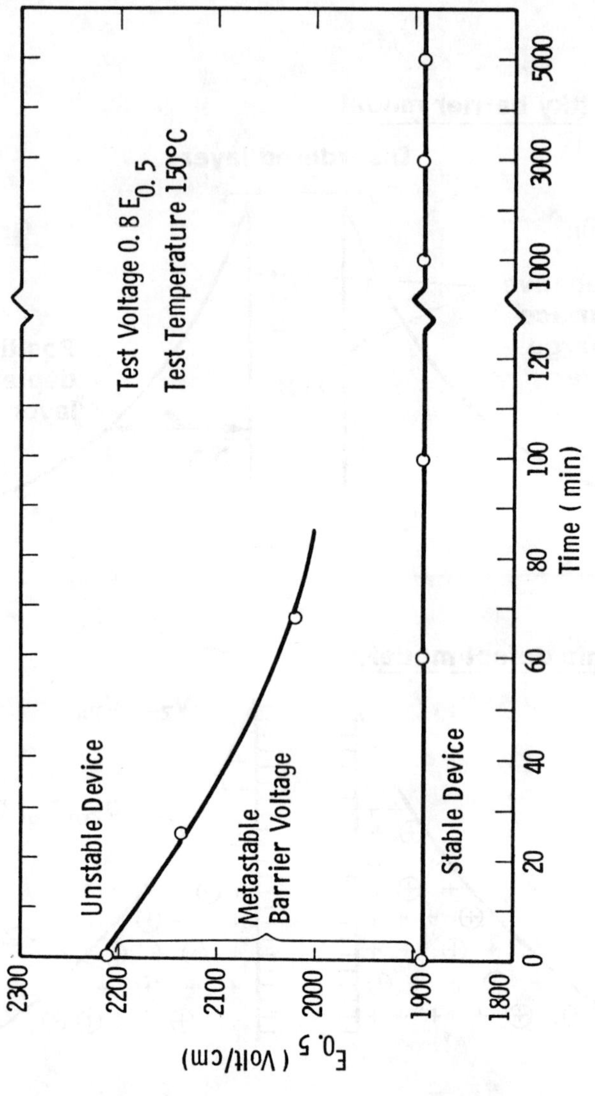

Fig. 2. The comparison of the barrier voltage ($E_{0.5}$) with time for a stable and an unstable device. Note the degradation of barrier voltage for the unstable device containing metastable barrier (Reference 6).

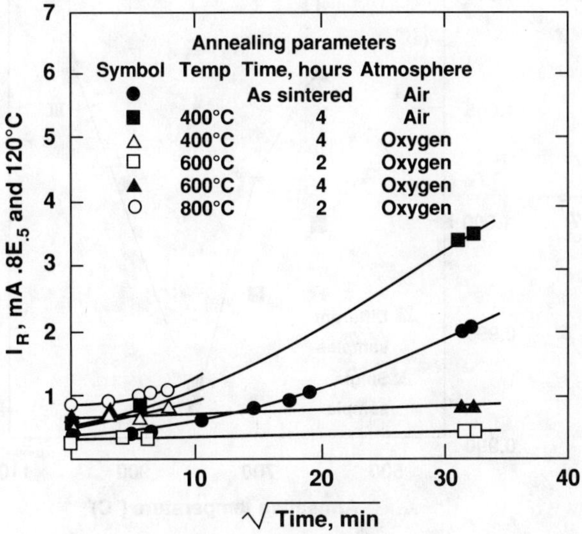

Fig. 3. Effect of annealing on varistor stability--a plot of resistive current I_R vs square root of time, $\sqrt{t}$. Note the stability of the device annealed at 600°C (Reference 10).

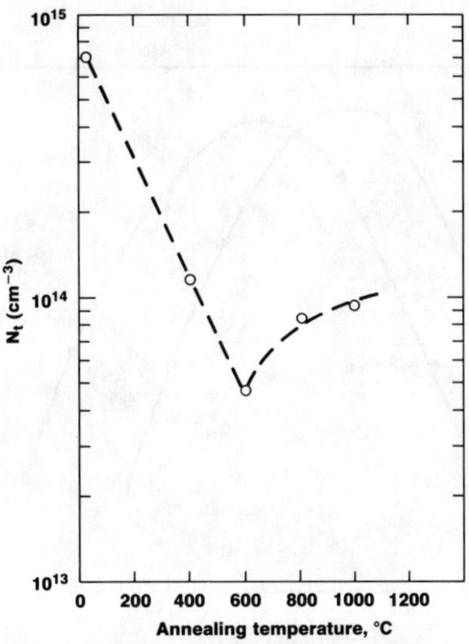

Fig. 4. The trap density L_2-level as a function of annealing temperature. Note the minimum trap density at 600°C anneal (References 9, 10).

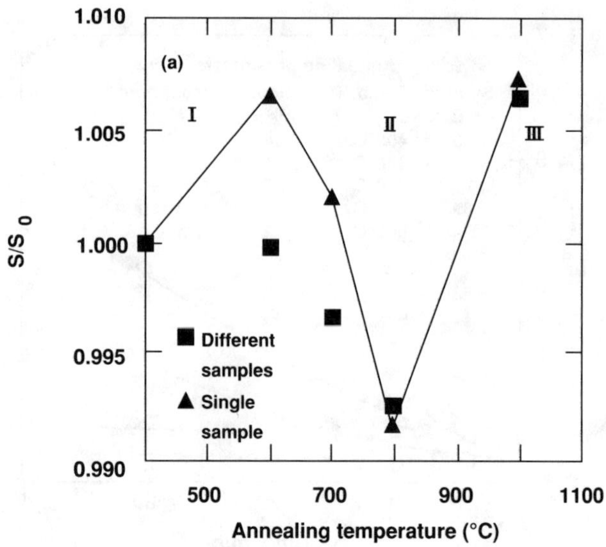

Fig. 5. Variation of normalized S-parameter with annealing temperature (the normalization was accomplished by dividing each S value by the S value at 400°C (S_0). Note the declining S-parameters between 600° and 800°C (Reference 11).

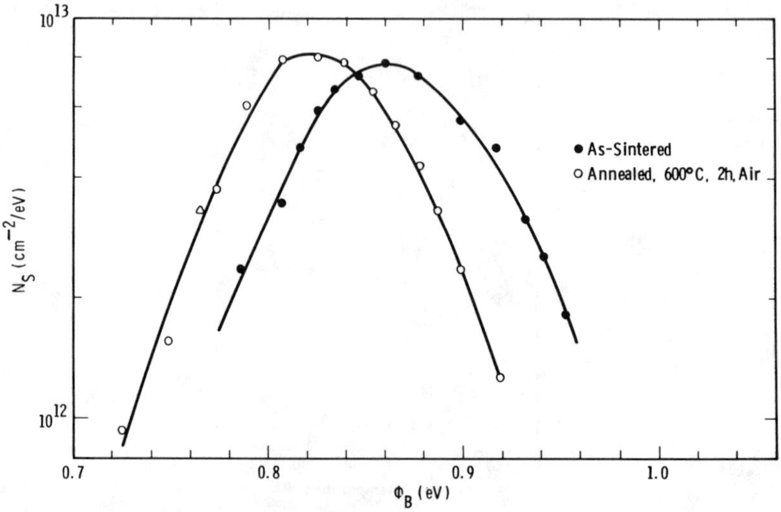

Fig. 6. The varistor surface state density (N_S) before and after annealing at 600°C. Note the shift in N_S to left as a result of annealing (Reference 14).

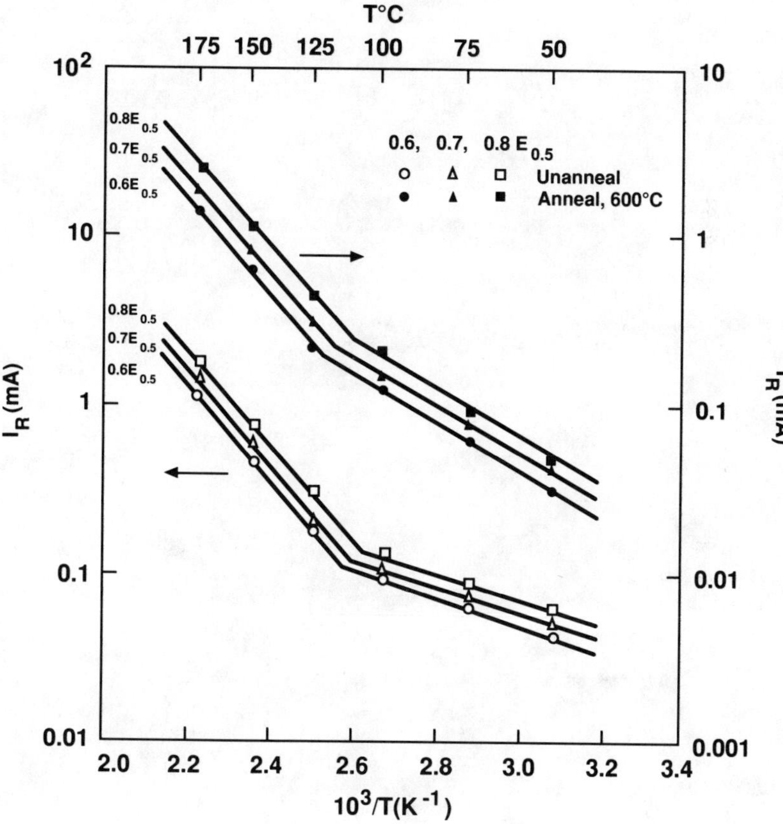

Fig. 7. Arrhenius plots of I_R-T curves for unannealed and 600°C-annealed samples. Note the change in slope for the annealed sample in the low temperature regime (Reference 15).

INDEX

Abouelleil, M., 457
Acrylic binders, 355, 375
Adhesion, 217, 685
Aging, 105
Alkoxy solution, 23
Alumina, 77, 167, 355, 365, 437, 575
Aluminum nitride, 91, 193, 217, 527, 583, 685
Analysis of variance, 51
Anisotropy, 503
Annealing, 297, 489, 711
Aqueous slurries, 355
Argon ion laser, 297
Atmosphere control, 375

Bailey, A.E., 105
Bailey, P.J., 167
Baking test, 365
Barium
  hexaferrite, 245
  magnesium tantalates, 153
  tetratitanate, 41
  titanate, 11, 23, 427, 489
Beall, G.H., 259
Berry, D.H., 489
Binders, 355, 375
Birnie, D.P., III, 503
Bonding glasses, 217
Borland, W., 555
Boron phosphate, 259
Bowen, H.K., 355
Brewer, J.A., 651
BSSZ ceramics, 313, 391
Burn, I., 375
Burton, L.C., 675

Canner, J.P., 313, 391
Capacitors, 427
Carbothermal reduction nitridation, 193
Carlson, R.O., 405
Characterization, 51, 69, 477
Charge transport, 675
Chemical
  processing, 41
  vapor deposition, 437
Chen, C.F., 193
Cherukuri, S.C., 217
Chip multilayer ceramic, 391
Co-fired ceramic multilayers, 611
Cold pressing techniques, 23
Color measurement, 193
Columbite precursor method, 105
Communication devices, 391
Compressive stresses, 11, 129
Concave solidification fronts, 297
Conductivity, 193, 685
Conopask, L., 457
Copper
  conductors, 375, 391, 555
  electrode, 313
Cordierite, 259
Crack growth, 129, 489
Cristobalite, 167
Crystallization, 217, 259

Degradation, 675
Densification, 279
Deposition, 437
Dey, S.K., 477
Di-phasic gel, 3
Diamond films, 3
Die attach, 217, 365
Dielectric
  aging, 105
  pastes, 217
  properties, 143, 153, 167, 675
Dimensional
  control, 611
  stability, 355
Direct nitridation, 193
Distortion analysis, 611
Doped
  barium titanate, 23
  PMN-PT ceramics, 105
Doughty, D.H., 179
Durability, 217

Eichelberger, C.W., 405

Electrical
conductivity, 217
field effects, 129
impedance, 611
properties, 69, 477
Electrodes, 313, 347, 603
Electromagnetic wave absorber, 11
Electrostrictive strain, 129
Emery, A.F., 297
Energy dispersive spectroscopy, 51
Enthalpy formation, 297
Ewsuk, K.G., 279
Externally nucleated, 259

Fatigue tests, 11
Ferroelectric/ferromagnetic composites, 711
Fibers, 11, 661
Fluormica, 259
Fox, B.H., 245
Freiman, S.W., 129
Fukunaga, O., 143
Furnace cooling, 143

Garino, T.J., 179
Giess, E.A., 167
Glasses, 217, 245, 259, 279, 347
Goo, E., 167
Goodnow, D., 179
Grain
boundaries, 23, 675, 711
growth, 23, 347
size, 23, 77, 91, 105, 503
Green tapes, 355
Grigoropoulos, C.P., 297
Gupta, T.K., 711

Hafnium titanates, 143
Hammetter, W.F., 179
Hanrahan, J.R., 489
Harada, H., 661
Hardness, 457
Hayashi, M., 427
Heat treatment, 143
Hernandez, C.L., 179
Heterocoagulation processing, 51

High
bias temperature test, 555
Density Interconnects (HDI), 405
frequency applications, 391
iron glasses, 245
Hot isostatic pressing, 11
Hung, C.C., 51
Hwang, C.S., 437
Hydrogen microfoams, 259
Hydrothermal-electrochemical synthesis, 427

Ichikawa, H., 661
Iio, S., 365
Ikawa, H., 143
Imai, H., 347
Impedance, 611
Impurity phases, 51
In situ preparation, 427
Indentation, 129, 457
Infrared radiation, 661
Interfaces, 685
Internal copper conductors, 375
Internally
cooled substrates, 527
nucleated, 259
Inuzuka, T., 153
Irrational strategies, 3
Ishizawa, N., 427

Jensen, R.H., 611, 629
Jiang, Q., 489
Johnson, D.L., 77

Kahn, M., 527
Kajiwara, T., 661
Kanda, A., 365
Kato, N., 365
Kato, S., 575
Kim, H.J., 437
Kolar, D., 69
Komatu, M., 91
Krasevec, V., 69
Krishnamani, K., 329
Kubo, H., 153

L-C filter, 391
Lapped substrates, 583

Laser
flash technique, 685
lithography, 405
melting, 297
Lateral migration resistance test, 555
Lead
magnesium niobate lead titanate, 105
oxide, 51
scandium tungstate, 69
zirconate titanate, 179, 477
Leakage current, 675
Levinson, L.M., 405
Liquid cooling, 527
Lithium disilicate, 259
Lougher, W., 583
Low
-field aging, 105
-K glass ceramics, 259

MacDowell, J.F., 259
Magnesium oxide, 105
Mandai, H., 313, 391
Manufacturing, 651
Martinez, S.L., 179
Materials synthesis, 3
Maxwell, J., 329
Mazdiyasni, K.S., 23
Mechanical properties, 11, 329
Metal housings, 217
Metallization, 217, 365, 583
Metallo-organic decomposition, 489
Mica composites, 259
Microchemistry, 685
Microfoams, 259
Microstructure, 69, 153, 167, 347, 477, 503, 685
Microwave device materials, 245
Mikeska, K.R., 611, 629
Mixedness, 51
Miyayama, M., 661
Modak, A.R., 477
Modeling,
electric field effect, 129
film hardness, 457
grain boundary defect, 711
low temperature defect, 179
sintering, 77
thermal, 297
Molten zone, 297

Molybdenum-manganese metallization, 365, 583
Mori, N., 661
Mullites, 167
Multicomponent powder mixtures, 51
Multilayer
ceramic capacitors, 313, 329, 375, 675
composites, 11, 347, 355
films, 503
Munson, M.J., 51
Muto, N., 661

Nahass, P., 355
Nakayama, Y., 365
Nano
composites, 3
-indentation hardness, 457
-size particles, 23
Narita, H., 143
Nebe, W.J., 417
Nelson, R., 527
New materials, 3
Newnham, R.E., 143
Nighan, W., 457
Niobium, 179
Nitridation, 193
Nitrogen pressure, 91
Non-
acqueous slurries, 355
reactive liquid phase sintering, 279
Notis, M.R., 685
Nozaki, S., 365
Numerical modeling, 297

Okamura, H., 391
Okazaki, K., 11
Organic slurries, 355
Orientation, 503
Overlay layers, 405
Oxide resistor pastes, 217

Packaging, 217, 259, 405, 527, 611, 629
Paste-substrate interaction, 575
PBN, 129
Pemberton, J.E., 41
Permeability, 217
Perovskite particle processing, 23, 69
Petric, M., 69
Phase assembly, 503
Photochemical machining, 259

Photoformable pastes, 417
Photolithography, 417
Phule, P.P., 41
Physical vapor deposition, 503
Pick-and-place considerations, 329
Piezoelectric substrates, 651
Pilgrim, S.M., 105
Platinum, 489
Pober, R.L., 355
Polymer layer overlays, 405
Porosity, 167, 503
Porter, W.C., 375
Post solder handling, 329
Powder processing, 51, 69
Precipitation, 51
Precursors, 105
Pressure-assisted sintering, 629
Price, D., 457
Printing, 583
Processing conditions, 503
Pull test, 583
Pyrochlore, 105, 575
PZT, 179, 477

QPLUS system, 555
QSIL system, 555
Qualitative model, 129
Quenching, 143

Radiation, 661
Rapid
  solidification technique, 245
  thermal annealing, 489
Rare earth oxides, 23
Rational strategies, 3
Rawal, B.S., 129, 329
Raynes, A.S., 129
Reactive electrode submerged arc process, 3
Recrystallization, 297
Relaxor materials, 11, 69, 105
Reliability, 675
Resonator materials, 153
Rhine, W.E., 355
Riman, R.E., 51
Risbud, S.H., 41
Robbins, W.L., 355
Roldan, J.M., 167
Roman, W., 457
Roy, R., 3
Ruthenium oxide, 575

Sakabe, Y., 313
Sanchez, E., 489
Santiago, J.J., 489
Sapphire, 575
Satellite systems, 153
Savrun, E., 193, 583
Sawhill, H.T., 611
Scaling law functions, 77
Schwartz, R.W., 179
Sealing glasses, 217
Seals, 365
Selective ion fixation, 3
Series Q materials, 555
Shrinkage rate, 77
Signal continuity, 611
Silica, 167
Silicon, 297, 437, 489, 661
Silver-glass electrodes, 347, 603
Sintering,
  aluminum nitride, 193
  electrodes, 347
  glasses, 279
  models, 77
  multilayers, 611, 629
  nitrogen pressure, 91
  non-reactive liquid, 279
  two-phase, 11
  viscous, 259
Sol-gel process, 259, 477, 503
Solder, 217, 329, 365, 603
Spin
  casting, 489
  coating, 179
Strain, 129
Strength, 217
Stress, 329
Strontium titanate, 105
Structural evolution, 41
Substrate materials, 91, 193, 217, 347, 365, 575, 583, 651, 685
Sugiyama, M., 153
Superconducting composites, 11, 51
Surface
  characterization, 51
  finish effects, 583
  land pattern, 611
  mount applications, 329, 391
Sutherland, A.E., 105

Suvorov, D., 69

Takada, N., 91
Takahashi, T., 91
Tape casting, 355, 611, 651
Termination effects, 329
Thermal
  annealing, 297, 489
  conductivity, 193, 685
  control, 527
  expansion, 217
  modeling, 297
  stress, 329
Thick films, 217, 347, 417, 555, 575, 603, 685
Thin films, 179, 217, 297, 427, 437, 457, 477, 489, 503, 685
Tissot, R.G., 179
Titanium dioxide, 437
Tuttle, B.A., 179
Two-phase mixed sintering, 11

Uhlmann, D.R., 503

Varistors, 711
Varshneya, A.K., 217
Vias, 417
Viscosity, 217

Wada, Y., 603
Wakino, K., 391
Wang, J.F., 51
Warner, D.A., 51
Wave absorber, 11
Wear resistance, 457
Westwood, A.D., 685
Wetting, 217, 365, 603
White, G.S., 129
Winzer, S.R., 105
Wire bondability, 365
Wojnarowski, R.G., 405

Yamaguchi, T., 347, 575, 603
Yanagida, H., 661
Yoo, S.E., 427
Yoshimura, M., 427
Yttrium, 583

Zero
  defect design, 329
  expanding material, 3
Zinc oxide, 51, 711
Zirconate titanate, 143, 651
Zybura, A., 245